המבנה הפנימי של חורים שחורים

וסינגולריות מרחב-זמן

INTERNAL STRUCTURE OF BLACK HOLES

AND SPACETIME SINGULARITIES

Annals of the Israel Physical Society Volume 13

INTERNAL STRUCTURE OF BLACK HOLES AND SPACETIME SINGULARITIES

An International Research Workshop
Haifa, June 29 – July 3, 1997

Edited on behalf of the Israel Physical Society by
Lior M. Burko and Amos Ori
Department of Physics – Technion

published by

Institute of Physics Publishing, Bristol and Philadelphia
and
The Israel Physical Society, Jerusalem

Published jointly by:
Institute of Physics Publishing, Dirac House, Temple Back, Bristol, BSI 6BE, UK;
and
The Israel Physical Society, P. O. B. 16105 Jerusalem 91160, Israel.
Editorial Office: c/o Physics Department, Technion, Haifa, 32000 Israel.

ISSN: 0309-8710 $2.00 + $0.50

British Library Cataloguing in Publication Data.
A catalogue record for this book is available from the British Library.

Library of Congress Cataloging-in-Publication Data are available.

ISBN: 0 7503 0548 7

Printed by Ayalon Offset Ltd., Haifa, Israel.

TABLE OF CONTENTS

PREVIOUS VOLUMES IN THIS SERIES

Vol. 1 ATOMIC PHYSICS IN NUCLEAR EXPERIMENTS
Proceedings of the International Workshop on Topics in Atomic Physics Related to Nuclear Experimentation held at Haifa.
Editors: Baruch Rosner and Rafael Kalish, Technion - Israel Institute of Technology, Haifa.
638 pages. 1977, ISBN 0-85274-355-6 ISSN 0309-8710

Vol. 2 STATISTICAL PHYSICS - "STATPHYS 13"
Proceedings of the 13th IUPAP Conference on Statistical Physics, held in Haifa.
Editors: Chanoch Weil (Executive Editor), Dario Cabib, Charles G. Kuper and Ilan Riess, Technion-Israel Institute of Technology, Haifa.
1,087 pages. 2 parts. 1978, ISBN 0-85274-356-4, ISSN 0309-8710.

Vol. 3 GROUP THEORETICAL METHODS IN PHYSICS
Proceeding of the VII International Colloquium on Group Theoretical Methods in Physics, held at Kiryat Anavim.
Editors: L. Horwitz, Y. Ne'eman, University of Tel Aviv.
417 pages, 1980. ISBN 0-85274-424-2. ISSN 0309-8710

Vol. 4 MOLECULAR IONS, MOLECULAR STRUCTURE AND INTERACTION WITH MATTER
Proceedings of the 38th Bat Sheva Seminar held at the Weizmann Institute of Science, Rehovoth, at Ein Bokek, and the Technion-Israel Institute of Technology, Haifa.
Editor: Baruch Rosner, Technion-Israel Institute of Technology, Haifa.
290 pages. 1981. ISBN 0-085274-441-2 ISSN 0309-8710

Vol. 5 PERCOLATION STRUCTURES AND PROCESSES
Invited Review Papers on the Theoretical and Experimental Aspects of Percolation Structures and Processes.
Editors: G. Deutscher, Tel Aviv University, R. Zallen, Xerox Corp. Rochester, N.Y., and J. Adler, Technion - Israel Institute of Technology, Haifa.
502 pages. 1982 ISBN 0-85274-477-3 ISSN 0309-8710

Vol. 6 VACUUM ULTRAVIOLET RADIATION PHYSICS – VUV VII
Proceedings of the 7th International Conference on Vacuum Ultraviolet Physics, held at Jerusalem.
Editors. A. Weinreb and A. Ron, Hebrew University, Jerusalem.
622 pages. 1984. ISBN 0-85274-760-8. ISSN 0309-8710.

Vol. 7 ELECTROMAGNETIC PROPERTIES OF HIGH SPIN NUCLEAR LEVELS
Proceedings of the Workshop held at the Weizmann Institute of Science, Rehovoth and at Ein Bokek.
Editors: Gvirol Goldring and Michael Hass, Weizmann Institute of Science, Rehovoth.
347 pages. 1984. ISBN 0-85274-775-6. ISSN 0309-8710.

Vol. 8 FRAGMENTATION FORM AND FLOW IN FRACTURED MEDIA
Proceedings of the F^3 Conference, held at Neve Ilan
Editors: R. Engelman and C.Yaeger, Sorek Nuclear Research Center.
628 pages, 1986. ISBN 0-85274-578-8. ISSN 0309-8710.

Vol. 9 DEVELOPMENTS IN GENERAL RELATIVITY, ASTROPHYSICS AND QUANTUM THEORY
A Jubilee Volume in Honour of Nathan Rosen
Editors: F. I. Cooperstock, Victoria, L. P. Horowitz, and J. Rosen, Tel Aviv University.
377 pages, 1990. ISBN 0-7503-0053-1. ISSN 0309-8710.

Vol. 10 CATACLYSMIC VARIABLES AND RELATED PHYSICS
Proceedings of the 2nd Technion Haifa Conference
Editors: O. Regev and G. Shaviv, Technion - Israel Institute of Technology.
337 pages, 1993. ISBN 0-7503-0282-8. ISSN 0309-8710.

Vol. 11 ASYMMETRICAL PLANETARY NEBULAE
Proceedings of the University of Haifa at Oranim Conference
Editors: A. Harpaz and N. Soker, University of Haifa at Oranim.
306 pages, 1995. ISBN 0-7503-0330-1. ISSN 0309-8710.

Vol. 12 THE DILEMMA OF EINSTEIN, PODOLSKY AND ROSEN – 60 YEARS LATER
An International Symposium in Honour of Nathan Rosen
Editors: A. Mann and M. Revzen, Technion - Israel Institute of Technology.
314 pages, 1996. ISBN 0-7503-0394-8. ISSN 0309-8710.

FOREWORD

The Israel Physical Society provides a forum for interaction amongst Israeli physicists. There are two types of membership: individual and corporate. Individual physicists and scientists in fields related to physics are eligible for individual membership while educational and research establishments, companies, foundations, etc. with an interest in the promotion of physics in Israel are eligible for corporate membership.

The Society organizes an annual national general conference. In addition it co-sponsors many international conferences of a more specialized nature, held in Israel. The Society publishes the "Annals" which primarily serve as a vehicle for rapid publication of the Proceedings of international conferences held under its sponsorship. The "Annals" also occasionally publishes an up-to-date review of a topical field in which many of our members are active.

Joint Series Editors

PREFACE

The workshop on "The Internal Structure of Black Holes and Spacetime Singularities" was held at the Technion—Israel Institute of Technology in Haifa, Israel from June 29 to July 3, 1997. To the best of our knowledge, this is the first meeting ever devoted primarily to this subject, and we hope that other meetings will follow. Thirty eight participants from thirteen countries attended the workshop. The main goal of the workshop was to encourage interaction between the participants, and to provide a stimulating and thought-provoking environment for new ideas. For this reason, much free time was made available, in addition to informal discussion sessions, where participants were free to raise open problems and ideas. We hope that these discussions will benefit the future research of the participants, and that new ideas and directions were born in the discussions.

At the workshop, twenty three participants presented their research formally, and most of them contributed written versions of their talks for this proceedings volume. Discussions and questions were not limited to the last few minutes of each talk. As the main goal of this workshop was the interaction between the participants, the latter were encouraged by the chairpersons of the sessions to present questions during the talks, and to initiate discussions on matters of principle. Indeed, we feel that important discussions followed the different talks, and we thank the participants for the active part they took in the workshop. Unfortunately, a few invited speakers were unable to be present at the workshop. However, the talks they planned to give at the workshop are of great importance, and therefore are included in the workshop's proceedings. During the workshop we also devoted time for excursions and sightseeing.

Since this volume is primarily the proceedings of a workshop, the different chapters are mainly written versions of the talks given at the workshop. However, the authors have written extended introductions to their contributions, so that a reader, well versed with Gravitation theory, but unaware of the current research in a specific sub-field, would be able to follow the text. Several topics are considered by more than one author. This brings into the text the different viewpoints various authors have, which were expressed during the discussion sessions in the workshop. As a result, this book is suitable for graduate students who wish to familiarize themselves with the various aspects of the structure of black holes, and also for researchers who wish to learn about advances in sub-fields other than their own.

The contents of this volume represent the various subjects the workshop focused on. Among these are the nature of the singularities (both null and spacelike) and the spacetime inside black holes coupled to various matter fields in classical and semi-classical gravity, as well as in alternative gravity theories; the cosmic censorship hypothesis and its connection to black hole interiors; cosmological singularities and their relation with black hole singularities; and black hole entropy and evaporation.

We thank the other members of the Organizing Committee, Éanna Flanagan, Tsvi Piran, Igor Novikov and Eric Poisson for helping us in the planning of the workshop. We also thank Ted Jacobson, Alfio Bonanno, Valeri Frolov, Matt Visser,

Tom Helliwell, Chris Chambers, Eric Poisson, Éanna Flanagan and Robert Mann for chairing the sessions of the workshop. We would like to thank the workshop secretary Liz Youdim for many hours of hard and devoted work, which helped us and facilitated the organization of the workshop considerably. We thank Jimmi Nishri, the Physics Department administrator for much good advice and for the administrative organization, and Gabi Schwarz for technical assistance. We thank Yossi Avron and Moshe Moshe for encouraging us to host the workshop at the Technion. The help of the Department of General Studies and Dafna Dressler, who let us use the theater in the Taub building is warmly acknowledged. We are indebted to the generous financial support of the Institute of Theoretical Physics and the Department of Physics of the Technion, which enabled us to organize the workshop. We would like to express our gratitude to the Israel Physical Society which facilitated the publication of this volume, and supported its production financially. We especially thank the series editors of the Annals of the Israel Physical Society, Joan Adler, Shulamit Eckstein, Barouch Rosner and Raoul Weil for their support and help. Finally, we thank Orna Burko for her invaluable help in the production and typing of this volume.

L. M. Burko and A. Ori

Haifa, December 1997

On the Cover: The cover figure is a schematic representation of an inter-universe wormhole, taken from the contribution to this volume by M. Visser and D. Hochberg. It represents a situation where the geometry of space is extremely curved, so much so that the universe folds up on itself and connects through to another universe. The region where the geometry is narrowest is called the "throat" or "neck" of the wormhole. The detailed behavior of the space-time geometry at and near the throat is what distinguishes a traversable wormhole from a non-traversable wormhole (black hole, Schwarzschild wormhole). If the throat is also an event horizon, then this diagram represents (part of) a black hole. If the throat is carefully engineered to prevent an event horizon forming the wormhole is traversable. These traversable wormholes are good models for studying semi-classical quantum gravity.

INTRODUCTION TO THE INTERNAL STRUCTURE OF BLACK HOLES

Lior M. Burko and Amos Ori

Department of Physics, Technion—Israel Institute of Technology, 32000 Haifa, Israel

Abstract

We briefly review the study of the internal structure of black holes. We describe in this introductory essay the main results which shape our understanding of the internal structure of black holes. We focus on the essential developments in our understanding of black hole interiors in classical General Relativity. In addition, we also describe in brief other directions in which interesting research has been conducted. We divide the progress made in this field into three categories, roughly parallel to three temporal periods: During the first period the internal structure of the exact Schwarzschild, Reissner-Nordström, and Kerr solutions were considered. During the second period the inner-horizon instability was studied mainly from linear analyses. During the third period nonlinear analyses replaced the linear ones. We briefly describe the main progress achieved during these three periods. Then, we describe nonlinear perturbation analysis of spinning black holes. We conclude by considering classically-stable Cauchy horizons and quantum-mechanical effects.

Introduction

The study of the internal structure of black holes is as old as the study of black holes themselves. From a naïve point of view, this investigation might look bizarre: by definition, no information can get out of a (classical) black hole, and therefore, no physical observation which tells us on its interiors can be made. However, the existence of a one-directional membrane-like hypersurface, known as the event horizon, through which no information can get out to external observers, should not constitute a barrier to human curiosity. It turns out that locally there is nothing special at the event horizon, and infalling observers will never tell (from local experiments that they can make) that they are passing through it. Therefore, there is no reason to assume that our physical laws are invalid inside the event horizon. The physical laws can still be implemented inside black holes.

Every acceptable physical theory has its own domain of applicability. That is, every theory has a range of parameters for which it is valid. Outside this range, the theory is invalid and gives false predictions. The frontier of physical research, therefore, is at the boundary of these ranges: of the greatest interests of scientists is to make experiments and observations at the most extreme conditions, and confront the predictions of the known theories with them. The theories, which are accepted because they were successfully tested at less extreme conditions, can thus be tested at new regimes. Important information on how new theories—if needed—should behave can be withdrawn from such experiments at extreme conditions. Another route to test existing theories is to examine their own physical and mathematical consistency under extreme conditions. As no observations and experiments regarding the interiors of black holes are available to us, this second route is very relevant. The theory of gravitation, Einstein's General Relativity, has been tested successfully in a certain range of field strengths. However, the theory has not been tested yet in the strong field regime.

General Relativity predicts, by virtue of the singularity theorems and under very plausible assumptions [1], the occurrence of spacetime singularities inside black holes. One approach to the singularities predicted by General Relativity is that the theory fails to describe nature at some point. The inevitability of singularities in General Relativity is viewed then as a warning that the theory predicts its own breakdown, and as a sign that another theory, perhaps Quantum Gravity, takes over [2]. According to this viewpoint, "proper physical variables do not and cannot go to infinity," [2] and consequently "... a theory that involves singularities and involves them unavoidably ... carries within itself the seeds of its own destruction" [3]. From this point of view, the existence of a singularity cannot be accepted , "[f]or a singularity brings so much arbitrariness into the theory that it actually nullifies its laws. ...Every field theory... must therefore adhere to the fundamental principle that singularities of the field are to be excluded" [4]. A somewhat weaker version of this approach is that even though the theory may not break down completely, the singularities signal that classical General Relativity is incomplete, in the sense that it does not provide boundary conditions for the field equations at the singular points [5].
A different approach to the predicted singularities is that singularities should be viewed "not as a warning of our ignorance, but as a source from which we can derive much valuable understanding" [6]. According to this second viewpoint, study of black hole interiors, and in particular study of the General Relativistic predictions about the occurrence and features of these singularities, may teach us something on what we should expect from a quantum theory of gravitation, or any other modification of General Relativity in the strong field regime. In particular, the present authors find it intriguing that a possible way in which General Relativity fails at one type of spacetime singularities is not through infinite destructive effects on physical objects, but through breakdown of predictability (i.e., a null weak singularity; see below), while at another type of spacetime singularities, it is the other way around (i.e., a strong singularity; see below). (A recent mathematical analysis of spacetime singularities can be found in [7]. For a recent review from a more philosophical point of view see

[8].)

A second motivation to the study of black hole interiors stems from the suggestion that a descent into black hole interiors might resemble the (time reverse of the) early universe. Indeed, the same arguments which lead to the inevitability of the initial cosmological singularity also lead to the inevitability of the singularity inside black holes. Therefore, the study of black hole interiors may be relevant to our understanding of the universe.

But there is also another strong motivation to the study of black-hole interiors: It may be possible that an object falling into a realistic spinning black hole would re-emerge into another universe (or into another region of our universe), through a white hole [9]. This possibility is demonstrated in the simplest solution for spinning black holes – the Kerr geometry. For many years it had been widely anticipated that the divergent blue shift at the inner horizon of Kerr would lead to the formation of a curvature singularity, which in turn would cause the ultimate destruction of any infalling physical object, due to unbounded tidal deformation and other effects. At present, however, the evidence is that no such unbounded deformation occurs at the inner horizon. Moreover, nothing in our present understanding of the theory of gravity indicates to the impossibility of the extension of geometry beyond the inner horizon. This issue will be considered in greater detail below. Therefore, a Kerr-type causal structure (i.e., a gravitational bounce through a wormhole) cannot be excluded. If indeed realistic spinning black holes admit such a Kerr-like causal structure , study of black hole interiors may have extremely important observational consequences.

The study of black hole interiors can be divided into three categories, roughly parallel to three temporal periods. At the first period, only the spacetime structure and the causal structure of the exact stationary black hole solutions (the Kerr-Newman family) were considered. The first black hole to be studied was naturally the Schwarzschild black hole. Then, also the interiors of Reissner-Nordström and Kerr black holes were studied. This period started in the 1960's, with the breakthrough in the understanding of the physical meaning of the coordinate singularity at the event horizon of the Schwarzschild black hole. Kruskal and Szekeres [12, 13] found (independently) regular coordinates which cross the event horizon and cover the entire Schwarzschild manifold [14]. This understanding revived the older insight of Oppenheimer and Snyder, that an infalling observer would cross the event horizon and enter the black hole, even though an external observer would never see him arriving at the event horizon [15]. Later, similar coordinates were found also for the Reissner-Nordström and for the Kerr black holes.
The second period is characterized by the study of linear perturbations of black holes. That is, the study of the evolution of various matter and radiation fields on the *fixed* background of the black hole. This period began in 1967 with the suggestion of Penrose [16]—that the inner horizon of Reissner-Nordström black holes is unstable to external perturbations which do not decay fast enough.
The third period began in 1981, with the first *nonlinear* analysis of black hole interi-

ors, made by Hiscock [17]. In the nonlinear analyzes the perturbations are coupled to the spacetime geometry, and consequently the geometry evolves dynamically under the perturbations.

In what follows, this Introduction will describe the main results obtained during these three periods, and which shape our understanding of the internal structure of black holes (see also [18]). We shall focus on the essential developments in our understanding of black hole interiors in classical General Relativity. In addition, we shall describe in brief other directions wherein interesting research has been conducted. First, we consider the exact Schwarzschild, Reissner-Nordström, and Kerr solutions, and review their main properties which are relevant to the internal structure. Then, we describe the inner-horizon instability, by means of the geometrical-optics approximation and also by linear analyses. Simplified nonlinear models of the spacetime dynamics, including the mass-inflation phenomenon and numerical analyses of a test model of a spherically-symmetric, self-gravitating scalar field, are reviewed next. Then, we describe nonlinear perturbation analysis of spinning black holes and their main results. We conclude by considering classically-stable Cauchy horizons and mention some semi-classical effects.

The authors are well aware of the fact that various points are in controversy among the different workers in the field. We have endeavored in this Introduction to present other viewpoints than our own as well. However, undoubtedly this Introduction is biased toward the viewpoint of the authors. Other interpretations and insights are presented elsewhere in this volume, and the reader is welcome to confront the various approaches.

The Schwarzschild, Reissner-Nordström, and Kerr solutions

Much of our present understanding of black holes' interiors is primarily based on three exact solutions: the Schwarzschild, Reissner-Nordström (RN), and Kerr solutions [19]. (For more information about the exact solutions and linear perturbation analysis see [20, 21].) Of these solutions, the simplest is the Schwarzschild solution, described by the line element

$$ds^2 = -f\,dt^2 + f^{-1}\,dr^2 + r^2\,d\Omega^2, \tag{1.1}$$

where

$$f = 1 - 2M/r, \tag{1.2}$$

and $d\Omega^2 = d\theta^2 + \sin^2\theta\,d\varphi^2$ is the line-element on the unit two-sphere. Here, r is the radial Schwarzschild coordinate, defined such that circles of radius r have circumference $2\pi r$. We also define t to be the temporal coordinate for a static observer in infinity. Namely, at large values of r the metric (1.1) is asymptotically Minkowskian, and in Minkowski spacetime t is the time of a static observer. [The metric (1.1,1.2) is independent of the coordinate t.]

This solution (1.1) together with the metric function (1.2) describe the unique spherically symmetric, static, vacuum black hole with mass M. The Penrose diagram in Fig. 1.1 displays the main properties of the Schwarzschild black hole. The event horizon (denoted EH) is located at $r = 2M$, where f vanishes. (Throughout this Introduction we use geometrized units $c = 1 = G$.) This horizon serves as a one-way membrane: physical objects (or observers) can only cross it from the past to the future. The black hole's interior, region II, is located to the future of the event horizon. As can be seen in Fig. 1.1, the spacelike $r = 0$ singularity completely blocks the future of region II. Once an observer has crossed the event horizon, he is doomed to hit this singularity. The spacetime curvature diverges at $r = 0$, as can be seen, for example, from the Kretschmann scalar $R_{\alpha\beta\gamma\delta}R^{\alpha\beta\gamma\delta}$, where $R_{\alpha\beta\gamma\delta}$ is the Riemann-Christoffel curvature tensor. For Schwarzschild, one obtains

$$R_{\alpha\beta\gamma\delta}R^{\alpha\beta\gamma\delta} = 48M^2/r^6. \tag{1.3}$$

Equation (1.3) implies that the tidal force becomes unbounded at $r = 0$. Any physical object that falls towards this singularity is completely torn apart by the infinite tidal force, which unboundedly compresses it in two directions ($d\varphi$ and $d\theta$) and at the same time stretches it in the radial direction [22].

The traditional view of the nature of the black hole's singularity was strongly influenced by the above features of the Schwarzschild geometry, as well as by similar features of other spherically symmetric solutions like Oppenheimer-Snyder [15] and Robertson-Walker. The later analysis by Belinsky, Khalatnikov, and Lifshitz (BKL) [23, 24, 25] provided additional support for this view. In that project, BKL investigated a class of inhomogeneous solutions that develop a succession of Kasner epochs in which the axes of contraction and expansion chaotically change directions and terminate at unbounded oscillations at a spacelike singularity—the so-called BKL singularity. (Cf. the "mixmaster" singularity [26].) The analysis by BKL suggests that this type of singularity is generic (namely, it depends on a sufficient number of "degrees of freedom"—arbitrary functions of the three spatial coordinates). Moreover, BKL argued that this is the only possible generic type of spacelike singularity. The BKL singularity is similar to Schwarzschild in that it is spacelike and destructive (although, unlike Schwarzschild, the BKL singularity is oscillatory).

A new phase in the investigation of the inner structure of black holes began in 1960, when Graves and Brill [27] discovered the causal structure of the RN solution (describing the unique spherically-symmetric, electrically-charged black hole). This solution is still described by the line-element (1.1), but this time with the metric function

$$f = 1 - 2M/r + Q^2/r^2, \tag{1.4}$$

where Q is the black hole's electric charge (we assume $0 < |Q| < M$). Graves and Brill realized that the inner structure of RN is dramatically different from that of Schwarzschild. This difference stems from the fact that in RN (1.4), the equation

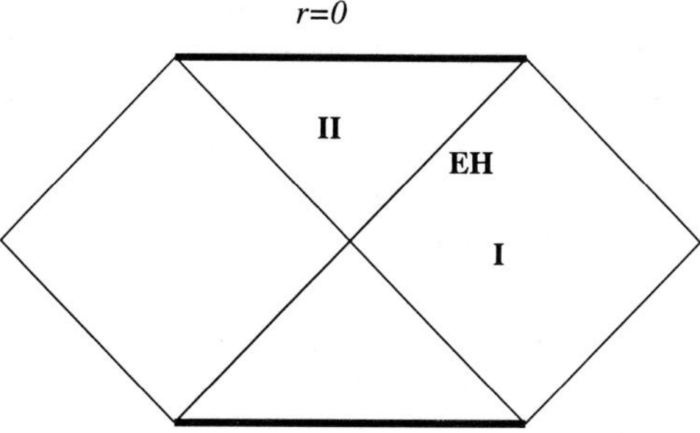

Figure 1.1: Penrose diagram of the extended Schwarzschild geometry. EH denotes the event horizon. The asymptotic region I is "our" external universe, and region II is the black hole's interior. Singularities are displayed by thick lines.

$f = 0$ has two roots:

$$r_{\pm} = M \pm \left(m^2 - Q^2\right), \tag{1.5}$$

[as opposed to the Schwarzschild case (1.2), where there is only one root, $r = 2M$, which corresponds to the event horizon]. The null hypersurfaces $r = r_+$ and $r = r_-$ are known respectively as the event horizon and the inner horizon of RN. The Penrose diagram of the (analytically extended) RN solution is shown in Fig. 1.2. The (double) $r = 0$ singularity is located to the future of the inner horizon, in regions III and III'. It is remarkable that this singularity is timelike, and does not block the way to the future. The geodesic equation may be easily solved, thereby showing that infalling (test) objects do *not* strike the $r = 0$ singularity: instead, after crossing the inner horizon, they arrive at a minimal $r > 0$ value, where r begins to increase (in a time-symmetric manner). Eventually, these objects are ejected through a "white hole" (region IV), into another asymptotically-flat universe (region V)—in sharp contrast with the behavior in the Schwarzschild geometry. A typical such worldline is shown in Fig. 1.2.

Motivated by this unusual causal structure of RN, Novikov [28] suggested that perhaps, due to deviations from spherical symmetry or quantum effects, in a realistic collapse scenario the collapsing matter (and also subsequent infalling bodies) will not crash into a central singularity: instead, it will emerge through a white hole into another universe. This scenario, known as gravitational bounce, was then farther investigated by de la Cruz and Israel [29]. An investigation of the collapse of charged dust spheres [28, 30] and of charged spherical thin shells [29] indicated that a transition from contraction to expansion can in principle occur in a completely regular manner.

(See, however Ref. [31]. The general solution of the Maxwell-Einstein equations for spherical configurations of charged dust is given in Ref. [32].)

The Kerr solution, describing a stationary, rotating, vacuum black hole, was discovered in 1963 [33]. In Boyer-Lindquist coordinates, it takes the form

$$\begin{aligned} ds^2 &= -\left(1-\frac{2Mr}{\rho^2}\right)dt^2+\frac{\rho^2}{\Delta}dr^2+\rho^2\,d\theta^2-\frac{4Mra}{\rho^2}\sin^2\theta\,d\varphi\,dt \\ &+ \left[r^2+a^2+2Mr\left(\frac{a}{\rho}\right)^2\sin^2\theta\right]\sin^2\theta\,d\varphi^2, \end{aligned} \tag{1.6}$$

where $\rho^2 \equiv r^2 + a^2\cos^2\theta$ and $\Delta \equiv r^2 - 2Mr + a^2$. This solution (1.6) depends on two parameters: the mass M, and the specific angular momentum (i.e., angular momentum per unit mass) a. The causal structure of Kerr is shown in Fig. 1.3. The similarity to the causal structure of RN is remarkable. In a manner analogous to Eq. (1.4), we may define for Kerr

$$f = \Delta/r^2 \equiv 1 - 2M/r + a^2/r^2. \tag{1.7}$$

As in RN, the event and inner horizons of Kerr are located at $r = r_+$ and $r = r_-$, respectively, where r_+ and r_- are the two roots of the equation $f = 0$, namely

$$r_\pm = M \pm \left(M^2 - a^2\right) \tag{1.8}$$

[cf. Eq. (1.5)]. We assume $0 < |a| < M$ (the equation $f = 0$ has no roots—and, correspondingly, there is no black hole—if $|a| > M$). Here, again, the curvature singularity at $r = 0$ is timelike and gravitationally-repulsive, and it does not block the way to the future. Typical infalling orbits will generally avoid the singularity and arrive at another external universe. Fig. 1.3 shows three such typical orbits. The main difference between the inner structure of RN and Kerr is that in the latter the singularity is in the shape of a ring (this is not evident from Fig. 1.3, because in this figure two spatial dimensions have been suppressed). Hence, infalling observers may travel through this ring and arrive at region VI—an asymptotically-flat universe that lies beyond the ring singularity (this region has no analogue in RN). This possibility is manifested by orbit 3 in Fig. 1.3. More details about the Kerr geometry can be found, e.g., in Ref. [34].

It is quite unlikely that realistic astrophysical black holes will be significantly charged: the electrostatic field induced by a net positive charge will act to balance itself by accreting negative charges, and vice versa. On the other hand, we do expect realistic black holes to have a significant amount of angular momentum: first, typical astrophysical objects are generally rotating. Even if a star originally rotates very slowly, conservation of angular momentum implies that it will speed up upon contraction (in gravitational collapse). Second, if the black hole has an accretion disc, the accretion process will presumably further increase the black hole's angular

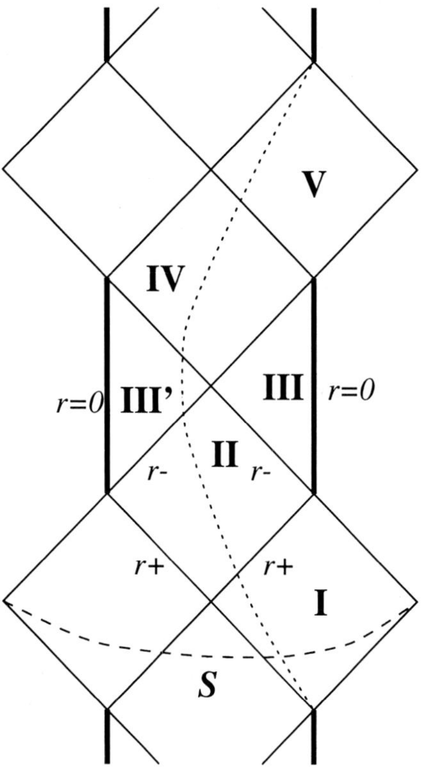

Figure 1.2: Penrose diagram of the extended Reissner-Nordström geometry. The event and the inner horizon are located at $r = r_+$ and $r = r_-$, correspondingly. The asymptotic region **I** is "our" universe. The surface **S** (dashed) is a typical initial hypersurface (partial Cauchy surface). The dotted curve denotes a typical timelike geodesic that falls into the black hole and then emerges into another external universe (region V). Singularities are depicted by thick lines

momentum, resulting in a ratio a/M that is very close to 1 [35]. (Note that $a = 0$ is the Schwarzschild limit, and $a = M$ is the maximal possible value for a Kerr black hole). In fact, Thorne [36] has found that for a typical black hole with an accretion disc, $a/M \approx 0.998$ may be anticipated [37]. Thus, of the above three simple black hole solutions (Schwarzschild, RN, and Kerr), the Kerr solution appears to be most interesting from the physical point of view. The exotic inner structure of Kerr should therefore be regarded, *a priori*, as the most natural candidate for the inner structure of realistic black holes. This observation yields important support for the idea of gravitational bounce. (We shall see, however, that this expectation is somewhat naïve.)

Although the RN solution is rather unrealistic, most previous attempts to explore the inner structure of black holes (in particular, the phenomenon of gravitational bounce; and the infinite blue shift that we discuss below) were based on the RN solution as a toy-model for Kerr. The reason is clear: the RN solution is much simpler than Kerr (it is spherically-symmetric), and yet its inner causal structure is fairly similar to that of Kerr. The simplest way to consider the instability of the inner horizon (see below) does not depend on the details of the spacetime geometry, but rather on the causal structure. It may therefore be hoped that the qualitative geometric aspects observed in a spherically-symmetric charged black hole will be relevant for uncharged, spinning black holes as well.

The inner-horizon instability

From the Penrose diagrams in Figs. 1.2 and 1.3 it is clear that in both the RN and Kerr solutions, the inner horizon is a Cauchy horizon for a typical initial hypersurface (a partial Cauchy surface) **S** in the external universe [38].

Consequently, specifying the initial data for some physical field (e.g., an electromagnetic or a gravitational field) on **S** will uniquely determine the evolution of that field up to the inner horizon—but the evolution in region III (or IV , V , etc.) is nonunique. Intuitively, this nonuniqueness occurs because region III is accessible to data propagating from the naked singularity at $r = 0$, or (in Kerr) from the asymptotically-flat universe VI. This breakdown of predictability is especially disturbing because the inner horizon itself is perfectly regular. This raises the following question: to what extent is the smooth inner horizon of RN or Kerr generic? Namely, would this feature be stable under small perturbations in the initial data (e.g., those specified on **S**)?

Penrose [16] was the first to point out that the inner horizon of RN is expected to be inherently unstable, and infalling radiation (whether electromagnetic or gravitational) is likely to convert it into a curvature singularity. His argument was based on the geometrical-optics approximation, according to which the infalling radiation is assumed to propagate along ingoing null rays. From the Penrose diagram 1.2 it is then evident that the radiation which falls into the RN black hole during the entire

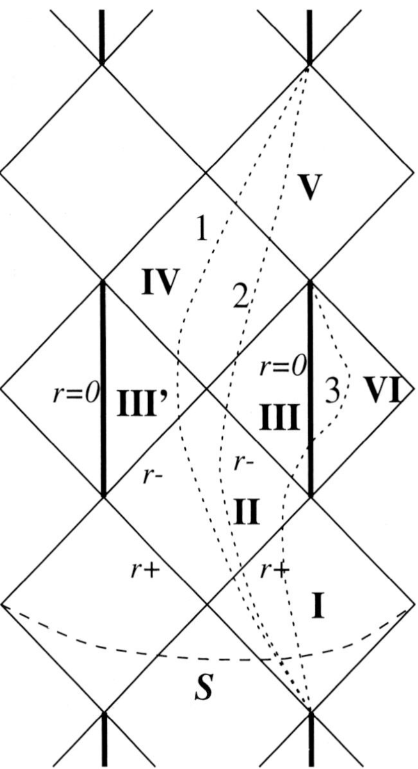

Figure 1.3: Penrose diagram of the extended Kerr geometry. The event and the inner horizon are located at $r = r_+$ and $r = r_-$, correspondingly. The asymptotic region **I** is "our" universe. The surface **S** is a typical initial hypersurface (partial Cauchy surface). The dotted curves numbered 1–3 denote three classes of timelike infalling orbits. While orbits 1 and 2 approach future timelike infinity of the external universe V, orbit 3 goes to the future timelike infinity of the external universe VI, located beyond the $r = 0$ ring singularity. Singularities are depicted by thick lines.

history of the external world (region I) accumulates at the inner horizon. A free-falling observer will reach the inner horizon at a finite proper time. He will thus "see" the infinitely long history of the external universe "flashing before his eyes" within a finite lapse of proper time. This means that the infalling radiation will be unboundedly blue-shifted at the inner horizon, with consequent divergence of the radiation's energy-density. In turn, such a divergence of the energy-momentum tensor may lead (via the Einstein equations) to a divergence of curvature, i.e., to the formation of a spacetime singularity. This phenomenon is known as the blue-sheet instability (or, in some papers, the blue-shift instability). Penrose [16] stressed that the place to look for a curvature singularity inside a black hole is specifically there, at the blue-shifted inner horizon. Hereafter, we shell refer to such a singularity formed due to the instability of the inner horizon as the inner-horizon singularity.

The above argument is not valid if the radiation influx from the external universe decays sufficiently quick. In discussing the influx rate, it is convenient to use the Eddington-like ingoing and outgoing coordinates, v and u, defined by $v = t + r_*$ and $u = t - r_*$ where r_* is the generalization of the Regge-Wheeler 'tortoise' coordinate defined by $r_*(r) = \int (r^2 + a^2)/\Delta \, dr$.

The blue-shift factor at the inner horizon grows exponentially with v. Thus, in order to prevent the blue-shift instability, the radiation influx must decay at least exponentially with v. In reality, one expects the radiation influx to decay much slowlier, because of two reasons: First, unless the collapsing object is strictly spherical (an unrealistic situation, especially because astrophysical objects are generally rotating), gravitational radiation must be produced during the collapse. Part of this radiation propagates away from the black hole, and is then scattered (off the spacetime curvature) down into the hole. This process has been analyzed by Price [39, 40] for the case of almost spherically-symmetric collapse. He found that the ingoing radiation decays like $v^{-(2l+2)}$, where l is the multipolar index of the mode under consideration. (See also more recent works in Refs. [41, 42, 43, 44]. For a generalization of Price's analysis to RN, see [45].) Since no monopole or dipole modes of gravitational radiation exist, at late time (i.e., large v) the ingoing perturbations will generally be dominated by the quadrupole mode ($l = 2$), and will thus decay like v^{-6}. Consequently, the blue-shift instability appears to be unavoidable. Second, black holes are not strictly isolated. Rather, they reside in the universe, and consequently constantly capture radiation of various sources. Of those sources, the slowliest decaying one may be the cosmic background radiation. It turns out that this radiation source may be considered as a constant-amplitude radiation (indeed, due to the expansion of the universe the cosmic background radiation cools down, but only on a time scale comparable to the age of the universe. For more details see below).

Linear perturbation analyses

The above simple argument for the divergence of the blue shift was based on the geometrical-optics approximation. The next obvious step is relaxing this approximation, and studying the behavior of dynamical fields (e.g., scalar, electromagnetic, or gravitational fields) near the inner horizon, using their appropriate wave equations. The first attempt in this direction was undertaken by Simpson and Penrose [46], who numerically studied the evolution of linear electromagnetic perturbations in RN. This analysis demonstrated the divergence of the electromagnetic field (and thereby of the associated energy-momentum tensor) at the inner horizon, for quite generic initial data outside the black hole. This confirmed the above-mentioned qualitative argument made by Penrose [16].

The numerical investigation by Simpson and Penrose [46] was later followed by an analytic calculation by Gürsel *et al.* [47, 48], who analyzed the evolution of perturbations in RN, from the event horizon to the inner horizon. They studied both scalar and combined electromagnetic-gravitational perturbations [49]. As initial data, they considered the inverse power-law decay at the event horizon, in accord with Price's [39] results. This analysis, too, demonstrated the occurrence of the blue-sheet instability: The electric field, as well as the scalar field's gradient, became unbounded at the inner horizon, thus indicating the divergence of the energy-momentum tensor (and the curvature) there. These results were later confirmed by an analytic calculation by Chandrasekhar and Hartle [50]. The technique used in both works ([47, 48, 50]) was basically similar and based on the following stages: (i) the initial perturbations at the event horizon are first decomposed into Fourier modes (and spherical harmonics); (ii) the evolution of each mode is then followed up to the inner horizon; and (iii) the perturbations at the inner horizon are recovered by a Fourier integration over all modes. Novikov and Starobinsky [51] also used a similar technique to analyze scalar-field perturbations in Kerr (assuming $v^{-(2l+2)}$ initial perturbations at the event horizon), and obtained similar results: the field's gradient becomes unbounded at the inner horizon. (See also on the instability of the inner horizon of RN Refs. [52, 53, 54, 55, 56] and on Kerr Ref. [57].)

It is remarkable that in both analyses ([47, 48, 50]), and for all types of fields investigated, the results were qualitatively the same: the fundamental fields themselves—i.e., the scalar field, the electromagnetic four-potential, and the metric perturbations—were all found to be well-behaved at the inner horizon. Moreover, at the inner horizon itself, these fields all vanish like $|u|^{-(2l+2)}$ as $u \to -\infty$ (i.e., at the past "edge" of the inner horizon) [47, 48]. It is only the gradients of these fields that diverge at the inner horizon. The same situation was found for a scalar field in Kerr [51]. This behavior is, in fact, in full harmony with the qualitative notion of "blue-shift instability": in terms of the fundamental fields, the shift toward the blue merely indicates an increase in the rate of variation (and hence in the gradients) of the fields, but usually not in their amplitude [58].

It is important to note that despite the finiteness of the fundamental fields, the divergence of gradients indicates a genuine curvature singularity—either directly (due to the infinite gradients of metric perturbations), or via the Einstein equations (due to the diverging energy-momentum tensor of, e.g., the electromagnetic field). The fact that the metric perturbations are finite at the inner horizon (moreover, they become arbitrarily small for large $|u|$) means, however, that the metric tensor itself is well-defined and non-singular there [59], despite the divergence of the Riemann tensor. Following Tipler [60], we refer to such a curvature singularity as a weak singularity. (A weak singularity may also be characterized by the rate of growth of the Riemann tensor components along a timelike or null geodesic that terminates at the singularity [60].) The weakness of the singularity may have important physical consequences: an evaluation of the deformation induced by the tidal force shows, that if the singularity is weak, then the deformation is finite (and, in some cases, extremely small and even undetectable). If, conversely, the singularity is strong (e.g., the Schwarzschild $r = 0$ singularity) then the tidal deformation is unbounded, which means that the infalling object is completely torn apart as it approaches the singularity.

Thus, if we were to regard the linear perturbations obtained in Refs. [47, 48, 50] as a genuine representative of the full (nonlinear) perturbations, we would arrive at the following picture: in a realistic black hole, due to ingoing perturbations, the inner horizon is converted into a curvature singularity; however, at least for large $|u|$, this singularity is null and weak—in remarkable contrast to the Schwarzschild singularity, which is spacelike and strong.

It is often stated that an infalling observer "measures" an infinite energy density upon arrival at the Cauchy horizon. Indeed, the linear analyses by Gürsel *et al* [47, 48] and by Chandrasekhar and Hartle [50] found that the projection of the energy-momentum tensor on the timelike worldline of the observer (or of the measurement device) diverges on the Cauchy horizon. More precisely, it was found that $T_{\mu\nu}u^{\mu}u^{\nu}$ diverges, where $T_{\mu\nu}$ is the energy-momentum tensor and u^{μ} is the observer's four-velocity. However, this does not guarantee a *measurement* of infinite energy. Whenever one speaks of a measurement, one actually considers the interaction of some measuring apparatus with the radiation: one should consider a specific apparatus, and analyze the interaction of this apparatus with the radiation fields. Of course, one could consider a plethora of different complicated devices. However, the essence of the measurement process can be captured by a simple model of a physical object and an analysis of its interaction with the Cauchy horizon singularity. We emphasize that unless such an analysis is made, there is only little sense in statements on measurements that observers make, and all that is meant is that $T_{\mu\nu}u^{\mu}u^{\nu}$ diverges.

A crucial question then is the following: how is an extended physical object affected by the divergent fields which arise from the linear analyses? In the Kerr spacetime there are also electromagnetic perturbations, which—like the scalar and gravitational perturbations—are also infinitely blue-shifted at the inner horizon. Because the RN geometry is much simpler than the Kerr geometry, the former has often been used as a simplified model. There are two sources for these electromagnetic perturbations. The first source for electromagnetic perturbations in RN results from the evolution of non-vanishing electromagnetic multipole moments (in the star) during the collapse, which are scattered off the spacetime curvature and captured by the black hole. It turns out that this process too leads to 'tails' at the event horizon, which decay at late times like $v^{-(2l+2)}$ (cf. the gravitational perturbations above) [45]. These perturbations are then infinitely blue-shifted at the inner horizon, which causes the electromagnetic field to diverge. However, an explicit calculation shows that the interaction of this electromagnetic field with physical objects is bounded, and is negligible for large $|u|$ [63, 64]. (This bounded interaction is closely related to the finiteness of the electromagnetic four-potential.) A second source, present for any realistic black hole, is the photons which originate from the cosmic background microwave radiation [65]. The energy of these photons indeed cools down as the universe expands, but the time-scale relevant for this cosmological cooling is much larger than the external time-scale of the captured photons which hit the infalling object until it arrives at distances comparable to the Planck scale from the inner horizon. Therefore, it is reasonable to consider constant-temperature photons, which are constantly captured by the black hole. This constant-temperature radiation source causes a non-integrable effects on infalling objects. However, if one introduces a cutoff at the Planckian scale, one finds again that the extent of this interaction is bounded, and quite small. Even though this interaction, although bounded, would most probably be fatal for a human astronaut who wishes to explore the black hole interiors, smaller and/or less sensitive objects may survive this interaction. In addition, the energy-momentum of these photons also act as a source term for the Einstein equations. However, with the Planckian cutoff, this energy density is also expected to be bounded. Therefore, despite the relative strength of this cosmological source, it is still possible for the inner horizon singularity to be weak.

A closely-related problem is the effect on extended objects by tidal distortions. The weakness of the singularity is manifested by finite (and for typical parameters—and large $|u|$—even negligible) tidal deformations. However, there are other effects which may be of importance. Herman and Hiscock [61] argued that although the distortions of the object are finite, their gradients diverge. Namely, that the internal velocities, and consequently also the absorbed kinetic energy and increase in temperature and entropy are unbounded. As a result, any extended object would be vaporized and destroyed upon approach to the Cauchy horizon. However, a closer examination [62] shows that at least up to the Cauchy horizon no such phenomenon occurs, and there is no reason to expect infinite energy absorption.

How much of this picture of a null and weak singularity remains valid when one considers nonlinear perturbations as well? Answering this question would require a fully nonlinear perturbation analysis. One would naturally try to employ the standard method of nonlinear perturbation expansion for that purpose. Unfortunately, the Fourier-decomposition method used in Refs. [47, 48, 50] is apparently rather difficult to generalize to nonlinear orders. This prevented any systematic nonlinear perturbation analysis for many years. In the meantime, though, it had been widely speculated that the second- and higher-order metric perturbations should diverge at the inner horizon, and thereby produce a strong, spacelike singularity. This expectation stemmed from the qualitative features of the field equations obtained in the nonlinear perturbation expansion. In that method, the original nonlinear field equation is converted into an infinite set of linear field equations, one for each order. The homogeneous part of these equations is the same for all orders, but the source term (the inhomogeneous part) is order-dependent. Typically, the source term for a given order is quadratic in first-order derivatives of lower-order perturbations (obviously, the field equation for the first order, i.e. the linear perturbation, has no source term). In particular, the source term for the second-order metric perturbations is quadratic in gradients of linear metric perturbations. This source term thus diverges at the inner horizon. It has often been assumed, for one reason or other, that this divergence of the source term will cause the divergence of the second-order metric perturbations as well. (See, e.g., the discussion in [50].) In the discussion below we shall refer to this argument as the "nonlinear divergence argument." (We point out, however, that this argument is incorrect; the divergence of the source terms does not imply the divergence of the second- or higher-order perturbations themselves. This has been illustrated in Ref. [66], and will be further verified in Refs. [67, 68] from the finiteness of nonlinear perturbations.)

Simplified nonlinear models and mass inflation

Should the second-order or higher-order perturbation terms be divergent at the inner horizon (as was widely expected during the 1980's) it would clearly signify the breakdown of the entire perturbation approach there [50]. Possibly because of this expectation, or because of the enormous technical difficulties, no systematic nonlinear perturbation analysis of the black hole interior has been carried out until recently [66]. There have been, however, important attempts to model these nonlinear perturbations by means of simplified models. Thus, Hiscock [17] suggested that in order to gain insight into the back-reaction gravitational effect of infalling radiation on the background geometry, one could mimic the highly blue-shifted radiation with a stream of massless particles (a null fluid). This led to a model consisting of a RN background perturbed by a spherically-symmetric radial inflow of massless particles. Very significantly, this situation can be described by an exact solution—the so-called charged Vaidya solution [69] (a generalization of the RN solution in which the mass becomes a function of the null coordinate v). Motivated by Price's results [39, 40], Hiscock considered an ingoing flux which at the event horizon decays like an inverse power of v. He found that the regular inner horizon of RN is indeed replaced, in the charged Vaidya solution, by a curvature singularity. (This singularity proves to be null and weak.) Hiscock's model [17] thus confirmed previous expectations, based on linear perturbation analyses. The singularity obtained in Hiscock's model, however, appears to be rather special, in the following sense: all curvature scalar polynomials (i.e., scalars obtained by contractions over products of the Riemann tensor) were found to be strictly finite at this singularity (i.e., a non-scalar singularity, or a "whimper" singularity [70]). Some researchers regarded this unexpected feature as an indication that Hiscock's model is an over-simplification: more realistic models would have to take into consideration also an *outgoing* flux, because of the scattering off the spacetime curvature *inside* the black hole (see below).

In order to obtain a more realistic description of the inner-horizon singularity, Poisson and Israel [71, 72, 73] constructed a model with two streams of spherically-symmetric null fluids—an ingoing one and an outgoing one. Such a two-fluid picture is very reasonable: the ingoing null fluid represents the ingoing gravitational radiation; the outgoing flux may represent the portion of this ingoing radiation that is scattered off the spacetime curvature (inside the black hole) and thus irradiates the inner horizon [74]. It is only the ingoing flux which is infinitely blue-shifted at the inner horizon. Yet, the outgoing flux has a remarkable effect on the inner-horizon singularity: in the presence of the outflux, the mass-function M (see definition in Ref. [73]) diverges at the singularity. (For comparison, note that this function is finite everywhere in Hiscock's model.) Although it was not possible at that time to provide an explicit expression for the metric functions near the singularity, Poisson and Israel were able to show that M grows exponentially in the ingoing null coordinate v (this coordinate is infinite at the inner-horizon singularity). Because of this exponential divergence of M, Poisson and Israel named this phenomenon "mass inflation." The diverging mass function indicates the divergence of some curvature scalar polynomials; For example, the Kretschmann scalar $R_{\alpha\beta\gamma\delta}R^{\alpha\beta\gamma\delta}$ diverges like M^2.

The mass-inflation model signified an extremely important milestone in the investigation of black holes' interiors—a field which was more or less dormant since the work of Hiscock [17] in 1981 and that of Chandrasekhar and Hartle [50] in 1982. For the first time, it was felt that we possess a reasonable (even if somewhat simplified), fully-nonlinear model of the singularity inside realistic black holes.

The exponential divergence of the mass function (and of curvature scalars) was naturally regarded by Poisson and Israel and others as an indication for the physical effectiveness and significance of the mass-inflation singularity. In the absence of an explicit solution for the metric functions in the mass-inflation model, however, it was difficult to evaluate the actual physical strength and other features of that singularity. This motivated Ori [75] to consider a simplified version of the mass-inflation model, in which the outgoing flux has the shape of a short burst (i.e., a delta-function in u). Such a short burst can be regarded as a null thin layer. The entire geometry can thus be described by matching two patches of the charged Vaidya solution along that null layer. By this method, Ori obtained an explicit expression for the metric functions (in double-null coordinates) near the inner-horizon singularity. In this model, too, the mass-function diverges exponentially at the singularity. Quite surprisingly, however, it turned out that the metric tensor has a well-behaved (i.e., finite and non-degenerate) limit at the mass-inflation singularity. In other words, the mass-inflation singularity is weak (in Tipler's terminology [60], which is employed throughout this paper). The tidal deformation experienced by an infalling object thus remains finite all the way up to the singularity. Although the model considered by Ori is only a simplified variant of the original mass-inflation model, it apparently captures the central features of the mass-inflation singularity [76]. A straightforward calculation (based on a simple iteration scheme) reveals that, for relevant fluxes of ingoing and outgoing radiation, the singularity in the original (smooth) mass-inflation model, as well, is null and weak [77]. Moreover, the singularity becomes arbitrarily weak as $u \to -\infty$.

Recently, there have been several interesting developments in this field. First, Bonanno *et al* [78] carried out an approximate analytical calculation of the asymptotic behavior at the inner-horizon singularity, for two variants of the spherically-symmetric mass-inflation model: (i) the original model of two null fluids, and (ii) the case in which the null fluids are replaced by a self-gravitating massless scalar field. In both cases, the asymptotic behavior at the singularity was found to be qualitatively similar to the simplified mass-inflation model in Ref. [75] and to the analysis in [77].

Despite these recent advances, our understanding of the null weak inner horizon singularity is still far from being complete. In particular, it is important to verify this picture by performing independent, *non-perturbative* analyses. This motivates one to employ numerical tools to study the structure of the inner horizon singularity. The numerical simulation of spinning black holes is difficult, as they are not spherically symmetric. Therefore, one is naturally led to study numerically, as a toy-model, the inner structure of spherically-symmetric charged black holes. As there are no radiative modes for the gravitational field in spherical symmetry (nor for the electromagnetic field), one usually mimics the field dynamics by the nonlinear evolution of a scalar field. This approach was first implemented by Gnedin and Gnedin [79, 80]. However, the analysis by Gnedin and Gnedin yielded conflicting results. They found that a spacelike singularity—which resembles the Schwarzschild singularity—forms, such that it completely precedes the inner horizon. This is in sharp contrast with the picture of a null and weak singularity. A closer look, however, reveals that the special null coordinates used in [79, 80] are not suitable for a numerical study of the region near the inner horizon: With these coordinates, even if one uses a very large grid (say, 10000×10000), there are no grid points in the relevant range $v \gg M$. However, the results by Gnedin and Gnedin raise the possibility that perhaps deep inside the black hole one would indeed find a spacelike singularity, in addition to the null and weak singularity at earlier times. The possibility that this might be the case was hinted in 1991 by Page [81], who studied the interiors of a spherical charged black hole with (ingoing and outgoing) radial null radiation, under the simplifying assumption of homogeneity. Page found that the singularity at $r = 0$ is spacelike rather than null, but could not determine if in inhomogeneous models the singularity at early times was null.

Brady and Smith [82] carried out a numerical simulation of the above scalar-field model (the same model used in Refs. [79, 80]). The numerical results seem to be consistent with the analyses in Refs. [75, 77] in that they show that the Cauchy horizon indeed becomes singular. Brady and Smith showed that under the outgoing flux of the scalar field the Cauchy horizon slowly contracts until it becomes spacelike at $r = 0$. In accord with Refs. [79, 80], Brady and Smith found that at some point the null singularity becomes spacelike. [This observation is not inconsistent with the perturbative approach presented in Ref. [77]: The perturbative approach is essentially based on an expansion around the past "edge" of the inner horizon (the point which, in the standard Penrose diagram (such as Figs. 1.2,1.3), intersects with future timelike infinity of the external universe I). This expansion is likely to be valid at some neighborhood of that point, but not far away. Thus, the perturbative approach does not provide significant information about the structure of the singularity (and spacetime, in general) deep inside the black hole, and in particular, the singularity may become spacelike (and strong) there.] Despite its remarkable achievements the analysis in Ref. [82] left several crucial questions unanswered. In particular, Brady and Smith did not analyze in detail the strength of the null portion of the singularity. In addition, they also indicated inconsistencies with the perturbative analyses, which are manifested by a non-vanishing deviation of the internal power-law indices from integer values. Brady and Smith estimated the value of their parameter σ (see [82]) to be non-zero beyond the numerical error. (The physical meaning of a non-zero value of σ is that the power-law index of physical fields inside the black hole is different from the powel-law index along the event horizon, even for an outgoing ray infinitesimally close to the event horizon.) However, perturbation theory predicts an identically vanishing value for this parameter σ. The validity of the predictions of the perturbative analyses is crucial, because they are the only tool we have at the moment for the study of the internal structure of spinning black holes. Consequently, another numerical analysis, which will study the strength of the null singularity in greater detail and which will question the value of σ, was needed. This was recently done by Burko [83]. Burko used the same model which was used by Refs. [79, 80, 82] and freely evolved the fields in double-null coordinates (see also [44]). This analysis found that the null singularity is indeed weak, and that in its early sections the metric perturbations are arbitrarily small, i.e., the values of the metric functions approach their RN counterparts at early times. In particular, Burko [83] found that the parameter σ of Ref. [82] identically vanish. Burko and Ori also considered the null and weak singularity analytically [84], and found exact analytical expressions for the blue-shift factors along the entire null singularity, in agreement with the numerical results in [83]. In particular, Ref. [84] shows analytically that σ must be constant along the CH. When combined with the linear results of [47] one can infer that this constant must equal zero, in accord with [83].

The emerging picture from the numerical analyses of [82, 83] is that the generators of the null singularity are monotonically focused to lower and lower values of r. Within a final lapse of affine parameter along the null singularity they are focused completely, namely, contract to $r = 0$, where the singularity becomes spacelike. The numerical analyses [82, 83] did not consider the details of the spacelike singularity. This was first studied by Burko, who was able to show that there indeed exists a generic solution for the model in question (under the simplifying assumption of homogeneity) with a spacelike singularity, with similar properties to those found numerically [85]. It is still an open question, however, whether a spacelike singularity would evolve also in more realistic models of black hole interiors.

Nonlinear perturbation analysis of Kerr

Despite this progress, an important question was still unresolved: Are the above-mentioned features of the mass-inflation singularity typical to realistic black holes, or just an artifact of the model? After all, the mass-inflation model is spherically-symmetric; the gravitational degrees of freedom cannot fully manifest themselves in such a model. Some researchers suspected that, in a generic (nonspherical) situation, the dynamics near the singularity would be dominated by the gravitational degrees of freedom, not by the matter fields. This expectation stemmed from the analysis of BKL [23, 24, 25], where it was found that usual matter-fields have a negligible effect on the BKL singularity. What then would be the effect of the pure gravitational degrees of freedom on the inner-horizon singularity, if strict spherical symmetry is relaxed? Would they leave the singularity null and weak, or convert it into a strong and spacelike one (presumably a BKL singularity)? Clearly, this question could not be answered in the framework of spherical symmetry. This forces one to generalize the mass-inflation model to a more realistic, nonspherical one.

Obviously, once we consider nonspherical (and nonlinear) gravitational perturbations, we do not need the null fluids any more. (In the previous spherically-symmetric models [17] and [71, 72, 73], the null fluids have been introduced in order to mimic the back-reaction effects of gravitational waves.) Also, since, in any event, we are forced to consider non-spherical dynamics, it is no longer so advantageous to use the unrealistic RN geometry as a background; the Kerr background may be employed as well. (In the previous models [17, 71, 72, 73, 75], it was crucial to consider the RN background, because the Kerr background does not admit spherically-symmetric perturbations.) Admittedly, even when dealing with nonspherical perturbations, the Kerr background is significantly more complicated than RN: in the former, the modes are spheroidal rather than spherical harmonics, the line-element is not diagonal, and the "dragging of reference frames" further complicates the analysis. On the other hand, the Kerr solution is conceptually simpler in that it is a vacuum solution. Thus, the natural perturbations of Kerr are the pure, vacuum, gravitational perturbations. (In RN, in contrast, there are no pure gravitational perturbations: all perturbation modes are combined gravitational–electromagnetic perturbations.) Beyond all these arguments of simplicity, the fact that in reality black holes are generally rotating (but are not charged) is critical, and should lead one to base the perturbation analysis on the Kerr background, rather than on that of RN.

The above discussion explained the need for a model of a black hole's interior which is non-spherical and at the same time fully nonlinear. Therefore, one would like to attempt a direct attack on a quite realistic model: the generically-perturbed vacuum Kerr black hole. Namely, we consider a vacuum rotating black hole, produced by the gravitational collapse of a generically-shaped spinning object. The prevalent view, succinctly expressed in the statement "a black hole has no hair," suggests that the external perturbations decay at late time. Since the initial data for perturbations inside the black hole are just the external perturbations, the presumed decay of the latter suggests that a small-perturbation approach might be applicable to the interior as well. One is thus led to the attempt of using the standard nonlinear perturbation expansion to analyze the black hole interior.

This approach is not self-evident: Since the above-mentioned "nonlinear divergence argument" suggests that the nonlinear perturbation terms are likely to diverge (and such a divergence would presumably signify the breakdown of the entire perturbative approach), it is not *a priori* certain that the standard nonlinear perturbation expansion would be applicable. However, the investigation of the mass-inflation model brings hope that the perturbation expansion might be well-behaved, after all. The mass-inflation field equations [73] are nonlinear, too. In the latter case, as well, it is natural to try an iteration scheme (which is, in some sense, analogous to a nonlinear perturbation expansion). The first term in this iteration scheme was calculated by Poisson and Israel [73], who found that the corresponding mass-function diverges (exponentially) at the singularity. Since the divergent mass function appears explicitly in the source terms for higher-order iterations, it was originally expected that the whole iteration scheme will be useless. It was later found, however, that when expressed in regular ("Kruskalized") double-null coordinates, the contribution of the first-iteration term to the metric functions is well-behaved, despite the divergence of the mass-function [75, 77]. Moreover, the contribution of higher-order terms was found to be finite, too (and even negligible) [77]. The whole iteration scheme thus appears to be well-behaved. This observation led Ori to apply a straightforward nonlinear perturbation expansion to the interior of a perturbed Kerr black hole [86].

Obviously, the implementation of this nonlinear perturbation expansion is not a simple task: it entails (a) generalizing the works of Simpson and Penrose [46], Gürsel *et al* [47, 48], Chandrasekhar and Hartle [50], and Novikov and Starobinsky [51] from linear to nonlinear perturbations; or (b) generalizing Hiscock's [17] and the mass-inflation [73] models from spherical to nonspherical perturbations, and from null fluids to metric perturbations; and (c) generalizing all these models (except that of Novikov and Starobinsky [51]) from RN to the more realistic Kerr background. The method used in previous linear perturbation analyses in order to evolve the perturbations from the event horizon to the inner horizon was based on a Fourier decomposition of the perturbations. This method is rather difficult to generalize to the nonlinear perturbations. This motivated Ori [56, 86] to design a simpler method, called the "late-time expansion." The main advantage of this method is that it is easily extended to the nonlinear orders (see [86]).

Thanks to the simplicity of the late-time expansion, Ori was able to implement the nonlinear perturbation expansion, and to obtain explicit expressions for the asymptotic form of linear as well as nonlinear metric perturbations near the inner horizon of Kerr. The main results were presented in Ref. [66]. These results confirmed the expectations based on the mass-inflation model: just like the linear term, the nonlinear metric-perturbation terms are all finite at the inner horizon (for $|u| \gg M$). Moreover, they all vanish at the limit $u \to -\infty$; the higher the order of a term in the perturbation expansion, the faster is its decay for large $|u|$. The curvature associated with the perturbations diverges at the inner horizon for all orders; but again, the higher the order, the weaker is the divergence. Thus, for large $|u|$ the whole expansion is dominated by the linear metric perturbation. (We note that curvature scalars, for example the Kretschmann scalar already diverged in the linear perturbation.) The singularity is weak and strictly null. It should be pointed out, though, that this singularity differs from the mass-inflation singularity in one important aspect: Whereas the mass-inflation singularity is monotonous, the inner horizon singularity of a generic spinning black hole is oscillatory (see [86]). We note that, although the calculations of [86] are fully analytic, they are not fully rigorous from the mathematical point of view. That is, the perturbations are given by an infinite series, in which indeed all terms vanish, but whose convergence has not been demonstrated. Nevertheless, this expansion *looks like* a well-behaved one: all the terms in this expansion are regular (that is, there are no diverging terms in the expansion). Moreover, all the expansion parameters tend to zero at the asymptotic limit in which we are interested (i.e., at the extreme past of the inner horizon). It is therefore conceivable (though not completely obvious) that, at least in some vicinity of this asymptotic limit, the expansion will indeed converge (at least asymptotically).

Recently, attempts have been made to examine the results of Ref. [66] from a non-perturbative, local, point of view. Thus, Brady and Chambers [87] found that the existence of a generic null weak vacuum singularity was consistent with the constraint equations. In addition, Ori and Flanagan [88] used a different construction and demonstrated that the vacuum null weak singularity was consistent with the full system of the Einstein equations—both with the constraints and evolution equations. This was more recently demonstrated also in the framework of plane-symmetric spacetrimes in Ref. [89].

According to the Authors' point of view, an important challenge now is to develop non-local, non-perturbative, mathematical methods for studying the formation of the null singularity, starting from regular, asymptotically-flat, initial data (corresponding to gravitational collapse of a spinning object), and follow the null singularity into the future to see whether it contracts to a spacelike singularity, or remains null throughout.

Classically stable horizons, and quantum mechanical effects

In the discussion above, we have considered the Reissner-Nordström and the Kerr black holes, where the inner horizon was shown to be unstable. One may wonder whether there exist configurations where the inner horizon would remain stable despite the perturbations. It can be readily shown that in the extreme Reissner-Nordström black hole (namely, a black hole with $Q = M$) the inner horizon is stable. This surprising result stems from the power-law divergence of the blueshift factor in the extreme case (compared with the exponential blueshift factor in the non-extreme case). However, extreme black holes constitute only a set of measure zero in parameter space, and would be very hard to create. It is intriguing therefore, that in non-asymptotically-flat models, parameters can be found for which the inner horizon is stable, for a non-zero measure in parameter space. That is, the inner horizon does not evolve into a spacetime singularity. Of course, these models are physically questionable, as they invoke the notion of a cosmological constant, and consequently they are not asymptotically-flat. (However, similar stability of the inner horizon may occur in non-asymptotically-flat spacetimes also with a vanishing cosmological constant. See, e.g., Ref. [90].) However, they are extremely interesting from the mathematical viewpoint; in addition, they enable us to consider a larger set in parameter space, which may deepen our understanding on the position General Relativity takes within all physically-sound relativistic theories of gravitation. In particular, if indeed in such theories the inner horizon does not evolve into a spacetime singularity, it bears much importance on the status of the cosmic censorship hypothesis, as there would be no obstacle for an infalling observer who is determined to probe the timelike singularity deep in the core of the black hole. The black hole-de Sitter spacetimes are dealt with exhaustively elsewhere in these proceedings.

In a series of papers [91, 92, 93, 94], it was shown that the inner horizon of the Reissner-Nordström-de Sitter black hole is classically stable, if the surface gravity at the inner horizon is smaller than the surface gravity at the cosmological horizon. This is physically possible in a universe with a very large cosmological horizon, if the black hole were very close to extremality. We have seen before, that astrophysically-likely black holes are indeed close to extremality [35, 36], which makes this possibility very interesting [95].

However, it has been argued that in situations where the inner horizon is classically stable, it would be quantum-mechanically unstable [96]. Namely, the classical treatment of the black hole interior would fail, and the realistic description of the internal structure of the black hole should be quantum mechanical. In the analysis of Marković and Poisson [96] the black hole is modeled by a 2-dimensional analog of the Reissner-Nordström-de Sitter black hole, and the quantum-mechanical backreaction was modeled by semi-classical effects. It was argued in [96], that the inner horizon turns into a singularity due to diverging vacuum polarization effects. If the semiclassical backreaction effects in the Marković-Poisson model [96] were formally extrapolated to arbitrary (higher than Planckian) curvature scales, one would obtain a strong singularity—that is, the total tidal deformations suffered by observers who hit the singularity would be unbounded—unlike singularities obtained from classical instabilities. However, the semiclassical approach is only valid as long as the curvature is sub-Planckian; at higher curvatures the treatment should be fully within the framework of the (as yet unknown) quantum theory of gravity. This is a situation analogous to the treatment of the electromagnetic effect of the cosmic background microwave photons, if extrapolated up to the inner horizon without introducing a cutoff at Planckian scales [65]. The effective strength of the singularity (in terms of tidal deformations) if a cutoff is introduced at Planck curvature scales seems to be weak. One might also worry to what extent the two-dimensional Marković-Poisson model is relevant for a four-dimensional spacetime. However, Ref. [96] raises an argument which suggests that that analysis may still be relevant for the four-dimensional case.

Quantum-mechanical effects may also be crucial for asymptotically-flat black holes. For a charged black hole Novikov and Starobinsky [97] showed that the effect of the electromagnetic pair production is to completely alter the spacetime geometry, well before the Planck regime. However, the situation for (uncharged) spinning black holes is different as there is no electromagnetic pair production, and the analogous process of gravitational pair production has a much weaker influence on the geometry of spacetime. This is the case as for electric and gravitational fields of comparable energy density, the number of gravitationally produced pairs is smaller than the number of electromagnetically produced pairs by a factor of $\sim L_p^2/(\alpha\mathcal{L}^2)$ where L_p is the Planck length, α is the fine structure constant, and $\mathcal{L}$ is the typical radius of curvature [97, 98, 99]. In addition, the charged pairs influence the geometry primarily indirectly through their strong influence on the electric field. Although the models of null weak singularities in charged black holes [73, 75, 100, 101] are thus unrealistic in that they ignore the above semi-classical effects (except for a certain range in the parameter space [97]), charged black holes can still be useful as a toy-model for the analysis of the gravitational pair production of neutral particles. In such analyses [100, 101] the model for the quantum-mechanical effects was the model of a free, linear, massless scalar test quantum field on mass inflation backgrounds. In order of magnitude, they find that quantum corrections to the geometry become important only near Planckian curvatures, where Quantum Gravity should take over.

Another semi-classical effect which should be accounted for is the back reaction of the evaporation of black holes through Hawking radiation [102]. Levin and Ori considered a very simplified toy-model [103], in which they modeled the evaporation by a flux of negative energy-density charged null fluid, that flows into a spherical charged black hole [104]. As charged null fluids bounce and then follow outgoing radial trajectories, it was shown that this effect was not expected to lead to a curvature singularity at the inner horizon. However, more realistic models and better understanding of quantum gravity are needed for a fuller understanding of this effect.

Bibliography

[1] S. W. Hawking and G. F. R. Ellis, *The large scale structure of space-time* (Cambridge University Press, Cambridge, 1973).

[2] J. A. Wheeler, in *Relativity, Groups, and Topology, Les Houches 1963*, edited by C. DeWitt and B. DeWitt (Gordon and Breach, New York, 1964).

[3] P. G. Bergmann, in *Some Strangeness in the Proportion*, edited by H. Woolf (Addison-Wesley, Reading, Massachusetts, 1980), p. 156.

[4] A. Einstein and N. Rosen, Phys. Rev. **48**, 73 (1935).

[5] S. W. Hawking, in *General Relativity: An Einstein Centenary Survey*, edited by S. W. Hawking and W. Israel (Cambridge University Press, Cambridge, 1979).

[6] C. W. Misner, Phys. Rev. **186**, 1328 (1969).

[7] C. J. S. Clarke, *The Analysis of Space-Time Singularities* (Cambridge University Press, Cambridge, 1993).

[8] J. Earman, *Bangs, Crunches, Whimpers, and Shrieks: Singularities and Acausalities in Relativistic Spacetimes* (Oxford University Press, Oxford, 1995).

[9] Note that unlike previous claims by Eardley [10], there is no indication for a local instability at the white-hole past horizon [11].

[10] D. M. Eardly, Phys. Rev. Lett. **33**, 442 (1974).

[11] A. Ori and E. Poisson, Phys. Rev. D **50**, 6150 (1994).

[12] M. D. Kruskal, Phys. Rev. **119**, 1743 (1960).

[13] G. Szekeres, Publ. Math. Debrecen **7**, 285 (1960).

[14] Actually, the conclusion that the singularity at the Schwarzschild radius is just a coordinate singularity and is removable by a proper coordinate transformation was first drawn by G. Lemaître, Annales de la Socièté Scientifique de Bruxelles

A 53, 51 (1933) [Gen. Relativ. Gravitation **29**, 641 (1997)]. However, Lemaître's insight was ignored at the time.

[15] J. R. Oppenheimer and H. Snyder, Phys. Rev. **56**, 455 (1939).

[16] R. Penrose, in *Battelle Rencontres, 1967 lectures in mathematics and physics*, edited by C. M. DeWitt and J. A. Wheeler (Benjamin, New York, 1968), p. 222.

[17] W. A. Hiscock, Phys. Lett. A **83**, 110 (1981).

[18] W. Israel, in *Black Hole Physics*, edited by V. De Sabbata and Z. Zhang (Kluwer, Dordrecht, 1992).

[19] We shall not consider the Kerr-Newman solution in this paper, for the following reasons (which will be further clarified below): it is neither realistic (we expect realistic black holes to be electrically neutral), nor simple (it is nonspherical). While the RN solution is often used as a toy-model for Kerr, the Kerr-Newman solution is not useful for that purpose, as it is obviously more complicated than Kerr.

[20] S. Chandrasekhar, *The Mathematical Theory of Black Holes* (Oxford University Press, Oxford, 1983).

[21] I. D. Novikov and V. P. Frolov, *Physics of Black Holes* (Kluwer, Dordrecht, 1989).

[22] The tidal deformation might be somewhat delayed by the internal forces (e.g., elasticity) in the case of a rigid body. But in view of the divergence of the tidal force at the singularity, the internal forces will eventually be negligible, and will not prevent the unbounded deformation at $r = 0$. (For, if the deformation were to remain finite, the internal forces would be finite, too; and the tidal forces are unbounded.) One may also hope that quantum effects will modify this picture, but these effects are only expected to occur at the Planck level (e.g., when the radius of curvature and/or the proper-time distance from the singularity is comparable to the Planck length). Therefore, even if, in principle, the quantum effects will prevent the singularity, in the Schwarzschild case these effects are unlikely to save a macroscopic extended object from complete breakup (presumably into sub-atomic debris) near $r = 0$.

[23] V. A. Belinsky and I. M. Khalatnikov, Zh. Eksp. & Teor. Fiz. **57**, 2163 (1969) [Sov. Phys.-JETP **30**, 1174 (1970)].

[24] I. M. Khalatnikov and E. M. Lifshitz, Phys. Rev. Lett. **24**, 76 (1970).

[25] V. A. Belinsky, I. M. Khalatnikov and E. M. Lifshitz, Usp. Fiz. Nauk. **102**, 463 (1970) [Advances in Physics **19**, 525 (1970)].

[26] C. W. Misner, Phys. Rev. Lett **22**, 1071 (1969).

[27] J. C. Graves and D. R. Brill, Phys. Rev. **120**, 1507 (1960).

[28] I. D. Novikov, Zh. Eksp. & Teor. Fiz. Pis'ma 3, 223 (1966) [Sov. Phys.-JETP Lett **3**, 142 (1966)].

[29] V. de la Cruz and W. Israel, Nuovo Cimento **51A**, 744 (1967).

[30] I. D. Novikov, Astron. Zh. **43**, 911 (1966) [Sov. Astron.-A. J. **10**, 731 (1966)].

[31] A. Ori, Phys. Rev. D **44**, 2278 (1991).

[32] A. Ori, Class. Quantum Grav. **7**, 985 (1990).

[33] R. P. Kerr, Phys. Rev. Lett. **11**, 237 (1963).

[34] B. O'Neill, *The Geometry of Kerr Black Holes* (A. K. Peters, Wellesley, Massachusetts, 1995).

[35] J. M. Bardeen, Nature **226**, 64 (1970).

[36] K. S. Thorne, Astrophys. J. **191**, 507 (1974).

[37] This value for a/M holds both for a stellar black hole which accretes matter from a stellar companion and for a super-massive black holes which accretes matter from a galaxy around it [36].

[38] In the analytically-extended geometry shown in Figs. 1.2 and 1.3 both branches of the inner horizon belong to the Cauchy horizon. We, however, shall actually be interested in black holes produced by gravitational collapse. In such black holes, the left-handed mirror image of region I does not exist; it is replaced by the geometry of the star's interior. Consequently, in a realistic black hole only the right-handed branch of the inner horizon is a Cauchy horizon.

[39] R. H. Price, Phys. Rev. D **5**, 2419 (1972).

[40] R. H. Price, Phys. Rev. D **5**, 2439 (1972).

[41] C. Gundlach, R. H. Price and J. Pullin, Phys. Rev. D **49**, 883 (1994).

[42] C. Gundlach, R. H. Price and J. Pullin, Phys. Rev.D **49**, 890 (1994).

[43] R. Gomez, J. Winicour and B.G. Schmidt, Phys. Rev. D **49**, 2828 (1994).

[44] L. M. Burko and A. Ori, Phys. Rev. D **56**, 7280 (1997).

[45] J. Bičák, Gen. Relativ. Gravit. **3**, 331 (1972).

[46] M. Simpson and R. Penrose, Inter. J. Theor. Phys. **7**, 183 (1973).

[47] Y. Gürsel, V. D. Sandberg, I. D. Novikov and A. A. Starobinsky, Phys. Rev. D **19**, 413 (1979).

[48] Y. Gürsel, I. D. Novikov, V. D. Sandberg and A. A. Starobinsky, Phys. Rev. D **20**, 1260 (1979).

[49] In a RN background, due to the nonvanishing electric field, the linearized electromagnetic and gravitational perturbations are coupled.

[50] S. Chandrasekhar and J.B. Hartle, Proc. R. Soc. Lond. A **384**, 301 (1982).

[51] I. D. Novikov and A. A. Starobinsky, Abstract of contributed papers of the 9th Intern. Conf. on General Relativity and Gravitation, Jena, DDR, p. 268 (1980).

[52] J. M. McNamara, Proc. R. Soc. London A **364**, 121 (1978).

[53] R. A. Matzner, N. Zamorano and V. D. Sandberg, Phys. Rev. D **19**, 2821 (1979).

[54] N. Zamorano, Phys. Rev. D **26**, 2564 (1982).

[55] A non-perturbative derivation of the instability of the interior of RN was given by F. J. Tipler, Phys. Rev. D **15**, 942 (1977). See, however, the related comments in F. J. Tipler, C. J. S. Clarke and G. F. R. Ellis, *General Relativity and Gravitation*, edited by A. Held (Plenum, New York, 1980), Vol. 2.

[56] A. Ori, Phys. Rev. D **55**, 4860 (1997).

[57] J. M. McNamara, Proc. R. Soc. London A **358**, 499 (1978).

[58] As a simple example, consider the usual Doppler blue-shift of a monochromatic plane electromagnetic wave in flat spacetime: it is only the electromagnetic field, not the vector-potential, which is amplified.

[59] This is because the actual metric tensor is (by definition) the sum of the background metric tensor and the metric perturbation, and both are finite. One still has to worry about $\det(g_{\mu\nu})$ being nonzero, but this is guaranteed if the metric perturbations are sufficiently small—which is indeed the case here (for sufficiently large $|u|$).

[60] F. J. Tipler, Phys. Lett. A **64**, 8 (1977).

[61] R. Herman and W. A. Hiscock, Phys. Rev. D **46**, 1863 (1992).

[62] A. Ori, these proceedings.

[63] L. M. Burko and A. Ori, Phys. Rev. Lett. **74**, 1064 (1995).

[64] L. M. Burko, Report No. gr-qc/9801018 (unpublished).

[65] L. M. Burko, Phys. Rev. D **55**, 2105 (1997).

[66] A. Ori, Phys. Rev. Lett. **68**, 2117 (1992).

[67] A. Ori, Gen. Relativ. Gravitation **29**, 881 (1997).

[68] A. Ori, to be published.

[69] B. Bonnor and P. C. Vaidya, Gen. Relativ. Gravit. **1**, 127 (1970).

[70] G. F. R. Ellis and A. R. King, Commun. Math. Phys. **38**, 119 (1974).

[71] E. Poisson and W. Israel, Phys. Rev. Lett. **63**, 1663 (1989).

[72] E. Poisson and W. Israel, Phys. Lett. B **233**, 74 (1989).

[73] E. Poisson and W. Israel, Phys. Rev. D **41**, 1796 (1990).

[74] Alternatively, the outgoing flux may represent the thermal radiation emitted from the surface of the hot collapsing object [73]. It turns out, however, that at the inner horizon this thermal radiation is exponentially small in $|u|$. In the early section of the inner horizon, which concerns us here, this thermal flux is therefore negligible compared to the scattered component.

[75] A. Ori, Phys. Rev. Lett. **67**, 789 (1991).

[76] This is true as long as one is interested in the dependence on v (which is the most crucial dependency for the inner-horizon singularity); Obviously, the discontinuous mass-inflation model [75] does not adequately describe the dependence of the inner-horizon singularity on u.

[77] A. Ori (unpublished)

[78] A. Bonanno, S. Droz, W. Israel, and S. M. Morsink, Proc. Roy. Soc. A **450**, 553 (1995).

[79] N. Y. Gnedin and M. L. Gnedin, Sov. Astron. **36**, 296 (1992).

[80] M. L. Gnedin and N. Y. Gnedin, Class. Quantum. Grav. **10**, 1083 (1993).

[81] D. N. Page, in *Black Hole Physics*, edited by V. De Sabbata and Z. Zhang (Kluwer, Dordrecht, 1992).

[82] P. R. Brady and J. D. Smith, Phys. Rev. Lett. **75**, 1256 (1995).

[83] L. M. Burko, Phys. Rev. Lett. **79**, 4958 (1997).

[84] L. M. Burko and A. Ori, submitted (also Report No. gr-qc/9711032).

[85] L. M. Burko, these proceedings.

[86] A. Ori, in preparation.

[87] P. R. Brady and C. M. Chambers, Phys. Rev. D **51**, 4177 (1995).

[88] A. Ori, and É. É. Flanagan, Phys. Rev. D **53**, R1754 (1996).

[89] A. Ori, to appear in Phys. Rev. D.

[90] G. T. Horowitz and H. J. Sheinblatt, Phys. Rev. D **55**, 650 (1997).

[91] F. Mellor and I. Moss, Phys. Rev. D **41**, 403 (1990).

[92] P. R. Brady and E. Poisson, Class. Quantum Grav. **9**, 121 (1992).

[93] F. Mellor and I. Moss, Class. Quantum Grav. **9**, L43 (1992).

[94] P. R. Brady, D. Núñez and S. Sinha, Phys. Rev. D **47**, 4239 (1993).

[95] Just before this volume was brought to the printer, a new paper appeared on the electronic archives. It turns out, according to the new analysis, that the inner horizon of Reissner-Nordström-de Sitter is classically *unstable* for the entire parameter space, in sharp contrast with earlier analyses. We emphasize that the papers in this book were written prior to the release of P. R. Brady, I. G. Moss, and R. C. Myers, Report No. gr-qc/9801032.

[96] D. Marković and E. Poisson, Phys. Rev. Lett. **74**, 1280 (1995).

[97] I. D. Novikov and A. A. Starobinsky, Zh. Eksp. Teor. Fiz. **78**, 3 (1980) [Sov. Phys. JETP **51**, 1 (1980)].

[98] Ya. B. Zel'dovich, JETP Lett. **12**, 307 (1970).

[99] Ya. B. Zel'dovich and A. A. Starobinsky, JETP Lett. **26**, 252 (1977).

[100] W. G. Anderson, P. R. Brady, W. Israel, and S. M. Morsink, Phys. Rev. Lett. **70**, 1041 (1993).

[101] R. Balbinot and E. Poisson, Phys. Rev. Lett. **70**, 13 (1993).

[102] S. W. Hawking, Commun. Math. Phys. **43**, 199 (1975).

[103] O. Levin and A. Ori, Phys. Rev. D **54**, 2746 (1996).

[104] Y. Kaminaga, Class. Quantum Grav. **7**, 1135 (1990).

THE CAUCHY HORIZON IN BLACK HOLE–DE SITTER SPACETIMES

Chris M. Chambers

Department of Physics, Montana State University, Bozeman, MT 59717-3840, USA

Abstract

The last seven years has produced a growing body of evidence which concludes that the Cauchy horizon in black hole de Sitter spacetimes is classically stable when the surface gravity at the cosmological event horizon is greater than that at the Cauchy horizon. That stability persists for a finite, but non-zero, region of the black hole's parameter space, (M, Q, J, Λ), suggests that black holes immersed in de Sitter space are counter-examples to the strong cosmic censorship hypothesis.

In this review we chronicle that body of evidence and describe the first steps of a program of numerical work aimed at better understanding the interior of black hole-de Sitter spacetimes. The review ends with a speculative account of the role that future work will take.

2.1 Introduction

Our discussions at Haifa have, among other things, emphasized the need for a correct formulation and proof of the *Strong Cosmic Censorship Hypothesis*. Stated in its simplest, physical form, the strong cosmic censorship hypothesis pronounces;

> All physically reasonable spacetimes are globally hyperbolic – Apart from a possible initial singularity, no spacetime singularity is ever visible to any observer.

Unfortunately cosmic censorship is an area which yields little return on much effort. We believe, however, we shall eventually be guided to a precise formulation and proof by examining example and counter-example to the hypothesis. While counter-examples are abundant in the literature, few can live up the claim of being 'reasonable' spacetimes. There are a few notable exceptions to this rule though, in particular the Ernst spacetime, a solution of the Einstein-Maxwell equations [1] and the focus of this contribution, black holes immersed in de Sitter space. The familiarity of black hole solutions make the de Sitter example a particularly attractive field of study.

Black Holes In de Sitter Space

The fate of an observer falling into a black hole is both an interesting and important problem within the framework of classical general relativity. The issue takes on a more notable significance if the black hole is charged or rotating. In this case the observer's journey appears to continue through the interior of the black hole and beyond, eventually emerging from a white hole into a new universe. The simplicity of this picture, however, belies the underlying physics. Deep within the black hole, concealed from the exterior by an event horizon, resides a spacetime singularity, characterized by infinite curvatures, and more importantly for our observer, infinite tidal forces. While fully able to avoid the crushing grip of the singularity, due in part to its timelike nature, the observer is unable to prevent the blatant violation of strong cosmic censorship that such a journey irrevocably promises – a sighting of the singularity. If the issues concerning how physics is going to describe such a *naked* singularity weren't enough, our observer is faced with an even more fundamental, though related, problem – a loss of predictability. It is well known that general relativity admits a well posed initial value problem, i.e., given some suitable initial data on a spacelike hypersurface, S say, the solution to the Einstein equations is uniquely determined everywhere within the domain of dependence $D(S)$ of S [2]. For some spacetimes though, $D(S)$ can often fail to cover the entire manifold M. So, even with suitable initial data on S, general relativity is unable to forecast the evolution of the spacetime beyond $D(S)$ in these cases. The boundary of $D(S)$, $H(S)$, is called the Cauchy horizon and marks the division between the region where general relativity is able to predict the evolution and the region where predictability of the field equations is lost. Black holes with charge or rotation, such as the Reissner-Nordström and Kerr solutions, are well known to possess such an horizon. For these black holes the Cauchy horizon coincides with an inner horizon that veils the singularity. In crossing the event horizon, all observer have unwittingly committed themselves to a journey that will inevitably encounter and attempt to cross this curious frontier that is the Cauchy horizon – an expedition that must contend with naked singularities, loss of predictability and, in some cases, causality violation. It may come as some relief to hear that physics (or nature if you prefer) appears to abhor this situation too and, indeed, conspires to prevent the passage through the interior, terminating it at the Cauchy horizon [3]. The conspiracy itself is rooted in an instability of the inner horizon to even the smallest time dependent perturbation, converting an initially regular horizon into a null, spacetime curvature singularity, effectively sealing off the tunnel to other worlds. The source of the instability can be understood quite simply as an *infinite proper time compression* effect. Because the causal past of the Cauchy horizon contains the entire universe external to the black hole, any observer approaching the horizon sees an infinite number of events in a finite proper time. Stated more dramatically; the observer sees the entire history of the external universe flash before their eyes, in the last few moments before they cross the Cauchy horizon. That this compression effect becomes larger as one approaches the Cauchy horizon, results in progressively larger energy densities being measured, until, at the horizon itself, they become infinite, and turn it into a spacetime singularity. Thus the Cauchy horizon instability restores predictability to the situation and provides an excellent example of a strong cosmic censorship obeying spacetime. So what if the causal past of the Cauchy horizon does not contain the entire external universe but just some part of that universe? In that case one can easily conceive a scenario in which the observer crossing the Cauchy horizon sees only a finite number of events in a finite proper time, possibly leading to no infinite blueshift effects and hence no infinite energy densities or curvature singularities at the horizon. In effect, such a spacetime could have a Cauchy horizon that is stable to time dependent perturbations and serve as a counter-example to strong cosmic censorship (reasonable or unreasonable). Does nature, in an attempt to prevent the embarrassment of naked singularities, plot to prevent such a spacetime occurring in a physically realistic situation? Amazingly it appears not. As we mentioned earlier, black holes immersed in de Sitter space prove to be notable counter-examples, for de Sitter black holes are part of a closed universe and play out much of the scenario discussed above. The familiarity of black hole solutions, combined with the knowledge that all the known black hole spacetimes of the Kerr-Newman family can be generalized to include a cosmological constant, establishes black hole de Sitter spacetimes as a favorable field of investigation.

During the last seven years a consistent picture of a classically stable Cauchy horizon in black hole-de Sitter spacetimes has emerged. Both linear studies and non-linear backreaction calculations indicate that the Cauchy horizon is stable. Although the scenario envisaged above, based on the infinite time compression effects, suggests the Cauchy horizon will always be stable, all analyses agree that stability persists for only a finite, but non-zero, measure on the parameter space (M, Q, J, Λ) of the black hole, infringing on the very spirit of the strong cosmic censorship hypothesis. Whether one views these spacetimes as reasonable counter-examples to the strong cosmic censorship hypothesis is a matter for personal tastes. In the past, much observational evidence was advanced to suggest that the cosmological constant, made famous by Einstein, is zero. Lately (within the last year), new evidence, based on improved gravitational lensing and cosmic microwave background observations, has surfaced and is currently challenging the established view of a vanishing cosmological constant [4]. At the least, since a cosmological constant does not violate any known law of physics, black hole-de Sitter spacetimes provide an excellent arena in which to examine the current form of the strong cosmic censorship hypothesis. The appeal of black hole-de Sitter spacetimes is also realized when comparing them with their asymptotically flat counterparts. While the Cauchy horizon in the Reissner-Nordström spacetime is well known to be unstable it is intriguing that the metric, expressed in well behaved coordinates, is regular there. Further to this, it has been demonstrated that while the tidal forces, due to the strong gravitational field of the Cauchy horizon singularity, on an observer grow without bound as the Cauchy horizon is approached the actual tidal deformation suffered by an observer does not grow without bound. These facts have re-opened the issues of traversability of the Cauchy horizon, and formed a major part of the discussions at Haifa. The answers to this debate are by no means clear. It has been proposed that the finite tidal deformations are irrelevant to the question of traversability [5] – any infalling object is completely destroyed at the Cauchy horizon since the energy absorbed by the object diverges as the horizon is approached. Our black hole-de Sitter model has none of these interpretational problems [6]. In this respect, the dilemmas that a traversable Cauchy horizon faces us with, are more clear cut in the case of black hole-de Sitter spacetimes than they are in the asymptotically flat case.

Arrangement

In accordance with the requirements of the organizers of the "Internal Structure of Black Holes and Spacetime Singularities" meeting at Haifa, this contribution has been written at a level and conciseness specifically intended for those who are conversant with General Relativity but are not familiar with the subject of Cauchy horizon stability, particularly graduate students for whom the subject literature, and logic behind it, is often confusing. That said, it is also hoped that those working in this and related fields will find this a worthy offering to the subject area. While every attempt has been made to present an impartial account of the subject, the narration undoubtedly suffers from my own personal bias.

In section 2.2 we present a mathematical description of black hole-de Sitter spacetimes. The section attempts to provide most of the basic mathematical material that is required for an understanding of later sections. Where this is not been possible the reader has been guided to compensating reference material. For the sake of simplicity, attention is focused exclusively on the spherically symmetric Reissner-Nordström-de Sitter solution. Much of what needs to be known about black hole-de Sitter spacetimes is most easily gained through studying the Reissner-Nordström-de Sitter spacetime. The different coordinate systems used throughout the spacetime, and the role they play, is explained and emphasized with the section concluding on a study of the behavior of radially free-falling geodesic observers. While the purpose of some of the concepts may not be immediately apparent, they will be become more clear on a second read through.

Section 2.3 chronicles an account of the relevant works performed to date. While each commentary is not exhaustive, every attempt has been made to produce a clear and concise summary of that particular study. Specific emphasis has been placed on the motivation and philosophy of the approach and a precise statement of the results obtained. Each account is followed by critical review of the method, and results, in an endeavor to provide the continuity required to go from one study to the next and to illustrate the logic in that step.

In Sec. 2.4 we provide additional material, in the form of comments, that have not, hereto, been published. Much of this material originates from the questions and comments that have arisen at previous conferences and from the many personal discussions that occurred during the Haifa meeting. With the knowledge of Sec. 2.3 at hand, these questions are explained and an attempt to address them is made. The conclusions of this section leads us nicely on to the next section.

The next section, Sec. 2.5, elaborates on the first steps of an ongoing program of numerical work aimed at understanding the interior of black hole-de Sitter spacetimes in more detail. The results and motivation of this first step, an investigation of the late time behavior of fields during gravitational collapse in de Sitter space, are presented. The conclusions of this study are discussed, with particular emphasis placed on their compatibility with the previous analytic studies of the interior.

Finally, we conclude in Sec. 2.6 with a summary of the results presented in this contribution. In keeping with the mood of the Haifa meeting we also take this opportunity to speculate on the role that future work might take in the study of the Cauchy horizon in black hole-de Sitter spacetimes.

We also include two appendices which provide additional material to the contribution, but need not be read in conjunction with the main body of text. An exhaustive bibliography is supplied at the end, which contains additional comments relating to each section.

Acknowledgments

It is a pleasure to thank the organizers of "The Internal Structure of Black Holes and Spacetime Singularities" workshop, at Haifa, for a pleasant meeting and for the generous hospitality offered during our stay. In particular my warm thanks go to Lior Burko, Amos Ori and Liz Youdim for making our visit to Haifa an extremely enjoyable one. I am also grateful for discussions with Alfio Bonnano, Patrick Brady, Eanna Flanaghan, Eric Poisson, Ian Moss and Amos Ori – all of which have aided this review.

Chris M. Chambers is a Fellow of The Royal Commission for the Exhibition of 1851, who's financial support is gratefully acknowledged. This work was supported in part by NSF Grant No. PHY-9722529 to North Carolina State University, and NSF Grant No. PHY-9511794 to Montana State University.

2.2 Reissner-Nordström-de Sitter Black Hole

2.2.1 The Spacetime

The generalization of the Reissner-Nordström black hole solution to include a cosmological constant has been provided by Carter [7]. In terms of the advanced Eddington-Finkelstein coordinates (v, r), the Reissner-Nordström de Sitter metric is

$$ds^2 = -f(r)dv^2 + 2dvdr + r^2 d\Omega^2 \ , \tag{2.1}$$

where

$$f(r) = 1 - \frac{2M}{r} + \frac{Q^2}{r^2} - \frac{r^2}{\alpha^2} \quad , \quad \alpha^2 = \frac{3}{\Lambda} \, . \tag{2.2}$$

In Eq. (2.2) M is the Bondi mass of the black hole, Q its electric charge and Λ is the cosmological constant. Throughout what follows, it will be assumed that Λ is a positive, non-zero, constant. The function $d\Omega^2 = d\theta^2 + \sin^2(\theta)d\varphi^2$ is the metric on the unit two sphere. The coordinate v is the standard advanced time coordinate, related to the Schwarzschild time coordinate t by

$$v = t + r_* \, , \tag{2.3}$$

with

$$\begin{aligned} r_* = \int \frac{dr}{f(r)} = & - \frac{1}{2\kappa_1} \ln\left|\frac{r}{r_1} - 1\right| + \frac{1}{2\kappa_2} \ln\left|\frac{r}{r_2} - 1\right| \\ & - \frac{1}{2\kappa_3} \ln\left|\frac{r}{r_3} - 1\right| + \frac{1}{2\kappa_4} \ln\left|\frac{r}{r_4} - 1\right| \, , \end{aligned} \tag{2.4}$$

where the arbitrary constant of integration has, for simplicity, been set to zero. The constants κ_j represent the surface gravity [8] at the corresponding j^{th} horizon, which are located at $r = r_j$. In general, the surface gravity is given by

$$\kappa_j = \lim_{r \to r_j} \sqrt{\frac{(\nabla_\mu |\ell^2|)(\nabla^\mu |\ell^2|)}{4|\ell^2|}} \, , \tag{2.5}$$

where ℓ is some suitably normalized Killing vector that is null on the jth horizon and $\ell^2 = \ell_\mu \ell^\mu$. In the case of a static, spherically symmetric spacetime with a line element similar to Eq. (2.1) [9], then Eq. (2.5) reduces to

$$\kappa_j = \frac{1}{2} \left| \frac{df}{dr} \right|_{r = r_j} \, , \tag{2.6}$$

which can be verified using Eq. (2.1) and $\ell = \partial_v$. The location of the horizons r_j is given by the roots to the equation

$$f(r) \equiv 0 \, , \tag{2.7}$$

which can be seen [via Eq. (2.2)] to reduce to a quartic equation in r. In general there will be four distinct roots [10] which we label r_1, r_2, r_3 and r_4. The lack of a cubic term in Eq. (2.7) forces at least one of the roots to be negative and unphysical in this spacetime [11]. We order the roots as follows

$$r_1 > r_2 > r_3 > r_4 \quad \text{where} \quad r_4 < 0 \, . \tag{2.8}$$

The root r_1 denotes the location of the *cosmological* event horizon, r_2 the location of the *black hole* event horizon and r_3 the location of the *inner* horizon (which from now on we shall refer to as the *Cauchy* horizon) of the black hole spacetime. We note that although r_4 does not correspond to a physical horizon in this spacetime, we are still able to define the constant κ_4 through Eq. (2.6). Though not strictly correct, κ_4 is often still referred to as a surface gravity. By factoring the metric function $f(r)$ as

$$f(r) = -\frac{(r-r_1)(r-r_2)(r-r_3)(r-r_4)}{\alpha^2 r^2}, \tag{2.9}$$

it is a simple task to show that the surface gravities are

$$\begin{aligned}
\kappa_1 &= \frac{(r_1-r_2)(r_1-r_3)(r_1-r_4)}{2\alpha^2 r_1^2}, \\
\kappa_2 &= \frac{(r_1-r_2)(r_2-r_3)(r_2-r_4)}{2\alpha^2 r_2^2}, \\
\kappa_3 &= \frac{(r_1-r_3)(r_2-r_3)(r_3-r_4)}{2\alpha^2 r_3^2}, \\
\kappa_4 &= \frac{(r_1-r_4)(r_2-r_4)(r_3-r_4)}{2\alpha^2 r_4^2}.
\end{aligned} \tag{2.10}$$

Using Eqs. (2.4), (2.9) and (2.10) it is not difficult to show that

$$\lim_{r\to r_j} f(r) = \pm 2(-1)^j r_j \kappa_j \, e^{2(-1)^j \kappa_j r_*}, \tag{2.11}$$

where $(+)$ indicates that we approach the jth horizon from above and the $(-)$ from below.

2.2.2 Coordinate Systems

The Penrose conformal diagram [12] for the Reissner-Nordström-de Sitter black hole spacetime is shown in Fig. 2.1. As well as the location of the three horizons r_1, r_2 and r_3, the timelike spacetime singularity at $r = 0$, and the path of radially infalling observer (see Sec. 2.2.3) AA$'$ are also shown. It is clear from the conformal diagram that the horizons divide the spacetime into four main regions, labeled I-IV in Fig. 2.1,

$$\begin{aligned}
&\text{Region I} &: \quad & r_1 < r < \infty \\
&\text{Region II} &: \quad & r_2 < r < r_1 \\
&\text{Region III} &: \quad & r_3 < r < r_2 \\
&\text{Region IV} &: \quad & 0 < r < r_3.
\end{aligned}$$

Figure 2.1 also displays the primed regions I$'$- IV$'$ which denote those regions of spacetime that are isometric to regions I-IV. In studying issues related to the Cauchy

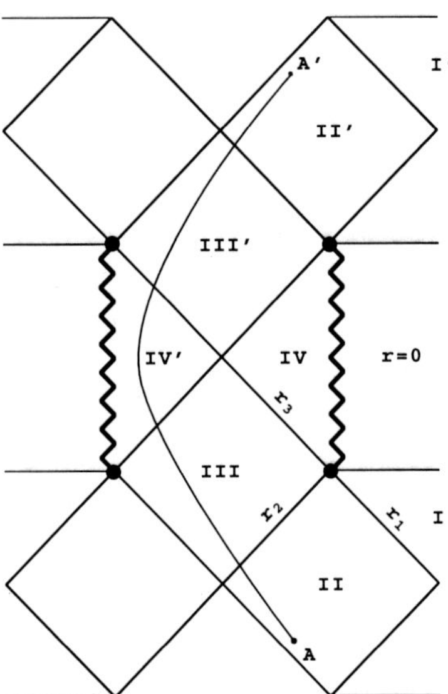

Figure 2.1: A portion of the Penrose conformal diagram for the Reissner-Nordström-de Sitter black hole spacetime. The locations of the cosmological event horizon at $r = r_1$, the black hole event horizon at $r = r_2$ and the inner horizon at $r = r_3$, which is also a Cauchy horizon for external initial value problems, are shown. The dark wavy line represents the timelike spacetime singularity at $r = 0$. Also shown are the path of an infalling observer AA′, the four main spacetime regions I-IV and their isometric counterparts labeled I′-IV′. The diagram can be extended in all direction to reveal an infinite lattice of asymptotically de Sitter universes.

horizon, we are predominantly interested in region II, which we shall frequently refer to as the exterior, and region III, which we shall refer to as the interior. In physical terms, we can consider region II as the *birth-site* for field fluctuations such as electromagnetic and gravitational waves [13]. These waves are continually scattered off the spacetime curvature, a fraction of which is transmitted across the cosmological event horizon into region I, while the remainder is scattered through the black hole event horizon into region III. The waves crossing the black hole event horizon eventually propagate throughout the interior region, perturbing the spacetime geometry there, and in particular the Cauchy horizon. It will, therefore, be beneficial if some time is devoted to each of these regions, paying special attention to the coordinate systems used there. For this purpose it proves useful to write the metric [Eq. (2.1)] in standard diagonal form using Eq. (2.3),

$$ds^2 = -f dt^2 + f^{-1} dr^2 + r^2 d\Omega^2 \ . \tag{2.12}$$

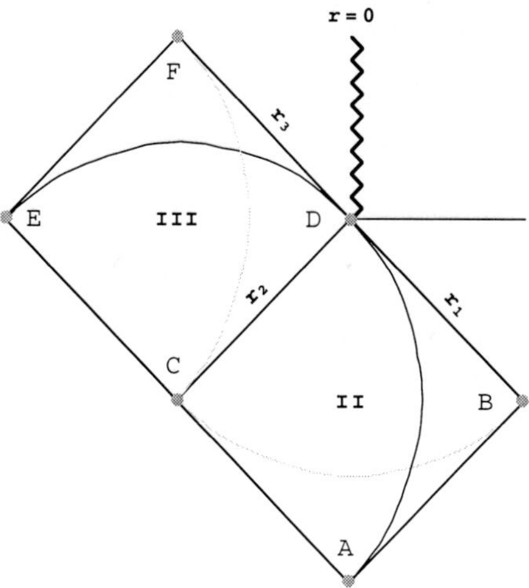

Figure 2.2: The conformal diagram of the exterior and the interior, regions II and III respectively, of the Reissner-Nordström-de Sitter black hole spacetime. The unbroken curved lines represent surfaces of constant r whilst the broken curved lines represent surfaces of constant t. For orientation with Fig. 2.1 the horizons and the singularity are also indicated.

Figure 2.2 shows enlarged versions of the regions II, and III, of Fig. 2.1. In each region, surfaces of constant r, the unbroken lines, and surfaces of constant t, broken lines, are shown. The corners of the regions are labeled A, B, C, D, E and F. These points can be seen to be singular points of the conformal diagram, in the sense that there is no one to one mapping between the corner points and the coordinates (t, r). For the sake of clarity we shall consider each region individually.

Region II

In region II, depicted in Fig. 2.3, we define two radial, null coordinates u and v by

$$v = t + r_* , \quad (2.13)$$

$$u = t - r_* \quad (2.14)$$

where v is defined exactly as in Eq. (2.3). In terms of (u, v) the spacetime metric [Eq. (2.1)] assumes the double null form

$$ds^2 = -f du dv + r^2 d\Omega^2 , \quad (2.15)$$

whose form will be useful in Sec. 2.5.2, when we study scalar wave propagation on a Reissner-Nordström-de Sitter background. Surfaces of constant u and v are shown in

Fig. 2.3 as lines at 45° to the horizontal, with $v =$ const. surfaces running parallel to the (future) cosmological event horizon BD and $u =$ const. surfaces running parallel to the (future) black hole event horizon CD. As indicated in the figure, we have defined the time coordinate t in Eq. (2.12) so that it is minus infinity on CAB and plus infinity on CDB. Using Eq. (2.4) one can determine that r_* takes the values minus infinity on ACD and plus infinity on ABD. With the definitions of u and v as above, [Eqs. (2.13) and (2.14)], it is not difficult to see that v assumes the values minus infinity on the (past) black hole event horizon AC and plus infinity on the (future) cosmological event horizon BD, whilst u is minus infinity on the (past) cosmological event horizon AB and plus infinity on the (future) black hole event horizon CD. In order to analytically continue the spacetime across the boundaries AB, BD, CD and AC one has to introduce new coordinates that are regular there (unlike u and v which diverge). The new coordinates are furnished by the dimensionless Kruskal-Szekeres coordinates, conventionally labeled U and V. In later sections we shall be interested in both the (future) cosmological and black hole event horizons. For this reason we shall consider the (U, V) coordinates for the boundaries BD and CD only, the extension to the other boundaries being self-evident. Near CD we define

$$U = -e^{-\kappa_2 u} , \tag{2.16}$$

$$V = e^{\kappa_2 v} \tag{2.17}$$

so that U tends to zero as u tends to plus infinity. In terms of these two coordinates, the metric, Eq. (2.15), close to the (future) black hole event horizon becomes

$$ds^2 \simeq \frac{2r_2}{\kappa_2} dU dV + r^2 d\Omega^2 , \tag{2.18}$$

which is regular at the black hole event horizon. Similarly, close to BD, we define

$$U = e^{\kappa_1 u} , \tag{2.19}$$

$$V = -e^{-\kappa_1 v} \tag{2.20}$$

so that V tends to zero as v tends to plus infinity. Near to the (future) cosmological event horizon the metric in these coordinates becomes

$$ds^2 = \frac{2r_1}{\kappa_1} dU dV + r^2 d\Omega^2 , \tag{2.21}$$

which is regular at the cosmological event horizon.

Region III

Figure 2.4 depicts region III of Fig. 2.1. We define two radial null coordinates (u, v) as before, defined by Eqs.(2.13) and (2.14). Surfaces of constant u and v in Fig. 2.4 are shown as lines at 45° to the horizontal, with v running parallel to the Cauchy

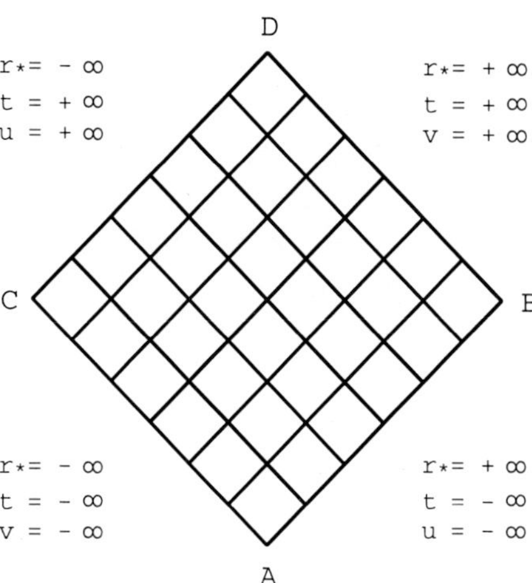

Figure 2.3: Part of the conformal diagram of Fig. 2.1 showing region II and details of its causal structure (see the text for details).

horizon DF and u running parallel to the black hole event horizon CD. It should be noted that the way we have defined u and v in region III is neither unique nor conventional [14]. Indeed, whilst r_* is constrained [via Eq. (2.4)] to take the values minus infinity along DCE and plus infinity along DFE, there is no such constraint on the parameter t. With u and v as defined, t assumes the values minus infinity along CEF and plus infinity along CDF, in order to preserve the requirement that both the Cauchy horizon and the cosmological event horizon be located at $v = +\infty$. One can equally well choose the more conventional definitions, $u = r_* + t$ and $v = r_* - t$, and have t be minus infinity on CDF and plus infinity on CEF [15]. No convention is strictly correct, but should be decided upon according to the problem at hand. The convention we choose is particularly well suited to the matching of waves across the event horizon, traveling from region II to region III. In looking at regular coordinate systems, near the boundaries of region III, we shall restrict our attention to DF, since the regular coordinates here play a crucial role when examining fluxes of energy crossing the Cauchy horizon, as we shall see in later sections. Near DF the regular coordinates (U, V) are defined by

$$U = e^{\kappa_3 u}, \tag{2.22}$$

$$V = -e^{-\kappa_3 v} \tag{2.23}$$

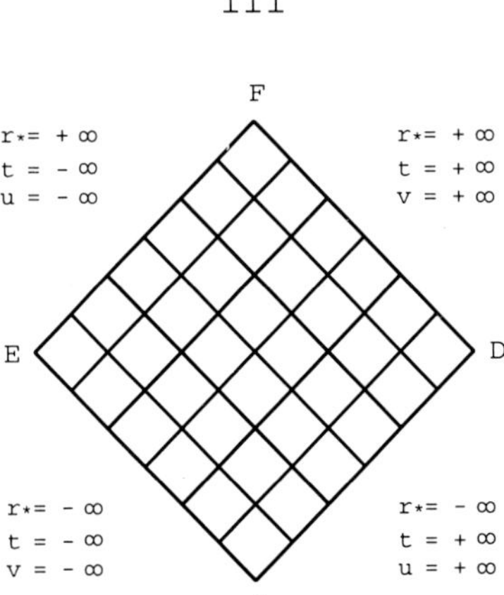

Figure 2.4: Part of the conformal diagram of Fig. 2.1 showing region III and details of its causal structure (see the text for details).

so that V tends to zero as v tends to plus infinity. The metric near DF, then becomes

$$ds^2 = -\frac{2r_3}{\kappa_3} dU\,dV + r^2 d\Omega^2 \,, \tag{2.24}$$

which is obviously regular at the Cauchy horizon.

2.2.3 Radial Geodesics

The role of geodesic observers, particularly what they measure, plays an important role in the study of any spacetime. In this section we shall consider certain concepts of geodesic observers that will be relevant to later sections. For the sake of both clarity and conciseness, we restrict our attention to radially free-falling observers ($d\theta = d\varphi = 0$). The first integrals of geodesic motion are

$$E^2 = \dot{r}^2 + f \,, \tag{2.25}$$

$$\dot{v} = \frac{E \pm \sqrt{(E^2 - f)}}{f} \tag{2.26}$$

where E is the constant of motion ('energy') associated with the timelike killing vector $\xi = \partial_v$, which is given by

$$E = -g_{\mu\nu}\xi^\mu u^\nu \,, \tag{2.27}$$

and $\mathbf{u} = d/d\tau$ is the observer's four velocity. The choice of sign in Eq. (2.26) depends upon whether or not the observer is traveling on a path of decreasing r, with respect to

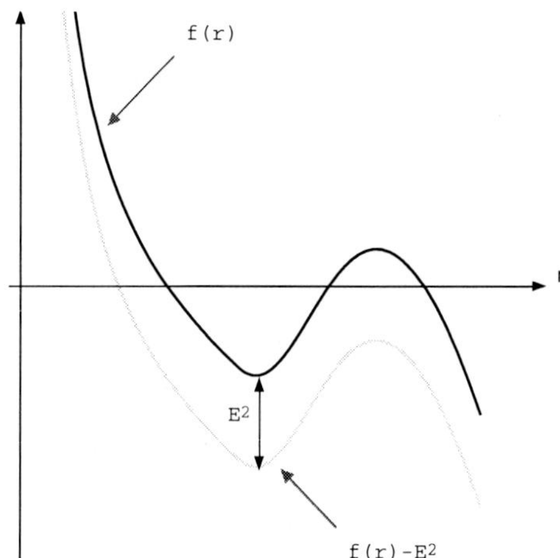

Figure 2.5: A graph of $f(r)$ vs. r and $f(r) - E^2$ vs. r. The three zeros in $f(r)$ (from right to left) correspond to the spacetime horizons r_1, r_2 and r_3. The zero in $f(r) - E^2$ at $0 < r < r_3$ corresponds to a turning point in the radial motion of a freely falling observer inside region IV′, as shown in Fig. 2.1.

τ, in which case the negative root is taken, or whether they are on a path of increasing r in which case the positive root is assumed. The path AA′ in Fig. 2.1 represents the typical world line of such an observer. In region IV′ the observer, having been on a path of decreasing r, reaches a turning point in her (or his) motion and proceeds to move along a path of increasing r. Eventually the observer reaches region II′, another asymptotically de Sitter universe isometric to region II. The occurrence of a turning point in the evolution for $0 < r < r_3$ can be understood by examining the radial geodesic equation (2.25). If a turning point exists then $\dot{r} = 0$. The problem is then to ascertain at what, if any, value of $r > 0$ does this occur. Setting $\dot{r} = 0$ in Eq. (2.25) yields

$$f(r) - E^2 = 0 \, , \tag{2.28}$$

the roots of which then define the turning points of the evolution. Figure 2.5 shows a plot of $f(r)$ vs. r for the generic case of three distinct spacetime horizons. The second plot is Eq. (2.28) vs. r, obtained by a vertical translation of $f(r)$ by an amount $-E^2$. For realistic observers there is one turning point in region IV′. Thus, there do exist paths for which radially infalling observers can enter the black hole, pass by the timelike spacetime singularity, and emerge into another asymptotically de Sitter universe.

We now turn our attention to observers that are in the vicinity of either the cosmological event horizon or the Cauchy horizon. The importance of these observers, to the study of Cauchy horizon stability, will become transparent in later sections.

For clarity in what comes later we use subscripts to denote which region we are considering. For radially free-falling observers approaching the cosmological event horizon, it is easy to show, using Eqs. (2.25) and (2.26), that

$$\dot{r}_{II} \simeq |E_{II}| , \tag{2.29}$$

$$\dot{v}_{II} \simeq \frac{2|E_{II}|}{f_{II}} , \tag{2.30}$$

The equation for $\dot{r}$ merely tells us that our observer is approaching the future cosmological event horizon (increasing r) at a rate that depends upon his initial energy. The equation for $\dot{v}$ is more interesting. As we approach the cosmological event horizon $\dot{v}$ diverges, since f tends to zero there. From Eq. (2.11), with $r = r_1$, we can see that [16]

$$\dot{v}_{II} \equiv \frac{dv_{II}}{d\tau_{II}} \simeq \frac{|E_{II}|}{r_1\kappa_1} e^{\kappa_1 v} \quad \text{as} \quad v \to \infty . \tag{2.31}$$

For observers near the Cauchy horizon, it can similarly be shown that

$$\dot{r}_{III} \simeq -|E_{III}| , \tag{2.32}$$

$$\dot{v}_{III} \simeq -\frac{2|E_{III}|}{f_{III}} . \tag{2.33}$$

Again we can see that $\dot{v}$ diverges as the observer approaches the Cauchy horizon, and one can show that

$$\dot{v}_{III} \equiv \frac{dv_{III}}{d\tau_{III}} \simeq \frac{|E_{III}|}{r_3\kappa_3} e^{\kappa_3 v} \quad \text{as} \quad v \to \infty . \tag{2.34}$$

Since $\dot{r}$ approaches a fixed constant value as we approach either horizon, the velocity of the observer will be dominated by the u^v component. Of course the divergence of $\dot{v}$ is due purely to the choice of coordinates, and one can choose coordinates, such as the Kruskal-Szekeres coordinates of Sec 2.2.1, that are regular at a particular horizon and in which the four velocity is well behaved there.

2.3 An Historical Overview

2.3.1 Mellor and Moss 1990

The first investigative study of Cauchy horizon stability in black hole-de Sitter spacetimes was performed by Felicity Mellor and Ian Moss in 1990 [17]. Confining their attention to the spherically symmetric Reissner-Nordström-de Sitter spacetime, Mellor and Moss considered the effect of gravitational perturbations, generated in the exterior of the black hole, on the Cauchy horizon in the interior. They examined the flux of radiation due to these perturbations, as seen by observers crossing the Cauchy horizon, and concluded that for the cases they studied, the Cauchy horizon is in fact stable to such perturbations [18].

Unfortunately a detailed exposition of the work of Mellor and Moss would lead to quite a *tour de force* in algebra and cause us to stray from picture we are trying to paint here. We will, therefore, sketch only an outline of their investigation.

Analysis

The method employed by Mellor and Moss is analogous to that used by Chandrasekhar and Hartle in their study of the Cauchy horizon instability in the Reissner-Nordström spacetime [19, 20]. The technique is that of linear perturbations about the background spacetime

$$\tilde{g}_{\mu\nu} = g_{\mu\nu} + \epsilon h_{\mu\nu} \ , \tag{2.35}$$

where $\tilde{g}_{\mu\nu}$ represents the components of the perturbed metric, $g_{\mu\nu}$ the components of the background metric and $h_{\mu\nu}$ is the perturbation about the background. The dimensionless quantity ϵ is just an expansion parameter. By constraining the perturbed metric to be a solution to the Einstein-Maxwell equations (since there is a non-zero background electromagnetic field) to linear order in ϵ, one obtains linearized equations for the metric perturbations $h_{\mu\nu}$. These perturbations fall into two distinct classes

Axial Perturbations : Change sign under a change of sign of the azimuthal angle φ

Polar Perturbations : Invariant under a change of sign of φ

With this definition it is easy to conclude, for a spherically symmetric background spacetime, that the polar perturbations have non-vanishing background values, whereas the axial perturbations vanish on the background. By considering the effect of the perturbations on the Ricci tensor, along with the linearized form of the Maxwell equations, it is quite remarkable that both the axial and polar perturbations can be cast into the form of a set of one-dimensional Schrödinger type wave equations, four in all,

$$\frac{d^2 Z_j^{\pm}}{dr_*^2} + (\sigma^2 - V_j^{\pm}) Z_j^{\pm} = 0 \quad \text{for} \quad j = 1, 2 \, . \tag{2.36}$$

Here the Z_j^- denote the two fields that describe the axial perturbations, Z_j^+ the polar perturbations and σ is the frequency of the perturbation modes with time dependence $e^{i\sigma t}$. The functions $V_j^{\pm}(r; \ell, M, Q, \Lambda)$ are the associated potentials, often referred to as the *effective potentials*, whose precise forms are too complicated to reproduce here but have been given by Mellor and Moss [17]. These equations are generalizations of the Regge-Wheeler and Zerilli equations, obtained from perturbations about asymptotically flat spacetimes [21]. One can, therefore, reduce the problem of linear perturbations to an ensemble of one dimensional scattering problems. In order to determine the whether or not the Cauchy horizon is stable, one examines the flux of radiation, due to the perturbations, encountered by an observer crossing the Cauchy horizon. The behavior of the flux near the Cauchy horizon is ultimately dictated by the scattering the fields have undergone in propagating from the exterior to the interior, which is in turn governed by the transmission (T) and reflection (R) coefficients. Close to the Cauchy horizon, the modes $Z \in \{Z_j^{\pm}\}$ have the asymptotic form

$$Z \to A(\sigma)e^{-i\sigma r_*} + B(\sigma)e^{i\sigma r_*} \quad \text{as} \quad r \to r_3 \,. \tag{2.37}$$

As an observer approaches the Cauchy horizon, their four-velocity $\mathbf{u}$ is proportional to ∂_V, where V is the Kruskal-Szekeres coordinate regular at the Cauchy horizon [see Sec. 2.2.2, Eq. (2.23)], and the flux of radiation he or she sees is proportional to

$$\partial_V z(t) = \frac{1}{\kappa_3} e^{\kappa_3 v} \partial_v z(t) \,, \tag{2.38}$$

where $z(t)$ is the Fourier transform of the product of the mode function $Z(\sigma, t)$ and the initial data function $W(\sigma)$, namely

$$z(t) = \frac{1}{2\pi} \int W(\sigma) Z(\sigma, t) e^{i\sigma t} d\sigma \,. \tag{2.39}$$

With the asymptotic form for the mode function, Eq. (2.37), the flux near the Cauchy horizon is proportional to

$$\partial_V z(t) = -\frac{i e^{\kappa_3 v}}{2\pi\kappa_3} \int \sigma W(\sigma) A(\sigma) e^{-i\sigma v} d\sigma \,. \tag{2.40}$$

It should be noted that the more conventional definition for the null coordinates, u and v, described in Sec. 2.2.2 have been used in obtaining Eq. (2.40). By implementing known results from one-dimensional scattering theory [19, 21, 22], and matching modes, scattered across the black hole event horizon from region II to region III, Mellor and Moss are able to show that

$$\overleftarrow{A}_{\rm III}(\sigma) = \frac{\overleftarrow{T}_{\rm II}(-\sigma)}{\overleftarrow{T}_{\rm III}(-\sigma)} \quad \text{and} \quad \overrightarrow{A}_{\rm III}(\sigma) = \frac{\overrightarrow{R}_{\rm II}(-\sigma)}{\overleftarrow{T}_{\rm III}(-\sigma)} \,, \tag{2.41}$$

where $\overleftarrow{Z}$ denotes an ingoing mode, originating from the (past) cosmological event horizon (AB in Fig. 2.3) in region II and $\overrightarrow{Z}$ denotes an outgoing mode, originating from the past black hole event horizon (AC in Fig. 2.3). The subscripts, of course, refer to the exterior and interior regions as defined in Sec. 2.2.2, and differ from those definitions in [17]. The analytic structure of the reciprocal transmission coefficients $T_{\rm II}^{-1}$ and $T_{\rm III}^{-1}$ are obtainable directly from the work of Chandrasekhar and Hartle [19]. Both $T_{\rm II}^{-1}$ and $T_{\rm III}^{-1}$ have poles along the imaginary σ axis at integer multiples of $i\kappa_+$

and $i\kappa_-$, where κ_- and κ_+ are the surface gravities (Sec. 2.2.1) of the horizons on the incident and transmission sides of the potential respectively. Due to zeros in $T_{\rm II}$ and $R_{\rm II}$ canceling the pole in $T_{\rm III}$, the leading pole in $A(\sigma)$ comes from a pole at $i\kappa_3$, whose residue results in a flux at the Cauchy horizon which is finite. The only other poles that could enter $A(\sigma)$ would come from $T_{\rm II}$ or $R_{\rm II}$. Using numerical techniques, Mellor and Moss provided convincing evidence that there are no poles in either $T_{\rm II}$ or $R_{\rm II}$ in the range $0 < {\rm Im}(\sigma) < \kappa_3$, thus concluding that the Cauchy horizon in a Reissner-Nordström-de Sitter black hole is stable to linear perturbations.

Remarks

There is one very subtle assumption in this analysis, which was only recognized later in the work of Brady and Poisson [23]. In studying the Reissner-Nordström-de Sitter interior, Mellor and Moss had not considered what constituted reasonable initial data in the exterior. We shall see that this is tantamount to ignoring the pole structure of the Fourier transform of the initial amplitude of the field modes $W(\sigma)$. By doing this there is an implicit assumption of vanishing flux at the cosmological event horizon which, on physical grounds, seems to be a rather strong restriction on the initial data. The more reasonable requirement of a finite, but non-zero, flux of energy at the cosmological event horizon still confirms that the Cauchy horizon can be stable, but only for a restricted family of solutions as we shall see next.

2.3.2 Brady and Poisson 1992

The question of Cauchy horizon stability in the Reissner-Nordström-de Sitter spacetime was revisited in 1992 by Patrick Brady and Eric Poisson [23]. Rather than contend with the formidable algebra of the gravitational perturbation method employed by Mellor and Moss, Brady and Poisson pursued a much simpler approach to the problem. Their model, a generalization of the study of perturbed a Reissner-Nordström interior due to Hiscock [24], mimics the perturbations propagating from region II to region III, across the event horizon, as a spherical inflow of null dust. Whilst an exact solution to the Einstein equations (in the presence of a cosmological constant) exists [25], Brady and Poisson instead treat the inflow as a linear perturbation, with the null dust propagating on a fixed Reissner-Nordström-de Sitter background. The beauty of the model is its simplicity together with its ability to capture the essential features of the Cauchy horizon stability issue. Whilst the model does not lend itself to a detailed investigation of the interior, it does incorporate the subtlety of the initial data problem missed in the Mellor-Moss analysis and allows us to make the following three, important, conclusions

- The infinite time compression effect, discussed in the introduction, is a *sufficient* requirement for an instability of the Cauchy horizon, but it is not a necessary requirement. A *necessary* condition is an infinite compression of the ratio of differential proper times.

- The requirement of a finite but non-zero flux of energy crossing the cosmological event horizon (initial data) leads to a significantly different stability condition than that obtained by Mellor and Moss. Namely the Cauchy horizon is stable provided that the surface gravity there is less than that at the cosmological event horizon,
$$\kappa_1 > \kappa_3 \ . \tag{2.42}$$
- If a backreaction calculation similar to that used by Poisson and Israel [26] to study the interior of Reissner-Nordström were performed, the model suggests the following conclusions
  1. If $\kappa_3 > 2\kappa_1$ then there would be a mass inflation type singularity.
  2. If $2\kappa_1 > \kappa_3 > \kappa_1$ then there would be a divergent flux at the Cauchy horizon but the internal mass function would approach some finite asymptotic value.

Analysis

The inflow of spherical null dust propagating on the fixed background, described by Eqs. (2.1) and (2.2), is characterized by the stress-energy tensor

$$T_{\mu\nu} = \frac{L(v)}{4\pi r^2} l_\mu l_\nu \ , \tag{2.43}$$

where $L(v)$ is the *luminosity* function and $l_\mu = -\partial_\mu v$ is a vector tangent to ingoing radial null geodesics. An observer crossing the inflow, with a four velocity $\mathbf{u}$, measures a flux of energy ρ given by

$$\rho \equiv T_{\mu\nu} u^\mu u^\nu \ . \tag{2.44}$$

A radially free-falling observer in the vicinity of the cosmological event horizon measures, therefore, a flux

$$\rho_{II} \simeq T_{vv} \dot{v}^2 = \frac{|E_{\rm II}|^2}{4\pi r_1^2} L(v) e^{2\kappa_1 v} \ , \tag{2.45}$$

where we have used Eq. (2.31) of Sec. 2.2.2. On physical grounds, we expect the flux of energy at the cosmological event horizon to be finite and in general non-zero. This requires that

$$L(v) = K(v) e^{-2\kappa_1 v} \ , \tag{2.46}$$

such that

$$\lim_{v \to \infty} K(v) \equiv K_\infty \neq 0 \ . \tag{2.47}$$

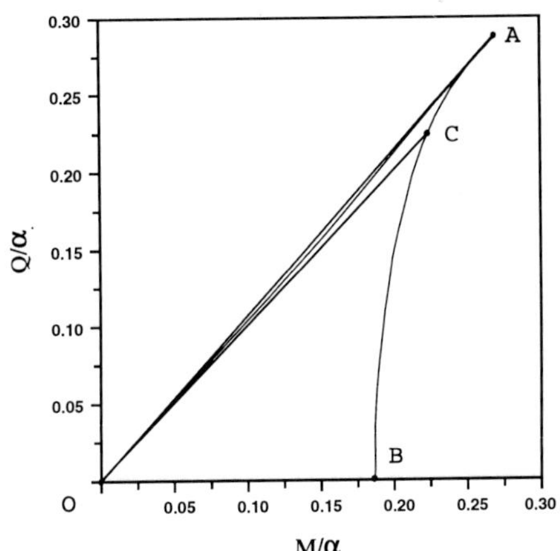

Figure 2.6: A plot of the black hole parameter space showing the region of stability defined by Eq. (2.42). The upper line OA denotes the condition $r_2 = r_3$ whilst the line AB denotes the condition $r_1 = r_2$. The lower line OA, just visible, denotes the condition $\kappa_1 = \kappa_3$. The narrow region between the lines OA is the valley of stability. The line OC denotes the condition $|Q| = M$ showing that the stability condition requires $Q > M$ which can always be achieved by adding charge to the black hole.

Brady and Poisson refer to this later condition as the *minimal requirement*, which ensures a non-zero flux is measured by the observer at the cosmological event horizon.

A radially infalling observer in the proximity of the Cauchy horizon measures a flux of energy ρ_{III}, due to the inflow, given by

$$\begin{aligned} \rho_{III} &= \frac{|E_{III}|^2}{4\pi r_3^2} L(v) e^{2\kappa_3 v} \\ &= \frac{|E_{III}|^2}{4\pi r_3^2} K(v) e^{2(\kappa_3 - \kappa_1)v} , \end{aligned} \tag{2.48}$$

using Eq. (2.46). Thus, at the Cauchy horizon, if $\kappa_3 > \kappa_1$ the energy density diverges, indicating an instability of the horizon [18]. On the other hand, if $\kappa_1 > \kappa_3$ then the energy density is finite, suggesting the horizon is stable.

Remarks

The stability condition of Eq. (2.42) offered by the Brady-Poisson model is more restrictive than the original proposal of Mellor and Moss, and one might ask whether or not this condition can be met in any realistic sense. Figure 2.6 shows the region in the parameter space (M, Q, Λ) for which the stability condition holds. Though small, the *valley of stability* is of finite, non-zero, measure on the parameter space.

An important point relating to this plot is that even if Λ is small, provided it is non-zero, one can always find solutions with a stable Cauchy horizon. There is absolutely no restriction to having an unacceptably large value of Λ [27]. Another point worth clarifying is that the stability condition requires $Q^2 > M^2$. Unlike the Reissner-Nordström black hole, there is no physical reason why one cannot keep putting charge on a Reissner-Nordström-de Sitter black hole to the point where the charge exceeds the mass [28]. In *practice* one could fine tune the black hole, by dropping in charge and/or mass, until it achieved a state lying in the stability region, allowing safe passage across the Cauchy horizon.

The stability condition also has the endearing quality that it can be understood intuitively, in terms of competing redshift and blueshift effects. At the cosmological event horizon there is an infinite redshift effect due to the cosmological expansion, whilst at the Cauchy horizon, due to its causal nature, there is an infinite blueshift effect. If the redshift outweighs the blueshift ($\kappa_1 > \kappa_3$) then the measured energy density at the Cauchy horizon is sufficiently diluted to render the horizon stable. Conversely, if the blueshift prevails over the redshift ($\kappa_3 > \kappa_1$) then the measured energy density is sufficiently dense to render the horizon unstable. Up until this juncture, the stability (or instability) of the Cauchy horizon had always been explained in terms of the time compression effects between the exterior and interior, as discussed in the Introduction. The Brady-Poisson analysis demonstrates that while we can consider this a sufficient condition, it is not a necessary condition. For $\kappa_1 < \kappa_3$, the Cauchy horizon is unstable, but there is no infinite proper-time compression since the black hole is part of a de Sitter, i.e., closed, universe. What is necessary is an infinite compression of the ratio of *differential* proper-times, since blueshift and redshift are precisely differential proper-time effects (See Appendix A.)

The existence of a region in the parameter space for which the Cauchy horizon is unstable, but there is no associated growth in the internal mass function, lead Brady and Poisson to speculate that if backreaction were taken into account [26], then the Cauchy horizon singularity might remain weak, in the sense that no scalars formed from the curvature tensors diverge there. Indeed, the Weyl scalar Ψ_2, that in spherical symmetry is characterized by the mass function, is regular at the Cauchy horizon. However, we shall in Sec. 2.3.4 that there is in fact a scalar curvature singularity at the horizon, characterized by the divergence of the Kretchsmann invariant $R_{\mu\nu\gamma\sigma}R^{\mu\nu\gamma\sigma}$.

2.3.3 Mellor and Moss 1992

In light of the work by Brady and Poisson, Mellor and Moss have reassessed their stability analysis [29]. By requiring that observers crossing the cosmological event horizon should measure a finite but non-vanishing flux of energy, it is relatively easy to show that the gravitational perturbation approach to the Cauchy horizon problem reproduces the stability condition, Eq. (2.42), in complete agreement with the Brady-Poisson model.

Analysis

The initial data requirement of a finite but non-vanishing flux of energy being measured at the Cauchy horizon is easily incorporated into the gravitational perturbation method by imposing pole structure on the initial data function $W(\sigma)$, defined in Eq. (2.39). Close to the cosmological event horizon, the modes have the asymptotic form

$$Z \rightarrow e^{i\sigma r_*} + R(\sigma)e^{-i\sigma r_*} .$$

The flux of energy [Eq. (2.38)] near the cosmological event horizon is thus proportional to

$$\partial_V z(t) = \frac{ie^{\kappa_1 v}}{2\pi\kappa_1} \int \sigma W(\sigma) e^{i\sigma v} d\sigma , \qquad (2.49)$$

Now, if $W(\sigma)$ has any poles in the range $0 < \sigma < i\kappa_1$, then the flux would diverge, indicating an unphysical instability of the cosmological event horizon. On the other hand, if there are no poles in the range $0 < \sigma \leq i\kappa_1$ the flux is vanishing at the cosmological horizon. From a physical point of view this is a very strong requirement on the initial data. If we impose the more reasonable requirement of a finite, non-vanishing flux at the cosmological event horizon then $W(\sigma)$ is forced to have a pole at $\sigma = i\kappa_1$. Using the results of Sec. 2.3.1, this implies that the Cauchy horizon will be stable provided that $\kappa_1 > \kappa_3$, as initially suggested by Brady and Poisson.

Remarks

From the analysis above it is clear that neglecting the relevant pole structure of the initial data, encoded in $W(\sigma)$, imposes an unreasonably strong restriction on the initial perturbations, namely the flux [Eq. (2.49)] at the cosmological event horizon, due to these perturbations, vanish. In relation to the Brady-Poisson model, this condition would require

$$\lim_{v\rightarrow\infty} K(v) = 0 ,$$

which can be seen, via Eq. (2.48), to lead to stability in all cases and clearly explains the original result of Mellor and Moss.

2.3.4 Brady, Núñez and Sinha 1993

Unlike the pure inflow model of Brady and Poisson, the perturbative content of the analysis by Mellor and Moss does incorporate the effects of field-scattering off the spacetime curvature. However, neither analysis takes account of non-linear backreaction effects produced by the field evolution. The first, and so far only, calculation to consider the effects of backreaction on the geometry was provided by Brady, Núñez and Sinha [30] in 1993.

Generalizing the backreaction model devised by Poisson and Israel [26], for their study of the Reissner-Nordström interior, Brady *et al.* arrived at the following important conclusions

- The Cauchy horizon in Reissner-Nordström-de Sitter is stable provided that
$$\kappa_1 > \kappa_3 .$$
- For $2\kappa_1 > \kappa_3 > \kappa_1$
  1. There is a divergent flux of energy at the Cauchy horizon but no corresponding growth of the internal mass function.
  2. The curvature singularity at the Cauchy horizon is *strong*, revealing itself as a divergence of the Kretchsmann scalar $R_{\mu\nu\gamma\sigma}R^{\mu\nu\gamma\sigma}$.
- For $\kappa_3 > 2\kappa_1$ the Cauchy horizon instability is qualitatively similar to that found in the Reissner-Nordström solution.

Analysis

The analysis of Brady *et al.*, whilst simple, does require a suitable knowledge of certain background material, which, for contextual reason, we have not presented here. In fact, the generalization of the Poisson-Israel model to include a cosmological constant is so straight forward that much of what one would learn from it can be gained from an understanding of the Poisson and Israel analysis [31]. For that reason we shall not attempt to reproduce an outline of the work by Brady *et al.* For the interested reader Appendix B details a simpler picture of the Brady-Núñez-Sinha backreaction model, originally due to Ori [32], which captures some of the essence of their analysis.

Remarks

The relevance of the work by Brady, Núñez and Sinha, to the study of Cauchy horizon stability in Reissner-Nordström-de Sitter spacetimes, cannot be over emphasized. As the only model, to date, to incorporate the effects of backreaction, it is both a natural and important extension to the preceding analyses. The agreement between the linear and non-linear studies is pleasing and, at present, is the best evidence we have toward the conjecture that the Cauchy horizon in black hole-de Sitter spacetimes is stable for $\kappa_1 > \kappa_3$. However, we should be careful not to paint too much of a rosy picture here. Two possible caveats to the work of Brady *et al.* that have direct bearing on the conjecture are

1. While the model includes backreaction, it does restrict these effects to a spherically symmetric spacetime.
2. The model imposes the minimal requirement, of Brady and Poisson, on the fields.

The first caveat is extremely difficult to address. To be more clear about this point, we have to ask what it is we are trying to prove in our studies of the Cauchy horizon in black hole-de Sitter spacetimes. We can pose this as a question;

> Can generic collapse, in de Sitter space, lead to a black hole with a stable Cauchy horizon?

It is difficult, if not impossible, to answer this question. Currently our best approach is to investigate more tractable models of the interior, such as those restricted to spherical symmetry, to guide our understanding and hopefully one day answer our questions. To what extent the backreaction will play a role in the stability issue is unclear. Indeed, since the curvature appears to be regular at the Cauchy horizon (in the case of stability) we might expect that the backreaction plays a minor role in the shaping of the spacetime interior. However, it may turn out that in a fully non-linear simulation of the collapse that the backreaction becomes important at early times, causing the interior to be significantly different to what our models predict. While we cannot provide definite answers to this and similar questions, we can gain some insight from our current models. With the conclusions of the linear model by Mellor and Moss in conjunction with the results of Brady *et al.* it is clear that the backreaction is not playing a significant role in the evolution of the spacetime close to the Cauchy horizon, at least within the confines of spherical symmetry and null dust flow. What about axisymmetric spacetimes? In Sec. 2.3.5 we shall discuss the details of a recent perturbation analysis of the Kerr-de Sitter spacetime, similar to that implemented by Mellor and Moss in their study of the Reissner-Nordström-de Sitter spacetime. The results of this analysis suggest that rotation has no effect on the stability conjecture, within the perturbative approach employed. Unfortunately no backreaction calculation, akin to that of Brady *et al.*, currently exist for Kerr-de Sitter spacetimes, against which we could compare the linear analysis. We shall have to wait to see whether backreaction, away from the confines of spherical, in anyway alters our current picture of stability.

The second caveat is more readily addressed. The requirement of a finite but non-vanishing flux at the Cauchy horizon leads to Eqs. (2.46) and (2.47) for the luminosity function $L(v)$. There is little doubt that this is the most physically reasonable condition one could impose on the initial perturbations. For the case of black holes in asymptotically flat black hole spacetimes there is no such requirement on the luminosity function [33]. In fact, requiring the perturbations to vanish at future null infinity ($v = r = \infty$) just requires the luminosity to be a decreasing function of advanced time. The exact form for the luminosity function is arrived at from an analysis of the late time behavior of fields in the exterior due to Price [34] and Bičák [35]. For black holes in de Sitter space, the minimal requirement is implicitly imposing the form on the late time behavior .of fields in the vicinity of the cosmological horizon. In Sec. 2.5.2 we will discuss the details of a numerical investigation into the late time behavior of fields in the Schwarzschild-de Sitter and Reissner-Nordström-de Sitter spacetimes and comment there on its compatibility with the minimal requirement of Brady and Poisson.

For $2\kappa_1 > \kappa_3 > \kappa_1$, the non-linear analysis shows the remarkable structure alluded to by Brady and Poisson [23] – a divergent flux but no corresponding mass inflation.

In this instance, the Weyl scalar Ψ_2 no longer acts as a measure of the divergence as it does in the asymptotically flat case, instead it is the Kretschmann invariant that signals the existence of a scalar (strong) singularity at the Cauchy horizon. It had been assumed that the existence of a divergent flux at the Cauchy horizon would always be met by an associated growth in the internal mass parameter. That mass-inflation does not occur in this parameter range makes the black hole-de Sitter case all the more interesting and worthy of study.

2.3.5 Chambers and Moss 1994

Until 1994, all analyses of the Cauchy horizon in black hole-de Sitter spacetimes had, for simplicity, confined their attention to the spherically symmetric Reissner-Nordström-de Sitter solution. However, in general, one expects a black hole formed in a realistic gravitational collapse situation to be rotating and uncharged. It is therefore, both natural and physically well motivated to consider how the inclusion of rotation might affect the current spherical picture of Cauchy horizon stability. The generalization of the stability analysis by Mellor and Moss to the case of a rotating but uncharged black hole was performed in 1994 by Chris Chambers and Ian Moss [36]. Studying linear perturbations of scalar, electromagnetic and gravitational fields on the Kerr-de Sitter spacetime, Chambers and Moss were able to conclude that the Cauchy horizon was stable, provided that $\kappa_1 > \kappa_3$, as in the spherical case. The method employed by Chambers and Moss is identical to that used by Mellor and Moss in their study of the Reissner-Nordström-de Sitter spacetime (Sec. 2.3.1). The essence of the approach is to transform the equations for the perturbations to a set of one dimensional Schrödinger-type wave equations, thus reducing the problem of linear perturbations to the more familiar problem of one dimensional scattering. The entire analysis is, as in the Mellor and Moss study, a *tour de force* in algebra. Consequently we sketch only an outline of the analysis.

Analysis

Chambers and Moss consider three types of field perturbation

- Scalar perturbations (spin = 0)
- Electromagnetic perturbations (spin = 1)
- Gravitational perturbations (spin = 2)

The scalar perturbations are represented as a massless, minimally coupled scalar field obeying $\Box\phi = 0$, propagating on a fixed background described by the Kerr-Newman-de Sitter spacetime. Using separable solutions of the form

$$\phi = R(r)S(\mu)e^{-i\omega t}e^{im\varphi} \tag{2.50}$$

where $\mu = a\cos(\theta)$ and a is the rotation parameter. Whilst no explicit solution to the angular equation exist, for $\Lambda = 0$ the solutions $S(\mu)$ are spheroidal wave functions. The equation for the radial function $R(r)$ is easily reduced to the form of a one dimensional scattering equation

$$\frac{d^2 Z}{dr_*^2} + VZ = 0 , \tag{2.51}$$

where $Z(r) = (r^2 + a^2)^{1/2} R(r)$, r_* is the analogous coordinate to Eq. (2.4) for Kerr-Newman-de Sitter and V is the effective potential.

The electromagnetic perturbations on a Kerr-de Sitter background (no background electromagnetic field), are governed by Maxwell's equations. Chambers and Moss resorted to the Newman-Penrose (NP) formalism to describe these equations [37]. In the NP formalism, the six independent components of the field strength tensor $F_{\mu\nu}$, are described by the three complex Maxwell scalars (ϕ_0, ϕ_1, ϕ_2). Trying separable solutions of the form Eq. (2.50), for each of the fields, enables one to write the equations for the radial functions as a set of one dimensional scattering equations similar to Eq. (2.51). The details of this transformation are quite involved and the can be found in [39].

The gravitational field in the NP formalism is described by the five complex Weyl scalars $(\Psi_0, \Psi_1, \Psi_2, \Psi_3, \Psi_4)$, representing the ten degrees of freedom of the gravitational field. Due to the nature of the Kerr-de Sitter spacetime [40], Ψ_2 is the only non-vanishing Weyl scalar on the background and only the remaining Weyl scalars can be used to describe the gravitational perturbations. While Ψ_0 and Ψ_4 are gauge invariant, and hence measurable, Ψ_1 and Ψ_3 are not and, with a judicious choice of gauge, can be made to vanish. With a significant amount of algebra [41], the equations for Ψ_0 and Ψ_4 can, by assuming separable solutions of the form Eq. (2.50), be cast in the form of Eq. (2.51). Again, for the details, one is guided to [39].

With the equations governing the perturbations transformed to a set of one dimensional scattering equations, much of the hard work is over. An investigation of the flux of energy, due to the perturbations, near the Cauchy horizon is performed in an identical way to that implemented by Mellor and Moss. Imposing the minimal requirement of Brady and Poisson and examining the pole structure of the transmission and reflection coefficients, one can show [36] that the Cauchy horizon in Kerr-de Sitter is stable provided that, as in the spherical case,

$$\kappa_1 > \kappa_3 ,$$

Remarks

The analysis of Chambers and Moss is the first to indicate that Cauchy horizon stability, in black hole-de Sitter spacetimes, is not an artifact of spherical symmetry. Indeed, this analysis provides the best evidence yet that generic collapse in de Sitter space can yield black holes with stable Cauchy horizons, violating the spirit (if not the letter) of the strong cosmic censorship hypothesis. Unfortunately, there is currently no calculation that takes into account the effects of backreaction on the spacetime. Thus we have no model against which we can test the predictions of the linear analysis. However, there is little evidence to suggest that a backreaction calculation would lead us to any other conclusions. The valley of stability in the Kerr-de Sitter case is similar to the charged case, shown in Fig. 2.6. One peculiar property of the Kerr-de Sitter spacetime is the existence of closed timelike curves, near the ring singularity at $r = 0$. These arise due to the timelike nature of the azimuthal angle φ for small values of r. The possibility of observers crossing into this region (IV) faces us with the problem of doing physics in the presence of closed timelike curves. There are many interesting aspects of the Kerr-de Sitter spacetime which have not received attention but provide thought provoking possibilities [39].

2.3.6 Marković and Poisson 1995

While we are primarily concerned with the classical stability of the Cauchy horizon in black hole-de Sitter spacetimes, we would not be presenting a fair picture if we did not include the results of a quantum analysis.

Classically we have seen that the Cauchy horizon can be stable for a certain region of the black hole parameter space (M, Q, J, Λ), which leads to problems associated with the loss of predictability that occurs beyond the Cauchy horizon. It is natural, therefore, to ask whether or not the Cauchy horizon can be quantum mechanically stable or whether quantum effects will restore predictability. So, how do the fluxes of quantum fields affect the spacetime in the vicinity of the Cauchy horizon? The intriguing answer to this question was provided by Dragoljub Marković and Eric Poisson [42] in 1995. Examining the quantum fluxes measured by an observer approaching the Cauchy horizon, Marković and Poisson were able to conclude that the horizon is quantum mechanically unstable, except for the set of zero measure solutions with

$$\kappa_1 = \kappa_3 ,$$

which are represented by the lower line OA in the parameter space plot of Fig. 2.6.

Analysis

To investigate the quantum stability of the Cauchy horizon in a black hole spacetime generally requires a knowledge of $\langle T_{\mu\nu} \rangle$, the renormalized expectation value of the stress-energy tensor associated with the quantum field. However, even within the confines of spherical symmetry, four dimensional calculations of $\langle T_{\mu\nu} \rangle$ are extremely difficult and, for the case of black hole-de Sitter spacetimes, the situation is especially difficult because one cannot choose from the standard vacuum states [43], such as the Hartle-Hawking or Unruh states. Marković and Poisson instead consider a simpler, but still instructive, approach to the problem, quantizing a conformally invariant scalar field on a two dimensional version of the Reissner-Nordström-de Sitter spacetime

$$ds^2 = -f(r)dudv \ , \tag{2.52}$$

where the pair (u, v) are the null coordinates defined in Sec. 2.2.2 and $f(r)$ is given by Eq. (2.2). A quantum state that is regular on both the cosmological horizons and the black hole event horizons, for the two-dimensional case, has been provided by Marković and Unruh [45]. For this state, the renormalized expectation value of the scalar field can be written as [46]

$$\langle T_{\mu\nu} \rangle = \theta_{\mu\nu} + t_{\mu\nu} + (48\pi)^{-1} R g_{\mu\nu} \ , \tag{2.53}$$

where R is the Ricci scalar associated with the two dimensional metric (2.52), $\theta_{\mu\nu}$ is a state independent object and $t_{\mu\nu}$ is a state dependent object. Examining the expectation value of the energy density measured by a freely falling observer crossing the Cauchy horizon, $\langle \rho \rangle = \langle T_{\mu\nu} \rangle u^\mu u^\nu$, in a regular coordinate frame reveals that

$$\langle \rho_{III} \rangle = \frac{|E|^2}{48\pi} (\kappa_1^2 - \kappa_3^2) e^{2\kappa_3 v} \ , \tag{2.54}$$

as v tends to infinity [cf. Eq. (2.48)]. Thus, unless $\kappa_1 = \kappa_3$, the Cauchy horizon in Reissner-Nordström-de Sitter is quantum mechanically unstable.

Remarks

While the calculation is restricted to a simple two-dimensional model, Marković and Poisson give a physical interpretation of Eq. (2.54), in terms of the thermal quanta emitted by the horizons and the gravitational redshifts and blueshifts such quanta undergo, that suggests the result should hold true in the four dimensional case and irrespectively of any particular quantum field.

The quantum mechanical instability of the Cauchy horizon is interesting for many reasons, but in particular it is that this situation demonstrates how, even in regions of spacetime where classical curvatures are not necessarily large, quantum effects can be important. One normally expects quantum effects to become important only when curvatures are sufficiently high (approaching planck scales). This result is similar to the situation in which a spacetime possesses regions containing closed timelike curves and regions that are free from them. The boundary between such regions, the *chronology* horizon, is sometimes stable classically but always quantum mechanically unstable. However, the quantum instability of the Cauchy horizon still standing, the existence of solutions to the classical Einstein equations with stable Cauchy horizons remains a disturbing issue.

2.4 Comments

In a classical setting, we can see that both linear and non-linear examinations of the interior of black hole-de Sitter spacetimes strongly suggest that the Cauchy horizon is stable for a finite, but non-zero, measure on the space of black hole parameters (M, Q, J, Λ). Moreover, simple but effective studies of the stability issues [23] yield intuitively compelling, and pleasing, insights into the mechanism at the heart of the Cauchy horizon stability condition, Eq. (2.42). Whilst the majority of researchers within the field are confident of the results and conclusions of the analyses in Sec. 2.3, some doubts, concerning the validity of the linear perturbation studies, have been raised. It is worthwhile devoting some time to a discussion of these doubts, addressing them directly with the results and conclusions of Sec. 2.3 at hand.

In Sec. 2.3.1 we briefly discussed the concept of a linear perturbation. For simplicity we shall consider a perturbation described by a scalar field. We can expand the scalar field as

$$\phi = \phi_0 + \sum_{n=1}^{\infty} \epsilon^n \phi_n \ , \tag{2.55}$$

where ϕ_0 is the background value of the field and ϕ^n represents the n-th order perturbation. Though not necessary here, we have introduced a dimensionless parameter ϵ which, in calculations, keeps track of the order of the perturbation. In the case of black holes the background field is, in general, zero. For linear perturbation studies one initially assumes that the scalar field is *sufficiently weak* , in the sense that the stress-energy associated with the field is negligibly small (no backreaction.) Formally this implies we consider the field equations to linear order in ϵ. The stress-energy tensor associated with the perturbation is quadratic in the field, and hence epsilon, so that to linear order the stress-energy is ignored. The spacetime is thus unaffected by the presence of the field and the field evolution takes place on a fixed background spacetime. If at any point of the evolution, the field becomes *large*, in the sense that its stress-energy becomes appreciable, then the linear theory is no longer valid and one must contemplate the construction of a model that takes account of non-linear effects. Indeed this is exactly what happens at the Cauchy horizon in the Reissner-Nordström spacetime [19]. The stress-energy tensor of the field becomes increasingly large as one approaches the Cauchy horizon. On the other hand, if the field remains *weak* everywhere in the region of interest then the linear theory, and its results, remain valid. As we discussed in the previous section, this is exactly what happens in the Reissner-Nordström-de Sitter spacetime. For $\kappa_1 > \kappa_3$ the fields stress-energy remains negligible and the results, conventionally, are taken to be representative of the full theory. Of course, for $\kappa_1 < \kappa_3$ the theory breaks down and is qualitatively similar to the case $\Lambda = 0$.

The doubts voiced about the results and conclusions of linear perturbation analyses concern themselves with precisely this conventional wisdom, that if a linear theory is finite then the full theory is finite too. Now, Eq. (2.55) involves an infinite sum over the field perturbations and although linear analyses show that ϕ_1 and its derivatives (appearing in the stress-tensor) are well behaved at the Cauchy horizon (for $\Lambda \neq 0$) this by no means guarantees that the infinite sum converges to a finite value [47]. In fact, not only should this sum be convergent of course but also those that appear in the stress-energy tensor. To prove that these sums converge to a finite value would be quite an undertaking, requiring a knowledge of the field behavior to all orders. A direct approach to the problem, showing convergence of the sums, is likely an impossible task. One possible approach, which would not directly prove convergence but could at least provide evidence of convergence, is based on an approach initially used by Ori to study non-linear perturbations in the Kerr spacetime [53]. This approach has been used to study the non-linear effects of a scalar field in the Reissner-Nordström-de Sitter spacetime [48], and predicts the correct linear behavior and allows one to obtain the behavior of each term in the expansion described by Eq. (2.55). By comparing the behavior of successive terms in the expansion, it may be possible to establish enough evidence to support the convergence of the sum. However, the work of Brady *et al.* [30] has, to some extent, answered this question already. Indeed, one can view this analysis as an alternative approach to addressing the convergence problem by directly attempting a non-linear study. That the results of this non-linear analysis agree with the results from the linear studies suggest that the conventional wisdom on linear perturbation theory is valid. This idea can even be taken one step further. Instead of assuming a model where null dust mimics the perturbations generated in the collapse, why not actually study the gravitational collapse of a body, which eventually forms a black hole? In this case one could attempt to follow the evolution of the collapse through the event horizon and in to the interior, paying special attention to spacetime near the Cauchy horizon. In the next section we shall address this idea in more detail.

2.5 Current Progress

One of the key ingredients to studying the interior of any black hole spacetime is a knowledge of the behavior of fields crossing the event horizon, due to scattering in the exterior of the black. In a realistic collapse situation these fields would always be present, as any initial perturbation (or any perturbation that forms during the collapse) in the collapsing body becomes dynamical after the onset of collapse and thus emits gravitational or electromagnetic waves. The waves, initially outgoing from the objects surface, will scatter due to their interaction with the spacetime curvature and a fraction of the wave will be reflected back to the surface. If an event horizon forms, there will always be a scattered component of the wave that crosses the event horizon and propagates through to the interior.

2.5.1 Radiative Tails

For black holes residing in asymptotically flat spacetimes, the behavior of these radiating fields at sufficiently late times after the collapse is well known now, both from analytic calculations [34, 35] and from numerical studies of collapse [49, 50]. The analytic and numerical studies both agree, that at late times ($v \to \infty$), along the event horizon, a physical field Ψ decays according to

$$\Psi \sim v^{-\gamma} \quad , \quad \gamma = 2\ell + 2 + P \tag{2.56}$$

where v is the standard advanced time coordinate of Sec. 2.2.1, ℓ is the multipole moment of the field ($\ell \geq s$) with spin s and P is a constant that assumes the value 0 if there is an initially static perturbation of the star and 1 otherwise. It is this late time decay that allows the gravitational field to relax to its final asymptotic state, parameterized by the three parameters associated with the non-radiatable multipoles ($\ell < s$) of the electromagnetic (Q) and gravitational fields (M, J) of the black hole. These late time fields are called the *tails* or *radiative tails* of collapse. Their importance lies in their role as initial data for studies of the interior, both analytically [26] and numerically [51, 52]. In the analytic studies the late time behavior of the field is imposed as a restriction on the stress-energy tensor of the matter that is flowing into the hole across the event horizon (i.e the field is supposed to mimic the late time component of the field scattered into the black hole).

For pure inflow models we have seen that the stress-energy has the form given by Eq. (2.43). The energy density measured by a freely falling radial observer, crossing the Cauchy horizon in an asymptotically flat black hole, is [Eq. (2.44)]

$$\rho_{CH} = \frac{|E^2|}{4\pi r_{CH}^2} L(v) e^{2\kappa_{CH} v} \ , \tag{2.57}$$

where to avoid confusion with the de Sitter case we have let $r = r_{CH}$ denote the location of the Cauchy horizon and κ_{CH} its surface gravity. Whereas in the de Sitter case the form of the luminosity function $L(v)$ followed the requirement of a finite but non-vanishing flux at the cosmological horizon, the situation in asymptotically flat spacetimes is different. The requirement that the flux as measured by an observer out at infinity approaches zero as v tends to infinity, just implies that $L(v)$ should tend to zero there. The form of $L(v)$ does not, therefore, follow naturally. To ascertain the form of $L(v)$ one examines the rate of change of the field, as measured by a freely falling observer

$$\dot{\Psi} \equiv \frac{d\Psi}{d\tau} = \Psi_{,\alpha} u^{\alpha} \ , \tag{2.58}$$

where τ is the observer's proper time. Now, close to the Cauchy horizon the four-velocity of the observer is dominated by the u^v component, $\dot{v}$. Therefore, as v tends to infinity

$$\dot{\Psi} \sim \Psi_{,v} \dot{v} \sim v^{-\gamma-1} e^{\kappa_{CH} v} \ .$$

The flux ρ_{CH} is proportional to $(\dot{\Psi})^2$, from which we can deduce that

$$L(v) \sim \alpha v^{-2(\gamma+1)} \quad \text{as} \quad v \to \infty , \tag{2.59}$$

where α is a constant.

To date, no analytic calculation of radiative tails in the Reissner-Nordström-de Sitter spacetime or the Schwarzschild-de Sitter spacetime exists, but it is surprising that the Brady-Poisson analysis did not actually require us to know any details of the form of the tails near either the black hole event horizon or the cosmological event horizon. Imposing a sufficiently general requirement on the observed flux at the cosmological horizon (finite and non-zero) allows one to ascertain [See Sec. 2.3.2, Eqn. (2.46)] that

$$L(v) \sim K(v) e^{-2\kappa_1 v} , \tag{2.60}$$

where $K(v)$ is some slowly varying function of v that tends to a finite, non-zero value, as v tends to infinity. Reversing the argument above for this form of the luminosity function we get that

$$\Psi \sim e^{-\kappa_1 v} \tag{2.61}$$

at late times, near the cosmological event horizon. Thus, one would claim that the radiative tails in black hole-de Sitter spacetimes are exponential with a folding time given by the surface gravity at the cosmological event horizon.

2.5.2 Radiative Tails In Black Hole-de Sitter Spacetimes

We commented above that, unlike the case of black holes in asymptotically flat spacetimes, no analytic work on radiative tails in black hole-de Sitter spacetimes exists. This is either due to a lack of interest, or more likely the comparative difficulty of working in such spacetimes. Even in the case of Schwarzschild-de Sitter the analytic work rapidly becomes difficult and tedious. The majority of the difficulties can be traced to the fact that

$$g^{\alpha\beta}\nabla_\alpha r \nabla_\beta r = f(r) = 1 - \frac{2M}{r} + \frac{Q^2}{r^2} - \frac{r^2}{\alpha^2} \quad , \quad \alpha^2 = \frac{3}{\Lambda} \tag{2.62}$$

has four roots and as r tends to infinity

$$f(r) \to -r^2/\alpha^2 .$$

The lack of any detailed analysis, combined with the difficulties of attempting an analytic investigation, led Brady, Chambers, Krivan and Laguna [54] to perform the first numerical investigation into the late time behavior of fields propagating in black hole-de Sitter spacetimes.

Brady, Chambers, Krivan and Laguna 1997

Brady *et al.* have studied, in some detail, the behavior of a massless, minimally coupled scalar field propagating on spherically symmetric spacetimes with a positive cosmological constant. Particular attention was focused on the late time behavior of the fields in three particular regions; (a) the cosmological event horizon, (b) the black hole event horizon and (c) future timelike infinity (point D in Fig. 2.3) – approached along surfaces of constant r between these two horizons. The methods they employ are similar to those used by Gundlach, Pullin and Price [49] in their numerical studies of radiative tails in asymptotically flat black hole spacetimes,

Linear Method The field propagates on the fixed background spacetimes of

- Schwarzschild-de Sitter
- Reissner-Nordström-de Sitter

Non-Linear Method The field is coupled to a general spherically symmetric spacetime through the Einstein-Klein-Gordon field equations

Linear Analysis

The idea behind a linear analysis is similar to that discussed in the considerations of linear perturbation theory. One assumes that the field is sufficiently weak that its effect upon the spacetime is negligible. This is tantamount to assuming that the scalar field is a linear perturbation on the spacetime, so that the stress-energy tensor which, for the field under attention, is

$$T_{\alpha\beta} = \phi_{,\alpha}\phi_{,\beta} - \frac{1}{2}g_{\alpha\beta}\phi_{,\gamma}\phi^{,\gamma} , \qquad (2.63)$$

is second order in the field and thus vanishes to linear order. Linear analyses are not just favorable for their mathematical simplicity. In cases where the spacetime curvature is small, as in the exterior region of a black hole, the results from linear analyses are fairly representative of the results from non-linear analyses as we shall demonstrate later. Indeed Price's original work on tails in the Schwarzschild spacetime, a linear perturbation analysis, has been verified numerically both by linear and non-linear evolutions [49, 50] to a high degree of accuracy.

In terms of the standard advanced and retarded times (u, v) the metrics for Schwarzschild-de Sitter and Reissner-Nordström-de Sitter are described by Eq. (2.15),

$$ds^2 = -f du dv + r^2 d\Omega^2 ,$$

with $f(r)$ given by Eq. (2.2), from which Schwarzschild-de Sitter is obtained simply by setting $Q = 0$. On this background, the massless minimally coupled scalar wave equation $\Box\phi = 0$ becomes

$$\Psi_{,uv} = -\frac{1}{4}V_\ell(r)\Psi , \qquad (2.64)$$

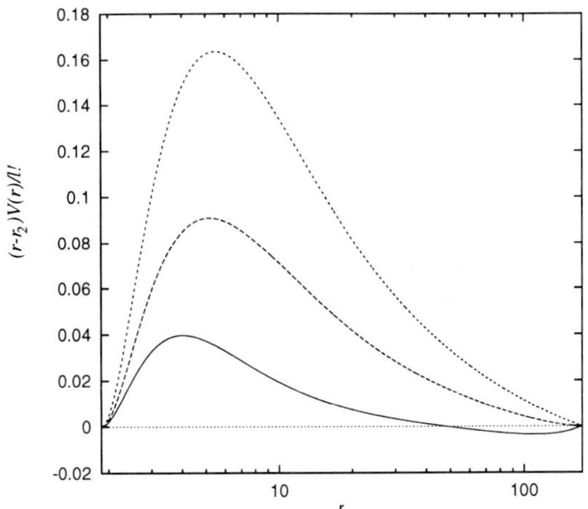

Figure 2.7: The effective potential experienced by a scalar field propagating on the fixed background spacetime of a Reissner-Nordström-de Sitter black hole with $Q = 0.5$, $M = 1$ and $\Lambda = 10^{-4}$. Shown are the potentials for the $\ell = 0$ (solid), $\ell = 1$ (dashed) and the $\ell = 2$ (dotted) ℓ-pole moments. The potential has been scaled by $(r - r_2)/\ell!$ to accentuate the nature of the potential when $\ell = 0$. Unlike the higher ℓ modes, $V_{\ell=0}$ shows a barrier followed by a well. For any ℓ-pole moment, the potential falls of exponentially as r approaches either horizon.

where the field has been decomposed into spherical harmonics

$$\phi = \sum_{\ell,m} \Psi(u,v) Y_{\ell m}(\theta,\psi) r^{-1}.$$

The function $V_\ell(r)$ is the effective potential for the scalar field (Sec. 2.3.1) and has the following form

$$V_\ell(r) = f(r) \left(\frac{\ell(\ell+1)}{r^2} + \frac{f'(r)}{r^2} \right) , \tag{2.65}$$

where $'$ denotes derivatives with respect to the function's argument. Plots of the potential, between the cosmological and black hole event horizons, are given in Fig. 2.7. Equation (2.64) is solved numerically by integrating it on the null grid defined by u and v. Initial data (characteristic) is given by the value of the field Ψ along two initial null surfaces $u = u_0$ and $v = v_0$. The details of the numerical method are adequately described in [49]. The results of the numerical integration are shown in Figs. 2.8, 2.9, 2.10 and 2.11. The initial data used to generate these figures was

$$\begin{aligned} \Psi(u=0,v) &= \exp\left[-\frac{(v-v_1)^2}{\sigma^2} \right] \\ \Psi(u,v=0) &= \Psi(u=0,v=0) , \end{aligned}$$

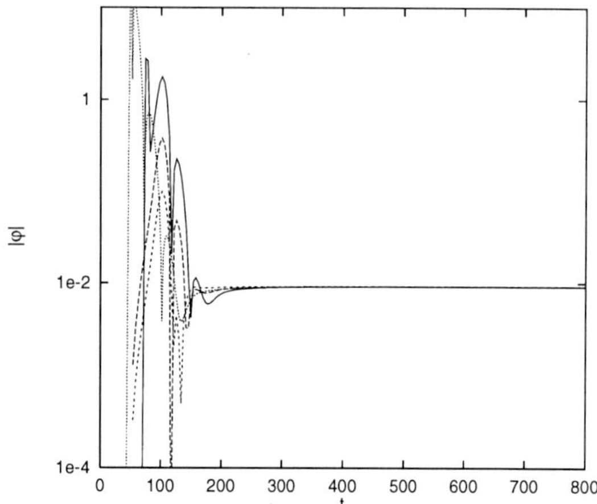

Figure 2.8: A plot of $|\phi_{\ell=0}|$ versus time for a Schwarzschild-de Sitter black hole spacetime. The field behavior is shown along four different surfaces (see text).

being representative of the other data sets employed by Brady *et al.*. All the graphs shown are for black holes with their mass (M) scaled to unity and $\Lambda = 10^{-4}$. Though this choice of Λ is arbitrary the results are qualitatively similar for any other value of Λ that is non-zero. The fields are shown plotted along four different surfaces,

- The cosmological event horizon
- The black hole event horizon
- Two surfaces of constant r approaching future timelike infinity

Since the late time behavior is identical along each surface we shall not distinguish between them in the figures. Figure 2.8 displays the monopole field behavior for a Schwarzschild-de Sitter black hole. At early times ($0 < t < 200$) the field behavior is dominated by quasi-normal ringing, associated with complex characteristic frequencies of the hole. At late times $t > 200$ the field approaches the same constant value on all four surfaces. Figure 2.9 shows the same results for a Reissner-Nordström-de Sitter black hole with $Q = 0.5$. A period of quasi-normal ringing is followed by a relaxation of the field to a constant value on all four surfaces. A detailed investigation of the field's late time behavior reveals

$$\phi_{\ell=0} \simeq \phi_0 + \phi_1(r)e^{-2\kappa_1 t} , \tag{2.66}$$

and that the constant field term ϕ_0 scales like Λ [54]. Figures 2.10 and 2.11 display the field behavior for $\ell = 1$ and the $\ell = 2$ modes for the case $Q = 0.5$. Again there is

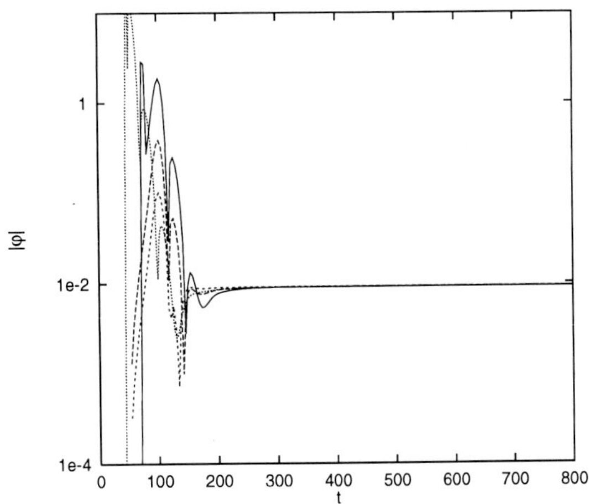

Figure 2.9: A plot of $|\phi_{\ell=0}|$ for a Reissner-Nordström-de Sitter black hole with $Q = 0.5$. The field behavior is shown along four different surfaces (see text).

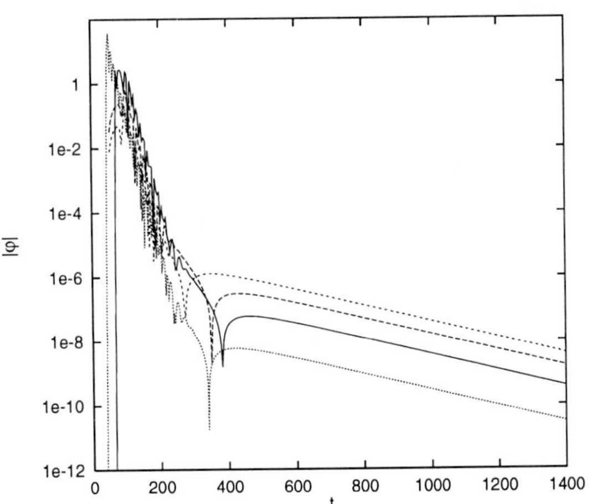

Figure 2.10: A plot of $|\phi_{\ell=1}|$ for a Reissner-Nordström-de Sitter black hole for $Q = 0.5$. The field is behavior is shown along four different surfaces (see text).

a period of quasi-normal ringing followed by a distinct exponential fall off. In general, an ℓ-pole mode of the field decays, at late times, like

$$\phi_\ell \sim e^{-\ell\kappa_1 t} \qquad (\ell > 0) . \tag{2.67}$$

Non-Linear Evolution

If the scalar field is allowed to couple to the spacetime via the stress-energy tensor, then one must contend with the Einstein-Klein-Gordon field equations

$$G_{\alpha\beta} \equiv R_{\alpha\beta} - \frac{1}{2}g_{\alpha\beta}R = 8\pi T_{\alpha\beta} - g_{\alpha\beta}\Lambda , \tag{2.68}$$

where $T_{\alpha\beta}$ is the stress-energy tensor for a massless, minimally coupled scalar field Eq. (2.63). For simplicity and tractability, attention is focused on spherically symmetric spacetimes, whose line element can be written as

$$ds^2 = -g\bar{g}du^2 - 2gdudr + r^2 d\Omega^2 . \tag{2.69}$$

The Einstein field equations, Eq. (2.68), then reduce to

$$(\ln g)_{,r} = 4\pi r^{-1}(h - \bar{h})^2 , \tag{2.70}$$
$$(r\bar{g})_{,r} = g(1 - \Lambda r^2) , \tag{2.71}$$
$$(r\bar{h})_{,r} = h \tag{2.72}$$

and the scalar wave equation, $\Box\phi = 0$, becomes

$$h_{,u} - \frac{\bar{g}}{2}h_{,r} = \frac{(h - \bar{h})}{2r}\left[g(1 - \Lambda r^2) - \bar{g}\right] , \tag{2.73}$$

where the auxiliary fields $(h, \bar{h})$ are defined by

$$\bar{h} = \frac{1}{r}\int h dr \equiv \phi . \tag{2.74}$$

Goldwirth and Piran [55] have devised a simple, but effective, numerical algorithm for integrating these equations on a (u, r) grid. This method has been implemented by Gundlach *et al.* to study radiative tails in asymptotically flat black hole spacetimes [49]. A refinement of this algorithm, which reduces numerical error near the $r = 0$ origin has been supplied by Garfinkle [56] and used by Brady *et al.* to integrate Eqs. (2.70)–(2.73). Initial data for the problem is given by the value of the field along some initial null cone centered at the origin of coordinates. Details of the numerical method can be found in [49] and references therein. In the graphs that follow, the initial data was Gaussian with

$$\phi = \phi_A \left(\frac{r}{r_0}\right)^2 \exp\left[-\frac{(r - r_0)^2}{\sigma^2}\right] , \tag{2.75}$$

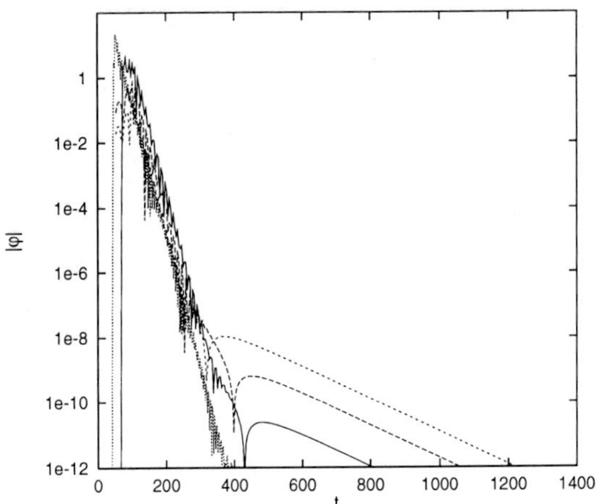

Figure 2.11: A plot of $|\phi_{\ell=2}|$ for a Reissner-Nordström-de Sitter black hole for $Q = 0.5$. The behavior of the field is shown along four surfaces (see text).

where ϕ_A is the amplitude of the field. The figures shown are, again, for $\Lambda = 10^{-4}$. The behavior of the field in this case is along three surfaces,

- The cosmological event horizon
- The black hole event horizon
- A surface of constant r approaching future timelike infinity

Again, because the qualitative behavior of the field at late times along each surface is similar, we do not explicitly distinguish each surface in the figures. The restriction to spherical symmetry implies that we gain only information about the $\ell = 0$ mode of the field, so the results plotted are for $\ell = 0$ only. In Fig. 2.12 the field behavior can be seen to be remarkably similar to that demonstrated by the linear analysis. At late times the field approaches the same constant value along all three surfaces, in confirmation of the test field results. Figure 2.13 displays the behavior of $\phi_{,r}$, which is proportional to $(\bar{h} - h)$ along surfaces of constant r. Brady *et al.* find that at late times

$$(\bar{h} - h) \sim e^{-2\kappa_1 t} \quad \text{as} \quad u \to \infty\,, \tag{2.76}$$

so that

$$\phi_{\ell=0} \simeq \phi_0 + \phi_1(r) e^{-2\kappa t} \quad \text{as} \quad t \to \infty\,, \tag{2.77}$$

in agreement with the linear perturbation analysis.

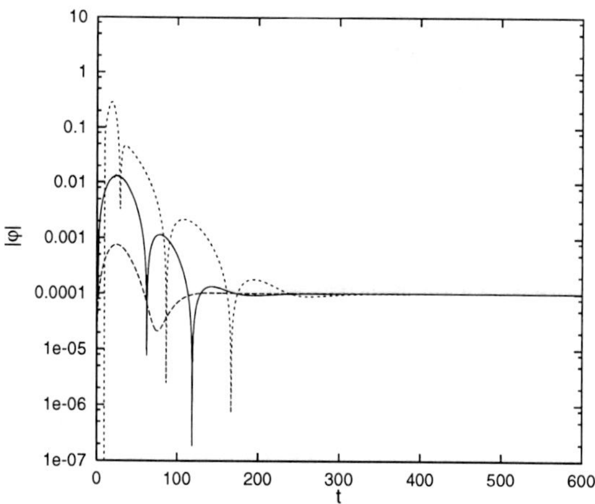

Figure 2.12: A plot of $|\phi|$ versus time t for a non-linear evolution. The field behavior is plotted for three the different surfaces explained in the text

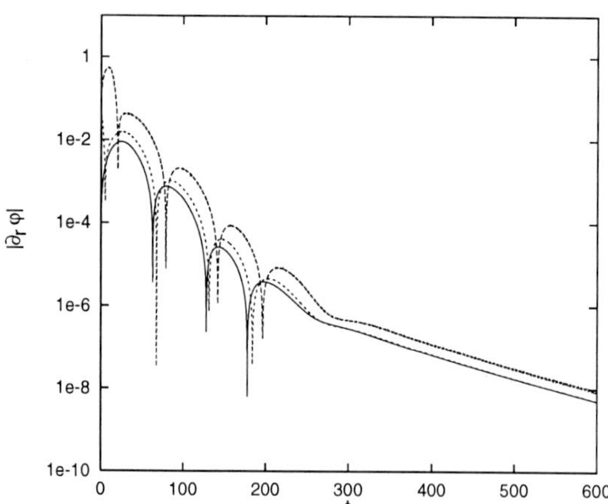

Figure 2.13: A plot of the derivative of the field $|\phi_{,r}|$ versus time t for a non-linear evolution. The field behavior is plotted for three the different surfaces explained in the text

2.5.3 Conclusions

For a massless, minimally coupled scalar field propagating on the fixed backgrounds of Schwarzschild-de Sitter and Reissner-Nordström-de Sitter, the late time behavior along the cosmological event horizon, the black hole event horizon and at future timelike infinity is given by

$$\phi \sim e^{-\ell\kappa_1 t} ,$$

for $\ell > 0$. For $\ell = 0$ behavior is slightly different,

$$\phi_{\ell=0} \sim \phi_0 + \phi_1 e^{-2\kappa_1 t} ,$$

asymptoting to a constant value at late times, which is ultimately confirmed by the non-linear evolution of a spherically symmetric scalar field coupled to general relativity.

The behavior of the $\ell = 0$ mode is, in some ways, unusual. As t tends to infinity, the field mode approaches a constant value, in fact the same constant value, at both the cosmological event horizon and the black hole event horizon. There is no analogue of this for the case of black holes in flat space. While Price's original work [34] demonstrates that there can be no static solutions to the scalar wave equation that are well behaved at infinity and the black hole event horizon, one can actually have the trivial static solution $\phi = constant$. The same holds true in black hole-de Sitter spacetimes. Even though the constant solution carries no stress-energy, it is intriguing that the appearance of a constant field value at the horizons does not occur in the flat space examples. An examination of the zero frequency reflection and transmission coefficients in the region exterior to a de Sitter black hole, shows they are non-zero, allowing the propagation of a constant mode to both horizons. In this respect, scattering in the exterior of a black hole-de Sitter spacetime is somewhat similar to scattering in the interior of a Reissner-Nordström black hole. However, in the exterior of a Reissner-Nordström black hole no such constant mode propagation is allowed. It is likely that the constant mode propagation in an $\ell = 0$ mode and the anomalous dip in the effective potential for that mode (Fig. 2.7) are not a mere coincidence, though no detailed investigation of this relation has been forthcoming. What is known is that for a conformally invariant scalar field the well in the potential disappears, and in this case the constant field is zero. A similar well in the effective potential occurs for the case of the gravitational perturbations detailed by Mellor and Moss [17]. It seems reasonable to suspect similar constant mode behavior will be observed in this case too [57].

At the start of this section on current work, we commented on how the minimal requirement of Brady-Poisson implicitly imposes form on the late time behavior of the perturbing field near the cosmological event horizon, described by Eq. (2.61). Since the minimal requirement is essential to the stability condition expressed in Eq. (2.42), it is important that the numerics support it. The numerical results have shown that we can, in general, expand the scalar field as

$$\phi(t,r) = \phi_0 + \sum_{n=1}^{n=\infty} \phi_n(r) e^{-n\kappa_1 t} ,$$

in the exterior at late times. Eq. (2.61) gives the form of the field we should expect from imposing the minimal requirement,

$$\phi \simeq a(r) + b(r) e^{-\kappa_1 t} .$$

We should note that this is a more general solution than that proposed in Eq. (2.61), where we dropped the constants of integration for simplicity. Making the identifications $a(r) = \phi_0$ and $b(r) = \phi_1(r)$ demonstrates that the numerical results confirm the minimal requirement of Brady and Poisson. It is worth noting that it is the $n = 1$, or $\ell = 1$ mode, that actually provides the required form for the radiative tails. The contribution from the other ℓ modes leads to a vanishing flux at the Cauchy horizon. In any generic perturbation we would expect all ℓ-pole moments to be present, thus assuring the correctness of the Brady-Poisson model.

2.6 Discussion

We began this review by stressing the need for a precise formulation and proof of the strong cosmic censorship hypothesis. It is almost twenty years since Penrose conjectured this stronger form of cosmic censorship [58], and today it still remains a cardinal, unsolved problem of general relativity. Instead of seeking a correct formulation of strong cosmic censorship, we have opted for a much simpler route – a search for reasonable counter-examples, in the belief that an understanding of these models will inevitably lead to a deeper comprehension of the censor issue. This hunt for counter-examples has lead us into realm of black hole-de Sitter spacetimes and to pose the question

> "Are black holes immersed in de Sitter space counter-examples to the current formulation of strong cosmic censorship?"

The answer to this appears, at least classically, to be yes. The linear perturbation analyses in Reissner-Nordström-de Sitter and Kerr-Newman-de Sitter and the back-reaction calculations, performed in spherical symmetry, agree – the Cauchy horizon in black hole-de Sitter spacetimes is stable provided that

$$\kappa_1 > \kappa_3 .$$

Quantum mechanically the answer appears to be no. It wasn't obvious (to me at least) that this divergence couldn't just be due to a bad choice of vacuum state. To prove quantum instability one actually needs to prove there are no quantum states that are regular on all three horizons. Eric Poisson has kindly pointed out to me that although Marković and Poisson did not give this proof, one does exist, at least in the two dimensional case. The proof is provided in Eric's contribution elsewhere in these proceedings. It might be interesting to see if this results is easily continued to the four dimensional case. Occasionally, the results of a two dimensional calculation do not concur with the results of a four dimensional analysis. An excellent demonstration of this occurs in the extreme Reissner-Nordström spacetime. In two dimensions it has been shown [59] that stress-energy of a quantized field will diverge on the event horizon of an extreme black hole, whereas a four dimensional calculation shows no sign of a divergence [60]. It maybe that a four dimensional calculation of quantum effects near the Cauchy horizon will lead to a somewhat different conclusion than that reached by the two dimensional analysis. The quantum instability is undoubtedly an important question. However, the fiery marriage between general relativity and quantum mechanics, that is *Quantum-Gravity*, has yet to be consummated. For that reason we have concentrated mainly on the classical stability of the Cauchy horizon in black hole-de Sitter spacetimes. The existence of solutions to the classical field equations exhibiting stable Cauchy horizons faces classical physics with many problems, including a description of the singularity and how to attempt physics in the presence of causality violating curves.

A look to the Future

We now turn our attention to the future. The following comments are based purely on personal speculation about future work and the role it will play in answering the question posed above. The purpose of this section is not just to seek the truth, but to try to encourage a more active participation in the study of black hole-de Sitter spacetimes.

With the results of Brady *et al.*, on the form of radiative tails in Reissner-Nordström-de Sitter [54], things are pretty much set up to allow a numerical investigation of the interior. An adequate amount of numerical algorithms and techniques now exist, which have been used in the study of the Reissner-Nordström spacetime [50, 51, 52], and should allow this problem to be confronted with some ease. It is hoped that the numerical results will provide further evidence toward the conjecture that the Cauchy horizon in black hole-de Sitter spacetimes can be stable for a fixed, non-zero, region of its parameter space. A numerical study will, at the least, allow us to gain some insight into the behavior of the spacetime in the vicinity of the Cauchy horizon and the behavior of fields propagating in the interior. These results can be used to gauge the adequacy of the linear analysis in describing the physics at the horizon and could provide information to initiate new analytic approaches for studying the interior.

A preliminary investigation of the numerical problem suggests that one should take an approach similar to that used by Brady and Smith [51] for studying the interior but implement the algorithm of Burko and Ori [50]. In this case the initial data is specified along some null surface that is taken to coincide with the event horizon. The alternative is to specify generic (Gaussian) initial data outside the hole (see [50, 52]) and let it evolve, through the event, to the interior. For Reissner-Nordström this is quite adequate and has the advantage that one does not have to know the precise form of the radiative tails. For the de Sitter case this is not really a satisfactory approach. As we saw in Sec. 2.5.2, the restriction to spherical symmetry [Eq. (2.69)] implies that if we start from generic initial data outside the hole then at the event horizon we will have only the radiative tail of the $\ell = 0$ mode. The form of this tail, Eq. (2.77), does not satisfy the minimal requirement of Brady and Poisson discussed in Sec. 2.3.2, for this mode there is a vanishing flux of energy at the cosmological event horizon. To obtain a non-vanishing flux requires the introduction of higher ℓ-modes, specifically the $\ell = 1$ mode. Specifying the initial data along the event horizon, as Brady and Smith did, allows, in some respects, the added flexibility of incorporating the effects of additional ℓ-modes.

One of the biggest problems facing anyone attempting to study the interior of a Reissner-Nordström-de Sitter black hole is the search for an initial data set that evolves into a solution with a stable Cauchy horizon. From Fig. 2.6, it is easy to see that the stability region is extremely small, in fact 0.6% of the entire parameter space by area. It might be possible, instead of hunting for initial data, to attempt some type of shooting method, with boundary conditions at the future event horizon and cosmological horizon. At present this is still a suggestion and may prove to be a dead end. Another approach which initially appeared sound was to fix regularity conditions, suggested by the analytic studies, on the spacetime and the fields at the Cauchy horizon and evolve them backward in time to the event horizon. The philosophy behind this idea being that one starts off with a solution that has a regular Cauchy horizon in order to see whether or not it can evolve from regular data at the event horizon. Of course we expect the data corresponding to the field to have the exponential form required by the tails. The downfall of this approach is that derivatives of the field, at the Cauchy horizon, vanish. It seems likely therefore that the evolution will end up showing no field propagating through the interior as a whole, i.e, vanishing field perturbation at the event horizon. Currently this remains the most difficult step on the path toward a numerical investigation of the interior. While the feeling is that this is not an insurmountable problem, it seems it will require some thought.

A Differential Proper Times

We present the arguments of Brady and Poisson [23] which demonstrate that a necessary condition for Cauchy horizon instability is that the ratio of *differential* proper times be divergent.

We consider two observers in the spacetime, an external observer in region II approaching the cosmological event horizon and an internal observer in region III

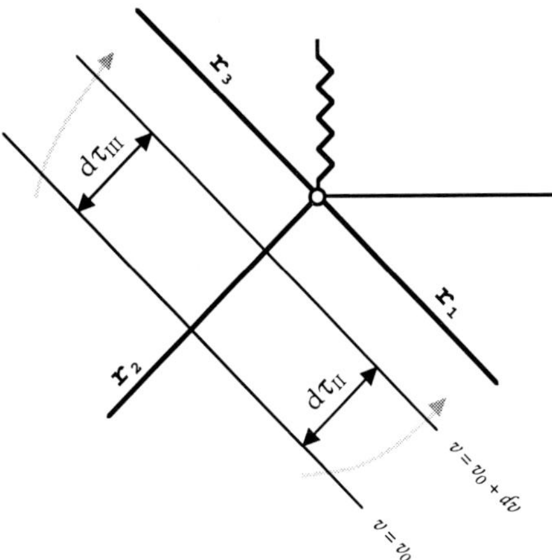

Figure 2.14: A schematic representation of the blueshift vs. redshift argument of Poisson and Brady. Shown are the inner horizon (r_3), the outer horizon (r_1) and the black hole event horizon (r_2). Also shown are the paths of the two observers (light arrows) approaching their respective horizons. v_0 and $v_0 + dv$ represent two radially ingoing null rays, meant to represent two successive wave crests. The external observer measures a proper time $d\tau_{II}$ between the crests whilst the internal observer measures a proper time $d\tau_{III}$

approaching the Cauchy horizon. Each observer measures the proper time, τ_{II} and τ_{III} respectively, between two successive wavefronts, labeled v_0 and $v_0 + dv$. The situation is shown schematically in Fig. 2.14. In region II, as the observer nears the cosmological event horizon, the u^v component of their four-velocity, $\dot{v}$, approaches

$$\dot{v}_{II} \equiv \frac{dv_{II}}{d\tau_{II}} \simeq \frac{|E_{II}|}{r_1\kappa_1} e^{\kappa_1 v} , \tag{2A.1}$$

where we have used the results of Sec. 2.2.2 [Eq. (2.31)]. Similarly, for observers in region III approaching the Cauchy horizon,

$$\dot{v}_{III} \equiv \frac{dv_{III}}{d\tau_{III}} \simeq \frac{|E_{III}|}{r_3\kappa_3} e^{\kappa_3 v} , \tag{2A.2}$$

[See Eq. (2.34)]. Since the observers are measuring proper time between the same successive wavefronts ($dv_{II} = dv_{III} \equiv dv$) and because both horizons are located at the same advanced time, then

$$\frac{\dot{v}_{II}}{\dot{v}_{III}} = \frac{d\tau_{III}}{d\tau_{II}} \sim e^{(\kappa_1 - \kappa_3)v} \quad \text{as} \quad v \to \infty . \tag{2A.3}$$

Therefore, if $\kappa_3 > \kappa_1$, the interior observer measures an increasingly smaller proper time between successive wave crests as she/he approaches the Cauchy horizon indicating a blueshift and hence an instability of the Cauchy horizon. If, on the other hand $\kappa_1 > \kappa_3$, the observer measures an increasingly larger proper time between the wavefronts as he/she approaches the Cauchy horizon, indicating a redshift. In this case the Cauchy horizon is stable. Even without performing the calculation, it is not hard to see that these observers take a finite proper time to reach there respective horizons. One can also see the inevitability of the blueshift instability of the Cauchy horizon in the Reissner-Nordström solution. Here, the cosmological event horizon is replaced by by future null infinity, $\mathcal{J}^+$, and $\kappa_1 \to 0$. In this case the internal observer will always measure an vanishingly small proper time between the wave crests as he approaches the inner horizon and so the Cauchy horizon is always unstable. This instability is due to the infinite time compression effects we discussed in the Introduction. Thus, we can conclude that while the infinite compression of the ratio of proper times is a sufficient condition for Cauchy horizon stability, it is not a necessary condition. A necessary condition is an infinite compression of the ratio of *differential* proper times expressed in Eq. (2A.3).

B Ori-Model

The Poisson-Israel model of mass inflation [26] models the fluxes of radiation in the interior, generated during the collapse to form a black hole, as a crossflow of lightlike particles moving radially outward and inward respectively. Whilst the form of the inflow is crucial to the analysis (modelling the late time behavior of the fields crossing the event horizon) the nature of the outflux is largely irrelevant, its presence is required only to precipitate a contraction of the generators of the Cauchy horizon. In the Ori model the outflux is modelled as a thin lightlike shell Σ, which allows an exact mass inflation solution. The generalization of this model to black holes in de Sitter space is simple [30]. Figure 2.15 depicts the conformal diagram of the situation. The null shell Σ divides the interior spacetime into two distinct regions, that to the past of Σ which we label $(-)$ and that to the future, which we label $(+)$. On either side of the shell the spacetime is that of Reissner-Nordström-Vaidya-de Sitter (RNVDS), with metrics

$$ds_{\pm}^2 = -f_{\pm} dv_{\pm}^2 + 2dv_{\pm} dr_{\pm} + r_{\pm}^2 d\Omega_{\pm}^2 \tag{2B.1}$$

$$f_{\pm} = 1 - \frac{2m_{\pm}(v_{\pm})}{r_{\pm}} + \frac{Q_{\pm}^2}{r_{\pm}^2} - \frac{r_{\pm}^2}{\alpha^2} \,. \tag{2B.2}$$

The stress-energy tensor is

$$T_{\alpha\beta}^{\pm} = \frac{L(v_{\pm})}{4\pi r_{\pm}^2} (\partial_\alpha v_{\pm})(\partial_\beta v_{\pm}) \,. \tag{2B.3}$$

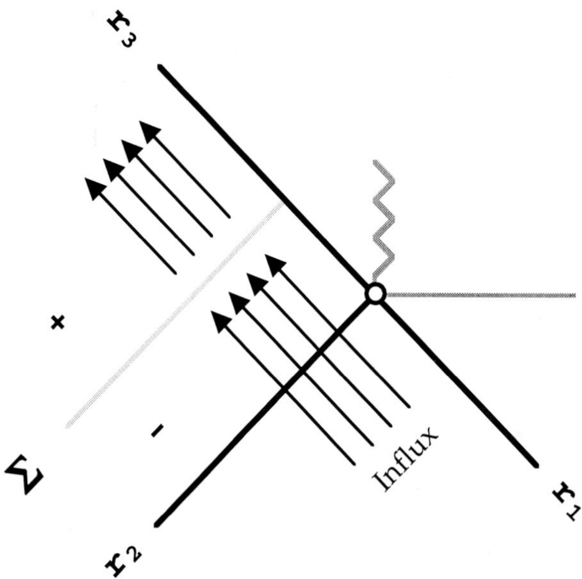

Figure 2.15: The conformal diagram for the Ori model in the Reissner-Nordström-de Sitter spacetime. Shown are the locations of the cosmological event horizon r_1, the black hole event horizon r_2 and the Cauchy horizon r_3. Also shown is an influx of radiation crossing the event horizon (arrows) and the outflow of radiation, idealized as a thin lightlike shell Σ, emanating from the collapsing star's surface to the left of the diagram (not shown). The outflow naturally divides the spacetime up into two distinct regions, that to its future (+) and that to its past (−).

The luminosity function $L(v)$ is related to the mass function $m(v)$ through the G_{vv} component of the Einstein equations

$$L(v) = \frac{dm(v)}{dv} \tag{2B.4}$$

Matching

The problem is to ascertain the nature of the spacetime to the future of Σ, the $(+)$ region. This is simply done by matching the two RNVDS spacetimes across the shell. The first requirement is that metric tensor $g_{\alpha\beta}$ be continuous across Σ, which requires that $r_+ = r_- \equiv r$ and $d\Omega_+^2 = d\Omega_-^2 \equiv d\Omega^2$. We also impose the physical conditions that Σ be electrically neutral, so that $Q_+ = Q_- \equiv Q$ and that it be pressureless, which requires that $\lambda_+ = \lambda_- \equiv \lambda$, where λ is the affine parameter along Σ. The two important matching conditions that can be derived from these are

- The null generators of Σ are the same on either side

$$2dr = f_+ dv_+ = f_- dv_- \tag{2B.5}$$

- Flux continuity across Σ

$$\frac{1}{f_+}\frac{dm_+}{d\lambda} = \frac{1}{f_-}\frac{dm_-}{d\lambda} \tag{2B.6}$$

The first condition is given by the continuity of the metric tensor and that null geodesics satisfy $dr/dv = f/2$. The second condition is the requirement that $\lambda_+ = \lambda_-$, which can be shown [61] to be equivalent to the requirement that $T^+_{\alpha\beta} m^\alpha_+ m^\beta_+ = T^-_{\alpha\beta} m^\alpha_- m^\beta_-$, where the $m^\alpha_\pm$ are the tangents to the null generators of Σ on either side. Thus the requirement of pressureless implies that Σ does not interact with the influx.

An Exact Solution

In the $(-)$ region we have that

$$L(v_-) = \frac{dm_-}{dv_-} = K(v_-)e^{-2\kappa_1 v_-} \ . \tag{2B.7}$$

Integrating this gives

$$m_-(v_-) = M - \frac{\alpha}{2\kappa_1} e^{-2\kappa_1 v_-} \ , \tag{2B.8}$$

where for convenience we have assumed that $K(v_-)$ is a constant α. In reality $K(v_-)$ is a slowly varying function of v_- and numerical analyses suggest that to first order $K(v_-)$ is constant. $M = m_-(\infty)$ is the final asymptotic mass. Equation (2B.5)) gives that the null geodesics in region $(-)$ obey

$$\frac{dr}{dv_-} = \frac{f_-}{2} \ . \tag{2B.9}$$

Expanding f_- about $r = r_3$ and letting v_- tend to infinity allows us to integrate this expression

$$\kappa_3(r - r_3) \simeq \beta e^{-\kappa_3 v_-} + \frac{\alpha\kappa_3}{2\kappa_1 r_3} \frac{e^{-2\kappa_1 v_-}}{(\kappa_3 - 2\kappa_1)} , \tag{2B.10}$$

where β is a constant of integration. Then we can write the asymptotic form of f_-

$$f_-(r, v_-) \simeq -2\beta e^{-2\kappa_3 v_-} - \frac{2\alpha}{r_3(\kappa_3 - 2\kappa_1)} e^{-2\kappa_1 v_-} , \tag{2B.11}$$

as $r \to r_3$ and $v_- \to \infty$. The second matching condition, Eq. (2B.7), allows us to obtain m_+ as a function of v_-,

$$\int \frac{dm_+}{f_+} = \int \frac{dm_-}{f_-} = \int \frac{dm_-}{dv_-} \frac{1}{f_-} dv_- . \tag{2B.12}$$

Using Eqs. (2B.2),(2B.8) and (2B.11) it is easy to show that

$$m_+(v_-) \simeq M + \frac{\gamma\beta M(\kappa_3 - 2\kappa_1)}{\alpha} + \gamma M e^{(\kappa_3 - 2\kappa_1)v_-} , \tag{2B.13}$$

where γ is another constant of integration. We can see immediately from this relation that the mass function to the future of the shell will inflate if and only if $\kappa_3 > 2\kappa_1$ which agrees with the work by Brady *et al.* [30]. We shall concentrate on the case that $\kappa_3 > 2\kappa_1$, the other cases follow the same line of reasoning. We can therefore estimate the leading behavior of the mass to be

$$m_+(v_-) \sim \gamma M e^{(\kappa_3 - 2\kappa_1)v_-} \quad \text{as} \quad v_- \to \infty . \tag{2B.14}$$

We require to know how v_+ and v_- are related, which we can do by using Eq. (2B.5)

$$\int dv_+ = \int \frac{f_-}{f_+} dv_- . \tag{2B.15}$$

Using Eqs. (2B.2),(2B.11) and (2B.14) we find that

$$v_+ \sim -\frac{\alpha}{\gamma M(\kappa_3 - 2\kappa_1)\kappa_3} e^{-\kappa_3 v_-} , \tag{2B.16}$$

where, without loss of generality, we have set the constant of integration to zero. If we remember that the regular coordinate at the Cauchy horizon is $V = -e^{-\kappa_3 v_-}$ [Eq. (2.23)] we can see that v_+ is Kruskal-like,

$$v_+ \sim \frac{\alpha}{\gamma M(\kappa_3 - 2\kappa_1)\kappa_3} V , \tag{2B.17}$$

then

$$v_- \sim -\frac{1}{\kappa_3} |-V| . \tag{2B.18}$$

The mass function $m_+(V)$ thus inflates as

$$m_+(V) \sim \gamma M(-V)^{\frac{2\kappa_1-\kappa_3}{\kappa_3}} \quad \text{as} \quad V \to 0\,. \tag{2B.19}$$

We define a new coordinate U such that

$$-\delta dU = m_+ dv_+ + r dr \quad \text{where} \quad \delta = \frac{\gamma M(\kappa_3 - 2\kappa_1) r_3}{\alpha \kappa_3}\,, \tag{2B.20}$$

so that close to the Cauchy horizon the metric becomes

$$ds_+^2 = -2e^{2\sigma} dU dV + r^2 d\Omega^2 \quad \text{where} \quad e^{2\sigma} = \frac{1}{\kappa_3^2}\frac{r_3}{r} \tag{2B.21}$$

From Eq. (2B.20) we can obtain the behavior of the radius

$$r^2 = r_3^2 - 2U\delta + \frac{\alpha}{(\kappa_3 - 2\kappa_1)\kappa_1}(-V)^{2\frac{\kappa_1}{\kappa_3}}\,, \tag{2B.22}$$

which reflects the slow contraction of the Cauchy horizon.

Bibliography

[1] G. T. Horowitz and H. J. Sheinblatt, Phys. Rev. D **55**, 650 (1997).

[2] Our notation for the causal structure follows Ref. [8], Chap. 8, p. 188.

[3] The phrase "terminating it at the Cauchy horizon" is probably a little strong. There are indications that the horizon is traversable. However, this point was the focus of much debate during the Haifa meeting and it is clear that not everybody agrees with the interpretations presented in favor of traversability. The issue is an important one and is adequately dealt with elsewhere in these proceedings, where the reader can make up his or her own mind.

[4] A brief search of the astro-ph archive at lanl, for the year 1997 to date, revealed 6 submissions offering evidence that challenges the 'establishment' view of a vanishing cosmological constant.

[5] R. Herman and W. A. Hiscock, Phys. Rev. D **46**, 1863 (1992).

[6] This is only true in a classical context. We shall see later that quantum mechanically, the Cauchy horizon in black hole-de Sitter spacetimes appears to be unstable.

[7] B. Carter, in *Black Holes*, edited by C. M. De Witt and B. S. De Witt (Gordon and Breach, New York, 1973).

[8] R. M. Wald, *General Relativity*, (The University of Chicago Press, Chicago and London, 1984).

[9] By *similar* we mean a spacetime for which $g^{\alpha\beta}\nabla_\alpha r \nabla_\beta r \equiv f(r)$.

[10] In this and what follows in subsequent sections, we shall implicitly assume that there exist four distinct roots to Eq. (2.7), and not concern ourselves with solutions possessing coincident horizons.

[11] For the Kerr-Newman-de Sitter spacetime the root r_4, though negative, does correspond to a physical horizon lying beyond the ring singularity at $r = 0$.

[12] R. Penrose, in *Relativity, Groups and Topology*, edited by C. M. De Witt and B. S. De Witt (Gordon and Breach, New York, 1963).

[13] This point assumes the collapse becomes dynamical in region II.

[14] The definitions of u and v in region II are not unique either, but they do follow the more usual convention. The main reason for introducing a less conventional definition of u and v here is to alert the reader to the fact it is possible and, in certain circumstances, it can be advantageous to define u and v differently.

[15] Again this is required so that both the Cauchy horizon, and cosmological event horizon, are located at the same advanced time $v = +\infty$.

[16] One may be surprised to see the argument of the exponential in Eq. (2.31) contains the advanced time coordinate v rather than the radial coordinate r_*. However, along a radial geodesic that crosses either horizon, it can be seen [from Eqs. (2.29) and (2.30) or Eqs. (2.32) and (2.33)] that $dr/dv = \frac{1}{2}f$. Integrating this, and using Eq. (2.4), one can easily show that $r_* \sim \frac{1}{2}v + k$ as the horizons are approached. In Sec. 2.2.3 we have, without loss of generality, set the integration constant k to zero.

[17] F. Mellor and I. Moss, Phys. Rev. D **41**, 403 (1990).

[18] Stability in this context requires that the flux, due to the perturbations, as measured by an observer crossing the Cauchy horizon be finite there.

[19] S. Chandrasekhar and J. B. Hartle, Proc. R. Soc. London **A384**, 301 (1982).

[20] A detailed and informative account of the method used by Chandrasekhar and Hartle can be found in [21].

[21] S. Chandrasekhar, *The Mathematical Theory of Black Holes* (Cambridge University Press, Cambridge, England, 1983).

[22] V. de Alfaro and T. Regge, *Potential Scattering* (North-Holland Press, Amsterdam, 1965).

[23] P. R. Brady and E. Poisson, Class. Quantum Grav. **9**, 121 (1992).

[24] W. A. Hiscock, Phys. Lett. **83A**, 110 (1981).

[25] This solution is generally referred to as the Reissner-Nordström-Vaidya-de Sitter solution. Some details of this spacetime are given in Appendix B.

[26] E. Poisson and W. Israel, Phys. Rev. D **41**, 1796 (1990).

[27] W. Israel, in *Black Hole Physics*, edited by V. de Sabbata and Z. Zhang (Kluwer Press, Amsterdam, 1992).

[28] Of course, like the Reissner-Nordström case, one can only keep putting charge on the black hole up to the point of extremality.

[29] F. Mellor and I. Moss, Class. Quantum Grav. **9**, L43 (1992).

[30] P. R. Brady, D. Núñez and S. Sinha, Phys. Rev. D **47**, 4239 (1993).

[31] The details of the Poisson-Israel model are dealt with elsewhere in these proceedings.

[32] A. Ori, Phys. Rev. Lett. **67**, 789 (1991).

[33] By strong we mean that we do not get a functional form for the luminosity function as we did in Sec. 2.3.2, Eq. (2.46).

[34] R. H. Price, Phys. Rev. D **5**, 2419 (1972); Phys. Rev. D **5**, 2439 (1972)

[35] J. Bičák, Gen. Relativ. Gravit. **3**, 331 (1972).

[36] C. M. Chambers and I. G. Moss, Class. Quantum Grav. **11**, 1034 (1994).

[37] A concise, but clear, discussion of the Newman-Penrose formalism can be found in [21] Chap. 1, p. 40. It transpires that the NP formalism is particularly well suited to the Kerr and Kerr-de Sitter spacetimes, with the perturbation equations showing a high degree of symmetry. Indeed, the first separation of the perturbation equations in the Kerr spacetime was completed by Teukolsky in 1973 [38] using the NP formalism.

[38] S. A. Teukolsky, Phys. Rev. Lett. **29**, 1114 (1972).

[39] C. M. Chambers, Ph.D thesis, University of Newcastle Upon Tyne, 1995.

[40] Kerr-de Sitter, like all black hole spacetimes, is Type D under the Petrov classification of spacetimes.

[41] For an idea of the amount of algebra required, one is guided to the comments made by Chandrasekhar in ref. [21], Chap. 9, p. 530.

[42] D. Marković and E. Poisson, Phys. Rev. Lett. **74**, 1280 (1995).

[43] For the standard vacuum states, the vacuum stress-energy diverges on the future black hole event horizon if the vacuum state is chosen so that the vacuum stress-energy is regular on the cosmological horizons [44] and vice-versa.

[44] W. A. Hiscock, Phys. Rev. D **39**, 1067 (1989).

[45] D. Marković and W. G. Unruh, Phys. Rev. D **43**, 332 (1991).

[46] R. Balbinot and R. Bergamini, Phys. Rev. D **40**, 372 (1989).

[47] Indeed, we know of many series like this whose successive terms decrease but whose sums do not converge. A particularly well known example is $\sum_1^\infty n^{-1}$.

[48] C. M. Chambers (unpublished).

[49] C. Gundlach, R. H. Price and J. Pullin, Phys. Rev. D **49**, 883 (1994); Phys. Rev. D **49**, 890 (1994).

[50] L. M. Burko and A. Ori, Phys. Rev. D **56**, 7820 (1997).

[51] P. R. Brady and J. D. Smith, Phys. Rev. Lett. **75**, 1256 (1995).

[52] L. M. Burko, Phys. Rev. Lett. **79**, 4958 (1997).

[53] A. Ori, Phys. Rev. Lett. **68**, 2117 (1992); Phys. Rev. D **55**, 4860 (1997).

[54] P. R. Brady, C. M. Chambers, W. Krivan and P. Laguna, Phys. Rev. D **55**, 7538 (1997).

[55] D. Goldwirth and T. Piran, Phys. Rev. D **36**, 3575 (1987).

[56] D. Garfinkle, Phys. Rev. D **51**, 5558 (1995).

[57] It is not clarified in ref. [17] which perturbation mode exhibits this behavior since only reference to a well in the effective potential for polar perturbation is made and it is not stated if this occurs for the lowest radiatable multipole or for others. It would not be too difficult to plot the potentials for each type of perturbation, mode, to ascertain this, but the constant mode is not important enough to warrant this in these proceedings.

[58] R. Penrose, in *General Relativity, an Einstein Centenary Survey*, edited by S. W. Hawking and W. Israel (Cambridge University Press, Cambridge, 1979).

[59] S. P. Trivedi, Phys. Rev. D **47**, 4233 (1993).

[60] P. R. Anderson, W. A. Hiscock and D. J. Loranz, Phys. Rev. Lett. **74**, 4365 (1995).

[61] C. Barrabès and W. Israel, Phys. Rev. D **43**, 1129 (1991).

BLACK–HOLE INTERIORS AND STRONG COSMIC CENSORSHIP

Eric Poisson

Department of Physics, University of Guelph
Guelph, Ontario, N1G 2W1, Canada

Abstract

Strong cosmic censorship holds that given suitable initial data on a spacelike hypersurface, the laws of general relativity should determine, completely and uniquely, the future evolution of the spacetime. Here it is argued that while strong cosmic censorship is enforced for all black holes residing in asymptotically flat spacetime, it is violated (within the classical formulation of general relativity) for some black holes residing in non asymptotically flat spacetime. It is suggested that the semi-classical formulation of general relativity might enforce strong cosmic censorship.

3.1 Introduction

The basic question underlying the theoretical study of black-hole interiors is "what is the structure of spacetime inside a realistic black hole?". A lot of progress has been made during the last few years toward answering this question, as will be obvious from the other contributions to these proceedings. The question I wish to consider in this contribution is the following, which is at once much more focused and much more fundamental: "Do the laws of general relativity uniquely determine the structure of spacetime inside the black hole, given suitable initial data placed at the onset of gravitational collapse?". I will argue that the evidence points to a negative answer in the case of the purely *classical* laws, but that there is hope for a positive answer in the framework of the *semi-classical* laws. This contribution is based mostly on a 1992 paper co-authored with Patrick Brady [1] and a 1995 paper co-authored with Draza Marković [2]. However, the point of view expressed here is entirely my own, and Patrick and Draza should not be held liable!

The scope of the question should be clarified before attempting to answer it. What is at stake here is the *global* structure of spacetime, that is, the full characterization of physical fields, including the metric, everywhere and at all times. The well-posedness of the initial value problem in general relativity [3] guarantees only that given suitable data on an initial surface, the solution to the Einstein equations will be unique (up to diffeomorphisms) everywhere within the *domain of dependence* of the initial surface.

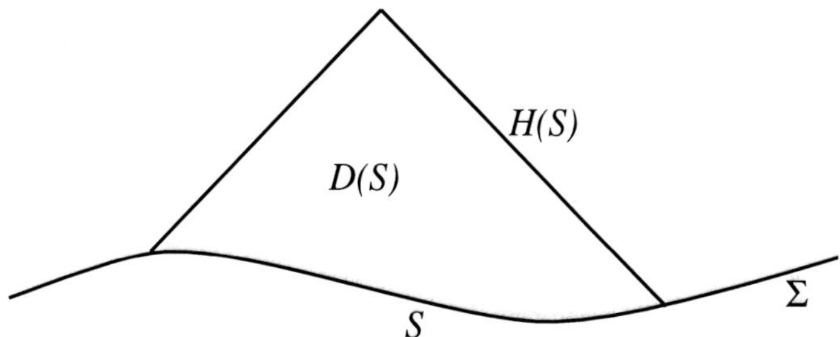

Figure 3.1: A spacelike hypersurface Σ (assumed without an edge) contains a set S. The evolution of initial data put on S is determined uniquely only within $D(S)$, the domain of dependence of S. The boundary of this region is $H(S)$, the Cauchy horizon of S. Strong cosmic censorship holds that as S is made to coincide with Σ, then $D(S)$ must coincide with M, the entire spacetime manifold. Then $H(\Sigma) = \emptyset$. In the diagram, only the future parts of $D(\Sigma)$ and $H(\Sigma)$ are shown.

Denoting this surface by Σ and its domain of dependence by $D(\Sigma)$, the question facing us is whether $D(\Sigma)$ coincides with M, the entire spacetime manifold. (See Fig. 3.1.) In other words, is there a region in the manifold which does not lie inside the domain of dependence of the initial surface? If we define the *Cauchy horizon* $H(\Sigma)$ of the initial surface to be the boundary of its domain of dependence, then the question is whether there exists a Cauchy horizon at all, and the answer is negative if $D(\Sigma) = M$.

This set of questions is usually grouped under the name *strong cosmic censorship.* The principle of strong cosmic censorship expresses the basic idea that starting from suitable initial data, general relativity should be able to predict, unambiguously, the complete future evolution of spacetime. As we have seen, a more technical way of expressing this idea is that the domain of dependence of the initial surface should be the entire spacetime manifold, or that the Cauchy horizon of the initial surface should be the empty set. A spacetime which satisfies these properties is said to be *globally hyperbolic.* A much more precise proposal for a strong cosmic censorship conjecture can be found in Wald's book [3].

A more frequently discussed form of cosmic censorship is the weak form, which states, roughly, that physically realistic spacetimes should not contain any globally naked singularities. A singularity is globally naked if it can be detected by observers at arbitrarily large distances. A singularity hidden behind an event horizon is not globally naked, and such a spacetime would therefore satisfy the weak form of cosmic censorship. The strong form asks for more: If an observer traverses the event horizon, will she encounter a (locally) naked singularity? If so, physical predictions made on the basis of the initial conditions will be upset by the presence of the singularity, and strong cosmic censorship will be violated. Strong cosmic censorship therefore holds that no singularity may be naked, even locally.

The question asked in this essay is whether the laws of general relativity enforce strong cosmic censorship.

3.2 Black holes in asymptotically flat spacetime

It is well known that the Reissner-Nordström, Kerr, and Kerr-Newman spacetimes contain timelike singularities inside their event horizons [4]. As these singularities are obviously naked, we must ask whether these spacetimes constitute a serious counter-example to strong cosmic censorship. I shall argue to the negative.

Figure 3.2 shows a conformal diagram of the Reissner-Nordström spacetime, whose metric is given by

$$ds^2 = -f dv^2 + 2dvdr + r^2(d\theta^2 + \sin^2\theta\, d\phi^2), \tag{3.1}$$

where

$$f = 1 - \frac{2M}{r} + \frac{Q^2}{r^2}. \tag{3.2}$$

Here, v is a null coordinate which is constant along radial ($d\theta = d\phi = 0$), ingoing (r decreasing) null geodesics; M denotes the mass of the black hole, and Q its charge. The spacetime contains two types of horizons, located where $f = 0$. The event horizon is at $r = r_e \equiv M + (M^2 - Q^2)^{1/2}$, while the inner horizon is at $r = r_i \equiv M - (M^2 - Q^2)^{1/2}$. The diagram shows clearly that the ingoing branch of the inner horizon is a Cauchy horizon for any spacelike hypersurface Σ preceding the formation of the event horizon. The physical origin of the Cauchy horizon is also clear: predictions made at any event P to the future of the Cauchy horizon would be upset by signals originating at the timelike singularity, where physics cannot be controlled.

Clearly, the Reissner-Nordström spacetime is a counter-example to strong cosmic censorship: the spacetime contains a Cauchy horizon, beyond which the evolution of physical fields becomes ambiguous. Furthermore, the same is true for the Kerr and Kerr-Newman spacetimes, which also contain Cauchy horizons. The real question, however, is whether these spacetimes constitute a *serious* counter-example to strong cosmic censorship. By this I mean that if these spacetimes could be shown to form a set of measure zero in some topological space of black-hole spacetimes, then they could be dismissed as inconsequential. On the other hand, if black-hole spacetimes with Cauchy horizons formed an open set, then we would have to conclude that strong cosmic censorship is not enforced by general relativity.

The construction of such a topological space would be a difficult undertaking which, however, will be necessary to settle the issue. (This point was forcedly made to me by Jim Isenberg.) Here I will argue that the Kerr-Newman spacetimes should form a set of measure zero in any reasonable topological space of black-hole spacetimes.

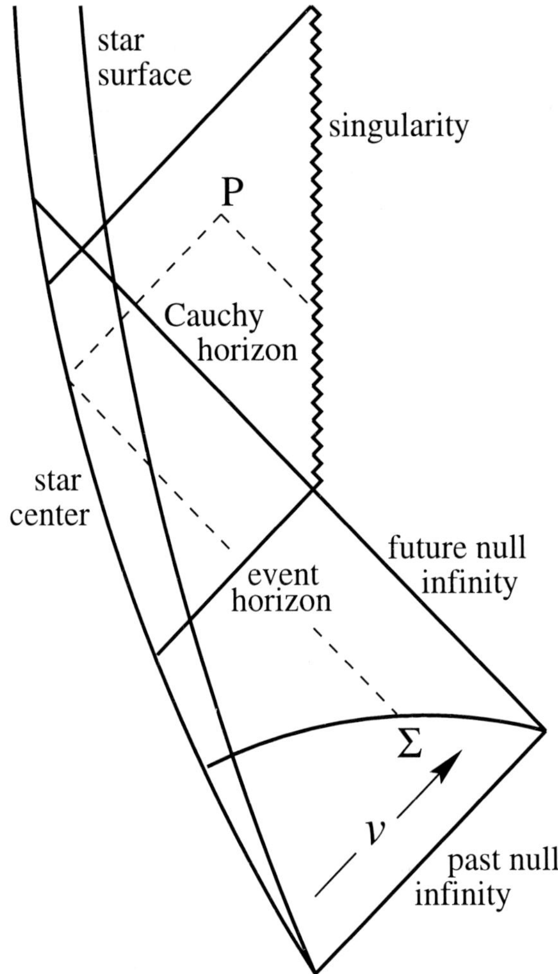

Figure 3.2: Conformal diagram of the Reissner-Nordström spacetime. The ingoing branch (left-going on the diagram) of the inner horizon is a Cauchy horizon for the hypersurface Σ. This is illustrated by the fact that light rays originating from event P and propagating backward in time run into the timelike singularity at $r = 0$. In this diagram, $v = \infty$ both at future null infinity and at the Cauchy horizon.

Consider first the Reissner-Nordström spacetimes. (There is one spacetime for each value of the parameters M and Q.) These spacetimes are very special, because they result from very special initial conditions: apart from being empty of any form of matter except for a static electric field, they are exactly spherically symmetric. However, it has long been known that slight deviations from these conditions, in the form of time-dependent matter fields or gravitational waves, produce large effects at the Cauchy horizon [5, 6, 7, 8, 9]. More precisely, physical quantities associated with the perturbations, such as the energy density measured by a free-falling observer, diverge at the Cauchy horizon. In other words, the Cauchy horizons of the Reissner-Nordström spacetimes are *unstable* to time-dependent perturbations. The same is true for the Kerr and Kerr-Newman spacetimes. This is an indication that these spacetimes must form a set of measure zero.

This, however, is not good enough, because the perturbation analysis involves only test fields in a fixed background spacetime. What must be understood, in a non-perturbative manner, is how the spacetime itself evolves under the slightly different choice of initial conditions. This is the question that Werner Israel and I started to examine in 1989 [10]. After a lot of work, carried out most notably by Israel's group in Canada and Amos Ori's group in Israel, the answer is now clear: Both for nonrotating and rotating black holes, the spacetime will develop a null curvature singularity at the Cauchy horizon. This singularity is not of a big-crunch type as in the Schwarzschild spacetime. Instead, it is characterized by an infinite growth of the internal mass function at the Cauchy horizon, whose area remains finite. This singularity is known as the *mass-inflation* singularity. In effect, the perturbations destroy the Cauchy horizon, and replace it with a null curvature singularity. The mass-inflation scenario is now firmly established in spherical symmetry, thanks to analytic [10, 11, 12, 13] and numerical [14, 15] calculations. The evidence is somewhat less firm, but still quite good, in the case of rotating black holes [16, 17, 18, 19].

One might ask the following question: "How is a null curvature singularity any better than a Cauchy horizon? After all, isn't predictability lost also at the singularity, because of the necessary breakdown of the classical laws there?" Eanna Flanagan provided the following answer during the workshop, which I fully endorse: The presence of a (nonsingular) Cauchy horizon inside a black hole is surprising because it signals the breakdown of the classical laws without any *local* indication that something may be wrong. For example, a free-falling observer would measure the curvature tensor to be well below Planckian values, and would never suspect that classical general relativity was about to lose predictive power. This, clearly, is not the case near a curvature singularity. In a sense, the loss of predictability occurring at a Cauchy horizon is much worse than the "mere" breakdown of the classical laws near a curvature singularity.

We may conclude that there is strong evidence that black-hole spacetimes with Cauchy horizons form a set of measure zero, because slight deviations in the initial conditions produce spacetimes whose causal structure is drastically different. In effect, slightly different initial conditions destroy the Cauchy horizon and replace it with a null curvature singularity. It is therefore tempting to suggest that the Kerr-Newman spacetimes do not constitute a serious counter-example to strong cosmic censorship. In fact, I suspect that the following statement is true: In the topological space of all asymptotically-flat black-hole spacetimes, the set of all spacetimes containing a Cauchy horizon has zero measure. Of course, no mathematically rigorous proof of this statement exists.

3.3 Cauchy-horizon instability

It is useful to have a clear understanding of the physical processes leading to the Cauchy-horizon instability. For simplicity, we restrict attention to the Reissner-Nordström spacetime.

We consider a simple model involving a test distribution of noninteracting massless particles. The particles originate from the region outside the black hole, move radially inward along curves of constant v, and eventually fall inside the black hole. They are described by the stress-energy tensor

$$T_{\alpha\beta} = \frac{L(v)}{4\pi r^2}(\partial_\alpha v)(\partial_\beta v), \tag{3.3}$$

where the luminosity function is given by $L(v) \sim v^{-p}$ (with p a positive constant) when $v \to \infty$, to correctly reproduce Price's inverse-power law decay of radiative fields outside the black hole [20, 21, 22, 23, 24]. We recall that $v = \infty$ designates both future null infinity and the Cauchy horizon (see Fig. 3.2).

The flux of particles is observed inside the black hole by a free-falling observer crossing the Cauchy horizon. This observer moves on a radial geodesic, has a four-velocity u^α, and measures the energy density of the particles to be $\rho = T_{\alpha\beta}u^\alpha u^\beta = L(v)\dot{v}^2/4\pi r^2$, where $\dot{v} \equiv u^v$. A simple calculation reveals that $\dot{v} \sim |\tilde{E}| \exp(\kappa_i v)$ when $v \to \infty$, where $\tilde{E} \equiv -u_v$ is the observer's energy parameter, and

$$\kappa_i = \frac{1}{2}\left|\frac{df}{dr}\right|_{r=r_i} \tag{3.4}$$

is the *surface gravity* of the inner horizon. Substitution yields

$$\rho \sim \frac{|\tilde{E}|^2}{4\pi {r_i}^2} v^{-p} e^{2\kappa_i v}, \qquad v \to \infty. \tag{3.5}$$

Thus, the measured energy density diverges as the observer reaches the Cauchy horizon. In terms of v, this divergence is exponential; in terms of τ, the amount of proper time left before reaching the Cauchy horizon, $\rho \sim 1/\tau^2$. The physical interpretation is that the particles pile up at the Cauchy horizon, which is a surface of infinite blueshift. Consequently, the energy density diverge there. It is perhaps not surprising that a full backreaction calculation reveals the existence of a null curvature singularity at the Cauchy horizon.

The extremal case ($|Q| = M$) requires a separate treatment, because $\kappa_i = 0$ for this spacetime. It turns out that in this case, the Cauchy horizon is *stable* to time-dependent perturbations [25]. The reason is that although an infinite blueshift still occurs at the Cauchy horizon, the blueshift factor diverges only as a power of v instead of exponentially. Because $L(v)$ decays with a larger power, the energy density stays finite. This observation should not affect our conclusions, as presumably, extremal black holes also form a set of measure zero of spacetimes.

3.4 Black holes in non asymptotically flat spacetime

While it appears highly plausible that strong cosmic censorship is enforced for black holes residing in asymptotically flat spacetime, the same cannot be said for black holes residing in asymptotically de Sitter spacetime.

Consider first the Reissner-Nordström-de Sitter spacetime [26], whose metric is given by Eq. (3.1) with

$$f = 1 - \frac{2M}{r} + \frac{Q^2}{r^2} - \frac{1}{3}\Lambda r^2, \tag{3.6}$$

where Λ is the cosmological constant, assumed to be positive. This spacetime contains three types of horizons, situated at the three positive roots of f. The cosmological horizon is located at $r = r_c$, the largest root. The black-hole event horizon is at $r = r_e$, and the inner horizon at $r = r_i$ is also a Cauchy horizon for any spacelike hypersurface Σ lying outside the black hole. A conformal diagram of this spacetime is presented in Fig. 3.3. We see from the diagram that the cosmological horizon plays here the same role that future null infinity played in the diagram of Fig. 3.2.

The major difference between this class of spacetimes and the Reissner-Nordström class is that here, the parameters $\{M, Q, \Lambda\}$ can be chosen so that the Cauchy horizon is *stable* to time-dependent perturbations. As was first shown by Brady and myself [1] on the basis of the simple argument presented in Sec. 3.5, and then confirmed by Mellor and Moss [27] on the basis of a complete perturbation analysis, this happens whenever

$$\kappa_i \leq \kappa_c, \tag{3.7}$$

where κ_i is the surface gravity of the inner horizon, defined by Eq. (3.4), and $\kappa_c = \frac{1}{2}|df/dr|(r_c)$ the surface gravity of the cosmological horizon. Equation (3.7) defines a small but finite region of the parameter space $\{M, Q, \Lambda\}$. This region is described in detail in Chambers' contribution to these proceedings. That the region must be small can be seen as follows (I thank Ian Moss for providing me with this argument): In situations close to realistic, the cosmological horizon would be at a very large radius,

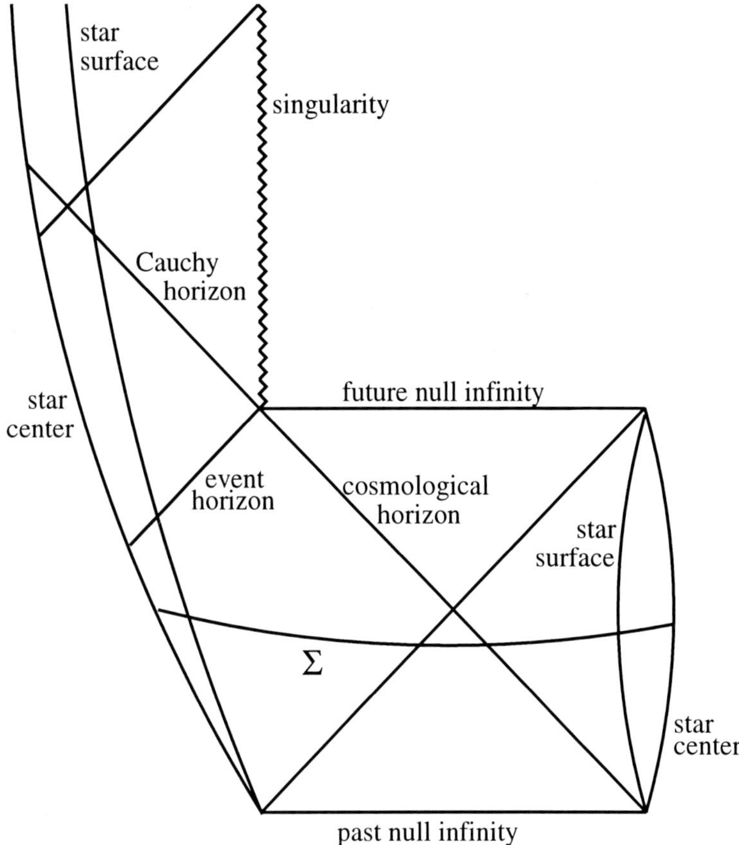

Figure 3.3: Conformal diagram of the Reissner-Nordström-de Sitter spacetime. The ingoing branch of the inner horizon is a Cauchy horizon for the hypersurface Σ. In this diagram, $v = \infty$ both at the cosmological horizon and at the Cauchy horizon.

making κ_c very small; to have κ_i smaller than this requires the black hole to very nearly extremal, $|Q| = M(1-\epsilon)$, with ϵ a very small positive number.

Equation (3.7) implies Cauchy-horizon stability not only for the Reissner-Nordström-de Sitter spacetimes, but also for the entire Kerr-Newman-de Sitter class. This was established by Chambers and Moss [28], who also showed that Eq. (3.7) defines a small but finite region of the parameter space $\{M, Q, a, \Lambda\}$, where a is the black hole's rotation parameter. Again, this is discussed in more detail in Chambers' contribution. Equation (3.7) was also shown to imply stability in a fully nonperturbative analysis restricted to spherical symmetry [29].

If the open region of parameter space happens to correspond to an open set in the topological space of black-hole spacetimes, then we would be forced to conclude that strong cosmic censorship is *not* enforced by general relativity. There is, of course, no rigorous proof that the spacetimes for which Eq. (3.7) is satisfied do indeed form an open set, but I would regard this as highly plausible.

This example of an apparent violation of strong cosmic censorship relies on the presence of a cosmological constant in the field equations, something which may be distasteful to some. However, another example was recently discovered by Horowitz and Sheinblatt [30], and it does not require a cosmological constant. (I thank Gary Horowitz for pointing out this work to me.) These authors consider two oppositely charged black holes, each uniformly accelerating in a background magnetic field. The solution to the Einstein-Maxwell equations describing this spacetime is known as the Ernst solution [31], and the causal structure of this spacetime is essentially identical to that of the Reissner-Nordström-de Sitter spacetime, except for the fact that the cosmological horizon is replaced by an acceleration horizon. Another similarity is that the spacetime is also not asymptotically flat. Horowitz and Sheinblatt show that the Cauchy horizon of the Ernst spacetime is *stable* whenever $\kappa_i \leq \kappa_a$, where κ_a is the surface gravity of the acceleration horizon. Again, this inequality defines a finite region of parameter space, and it is tempting to suggest that this corresponds to an open set in the topological space of black-hole spacetimes. If this were true, then strong cosmic censorship would be violated also for this class of spacetimes.

As we see, the fact that strong cosmic censorship might be violated for certain black-hole spacetimes is not necessarily due to the presence of a cosmological constant in the field equations. Instead, the essential features seem to be the presence of a "large" horizon and the absence of asymptotic flatness. I would conjecture the following statement: In the topological space of all black-hole spacetimes, including those which are not asymptotically flat, the set of all spacetimes containing a Cauchy horizon is open. If this statement is true, then strong cosmic censorship is not enforced by general relativity.

3.5 Cauchy-horizon stability

It is easy to understand why the Cauchy horizon of the Reissner-Nordström-de Sitter spacetime is stable when $\kappa_i \leq \kappa_c$. We consider once again the simple model of Sec. 3.3, involving noninteracting massless particles described by the stress-energy of Eq. (3.3).

In Sec. 3.3, the luminosity function $L(v)$ was taken to decay as an inverse-power law in order to correctly reproduce the behavior of radiative fields in the Reissner-Nordström spacetime. This choice, however, is not appropriate for the Reissner-Nordström-de Sitter spacetime [32]. To determine the correct behavior of $L(v)$ we shall calculate ρ_c, the energy density of the infalling particles as measured by a free-falling observer crossing the cosmological horizon. We recall that $v = \infty$ designates both the cosmological horizon and the Cauchy horizon (see Fig. 3.3). The steps are the same as in Sec. 3.3, and we find

$$\rho_c \sim \frac{\tilde{E}_c^2}{4\pi {r_c}^2} L(v) e^{2\kappa_c v}, \qquad v \to \infty, \tag{3.8}$$

where $\tilde{E}_c$ is the observer's energy parameter. We demand that the observer measure a finite, nonvanishing energy density as she crosses the cosmological horizon. This requires $L(v)$ to decay exponentially:

$$L(v) \sim K e^{-2\kappa_c v}, \qquad v \to \infty, \tag{3.9}$$

where K is a constant. While Eq. (3.9) serves mostly to remedy the bad behavior of the coordinates v and r at the cosmological horizon, it also expresses the fact that this horizon is a surface of infinite redshift.

Substituting Eq. (3.9) into the calculation of Sec. 3.3 reveals that the energy density of the particles, as measured by an observer crossing the *Cauchy* horizon, is given by

$$\rho_i \sim \frac{|\tilde{E}_i|^2}{4\pi {r_i}^2} K e^{2(\kappa_i - \kappa_c)v}, \qquad v \to \infty. \tag{3.10}$$

We see that this quantity diverges when $\kappa_i > \kappa_c$, but that it stays finite (in fact, goes to zero) when $\kappa_i \leq \kappa_c$. This is just the condition expressed by Eq. (3.7). The interpretation is clear. The factor $\exp(2\kappa_i v)$ comes from the infinite blueshift occurring at the inner horizon, while the factor $\exp(-2\kappa_c v)$ comes from the infinite redshift occurring at the cosmological horizon. When $\kappa_i > \kappa_c$ the blueshift wins over the redshift, and the Cauchy horizon is unstable. On the other hand, when $\kappa_i \leq \kappa_c$ the redshift wins, and the Cauchy horizon is stable.

3.6 Quantum effects

We now consider the quantum stability of the Reissner-Nordström-de Sitter spacetime. This question is addressed by admitting the existence of quantized matter fields in the spacetime, and examining the behavior of $\langle T^{\alpha\beta} \rangle$, the renormalized expectation valuc of their stress-energy tensor, near the Cauchy horizon. This is a difficult question, even when no attempt is made to take into account the spacetime's response to the quantum stress-energy tensor. The difficulty resides in the calculation of $\langle T^{\alpha\beta} \rangle$, even in a fixed spacetime possessing spherical symmetry. The difficulty is even more acute in our case, because (as will become clear below) the quantum state cannot be chosen among the standard ones, such as the Hartle-Hawking or Unruh vacua.

We shall therefore consider a simpler problem, that of quantizing matter fields in a two-dimensional version of the Reissner-Nordström-de Sitter spacetime, with metric

$$ds^2 = -f\, dudv, \tag{3.11}$$

where f is given by Eq. (3.6) and the null coordinate u is defined by $du = dv - 2f^{-1}dr$. (It should be noted that the coordinates u and v cannot be extended across the event horizon. This will not be a problem for this calculation.) In two dimensions, $\langle T^{ab} \rangle$ can be computed explicitly [33]; we will do so for a conformally invariant scalar field.

The calculation of $\langle T^{ab} \rangle$ starts with the trace-anomaly equation [34],

$$\langle T \rangle = \alpha R, \tag{3.12}$$

where $\alpha = (24\pi)^{-1}$ and $R = -f''$ is the Ricci scalar associated with the metric of Eq. (3.11); primes indicate differentiation with respect to r. This equation immediately gives us one component of the stress-energy tensor,

$$\langle T_{uv} \rangle = \frac{\alpha}{4} f f''. \tag{3.13}$$

The others are obtained by integrating the conservation equations, $\langle T^{ab} \rangle_{;b} = 0$. A straightforward calculation yields

$$\langle T_{uu} \rangle = -\frac{\alpha}{2} \Big[F(r) - A(u) \Big] \tag{3.14}$$

and

$$\langle T_{vv} \rangle = -\frac{\alpha}{2} \Big[F(r) - B(v) \Big], \tag{3.15}$$

where

$$F(r) = \frac{1}{4} \Big(f'^2 - 2ff'' \Big), \tag{3.16}$$

while $A(u)$ and $B(v)$ are arbitrary functions which serve to define the state of the quantum field. We demand that this state be regular on the cosmological and event horizons, so that the quantum stress-energy tensor will also be regular there.

To see what requirements must be made on $A(u)$ and $B(v)$, we construct $\langle \rho \rangle \equiv \langle T_{ab} \rangle u^a u^b$, the energy density as measured by a free-falling observer with two-velocity u^a. This has components

$$u^v = \frac{\tilde{E} \pm (\tilde{E}^2 - f)^{1/2}}{f} \tag{3.17}$$

and

$$u^u = \frac{\tilde{E} \mp (\tilde{E}^2 - f)^{1/2}}{f}, \tag{3.18}$$

where $\tilde{E}$ is the observer's energy parameter, and the upper (lower) sign is chosen if r is increasing (decreasing) along the world line. We obtain

$$\langle\rho\rangle = -\frac{\alpha}{2f^2}\Big[(2\tilde{E}^2 - f)(2F - A - B) \pm 2\tilde{E}(\tilde{E}^2 - f)^{1/2}(A - B) - f^2 f''\Big]. \tag{3.19}$$

We now want to evaluate $\langle\rho\rangle$ near the horizon $r = r_j$, where r_j stands for either r_c (cosmological horizon), r_e (event horizon), or r_i (Cauchy horizon). A straightforward calculation shows that near $f = 0$, Eq. (3.19) reduces to

$$\langle\rho\rangle = -\frac{\alpha}{2f^2}\Big\{\big[2{\kappa_j}^2 - (1 \mp \epsilon)A - (1 \pm \epsilon)B\big](2\tilde{E}^2 - f) + O(f^2)\Big\}, \tag{3.20}$$

where $\kappa_j = \frac{1}{2}|f'(r_j)|$ is the surface gravity of the horizon under consideration, and $\epsilon \equiv \mathrm{sign}(\tilde{E})$. Notice that we have never used the detailed form of the function $f(r)$ in this calculation.

We first evaluate $\langle\rho\rangle$ at the cosmological horizon, where $r = r_c$ and $v = \infty$. An observer crossing this horizon has a positive energy parameter ($\epsilon = +1$) and moves in the direction of increasing r (upper sign). The term within the square brackets therefore reduces to $2{\kappa_c}^2 - 2B(\infty)$. If we demand

$$B(v \to \infty) = {\kappa_c}^2 + O(e^{-2\kappa_c v}), \tag{3.21}$$

then this term will be $O(f^2)$ and $\langle\rho\rangle$ will be well-behaved at the cosmological horizon. Thus, our choice of quantum state is restricted by Eq. (3.21).

Next, we evaluate $\langle\rho\rangle$ at the event horizon, where $r = r_e$ and $u = \infty$. An observer crossing this horizon has a positive energy parameter ($\epsilon = +1$) and moves in the direction of decreasing r (lower sign). The square-brackets term becomes $2{\kappa_e}^2 - 2A(\infty)$. If we demand

$$A(u \to \infty) = {\kappa_e}^2 + O(e^{-2\kappa_e u}), \tag{3.22}$$

then this term will be $O(f^2)$ and $\langle\rho\rangle$ will be well-behaved at the event horizon. Thus, our choice of quantum state is further restricted by Eq. (3.22). It is interesting to note that only the asymptotic behaviors of $A(u)$ and $B(v)$ are restricted by the regularity conditions. Otherwise, these functions are completely arbitrary, allowing for much freedom in the choice of the quantum state. A *particular* quantum state meeting the regularity requirements is the Marković-Unruh vacuum [35].

Finally, we evaluate $\langle\rho\rangle$ at the Cauchy horizon, where $r = r_i$ and $v = \infty$. Since an observer crossing this horizon moves with a negative energy parameter in the direction of decreasing r, we must use $\epsilon = -1$ and the lower sign. The square-brackets term becomes $2{\kappa_i}^2 - 2B(\infty)$ which, by virtue of Eq. (3.21), is $2({\kappa_i}^2 - {\kappa_c}^2)$. This gives

$$\langle \rho \rangle \sim 2\alpha \tilde{E}^2({\kappa_i}^2 - {\kappa_c}^2)\,\frac{1}{f^2}. \tag{3.23}$$

Thus, $\langle \rho \rangle$ diverges at the Cauchy horizon.

We have proved the following theorem:

> In two-dimensional Reissner-Nordström-de Sitter spacetime, for *any* quantum state regular on both the cosmological horizon and the event horizon, the renormalized expectation value of the stress-energy tensor of a conformally invariant scalar field diverges at the Cauchy horizon, except when $\kappa_i = \kappa_c$.

The theorem implies that the two-dimensional spacetime is quantum mechanically unstable, except for the set of measure zero of spacetimes for which $\kappa_i = \kappa_c$. A similar theorem is believed to hold also for the Ernst spacetime [30].

What does this theorem tell us about the four-dimensional world? Although we shall not go into these details here [2], physical intuition suggests that the four-dimensional spacetime must also be quantum mechanically unstable. Indeed, it appears that the quantum physics of the Cauchy-horizon instability, which is revealed by the two-dimensional calculation, is robust and does not depend on the dimensionality of spacetime nor the nature of the quantum field (its spin, conformal invariance, etc.). The quantum mechanical instability can be intuitively explained in terms of fundamental processes such as the creation of thermal quanta near horizons, and the gravitational redshifts and blueshifts that these quanta undergo. Since these processes take place equally well in four as in two dimensions, I am quite confident that the quantum stress-energy tensor will diverge also at the Cauchy horizon of the four-dimensional spacetime.

3.7 Conclusion

I will conclude with these two statements:

- Black-hole Cauchy horizons may be classically stable if the black hole does not reside in asymptotically flat spacetime. This suggests that strong cosmic censorship is not enforced by the classical formulation of general relativity.

- Black-hole Cauchy horizons are always quantum mechanically unstable, except possibly for a set of measure zero of spacetimes. This suggests that strong cosmic censorship might be enforced by the semi-classical formulation of general relativity.

Although these statements are clearly not supported by rigorous mathematical proofs, I regard the evidence for their validity as quite compelling. (Of course, the second part of the second statement is pure speculation.)

To the extent that these statements are true, it is most intriguing that quantum physics must be invoked in order to restore the full predictive power of general relativity. Here we may notice an interesting similarity with the physics of chronology horizons [36].

And to the extent that the first statement is true, the following question remains: "Does the semi-classical formulation of general relativity really enforce strong cosmic censorship?"

Acknowledgments

This work was supported by the Natural Sciences and Engineering Research Council of Canada. It is a pleasure to thank Lior Burko, Amos Ori, and Liz Youdim for having organized such a wonderful workshop.

Bibliography

[1] P. R. Brady and E. Poisson, Class. Quantum Grav. **9**, 121 (1992).

[2] D. Marković and E. Poisson, Phys. Rev. Lett. **74**, 1280 (1995).

[3] R. M. Wald, *General relativity* (University of Chicago, Chicago, 1984).

[4] S. W. Hawking and G. F. R. Ellis, *The large scale structure of space-time* (Cambridge University Press, Cambridge, 1973).

[5] M. Simpson and R. Penrose, Int. J. Theor. Phys. **7**, 183 (1973).

[6] J. M. McNamara, Proc. R. Soc. London **A358**, 499 (1978); **A364**, 121 (1978).

[7] Y. Gürsel, V. D. Sandberg, I. D. Novikov, and A. A. Starobinsky, Phys. Rev. D **19**, 413 (1979); **20**, 1260 (1979).

[8] R. A. Matzner, N. Zamorano, and V. D. Sandberg, Phys. Rev. D **19**, 2821 (1979).

[9] S. Chandrasekhar and J. B. Hartle, Proc. R. Soc. London **A 284**, 301 (1982).

[10] E. Poisson and W. Israel, Phys. Rev. D **41**, 1796 (1990).

[11] A. Ori, Phys. Rev. Lett. **67**, 789 (1991).

[12] A. Bonanno, S. Droz, W. Israel and S. M. Morsink, Phys. Rev. D **50**, 7372 (1994).

[13] A. Bonanno, S. Droz, W. Israel and S. M. Morsink, Proc. R. Soc. London **A450**, 553 (1995).

[14] P. R. Brady and J. D. Smith, Phys. Rev. Lett. **75**, 1256 (1995).

[15] L. M. Burko, Phys. Rev. Lett. **79**, 4958 (1997).

[16] A. Ori, Phys. Rev. Lett. **68**, 2117 (1992).

[17] P. R Brady and C. M. Chambers, Phys. Rev. D **51**, 4177 (1995).

[18] A. Ori and É. É. Flanagan, Phys. Rev. D **53**, 1754 (1996).

[19] S. Droz, Phys. Rev. D **55**, 3575 (1997).

[20] R. H. Price, Phys. Rev. D **5**, 2419 (1972); 2439 (1972).

[21] C. Gundlach, R. H. Price, and J. Pullin, Phys. Rev. D **49**, 883 (1994); 890 (1994).

[22] L. M. Burko and A. Ori, Phys. Rev. D **56**, 7820 (1997).

[23] W. Krivan, P. Laguna, and P. Papadopoulos, Phys. Rev. D **54**, 4728 (1996).

[24] W. Krivan, P. Laguna, P. Papadopoulos, and N. Andersson, Phys. Rev. D **56**, 3395 (1997).

[25] E. Poisson (unpublished).

[26] B. Carter, in *Black Holes*, edited by C. DeWitt and B. S. DeWitt (Gordon and Breach, New York, 1973).

[27] F. Mellor and I. Moss, Class. Quantum Grav. **9**, L43 (1992).

[28] C. M. Chambers and I. Moss, Class. Quantum Grav. **11**, 1035 (1994).

[29] P. R. Brady, D. Núñez and S. Sinha, Phys. Rev. D **47**, 4239 (1993).

[30] G. T. Horowitz and H. J. Sheinblatt, Phys. Rev. D **55**, 650 (1997).

[31] F. Ernst, J. Math. Phys. (N.Y.) **17**, 54 (1976).

[32] P. R. Brady, C. M. Chambers, W. Krivan, and P. Laguna, Phys. Rev. D **55**, 7538 (1997).

[33] P. C. W. Davies, S. A. Fulling, and W. G. Unruh, Phys. Rev. D **13**, 2720 (1976).

[34] N. D. Birrell and P. C. W. Davies, *Quantum fields in curved space* (Cambridge University Press, Cambridge, 1982).

[35] D. Marković and W. G. Unruh, Phys. Rev. D **43**, 332 (1991).

[36] M. Visser, *Lorentzian wormholes, from Einstein to Hawking* (American Institute of Physics, Woodbury, 1995).

GRAVITATIONAL COLLAPSE WITH MASSIVE SCALAR FIELDS

Ian G. Moss

Department of Physics, University of Newcastle–upon–Tyne, NE1 7RU, U.K.

Abstract

Naked singularities are a severe problem for classical relativity. We now have examples of spacetimes and matter fields that develop naked singularities whilst not being too sensitive to variations of the initial conditions. The black hole de Sitter models are discussed here. These models have pre-existing black holes, but the possibility exists for forming these black holes by the collapse of a charged star. This can be modeled by considering including a massive scalar field in the theory.

4.1 Introduction

One of the goals of modern physics is the unification of classical gravity with quantum mechanics. We might hope to make some progress towards this goal by studying those situations where the classical theory is pushed into making an unreasonable prediction. One such situation is the classical behavior of matter collapsing under its own gravity. This often leads, so theory would predict, to the formation of singularities. A singularity is said to form when there are geodesics which come to an end and cannot be extended further.

The singularity theorems of Hawking and Penrose [1] tell us that the occurrence of singularities is particularly robust, in the sense that when they occur they are insensitive to small perturbations in the initial conditions, but these theorems tell us nothing about the nature of the singularity. Three types of singularity that might form in gravitational collapse are shown in figure 4.1. Timelike singularities are especially interesting because they are visible in the sense that they intersect the past of some world-lines.

Timelike singularities are a feature of the Reissner-Nordstrom and Kerr black hole solutions. They are visible from any point in the fully extended spacetimes, but this is now believed to be unphysical. If the black holes formed by gravitational collapse, then the singularities would form inside the event horizon and be invisible from the outside. We also know that the singularities are unstable and are converted into null or spacelike singularities by small perturbations (as described in many places in

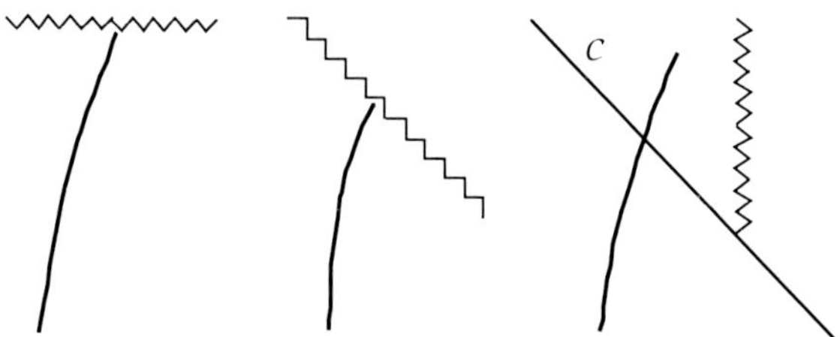

Figure 4.1: Spacetime diagrams of three singularities. In the first two cases the world-line hits the singularity without warning. In the third example the singularity comes into view from the world-line.

this book). It used to be believed that these features where typical of all timelike singularities. This forms the basis of the cosmic censorship conjecture.

Cosmic censorship is concerned with the development of singularities to the future of an initial hypersurface $\mathcal{L}$ in a spacetime with a metric that satisfies the Einstein field equations. We talk about generic singularities to indicate that their form is stable under small perturbations. The spacetime is usually assumed to be asymptotically flat or at least de Sitter. It may have event horizons bounding regions which cannot be seen from infinity.

There are two main divisions of cosmic censorship called the weak and strong cosmic censorship conjectures. Singularities that are visible to the asymptotic region are known as naked singularities. According to the weak cosmic censorship conjecture, generic singularities are never naked. They may exist hidden inside event horizons. This form of the conjecture is the original form [2] and it is assumed in many results about black holes, including the area increase theorems.

The strong version of cosmic censorship states that generic singularities are never visible. If the conjecture were true then it would be possible to integrate the Einstein equations as far as any geodesic can extend. Such spacetimes are said to be globally hyperbolic. If the conjecture is violated, then there exists a future boundary to the Cauchy development of the initial data, called the Cauchy horizon.

The status of these conjectures will obviously depend on the nature of the stress-tensor. Spacetimes with visible singularities are known for many models with free particles [3, 4] and one with perfect fluids [5]. The singularities in these models have never been shown to be generic. There is only one counter-example known so far in which the singularities are stable [7, 8, 9]. This consists of a charged or rotating black hole in de Sitter space. The stress-energy tensor includes a Maxwell term and a cosmological constant.

The original idea for this counter example came from considering what an observer passing through the Cauchy horizon would see. In the Reissner-Nordstrom solution,

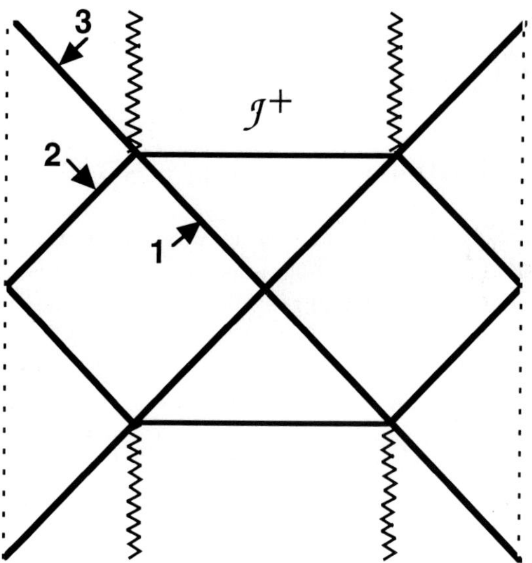

Figure 4.2: The global structure of a charged black hole-de Sitter spacetime. There are cosmological horizons $\mathcal{S}_1$, event horizons $\mathcal{S}_2$ and Cauchy horizons $\mathcal{S}_3$. Light reaching the Cauchy horizon need not have infinite blue-shift.

an observer sees the entire development of the outside world before crossing the event horizon. The compression of all this information into a finite time forces the incoming radiation to be shifted to arbitrary high frequencies and this destabalises the Cauchy horizon. In de Sitter space the observer crossing the Cauchy horizon only sees the external world up to a particle horizon. The frequency shift of the incoming light depends only on the local geometry of each of these horizons.

Both the charged and rotating black hole solutions are stable at the Cauchy horizon for a range of charges and angular momenta. The parameter range is small, but finite at any non-zero value of the cosmological constant. The cosmological constant allows a spatially compact exterior region, but there seems to be no reason why similar examples could not be constructed with other types of matter.

The black hole-de Sitter solution shows that stable Cauchy horizons can form from regular initial data and therefore Einstein's theory of gravity is an incomplete scheme for prediction. However, the initial data in this example already contains black holes. It is natural to ask if a spacetime of this sort can form from the gravitational collapse of matter in a closed (three-sphere topology) universe. This does seem to be the case, as we shall see, both with collapsing dust and approximately for a charged scalar field.

4.2 Black Holes in de Sitter Space

The metric of a stationary black hole in de Sitter space was first given by Carter [6]. The metric depends on the black hole mass M, charge Q and rotation parameter a. For a suitable range of parameters, the fully extended spacetime is periodic in space and time with many separate asymptotic regions. One fundamental cell contains a cosmological horizon $\mathcal{S}_1$, a black hole event horizon $\mathcal{S}_2$ and a Cauchy horizon $\mathcal{S}_3$, as shown in figure 4.2. These are located at the positive zeros of the polynomial

$$\Delta = (r^2 + a^2)(1 - \tfrac{1}{3}\Lambda r^2) - 2Mr + Q^2/(4\pi) \tag{4.1}$$

The local geometry of each horizon can be characterized by its surface gravity κ_i, given by

$$\kappa_i = \tfrac{1}{6}\Lambda\chi^{-2}\sigma^{-2}\prod_{j\neq i}|r_i - r_j| \tag{4.2}$$

where

$$\chi^2 = 1 - \tfrac{1}{3}\Lambda a^2 \qquad \sigma^2 = r^2 + a^2 \tag{4.3}$$

Figure 4.3 shows the parameter space divided up by the values of the surface gravities. More details about the metric, including the way to construct this diagram, are described in reference [9].

The surface gravities can be used to quantify the amount of red-shifting that is seen on light rays close to the horizon. There is a family of null geodesics that have $v = t + r^*$ fixed, where r^* is a new radial coordinate

$$dr^* = \chi^2\Delta^{-1}\sigma^2\, dr. \tag{4.4}$$

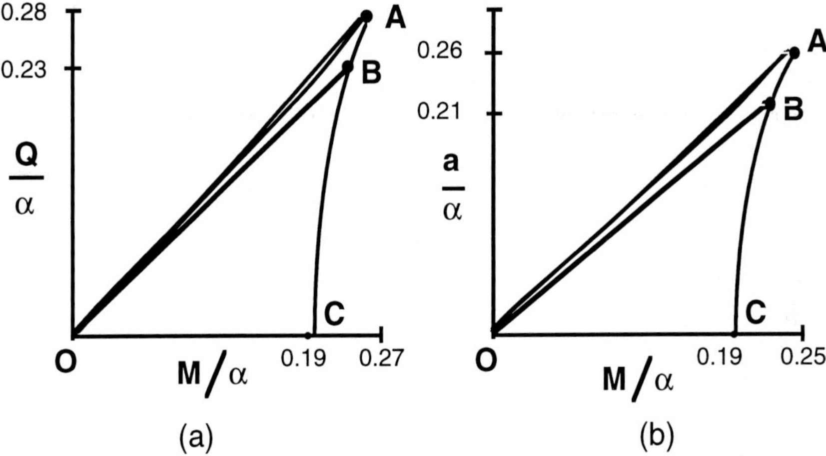

Figure 4.3: The black hole parameter space scaled by $\alpha^2 = 2/\Lambda$ and divided up by surface gravity. In the interior of the triangular region there are three horizons. Two surface gravities coincide along the lines crossing the triangle.

It follows from the definition of r^* and the surface gravity that close to horizon $\mathcal{S}_i$, where $|r^*| \to \infty$,

$$|r - r_i| \sim r_i \exp(-2\kappa_i |r^*|) \tag{4.5}$$

The four-velocities of timelike geodesics with energy E crossing the Cauchy or the cosmological horizons approach $r^2 E/(2\Delta)\mathbf{e}_v$, where $\mathbf{e}_v$ is a coordinate vector. An ingoing wave $\propto \exp(i\omega v)$ appears to have frequency $\omega_i = \omega \exp(\kappa_i v)$. The relative blue-shift factor seen by the observer at the Cauchy horizon is $\exp(\kappa_3 - \kappa_1)v$ as $v \to \infty$ [8]. There is an infinite flux of energy when $\kappa_3 > \kappa_1$. The alternative occupies a small, but non-zero range of the parameter space provided that $\Lambda > 0$. This is the parameter range to look for for stability.

Proving stability of the Cauchy horizon requires an analysis of the full system of equations. The linear analysis was performed for the charged case in reference [7] and for the rotating case in reference [9]. The combination of charge and rotation remains an open problem. We shall outline some features of the analysis for rotating holes.

The perturbation analysis falls into two parts. The first stage involves separation of the field equations to reduce them to a scattering problem for the radial modes. The second stage involves analysis of the scattering problem to prove stability. Fortunately, both parts of the analysis can be performed by tracing the steps of the perturbation analysis presented in detail by Chandrasekhar [10] for vanishing cosmological constant.

The field strengths can be converted into fully symmetric SL(2,C) tensors $\Phi^{AB\ldots C}$ [10]. There are $2s$ indices for a spin-s tensor, and therefore $2s+1$ independent components of the symmetric tensor, which can be labeled Φ_n, $n = -s, \ldots, s$. The separable solutions to the field equations have the form

$$\Phi_s = \mathcal{R}(r)\mathcal{S}(\mu)e^{-i\omega t + im\phi}. \tag{4.6}$$

The equations for the radial modes $\mathcal{R}(r)$ can be put into a Starobinski-Teukolsky form,

$$\left(\mathcal{D}_{-s/2}\Delta\mathcal{D}^{\dagger}_{s/2} + 2(2s-1)i\omega\chi^2\right)\mathcal{R} = \lambda_l\mathcal{R}. \tag{4.7}$$

where λ_l is an eigenvalue of the angular equations. The gravitational case includes an additional cosmological term,

$$\left(\mathcal{D}_{-1}\Delta\mathcal{D}^{\dagger}_{1} + 6i\omega\chi^2 - 2\Lambda r^2\right)\mathcal{R} = \lambda_l\mathcal{R}, \tag{4.8}$$

The radial derivatives appearing in these equations are

$$\mathcal{D}_n = \frac{\partial}{\partial r} + \frac{iK}{\Delta} + n\frac{\Delta'}{\Delta}, \qquad K = \chi^2 a\sigma^2(m\Omega - \omega). \tag{4.9}$$

where $\Omega = a/\sigma^2$.

The radial equation can be analyzed by scattering theory in the r^* coordinate. When rewritten in terms of r^*, equation (4.7) becomes

$$\left(\frac{d^2}{dr^{*2}} + V(r)\right)\sigma\mathcal{R} = 0 \tag{4.10}$$

The potential resembles a potential barrier and approaches constant values at the horizons. It cannot in general be expressed simply in terms of the r^* coordinate, but it has convergent series expansions in $\exp(-2\kappa_i|r^*|)$. Such potentials are called Yukawian and lead to scattering coefficients with simple anyliticity properties that prove crucial to the stability analysis.

We can impose initial data on the time-symmetric Cauchy hypersurface in the black hole exterior region. The linear perturbations propagate through the event horizon and up to the Cauchy horizon. Both the exterior and interior region produce a set of scattering coefficients and we multiply the two together.

The original fields can be recovered from the modes by integration,

$$\Phi_s = \sum_{l,m}\int \frac{d\omega}{2\pi} W(lm\omega)\mathcal{R}(lm\omega; r)\mathcal{S}(lm\omega;\mu)e^{im\phi - i\omega t} \tag{4.11}$$

The functions $W(lm\omega)$ depend on the choice of initial data. The actual field components measured by an observer close to the Cauchy horizon are not Ψ_s, but actually

$\Phi^{(obs)} = e^{\kappa_3 v}\Phi_s$. (For the Maxwell and Weyl fields the difference is due to a Lorentz boost between the tetrad frame defining Ψ_s and the observer's frame). Close to the Cauchy horizon, where $v \to \infty$, the asymptotic forms of the mode functions can be used, which depend on scattering coefficients A and B,

$$\Phi_s^{(obs)} \sim e^{\kappa_3 v} \sum_{l,m} \int \frac{d\omega}{2\pi} W(lm\omega) S(lm\omega;\mu) \times (A\, e^{-ik(\omega)v} + B\, e^{ik(\omega)u}) e^{im(\phi - i\Omega(r_3)t)} \tag{4.12}$$

Although the exponential factor in front diverges at the Cauchy horizon, the actual integral decreases exponentially. This is due to the analyticity properties of the scattering coefficients, which allows movement of the ω integration contour to imaginary values of k. In fact, poles in $W(lm\omega)$ at $m\Omega(r_1) - i\kappa_1$ determine the leading order behavior of the integral. Consequently, $\Phi_s^{(obs)}$ diverges as $\exp(\kappa_3 - \kappa_1)v$, and it is finite if $\kappa_1 > \kappa_3$.

4.3 Gravitational Collapse

The black hole solutions in de Sitter space do not answer the question of whether a stable Cauchy horizon can form as a result of gravitational collapse. In order to examine issue we can consider the spherically symmetrical collapse of charged matter in a de Sitter universe. The first step will be to look for solutions to the field equations for simple types of matter. It is also important to discuss the stability of these solutions. However, if the matter density happens to be zero in the neighborhood of the Cauchy horizon, then the linear perturbations of the matter field will decouple from the gravitational perturbations in this region and simplify the problem.

Analytic solutions can usually be found for collapsing spherical dust clouds. Oppenheimer and Schneider performed the original calculations on uncharged dust with a vanishing cosmological. Tolman and Bondi found the general class of metrics of this type and Amos Ori has given the solution to the Einstein equations for charged dust [11]. We shall review the charged dust solution in a form that allows us to find an approximate solution to the Einstein equations for a charged scalar field.

The spherically symmetric metric will be written in the following form,

$$ds^2 = -e^{2\Psi} dt^2 + e^{2\Lambda} dr^2 + R^2(d\theta^2 + \sin^2\theta\, d\phi^2) \tag{4.13}$$

where $\Psi(r,t)$, $\Lambda(r,t)$ and $R(r,t)$ are the only free functions. We also take a Maxwell field

$$F_{rt} = \frac{e^{\Psi+\Lambda} q}{4\pi R^2} \tag{4.14}$$

with $q(r,t)$. The collapsing matter will have a stress-energy tensor T_{ab} and current vector J^a.

The relevant components of the Einstein tensor in an orthonormal frame are given below, where $\hat{\partial}_t$ denotes $e^{-\Psi}\partial/\partial_t$ and $\hat{\partial}_r$ denotes $e^{-\Lambda}\partial/\partial_r$.

$$
\begin{aligned}
G^{\hat{r}}{}_{\hat{t}} &= 2R^{-1}\hat{\partial}_t R\hat{\partial}_r\Psi - 2R^{-1}\hat{\partial}_t\hat{\partial}_r R && (4.15)\\
G^{\hat{r}}{}_{\hat{r}} &= -2R^{-1}\hat{\partial}_t\hat{\partial}_t R + 2R^{-1}\hat{\partial}_r R\hat{\partial}_r\Psi \\
&\quad + R^{-2}\left((\hat{\partial}_r R)^2 - (\hat{\partial}_t R)^2 - 1\right) && (4.16)\\
G^{\hat{t}}{}_{\hat{t}} &= 2R^{-1}\hat{\partial}_r\hat{\partial}_r R - 2R^{-1}\hat{\partial}_r R\hat{\partial}_r\Lambda \\
&\quad + R^{-2}\left((\hat{\partial}_r R)^2 - (\hat{\partial}_t R)^2 - 1\right) && (4.17)
\end{aligned}
$$

The Tolman-Bondi solution for collapsing dust suggests that the field equations will simplify if we introduce functions $m(r,t)$ and $u(r,t)$,

$$
\begin{aligned}
u(r,t) &= \mathrm{e}^{-2\Lambda}R'^2 && (4.18)\\
m(r,t) &= \tfrac{1}{2}R\left(e^{-2\Psi}\dot{R}^2 - u + 1 + q^2/(4\pi R^2) - \tfrac{1}{3}\Lambda R^2\right), && (4.19)
\end{aligned}
$$

The Einstein-Maxwell equations can now be written

$$
\begin{aligned}
\dot{m} &= -4\pi R^2\dot{R}T^r{}_r + 4\pi R^2 R'T^r{}_t + R^{-1}(q^2/8\pi)^{\cdot} && (4.20)\\
m' &= -4\pi R^2 R'T^t{}_t + 4\pi R^2 R'T^t{}_r + R^{-1}(q^2/8\pi)' && (4.21)\\
\dot{u} &= -8\pi RR'T^r{}_t + 2R'\dot{R}e^{-2\Lambda}\Psi' && (4.22)\\
\dot{q} &= 4\pi R^2 e^{\Psi+\Lambda}J^r && (4.23)\\
q' &= 4\pi R^2 e^{\Psi+\Lambda}J^t && (4.24)
\end{aligned}
$$

In addition, we must impose the conservation laws $\nabla\cdot\mathbf{T} = 0$ on the matter fields. The cosmological constant and electric fields only enter through equation (4.19).

The simplest situation is a collection of charged particles moving with four-velocity $\mathbf{u}$, then

$$
\begin{aligned}
\mathbf{T} &= \rho\mathbf{u}\otimes\mathbf{u} && (4.25)\\
\mathbf{J} &= \epsilon\rho\mathbf{u} && (4.26)
\end{aligned}
$$

It is possible to choose our coordinate system to have the particles moving in the surfaces with r constant. Equations (4.21) and (4.24) suggest that m and q are interpreted as the mass and charge within a radius r. The equations (4.20) and (4.23) imply that

$$q \equiv q(r) \qquad m \equiv m(r) \tag{4.27}$$

The conservation of energy gives

$$\epsilon \equiv \epsilon(r) \tag{4.28}$$

together with

$$\rho R^4 \Psi' = \frac{(q^2)'}{32\pi^2}. \tag{4.29}$$

If we eliminate the q' using equation (4.24), this gives

$$\Psi' = e^{\Lambda} \frac{q\epsilon}{4\pi R^2} \tag{4.30}$$

The physical interpretation of this equation is that, in the comoving frame, the gravitational (inertial) forces exactly balance the electrostatic forces on the particles.

We can substitute equation (4.30) into equation(4.22) to obtain an equation for u, which can be integrated to give

$$u = \left(E - \frac{q\epsilon}{4\pi R}\right)^2 \tag{4.31}$$

where $E = m'\epsilon/q'$ is a function of r.

The behavior of the solutions can now be seen from equation (4.19), which can be re-organized into a more suggestive form,

$$e^{-2\Psi}\dot{R}^2 + V(R, r) = 0 \tag{4.32}$$

where

$$V(R, r) = 1 - \frac{2m}{R} + \frac{q^2}{4\pi R^2} - \tfrac{1}{3}\Lambda R^2 - u(R, r) \tag{4.33}$$

For a collapsing star, we can suppose that the density is non-zero and positive up to a fixed value of r. The values of m and q should rise monotonically to the total mass M and charge Q of the star. We will also suppose that there are four roots $R = R_i(r)$ where V vanishes, with $R_1 > R_2 > R_3 > 0 > R_4$. If the star starts from rest, with $R = r$, then $R_2 = r$ and the parameters are located in the triangular region of Figure 4.3.

If Ψ is well-behaved, then each spherical shell that makes up the star will contract and bounce when $R = R_3$. Before the bounce occurs, there is a shell-crossing singularity where $R' = 0$ and the density diverges. The existence of the singularity can be seen from the radial derivative of equation (4.32) at the point $\dot{R} = 0$, (using 4.30)

$$\frac{Q\epsilon V}{4\pi R^2 u^{1/2}} R' + \frac{\partial V}{\partial r} + \frac{\partial V}{\partial R} R' = 0 \tag{4.34}$$

With our assumptions about the roots of V, both derivatives of V can be shown to be negative. Therefore R' is negative, but $R' = 1$ initially and R' has to vanish an an intermediate point.

The solution proceeds by introducing a new function $f = e^{-\Lambda-\Psi}\dot{R}$, and showing that it satisfies

$$\dot{f} = f^2 \left|\frac{1}{4uV^3}\right|^{1/2} \frac{\partial V}{\partial r} \dot{R} \tag{4.35}$$

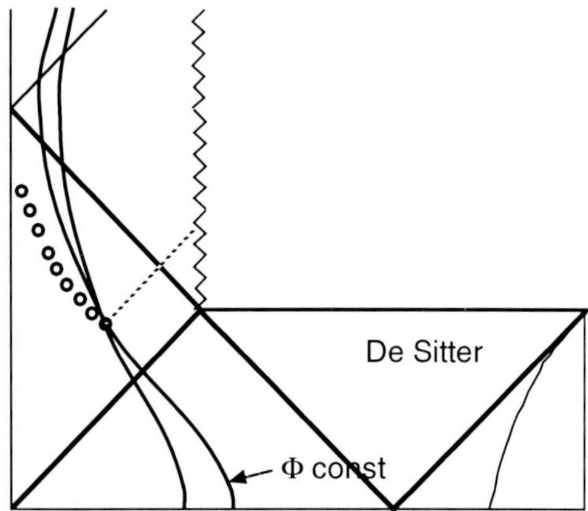

Figure 4.4: Spacetime diagrams with a collapsing star. Shell-crossings are shown schematically by circles. The dust approximation is valid up to the null geodesic (dotted).

The integral of this equation gives $f(R, r)$ and with (4.32) leads to an ordinary differential equation $R' = |uV|^{1/2}/f$.

The solution can only be applied up to the shell-crossing singularity. If all of the shell crossing singularities occur inside of the event horizon, as in figure 4.4, then then solution close to at least part of the Cauchy horizon would not be affected. This is an interesting situation in which a Cauchy horizon would form from gravitational collapse. The red-shift argument of the previous section suggests that the singularity would be generic in the appropriate parameter range.

The dust models have limited practical appeal because of breakdown of the solution at the shell-crossing singularities. We will now consider another model based upon a charged scalar field. We will see that the solution up to the shell crossing can be very similar to the dust solution just examined. In order to follow the solution beyond this point we would have to resort to numerical methods, but at least we expect the solution to continue consistently.

The stress-energy tensor components for the scalar field ϕ with mass μ and charge e include

$$T^t{}_t = -e^{-2\Psi}|D\phi|^2 - e^{-2\Lambda}|\phi'|^2 - \mu^2|\phi|^2 \tag{4.36}$$

$$T^r{}_r = +e^{-2\Psi}|D\phi|^2 + e^{-2\Lambda}|\phi'|^2 - \mu^2|\phi|^2 \tag{4.37}$$

where $D\phi = \dot{\phi} - ie\psi\phi$ and ψ is the Coulomb potential, $\psi' = F_{rt}$. The current is given by

$$j_t = -ie(\phi^* D\phi - \phi D\phi^*) \tag{4.38}$$
$$j_r = -ie(\phi^* \phi' - \phi\phi'^*) \tag{4.39}$$

We will find an approximate solution for the gravitational collapse of a body whose radius λ is much larger than the Compton wavelength of the scalar field,

$$\lambda\mu \gg 1 \tag{4.40}$$

The uncharged case was described in reference [12], where we introduced an asymptotic expansion in $1/(\lambda\mu)$. Here we will only consider the leading term.

The scalar field is written in adiabatic form with slowly varying amplitude,

$$\phi(r,t) = \tfrac{1}{\sqrt{2}}\mu^{-1}\Phi(r,t)e^{i\mu t} \tag{4.41}$$

This is substituted into the field equations, keeping only the leading order terms.

Before proceeding, we compare the Maxwell field (4.14) with equation (4.30), to see that if $\epsilon = \mu/e$, the Coulomb potential is

$$\psi = \frac{\mu}{e}(e^{\Psi} - 1). \tag{4.42}$$

In the comoving coordinate frame the gravitational red-shift cancels the work done against the electromagnetic forces, and we get

$$D\phi = \tfrac{i}{\sqrt{2}}\Phi + O(\mu^{-1}) \tag{4.43}$$

Consequently, from (4.37), $T^r{}_r = O(\mu^{-2})$ and

$$\mathbf{T} = \Phi^2 \mathbf{u} \otimes \mathbf{u} + O(\mu^{-2}) \tag{4.44}$$
$$\mathbf{J} = \frac{\mu}{e}\Phi^2 \mathbf{u} + O(e\mu^{-1}) \tag{4.45}$$

To leading order, all of the equations reduce to the previous equations for charged dust. The approximation breaks down when Φ becomes large, which is where the shell-crossing singularities occur, but this can be inside the event horizon [12].

These two models, charged dust and a massive scalar field, both give examples of gravitational collapse with naked singularities. Furthermore, when the parameters are in the stability regime for the black hole-de Sitter Cauchy horizon, it seems likely that the the singularity will be generic and that a linear stability analysis may be feasible.

4.4 Conclusions

General Relativity does not preclude the possibility of visible singularities. According to the linear perturbation analysis of the black hole-de Sitter spacetime, they can also be generic. This is an important development in that all other examples could be unstable. The believers in Strong Cosmic Censorship can hold on to the hope that the linear analysis has zero radius of convergence or they can insist on stronger energy conditions on the matter. In my opinion, the black hole-de Sitter model is only the one of many more visible singularities around if we want to look for them.

The two matter models described in this article are incomplete, but they are steps towards more realistic models. The stability analysis of these models may be possible by standard methods. More complicated models may require new methods of stability analysis.

Physics in the vicinity of the Cauchy horizon is necessarily quantum physics, even though the curvature might be classically small compared to the Planck scale. One view would be that quantum effects push the curvature at the Cauchy horizon to Planckian values, effectively converting the timelike singularity into a null singularity. Another possibility would be that the naked singularity becomes a spacelike singularity, similar to a localized version of the big bang. Recent years have seen interesting developments in the quantum theory of the big bang singularity and it may be the right time to start to investigate the quantum theory of other types of singularity.

Bibliography

[1] S. W. Hawking and G. F. R. Ellis, *The large scale structure of space-time* (Cambridge University Press, Cambridge, 1973).

[2] R. Penrose, Rivista Nuovo Cimento **1**, 252 (1969).

[3] S. L. Shapiro and S. A. Teukolski, Phys. Rev. Lett. **66**, 994 (1991).

[4] V. Gorini, G. Grillo and M. Pelizza, Phys. Lett. **A 135**, 154 (1989).

[5] A. Ori and T. Piran, Gen. Relativ. Gravitation **20**, 7 (1988).

[6] B. Carter, Commun. Math. Phys. **10**, 280 (1968).

[7] F. Mellor and I. G. Moss, Phys. Rev. D **41**, 403 (1990); Class. Quantum Grav. **9**, L43 (1992).

[8] P. R. Brady and E. Poisson, Class. Quantum Grav. **9**, 121 (1992).

[9] C. M. Chambers and I. G. Moss, Class. Quantum Grav. **11**, 1035 (1994).

[10] S. Chandrasekhar, *The mathematical theory of black holes* (Oxford University Press, Oxford, 1983).

[11] A. Ori, Class. Quantum Grav. **7**, 985 (1990).

[12] S. M. C. V. Gonçalves and I. G. Moss, Class. Quantum Grav. **14**, 2607 (1997).

ENCOUNTER OF A STABILITY CONJECTURE WITH BLACK–HOLE CAUCHY HORIZONS

T. M. Helliwell [a] and D. A. Konkowski [b]

a) Department of Physics, Harvey Mudd College, Claremont, California 91711
b) Department of Mathematics, U.S. Naval Academy, Annapolis, Maryland 21401

Abstract

The stability of black-hole Cauchy horizons is reviewed in the context of a previously-developed stability conjecture which judges stability using the behavior of test fields in black hole spacetimes. In several cases the conjecture can be tested against exact solutions incorporating various added fields. Cauchy horizons are studied in Reissner-Nordström, Kerr, Reissner-Nordström deSitter, and (1+1) and (2+1) dimensional dilaton black holes. In some cases the Cauchy horizon is converted by the field into a scalar curvature singularity, and in other cases it is converted into a non-scalar curvature singularity. Several challenges to the conjecture are presented, and the limitations they place upon the applicability of the conjecture are discussed.

5.1 Introduction

Realistic black holes contain Cauchy horizons. A black-hole Cauchy horizon is the boundary of the causal development of information given on a spacelike hypersurface outside the event horizon of the black hole. A Penrose diagram of a charged, non-rotating black hole (or of the equatorial plane of a rotating black hole) is shown in Fig. 5.1. As a star collapses, an event horizon (EH) is created, and within the event horizon a Cauchy horizon (CH) is formed. Light moves at 45° in the diagram, so it is clear that a hypothetical observer traveling through the EH and then through the CH would enter a region whose properties could not be predicted entirely from data entered on a spacelike surface outside the black hole. In principle the observer could see and be influenced by the timelike curvature singularity at $r = 0$, for example, a violation of strong cosmic censorship [1].

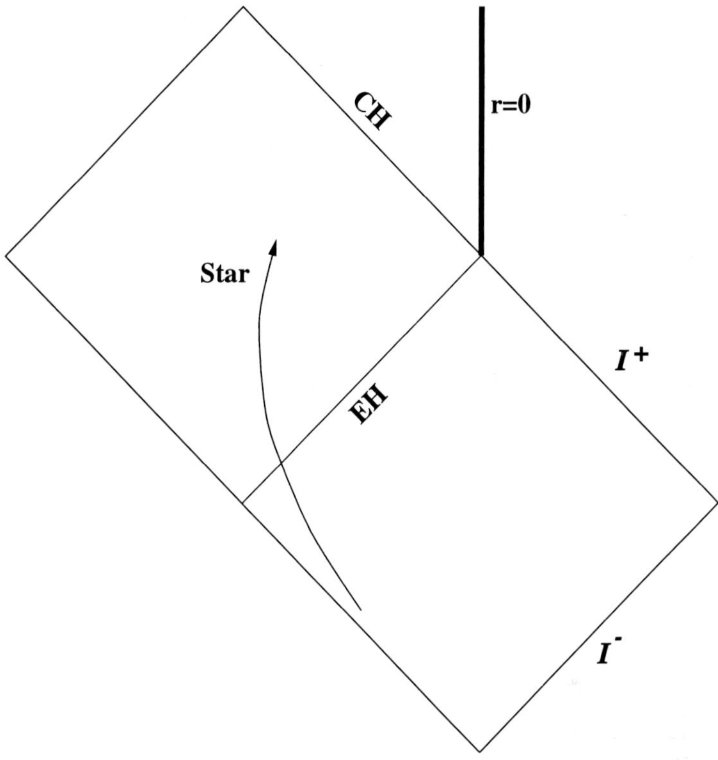

Figure 5.1: Penrose diagram of a charged, nonrotating (Reissner-Nordström) black-hole spacetime.

It has long been suspected that black hole Cauchy horizons are unstable. If the instability renders the CH impenetrable, strong cosmic censorship would be protected. Already in 1968, Roger Penrose noted that light falling into a spherically-symmetric charged black hole would be infinitely blue-shifted from the point of view of an observer falling toward the Cauchy horizon [2]. Suppose for example there is an external source which emits periodic inward-moving pulses of light. An observer falling toward the CH would reach it in a finite proper time, during which an infinite number of the external pulses would be received, piled up in the near vicinity of the CH. This infinite compression might be expected to destabilize the CH; many authors have presented evidence that this is indeed the case [3, 4, 5, 6].

Here we review the stability of various black-hole Cauchy horizons in the context of a stability conjecture we have developed. The conjecture judges the stability of CHs when various added fields are imposed. If the conjecture indicates that a horizon is unstable, it also indicates what type of singularity is created. The encounter of the stability conjecture with black-hole Cauchy horizons is an interesting one, in part because stability or instability can be predicted in a number of situations in which exact calculations have not been carried out. Also, in some other situations exact results are known, which can serve as a test of the conjecture. In this way insights have been gained about the strengths and limitations of the conjecture. In Sec. 5.2 we review the classification of singularities and a singularity stability conjecture. In Sec. 5.3 we introduce a Cauchy-horizon stability conjecture, and test it in the Reissner-Nordström spacetime. In Sec. 5.4 we discuss mass inflation in the context of the stability conjecture. In Sec. 5.5 we discuss other types of black holes. In Sec. 5.6 we present several challenges to the stability conjecture, and in Sec. 5.7 we assess the conjecture's strengths and limitations.

5.2 Singularity classification and a singualrity stability conjecture

According to a classification scheme proposed by Ellis and Schmidt [7], there are three basic types of singularities possible in maximal spacetimes:

(1) scalar curvature singularities;

(2) nonscalar curvature singularities;

(3) quasiregular singularities.

(1) All observers approaching a scalar curvature singularity (SCS) find that physical quantities such as energy density and tidal forces diverge.
Examples of SCSs include the center of Schwarzschild black holes and the beginning of Friedmann-Robertson-Walker cosmologies.

(2) Some free-falling observers approaching a nonscalar curvature singularity (NSCS) experience divergent tidal forces, but observers moving along other curves passing arbitrarily close to a NSCS experience perfectly regular tidal forces.
Examples of NSCSs include tilted homogeneous cosmologies and singular gravitational plane waves.

(3) No observers approaching a quasiregular singularity find that physical quantities diverge, although their world lines suddenly end at the singularity in a finite proper time.
Examples of quasiregular singularities occur in idealized cosmic strings, the Taub-NUT spacetime, and the fold-like singularities in colliding gravitational plane waves.

Our version of the Ellis and Schmidt classification scheme is as follows [8].

Define singular points as the endpoints of incomplete geodesics in maximal spacetimes. Such a singular point q is a quasiregular singularity if all components of the Riemann tensor R_{abcd} evaluated in a parallel-propagated orthonormal (PPON) frame along an incomplete geodesic ending at q are C^0 (or C^{0-}). In other words, the Riemann tensor components tend to finite limits (or are bounded). A singular point q is a curvature singularity if some components are not bounded in this way. If all scalars in the metric tensor g_{ab} , the antisymmetric tensor η_{abcd} , and R_{abcd} nevertheless tend to a finite limit (or are bounded), the singularity is a NSCS, but if any scalar is unbounded, the point q is a SCS. The stability conjecture for mild (i.e., quasiregular or nonscalar curvature) singularities is as follows.

If a test field stress-energy tensor evaluated in a PPON frame mimics the behavior of the Riemann tensor components which indicate a particular type of singularity, then a complete nonlinear back-reaction calculation would show that this type of singularity occurs.

For example, if a test-field stress-energy scalar $T^{\mu}{}_{\mu}$ or $T^{\mu\nu}T_{\mu\nu}$ diverges at a mild singularity in a background spacetime, according to the conjecture the singularity will be converted into a SCS in an exact back-reaction spacetime, in which the field is allowed to influence the geometry. For a summary of special cases in which the conjecture has been applied, see reference [8].

5.3 Cauchy horizon stability conjecture

We have extended the conjecture for mild singularities to cover Cauchy horizons as well [8]. In this case the conjecture is as follows.

For all maximally extended spacetimes with Cauchy horizons, the back reaction due to a field (whose test-field stress-energy tensor is $T_{\mu\nu}$) will affect the horizon in the following manner.

(1) If both $T^{\mu}{}_{\mu}$ and $T^{\mu\nu}T_{\mu\nu}$ are finite and if the stress-energy tensor $T_{(ab)}$ in all parallel-propagated orthonormal frames (PPON frames) is finite, then the Cauchy horizon remains nonsingular.

(2) If $T^{\mu}{}_{\mu}$ or $T^{\mu\nu}T_{\mu\nu}$ are finite but $T_{(ab)}$ diverges in some PPON frame, then a NSCS will be formed at the Cauchy horizon.

(3) If $T^{\mu}{}_{\mu}$ or $T^{\mu\nu}T_{\mu\nu}$ diverges, then a SCS will be formed at the Cauchy horizon.

It is usually not possible to consider *all* PPON frames in (2), so in practice the conjecture indicates only instability, not stability, in the formation of NSCSs.

As an example of using the conjecture, consider the CH of Reissner-Nordström under the influence of a null dust test-field. The Reissner-Nordström EH is located at $r = m + \sqrt{m^2 - e^2}$, where m and e are the hole's mass and charge, and the CH is located at $r = m - \sqrt{m^2 - e^2}$. The metric is

$$ds^2 = -f(r)dv^2 + 2dvdr + r^2 d\Omega^2 \tag{5.1}$$

in terms of the Eddington-Finklestein advanced time v, where

$$f(r) = 1 - \frac{2M}{r} + \frac{e^2}{r^2}.$$

Null dust is a pressureless null fluid which can be a good approximation to electromagnetic radiation in the short-wavelength limit, an entirely appropriate limit for the highly blue-shifted radiation falling into the black hole along the CH. The stress-energy tensor for null dust is $T^{\mu\nu} = \rho l^\mu l^\nu$, where ρ is the null-dust density (a scalar) and l^μ is the null vector along the ray. The stress-energy scalars $T^\mu{}_\mu = T^{\mu\nu}T_{\mu\nu} = 0$ for null dust, because l^μ is a null vector. Therefore, according to the conjecture no SCS should be formed [8]. To see whether a NSCS should be formed instead, we need to evaluate elements of the stress-energy tensor in a PPON frame approaching the CH. In a radial PPON frame one finds that the stress-energy tensor is

$$T_{(ab)} \sim \frac{F(v)}{f^2} \sim F(v)e^{2k-v} \tag{5.2}$$

where $k_- = -(1/2)df/dr|r_-$ is the "surface gravity" at r_- [9]. If F(v) falls off asymptotically as a power-law in v, the typical form for scattering into a black hole [10], then $T_{(ab)}$ diverges at the CH. Therefore the conjecture indicates the CH will be converted into a NSCS. In fact, the corresponding exact back-reaction solution was constructed long ago by Hiscock [11]. It is a so-called Reissner-Nordström-Vaidya spacetime in which null dust falls into a charged, nonrotating black hole. The exact Reissner-Nordström-Vaidya solution has the same metric as that of Eq. (5.1), except that now

$$f = f(r, v) = 1 - \frac{2M(v)}{r} + \frac{e^2}{r^2}. \tag{5.3}$$

That is, the mass has become a function of the advanced time v . For a power-law tail of infalling radiation,

$$\begin{aligned} M(v) &= m_o - \delta \ (v < v_0) \\ &= m_o - \delta(\tfrac{v_o}{v})^n \ (v > v_o). \end{aligned} \tag{5.4}$$

Curvature scalars converge at $r = r_-$: That is, $R = 0 = R_{\mu\nu}R^{\mu\nu}$, and

$$R_{\mu\nu\lambda\sigma}R^{\mu\nu\lambda\sigma} = \frac{48M^2(v)}{r^6} - \frac{80M(v)e^2}{r^7} + \frac{56e^4}{r^8}. \tag{5.5}$$

Therefore the Cauchy horizon does not become a SCS. It forms instead a NSCS; Hiscock shows that the local tidal force diverges for an observer falling to the Cauchy horizon of the Reissner-Nordström-Vaidya geometry. The stability conjecture is therefore in agreement with the exact back-reaction calculation.

5.4 Mass inflation

Poisson and Israel [9] have shown that if outgoing radiation is added to the infalling radiation within a Reissner-Nordström black hole, the local mass from the point of view of an observer falling toward the CH will increase exponentially. Instead of a NSCS, the CH is converted into a null scalar curvature singularity. The null-dust stress-energy tensor is now

$$T^{\mu\nu} = \rho_{in} l^\mu l^\nu + \rho_{out} n^\mu n^\nu \tag{5.6}$$

where n^μ is an outwardly-directed null vector. Poisson and Israel show that the local mass obeys the wave equation [9]

$$\Box m = -(4\pi)^2 r^3 T_{\mu\nu} T^{\mu\nu}. \tag{5.7}$$

There is no exact *general* solution of this equation, but they show that $m \sim v^{-p} e^{K_- v}$ as $v \to \infty$, i.e., that the local mass diverges exponentially as the CH is approached, which is the phenomenon of "mass inflation". The Coulomb component of the Weyl scalar diverges, and a null SCS is formed at the CH. (We discuss a special exact solution in Section 5.6.)

The problem of both infalling and outgoing null dust can also be studied using the test-field stability conjecture [8]. A test field with stress-energy as given by Eq. (5.6) has null vectors

$$l^\mu = (0, -1, 0, 0) \text{ and } n^\mu = (-2/f(r), -1, 0, 0), \tag{5.8}$$

where $f(r)$ is the metric function of the background Reissner-Nordström solution. The continuity equation $T^{\mu\nu}{}_{;\nu} = 0$ shows that ρ_{in} and ρ_{out} have the forms

$$\rho_{in} = \frac{F(v)}{4\pi r^2} \text{ and } \rho_{out} = \frac{G(u)}{4\pi r^2}, \tag{5.9}$$

where v and u are the advanced and retarded times, both null coordinates. The corresponding stress-energy scalars are $T^\mu{}_\mu = 0$ and

$$T_{\mu\nu} T^{\mu\nu} = 2\rho_{in}\rho_{out}[l^\mu n_\mu]^2 = 8\frac{F(v)G(u)}{r^4 f^2}, \tag{5.10}$$

which diverges as $r \to r_-$ if

(i) $G(u) \neq 0$ (That is, if there is at least some outgoing radiation)

(ii) $F(v)$ falls off as a power law in v .

The scalar $T_{\mu\nu}T^{\mu\nu}$ diverges because the quantity $1/f^2$ diverges exponentially as $v \to \infty$. The conjecture therefore indicates that a SCS will be formed at the CH, in agreement with the back-reaction calculations.

Various authors have also studied the behavior of scalar and electromagnetic test-fields in the background Reissner-Nordström geometry [4, 5, 6]. For example, solutions of the massless scalar wave equation $\Box\Phi = 0$ have the form [4, 6]

$$\Phi(t, r, \theta, \phi) = \sum_{lm} Y_{lm}(\theta, \phi) \int_{-\infty}^{\infty} dk e^{ikt} (\frac{1}{r}) \psi_{lmk}(r). \tag{5.11}$$

For purely infalling waves at the event horizon, scattering in the interior creates outgoing waves as well. Near the CH, field modes have the form

$$\Phi_{lm}(u, v, \theta, \phi) \sim [\frac{A_{lm}}{v^{2l+2}} + \frac{B_{lm}}{u^{2l+2}}] Y_{lm}(\theta, \phi). \tag{5.12}$$

The stress-energy scalars are $T^{\mu}{}_{\mu} = -S$ and $T^{\mu\nu}T_{\mu\nu} = S^2$, where $S = g^{\alpha\beta}\Phi_{,\alpha}\Phi_{,\beta}$. The dominant term of S_{lm} is

$$S_{lm} \sim \frac{e^{k(u+v)/2}}{(uv)^{2l+3}}, \tag{5.13}$$

which diverges as v $\to \infty$ (The CH), so the conjecture indicates that such scalar waves will convert the CH into a SCS, consistent with the effects of null dust [9]. Electromagnetic waves behave similarly, and should also convert the CH into a SCS.

5.5 Other black holes

The conjecture has also been applied to several other types of black holes.

(a) Kerr

Rotating black holes have the Kerr geometry, with event and Cauchy horizons at $r_{\pm} = m \pm \sqrt{m^2 - a^2}$, where a is the angular momentum/mass. An observer falling toward the Cauchy horizon would experience the same blue-shift effect of infalling radiation as in the Reissner-Nordström case. In the case of an inward-moving principal null congruence of null dust, the stress-energy tensor is $T_{in}{}^{\mu\nu} = \rho_{in} l^{\mu} l^{\nu}$ with

$$l^{\mu} = (\frac{r^2 + a^2}{\Delta}, -1, 0, \frac{a}{\Delta}) \tag{5.14}$$

where $\Delta = r^2 + a^2 -.2mr$ [12]. The density has the form

$$\rho_{in}(v, r, \theta) = \frac{A(\theta)F(v)}{r^2 + a^2 \cos^2 \theta} \tag{5.15}$$

where $v = r_* + t$, with r_* the tortoise coordinate

$$r_* = r + [\frac{r_+ + a^2}{r_+ - r_-}] \ln |r - r_+| - [\frac{r_-{}^2 + a^2}{r_+ - r_-}] \ln |r - r_-|. \tag{5.16}$$

The stress-nergy scalars are zero, and in an equatorial-plane PPON frame [12],

$$T_{(00)} \sim \frac{F(v)}{\Delta^2} \sim F(v)e^{2k-v} \tag{5.17}$$

which diverges if F(v) falls off as a power law in v. Therefore a NSCS is predicted by the conjecture for purely inflowing null dust, just as for Reissner-Nordström black holes.

If outgoing null dust is introduced as well, the added stress-energy is ${T_{out}}^{\mu\nu} = \rho_{out} n^\mu n^\nu$, with

$$n^\mu = (-\frac{r^2+a^2}{\Delta}, -1, 0, -\frac{a}{\Delta}) \tag{5.18}$$

and

$$\rho_{out}(u,r,\theta) = \frac{B(\theta)G(u)}{r^2+a^2\cos^2\theta} \tag{5.19}$$

where $u = r^* - t$. The scalar $T_{\mu\nu}T^{\mu\nu} = 8\frac{F(v)G(u)}{\Delta^2}$ diverges at the CH, so the conjecture predicts the horizon will become a SCS if $G(u) \neq 0$ and F(v) falls off as a power law in v. This behavior is similar to that in Reissner-Nordström black holes.

(b) Reissner-Nordström-deSitter

A charged non-rotating black hole is embedded in an exponentially expanding deSitter universe. The metric again has the form of Eq. (5.1), where now [13]

$$f = f(r,v) = 1 - \frac{2M(v)}{r} + \frac{e^2}{r^2} - \frac{1}{3}\Lambda r^2. \tag{5.20}$$

There are three horizons: a cosmological horizon at r_c , an event horizon at r_+, and a CH at r_-, as shown in Fig. 5.2. Radiation falling into the black hole along the CH is first highly redshifted along the cosmological horizon, which can compensate for the blue shift along the Cauchy horizon.

Various authors have included both infalling and outgoing null dust and checked for mass inflation [13]. There are three cases, depending upon the relative sizes of the surface gravities at the Cauchy horizon (k_-) and at the cosmological horizon (k_c):

(i) $k_- \leq k_c$ the Cauchy horizon is stable

(ii) $k_c < k_- < 2k_c$ there is a divergent influx but finite mass (there is no mass inflation)

(iii) $2k_c < k_-$ there is both a divergent influx and mass inflation

What type of singularity is formed in cases (ii) and (iii) ? In case (iii), the Weyl scalar $\Psi_2 \sim m/r^3 \to \infty$, so a SCS forms. In case (ii), Ψ_2 is finite, but both $R_{\mu\nu}R^{\mu\nu}$ and $R^{\mu\nu\lambda\rho}R_{\mu\nu\lambda\rho}$ diverge *if* there is continuous outflow. So a SCS is formed if there is continuous outflow. The stability conjecture can be tested using these results. Cai

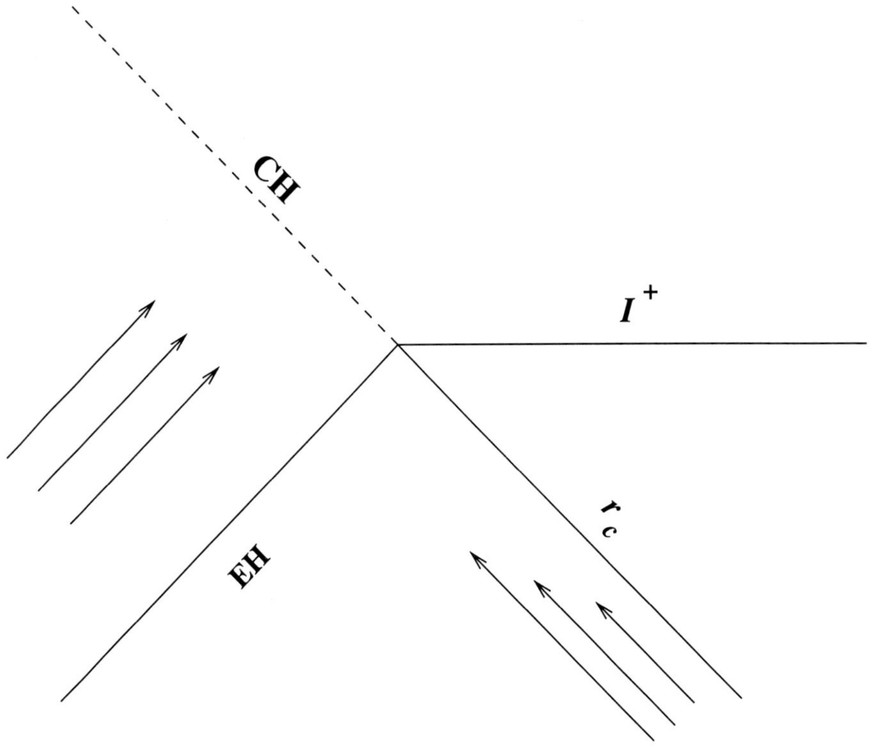

Figure 5.2: Penrose diagram for the Reissner-Nordström-de Sitter Black Hole

and Su [14] have shown that if one considers infalling null dust only, the conjecture indicates the CH is stable if $k_- \leq k_c$, and a NSCS is indicated if $k_- > k_c$. They show that these results agree with exact back-reaction Reissner-Nordström-deSitter-Vaidya. In the crossflow case, with both infalling and outgoing null dust, they find that the test-field stress-energy tensor is

$$T_{\mu\nu}T^{\mu\nu} = \frac{F(v)G(u)}{2\pi^2 f^2} \sim \exp 2(k_- - k_c)v \tag{5.21}$$

if G(u) $\neq 0$, where $\rho_{out} = G(u)/4\pi^2 r^2$. So according to the conjecture the CH should be stable if $k_- \leq k_c$, and a SCS should be formed if $k_- > k_c$, assuming continuous outflow. This agrees with the exact back-reaction case. We take up later the case of impulsive outflow.

(c) Lower-dimensional black holes

Mass inflation has also been demonstrated in (1+1)-dimensional dilaton black holes [15]. A novel feature is their multihorizon structures, in which CHs may have either vanishing or nonvanishing surface gravity. For CHs with non-vanishing surface gravity, the results are similar to Reissner-Nordström: A NSCS is formed for purely infalling radiation, and a SCS is formed if there is crossflow. For CHs with vanishing surface gravity, and with a power-law tail of infalling radiation $\sim v^{-(p+1)}$, if $p \geq 1+2/b$ the CH is stable; if $1 + 1/b \leq p < 1 + 2/b$ there is a SCS at the CH, but no mass inflation; and if $p < 1 + 1/b$ there is a SCS *with* mass inflation. Here b is an integer related to the number of degenerate roots of the equation determining the horizon. Stability conjecture calculations have been carried out by Cai [16]. Results agree with the exact solution.

Cai and Zhao [17] have considered (2 + 1)-dimensional dilaton black holes, and have shown that again the stability conjecture agrees with back-reaction calculations.

5.6 Challenges to the stability conjecture

In every case discussed up to now, the predictions of the Cauchy horizon stability conjecture agree with the exact back-reaction calculations which have been carried out. There are, however, some interesting challenges to the conjecture.

(a) Shell-focusing singularities

Consider an imploding shell of null dust in a background anti-de Sitter spacetime [18]. According to the stability conjecture, no SCS can form, because the stress-energy scalars are identically zero for null dust. In fact, the conjecture predicts that a NSCS will form, but only at the collapse point r = 0 . An exact anti-de Sitter-Vaidya solution shows that the curvature scalars are [18]

$$R = -12\alpha^2, R_{\mu\nu}R^{\mu\nu} = 36\alpha^4, \text{ and } R_{\mu\nu\lambda\sigma}R^{\mu\nu\lambda\sigma} = 24(\alpha^4 + \frac{2H^2(v)}{r^6}), \tag{5.22}$$

where α = constant and H(v) is arbitrary. The latter (Kretschmann) scalar diverges as $r \to 0$, so that r = 0 is a SCS instead of the NSCS predicted by the conjecture.

Why does the conjecture fail to indicate the correct type of singularity in this case? The reason is clearly related to that fact that even though the null-dust density diverges at the collapse point, scalars constructed from $T_{\mu\nu}$ are zero. If the shell were constructed of realistic electromagnetic waves instead of the hypothetical null dust, the scalar $T^{\mu}{}_{\mu}$ would still vanish , but $T^{\mu\nu}T_{\mu\nu}$ would diverge as $r \to 0$. So with these more physical waves, the conjecture would be valid as is. The problem is with the unphysical null dust, which is pressureless even when it piles up. The failure of the conjecture is also related to the fact that the curvature singularity at r = 0 is only in the Weyl tensor $C_{\mu\nu\lambda\sigma}C^{\mu\nu\lambda\sigma} = 48H^2(v)/r^6$, and not in the Ricci tensor. Because the conjecture inspects test-field stress-energy tensors, which can be related to the Ricci tensor through the field equations, it is not surprising that the conjecture can only predict divergences in the Ricci tensor and not the Weyl tensor portion of the curvature.

It is straightforward to modify the conjecture to take account of null dust shell-focusing singularities, simply by adding the divergence of any scalar null-dust density to the divergence of a stress-energy scalar as a criterion for the creation of an SCS. The resulting inclusive statement of the conjecture is as follows [19].

1. If $T^{\mu}{}_{\mu}$, $T^{\mu\nu}T_{\mu\nu}$, and the density of any null dust are finite, and the stress-energy tensor $T_{(ab)}$ in all parallel-propagated orthonormal frames (PPON frames) is also finite, then the Cauchy horizon remains nonsingular.

2. If $T^{\mu}{}_{\mu}$, $T^{\mu\nu}T_{\mu\nu}$, and the density of any null dust is finite, but $T_{(ab)}$ diverges in some PPON frame, then a NSCS will be formed at the Cauchy horizon.

3. If $T^{\mu}{}_{\mu}$ or $T^{\mu\nu}T_{\mu\nu}$ diverges, or if the density of any null dust diverges, then a SCS will be formed at the Cauchy horizon.

(2) beam of null dust incident on an impulsive gravitational plane wave

Consider an impulsive gravitational plane wave, moving with advanced time v = 0, shown in Fig. 5.3. In region I, to the lower left, the spacetime is flat with the double-null metric

$$ds^2 = 2dudv + dx^2 + dy^2 \text{ (Region I)} \tag{5.23}$$

In region II, to the upper right, the spacetime is also flat, with metric

$$ds^2 = 2dudv + (1+v)^2dx^2 + (1-v)^2dy^2 \text{ (Region II)} \tag{5.24}$$

There is a CH along v = 1 in region II [20].

Suppose that a beam of null dust originates in region I and strikes the gravitational wave head on. The beam is distorted as it passes through the gravitational wave, so that an originally circular beam will double its width in the x direction, while narrowing its width to zero in the y direction, as it reaches the Cauchy horizon at v =

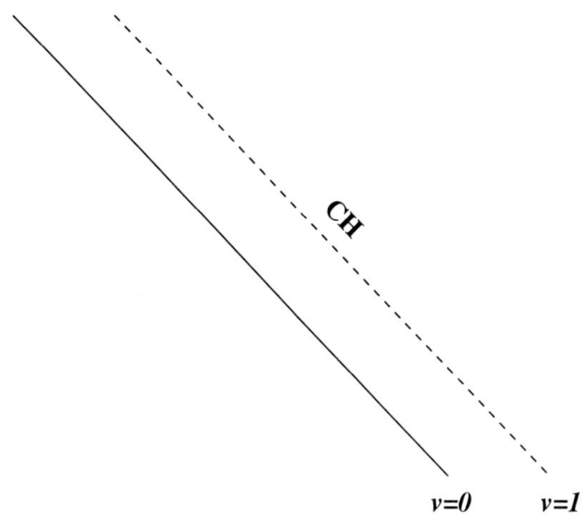

Figure 5.3: An impulsive gravitational plane wave at $v = 0$. The wave has a Cauchy horizon at $v = 1$.

1. The null-dust density $\rho \to \infty$ as $v \to 1$, so as with the shell-focusing singularity, a scalar curvature singularity is expected to form at the Cauchy horizon, even though no scalar constructed from $T_{\mu\nu}$ diverges there. The original conjecture fails to catch the SCS in this case (it predicts a NSCS instead), but as in the shell-focusing singularity of part (a), the revised conjecture correctly predicts a SCS.

Other test fields also indicate that a SCS is generally formed at the CH [20]. For example, the scalar wave equation $\Box\Phi = 0$ can be solved in the background spacetime. If $\Phi = f(u)$ in region I, then $\Phi = \frac{f(u)}{\sqrt{1-v^2}}$ in region II. The stress-energy scalars also diverge: $T^{\mu}{}_{\mu} \sim (1-v)^{-2}$ and $T_{\mu\nu}T^{\mu\nu} \sim (1-v)^{-4}$. Electromagnetic waves generally cause a SCS to form as well according to the conjecture, since in this case $T^{\mu}{}_{\mu} \sim 0$ and $T_{\mu\nu}T^{\mu\nu} \sim (1-v)^{-4}$. In summary, as with spherical shell-focusing, the conjecture is valid provided the null-dust $\rho \to \infty$ is added as a criterion for a SCS.

(3) Reissner-Nordström with power-law infalling and impulsive outgoing null dust

There is an exact mass-inflation solution for Reissner-Nordström black holes due to Ori [21]. The spacetime has four regions as shown in Fig. 5.4, separated by the beginning of the continuous inflow of null dust and by the impulsive outgoing shell.

There is mass inflation along the CH everywhere to the future of the impulsive outgoing shell. Along this portion there is also a SCS, because the Weyl "Coulomb" scalar $\Psi_2 \sim m/r^3$ diverges everywhere along it. There is only a Weyl SCS ($C_{\mu\nu\lambda\sigma}C^{\mu\nu\lambda\sigma} \to \infty$) on the CH; the scalars R and $R_{\mu\nu}R^{\mu\nu}$ made from the Ricci tensor stay finite, except where the impulsive outgoing shell intersects the CH. This contrasts with the case of continuous outflow (Poisson and Israel), for which there is a divergence of $R_{\mu\nu}R^{\mu\nu}$ as well as the Weyl scalar singularity. The stability conjecture correctly indi-

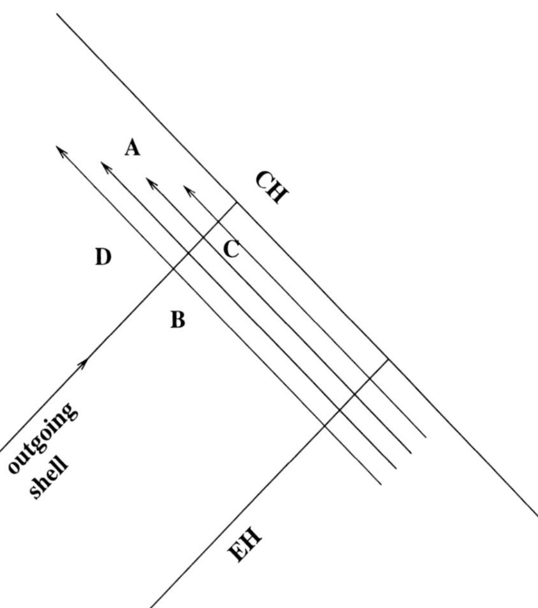

Figure 5.4: Reissner-Nordström with continuous inflow and an impulsive outflow and a line representing a single impulsive outgoing beam.

cates that the CH is singular, but that it is a NSCS instead of the actual SCS to the future of the point at which the outgoing null dust impacts the CH. This is because the test-field stress-energy scalars are $T^{\mu}{}_{\mu} = T^{\mu\nu}T_{\mu\nu} = 0$ along the future CH.

(4) Reissner-Nordström-de Sitter with power-law infalling and impulsive outgoing null dust

A continuous outflow of radiation in the Reissner-Nordström-de Sitter black hole can likewise be replaced by an impulsive outflow. In the mass-inflation regime ($2k_c < k_-$) the results are the same as for Reissner-Nordström: The conjecture indicates a NSCS, but there is actually a SCS in back reaction. In the intermediate regime ($k_c < k_- < 2k_c$) , which exhibits an infinite blue shift along the CH but no mass inflation, the conjecture and back-reaction solution agree upon a NSCS.

Why does the conjecture fail to indicate the correct type of singularity with impulsive outflow, both in the RN and RNdS spacetimes? In exact back reaction, the outgoing radiation creates a Weyl curvature singularity, which persists along the CH to the future of its point of origin. The stability conjecture can miss such purely Weyl singularities, because the stress-tensor scalars $T^{\mu}{}_{\mu}$ and $T^{\mu\nu}T_{\mu\nu}$ are related through the field equations to scalars in the Ricci tensor portion of the curvature only. (The SCS can't be detected by looking at the local null-dust density in these cases, since ρ remains finite.)

The persistence in mass inflation (and the singularity in Weyl curvature $\Psi_2 \sim m/r^3$) after the impulsive outflow is clear from Ori [21]. One can also prove the persistence from the Poisson-Israel mass equation [9]

$$\Box m = -(4\pi)^2 r^3 T_{\mu\nu} T^{\mu\nu}. \tag{5.25}$$

Use Kruskalized coordinates U, V in the cross-flow region [9], i.e. in the upper quadrant of Fig. 5.4, with $U = -\infty$ on the event horizon, $V = -\exp -k_- v$,

$$4\pi r^2 \rho_{out} = L_{out}(U), \tag{5.26}$$

and

$$4\pi r^2 \rho_{in} = L_{in}(V) = (k_o V)^{-2} dm_{in}(v)/dv. \tag{5.27}$$

Suppose the inward and outward fluxes are turned on at V_1, U_1, respectively. Then in the cross-flow region the mass function is

$$\begin{aligned} m(U,V) &= \int_{U_1}^{U} \int_{V_1}^{V} (-r' g'_{UV})^{-1} L_{out}(U') L_{in}(V') dU' dV' \\ &+ m_{in}(v) + m_{out}(u) - m_1. \end{aligned} \tag{5.28}$$

Let $m_{in}(v) = m_o - \alpha v^{-p}$, corresponding to the expected power-law tail, so

$$L_{in}(V) = (k_o V)^{-2} \frac{dm_{in}(V)}{dv} \sim V^{-2} ln^{-(p+1)} |-V|, \tag{5.29}$$

which diverges as $V \to 0$. If L_{out} is an impulsive shell confined between U_1 and $U_1 + \varepsilon$, then near the CH to the future of the shell,

$$m(U,V) = L_{out} \varepsilon \int_{V_1}^{V} (-r' g'_{U_1 V})^{-1} L_{in}(V') dV' \tag{5.30}$$

which diverges as $V \to 0$ in a manner independent of U. That is, mass inflation (and the Weyl scalar singularity) persist along the future CH, after the outward shell has already reached the CH. The outgoing shell creates both a scalar Ricci singularity and a Weyl singularity at the CH. The Weyl singularity persists along the future CH, but the scalar Ricci singularity does not. The conjecture is sensitive only to Ricci singularities, so it misses the persistence of the SCS along the future CH, and sees only a NSCS.

5.7 Assessment of the Cauchy horizon stability conjecture

The stability conjecture is, of course, no substitute for exact back-reaction calculations. It can, however, indicate whether singularities should form, and what type they should be, in the multitude of cases where exact calculations have not been carried out.

The conjecture is limited in that it cannot handle the influence of gravitational waves on a Cauchy horizon, because gravitational waves have no stress-energy tensor. The conjecture can also miss Weyl tensor singularities, so the conjecture can incorrectly judge the singularity type in some cases.

When supplemented with the behavior of null-dust density, the conjecture has so far been found to be perfectly reliable in indicating Ricci-tensor singularities. Electromagnetic-wave solutions of Maxwell's equations studied up to now, which in the case of black holes include the naturally continuous backscattering inside, are all consistent with the conjecture.

A full understanding of any limitations of the conjecture will involve a better understanding of the evolution of Weyl-tensor singularities. We are currently studying this question.

Acknowledgements

The authors are grateful to the workshop organizers Amos Ori, Lior Burko, and Liz Youdim, and to the Technion for its hospitality. One of us (D.A.K.) was partially funded by NSF grants Nos. PHY-9312448 and PH4-9602068.

Bibliography

[1] F. J. Tipler, C. J. S. Clarke, and G. F. R. Ellis, in *General Relativity and Gravitation*, edited by A. Held (Plenum, New York, 1980).

[2] R. Penrose, in *Battelle Rencontres*, ed. C. M. DeWitt and J. A. Wheeler (W. A. Benjamin, New York, 1968).

[3] J. M. McNamara, Proc. R. Soc. London **A 358**, 449 (1978); **A 364**, 121 (1978); S. Chandrasekhar and J. B. Hartle, Proc. R. Soc. London **A 284**, 301 (1982); N. Zamorano, Phys. Rev. D. **26**, 2564 (1982).

[4] Y. Gürsel, V. D. Sandberg, I. D. Novikov, and A. A. Starobinski, Phys. Rev. D. **19**, 413 (1979).

[5] R. Matzner, N. Zamorano, and V. Sandberg, Phys. Rev. D. **19**, 2821 (1979).

[6] Y. Gürsel, I. D. Novikov, V. D. Sandberg, and A. A. Starobinski, Phys. Rev. D. **20**, 1260 (1980)

[7] G. F. R. Ellis and B. G. Schmidt, Gen. Relativ. Grav. **8**, 915 (1977).

[8] T. M. Helliwell and D. A. Konkowski, Phys. Rev. D. **47**, 4322 (1993).

[9] E. Poisson and W. Israel, Phys. Rev. Lett. **63**, 1663 (1989); Phys. Lett. **B 233**, 74 (1989); Phys. Rev. D. **41**, 1796 (1990).

[10] R. H. Price, Phys. Rev. D. **5**, 2419 (1972); **5**, 2439 (1972).

[11] W. A. Hiscock, Phys. Lett. **A 83**, 110 (1981).

[12] D. A. Konkowski and T. M. Helliwell, Phys. Rev. D **50**, 841 (1994).

[13] P. R. Brady and E. Poisson, Class. Quantum Grav. **9**, 121 (1992); F. Mellor and I. Moss, Class. Quantum Grav. **9**, L43 (1992); P. R. Brady, D. Nunez, and S. Sinha, Phys. Rev. D **47**, 4239 (1993).

[14] R.-G. Cai and R.-K. Su, Phys. Rev. D **52**, 666 (1995).

[15] S. Droz, Phys. Lett. **A 191**, 211 (1994); R. Balbinot and P. R. Brady, Class. Quantum Grav. **11**, 1763 (1994); J. S. F. Chan and R. B. Mann, Phys. Rev. D **50**, 7376 (1994).

[16] R.-G. Cai, Phys. Rev. D **53**, 5696 (1996).

[17] R.-G. Cai and D.-F. Zhao, Phys. Lett. **A 208**, 281 (1995).

[18] T. M. Helliwell and D. A. Konkowski, Phys. Rev. D **51**, 5517 (1995).

[19] D. A. Konkowski and T. M. Helliwell, Phys. Rev. D **54**, 7898 (1996).

[20] D. A. Konkowski and T. M. Helliwell, Class. Quantum Grav. **6**, 1847 (1989).

[21] A. Ori, Phys. Rev. Lett. **67**, 789 (1991).

APPLYING A STABILITY CONJECTURE TO PLANE–WAVE CAUCHY HORIZONS

D. A. Konkowski [a] and T. M. Helliwell [b]

a) Department of Mathematics, U.S. Naval Academy, Annapolis, Maryland 21401, USA
b) Department of Physics, Harvey Mudd College, Claremont, California 91711, USA

Abstract

Plane-wave spacetimes and their accompanying Cauchy horizons are discussed. The behavior of test fields (scalar, electromagnetic, and null dust) is investigated along with corresponding exact solutions to the field equations. These results are used to test a previously-developed stability conjecture and a stability theorem given. It is shown that the stability conjecture reliably indicates Ricci-tensor singularities, but that it can miss Weyl-tensor singularities. This is not a problem in the case of plane waves as long as no Weyl-tensor singularities are inadvertantly introduced on the initial-value surface.

6.1 Introduction

In Einstein-Maxwell theory plane waves focus light [1, 2]. A gravitational wave will focus a light beam to a line, while a purely electromagnetic wave will focus it to a point. The future light cone of a point in front of the wave will be focused and appear as the past light cone of this line or point. This unusual behavior led Penrose [2] to discover a remarkable property of plane wave spacetimes: they have no global Cauchy hypersurfaces. If you try to set up initial values for a plane wave, no global spacelike hypersurface will work. As Griffiths [1] explains, the spacelike hypersurface would need to lie completely in the past of any future null cone from any point on the spacelike surface, but the plane wave focuses these light cones and turns them over, which bends the prospective Cauchy surface so it cannot extend to infinity. (An excellent illustration of this phenomenon is given in Penrose [2] and reproduced in Griffiths [1].) Therefore general relativistic plane wave spacetimes are not globally hyperbolic. Some are free of curvature singularities and have Cauchy horizons behind the propagating wave, while others have non-scalar curvature singularities [3]. Here we study the stability of plane-wave Cauchy horizons in the context of a previously developed Cauchy horizon conjecture (see the paper by Helliwell and Konkowski in this volume). In this paper, various fields, all of which propagate in the same direction as the gravitational wave, are added to plane-wave spacetimes with Cauchy horizons. Yurtsever [4] showed that plane waves with Cauchy horizons remain nonsingular when perturbed by either linear or nonlinear plane-symmetric radiation propagating parallel to the original wave. Here we use test fields and our stability conjecture to indicate Cauchy-horizon stability in an exact back-reaction solution. The beauty of these spacetimes from our viewpoint is that the back reaction can be solved in each case and further insight gained into the applicability of the stability conjecture. The plan of this paper is as follows: In Section 6.2 we define singularity types and briefly review the stability conjecture. In Section 6.3 we review the properties of plane-wave spacetimes, including purely gravitational waves. In Section 6.4 we introduce test fields, their stress-energy tensors, and the corresponding exact back-reaction solutions. In Section 6.5 we present a stability theorem for plane-wave spacetimes, and in Section 6.6 we give our conclusions.

6.2 Singularity classification and stability conjectures

We use a singularity classification scheme based on one devised by Ellis and Schmidt [3]. They classified singularities in maximal spacetimes into three basic types: quasiregular, nonscalar curvature, and scalar curvature. The mildest singularity is quasiregular and the strongest is scalar curvature. At a scalar curvature singularity, physical quantities such as energy density and tidal forces diverge in the frames of all observers who approach the singularity. At a nonscalar curvature singularity, many observers who approach the singularity also experience divergent tidal forces, but there exist curves through each point arbitrarily close to the singularity such that observers moving on these curves experience perfectly regular tidal forces [3]. For a quasiregular singularity, no observers see physical quantities diverge, even though their world lines end at the singularity in a finite proper time. Our version of the Ellis and Schmidt classification scheme can be expressed mathematically. We define singular points simply as the end points of incomplete geodesics in maximal spacetimes; Ellis and Schmidt use instead a b-boundary construction to define the singular points. In our scheme a singular point q is a quasiregular singularity if all components of the Riemann tensor R_{abcd} evaluated in an orthonormal frame parallel propagated along an incomplete geodesic ending at q are C^0 (or C^{0-}). In other words, the Riemann tensor components tend to finite limits (or are bounded). On the other hand, a singular point q is a curvature singularity if some components are not bounded in this way. If all scalars in g_{ab}, the antisymmetric tensor η_{abcd}, and R_{abcd} nevertheless tend to a finite limit (or are bounded), the singularity is nonscalar, but if any scalar is unbounded, the point q is a scalar curvature singularity. We have investigated whether mild (quasiregular and nonscalar curvature) singularities are a stable feature of spacetime, and have formulated a conjecture to assist in answering the question. The conjecture, its predictions, and comparisons with those few cases in which back-reaction calculations have been carried out, are summarized in reference [5]. The stability conjecture predictions agree with all of the exact solutions which have been found. We have extended our techniques to study also the stability of Cauchy horizons. The Cauchy horizon stability conjecture [6] is as follows.

For all maximally extended spacetimes with Cauchy horizons, the back reaction due to a field (whose test-field stress-energy tensor is $T_{\mu\nu}$) will affect the horizon in the following manner. (1) If $T^{\mu}{}_{\mu}$, $T^{\mu\nu}T_{\mu\nu}$, and any null-dust density ρ are finite, and if the stress-energy tensor $T_{(ab)}$ in all parallel-propagated orthonormal frames (PPON frames) is finite, then the Cauchy horizon remains nonsingular. (2) If $T^{\mu}{}_{\mu}$, $T^{\mu\nu}T_{\mu\nu}$, and any null-dust density ρ are finite, but $T_{(ab)}$ diverges in some PPON frame, then a nonscalar curvature singularity will be formed at the Cauchy horizon. (3) If $T^{\mu}{}_{\mu}$, $T^{\mu\nu}T_{\mu\nu}$, or any null-dust density diverges, then a scalar curvature singularity will be formed at the Cauchy horizon.

The conjecture predicts not only whether a singularity will form if the applied field is allowed to influence the geometry; it predicts also which *type* of singularity will form. It is usually not possible to consider all PPON frames in (2), so in practice the conjecture indicates only instability, not stability, in the formation of nonscalar curvature singularities. The conjecture and its applications are discussed more fully in the paper by Helliwell and Konkowski in this volume.

6.3 Plane waves

Plane-wave spacetimes can be described by the metric [1]

$$ds^2 = -2dudv + dx^2 + dy^2 + H_{ij}(u)x^i x^j du^2. \tag{6.1}$$

Here $i, j = 1, 2$ with $x^1 = x$ and $x^2 = y$; also $H_{ji} = H_{ij}$. The entire spacetime manifold is covered in a 1-1 fashion by a single coordinate patch with all four coordinates ranging from $-\infty$ to $+\infty$. The possibly nonzero elements of the Riemann tensor are (in coordinates u, v, x, y)

$$R_{0202} = -H_{xx},\ R_{0303} = -H_{yy},\ R_{0203} = -H_{xy}, \tag{6.2}$$

and others obtained from these by the usual symmetry operations. The only possibly nonzero component of the Ricci tensor is

$$R_{00} = -(H_{xx} + H_{yy}). \tag{6.3}$$

All scalar curvature invariants of the Riemann tensor, including higher derivative curvature invariants, are zero, so plane-wave spacetimes cannot have scalar curvature singularities [3].

Linearized plane waves have exactly the same freedom in their construction as full general relativistic plane waves, and have exactly the same degree of symmetry. Plane waves admit a five-parameter group of motions which acts transitively on all null geodesics except for those which are parallel to the propagation direction of the wave [3]. The invariance of the five-parameter group of isometries means that the timelike geodesic $\{x = y = 0,\ u = -v = \frac{1}{\sqrt{2}}s\}$ is equivalent to any other timelike geodesic. Without loss of generality we can therefore choose a PPON frame with $x = y = 0$ and components

$$\begin{gathered} e^{\mu}_{(0)} = (\dot{u}, \dot{v}, \dot{x}, \dot{y}) = (\alpha, \alpha/2, 0, 0)\ e^{\mu}_{(1)} = (\alpha, -\alpha/2, 0, 0), \\ e^{\mu}_{(2)} = (0, 0, 1, 0),\ e^{\mu}_{(3)} = (0, 0, 0, 1) \end{gathered} \tag{6.4}$$

where α is an arbitrary positive constant. The nonzero elements of the Riemann tensor in this PPON frame are

$$\begin{gathered} R_{(0202)} = R_{(0212)} = R_{(1212)} = -\alpha^2 H_{xx} \\ R_{(0303)} = R_{(0313)} = R_{(1313)} = -\alpha^2 H_{yy} \\ R_{(0203)} = R_{(0213)} = R_{(1203)} = R_{(1213)} = -\alpha^2 H_{xy} \end{gathered} \tag{6.5}$$

and others obtained by symmetry operations; these describe the tidal-force components experienced by an observer falling freely in the spacetime. A nonscalar curvature singularity (NSCS) clearly results if and only if at least one of the $H_{ij}(u)$ diverges at a point. A plane-wave spacetime which does not contain a NSCS will contain a Cauchy-Killing horizon. As mentioned in the Introduction, Penrose has shown [2] that no spacelike Cauchy surface exists in plane-wave spacetimes, so one cannot imbed a plane wave globally in any hyperbolic normal pseudo-Euclidean space.

Now we consider the special case of plane-wave spacetimes with constant polarization. These are waves in which $H_{xy}(u) = 0$, so in Brinkmann form the metric is [1, 7, 8]

$$ds^2 = -2du dv + dx^2 + dy^2 + [H_{xx}x^2 + H_{yy}y^2]du^2. \tag{6.6}$$

The Rosen form of the metric is [1, 9]

$$ds^2 = -2du d\bar{v} + F^2(u) d\bar{x}^2 + G^2(u) d\bar{y}^2 \tag{6.7}$$

where the coordinate transformation between the two metrics is

$$\bar{v} = v - \frac{1}{2}\left(x^2\frac{F'}{F} + y^2\frac{G'}{G}\right),\quad \bar{x} = \frac{x}{F},\quad \bar{y} = \frac{y}{G}, \tag{6.8}$$

with $H_{xx}(u) = F''/F$, $H_{yy}(u) = G''/G$. Here F'' is the second derivative of $F(u)$, etc. The null surface $S = \{u = f\}$ on which either $F(u)$ or $G(u)$ vanishes is a Cauchy horizon for the spacetime region $\{u < f\}$. The surface S is also a Killing horizon: At least one of the plane-symmetric generators (depending upon whether $F(u)$ or $G(u)$ vanishes at $u = f$) becomes null on S. The null surface S is regular as long as both

H_{xx} and H_{yy} are bounded at $u = f$. If either H_{xx} or H_{yy} diverges as $u \to f$, then S is a null nonscalar curvature singularity, with all scalar curvature invariants equal to zero.

If the plane wave is purely gravitational, then $R_{00} = -(H^g_{xx} + H^g_{yy}) = 0$, so the metric coefficients obey $H^g_{yy} = -H^g_{xx}$. In Rosen coordinates the gravitational wave has metric coefficients F(u) and G(u), where $F''/F = G''/G$. The null surface on which $F(u)$ or $G(u)$ vanishes is a Cauchy-Killing horizon; we choose this to occur at $u = l$. If the plane wave is impulsive, for example, with a delta-function pulse at $u = 0$, the spacetime region behind the wave has the metric

$$ds^2 = -2dud\bar{v} + (1+u)^2 d\bar{x}^2 + (1-u)^2 d\bar{y}^2 \text{ (Rosen form)} \tag{6.9}$$

$$ds^2 = -2dudv + dx^2 + dy^2 \text{ (Brinkmann form)} \tag{6.10}$$

This region $u > 0$ of the spacetime is flat with $H^g_{xx} = F''/F = 0$ and $H^g_{yy} = G''/G = 0$; A Cauchy-Killing horizon occurs at $u = 1$.

6.4 Test fields and exact solutions

In this section we add scalar fields, electromagnetic fields, and null dust to a background gravitational plane wave. The added fields all propagate parallel to the gravitational wave, so the exact back-reaction solutions still have the plane wave metric, but with different metric coefficients. That is, $F(u)$ and $G(u)$ will be different than before, although each will reduce to the gravitational-wave F and G if the field is turned off. The same is true for H_{xx} and H_{yy} in the Brinkmann form. Predictions of the stability conjecture will be worked out using test fields, and then compared with exact solutions of Einstein's equations. The curvature scalar is zero for plane waves , so the Einstein equations for the exact solutions are $R_{(ab)} = 8\pi T_{(ab)}$ in a PPON frame. The PPON-frame Ricci tensor is

$$R_{(ab)} = e^{\mu}_{(a)} e^{\nu}_{(b)} R_{\mu\nu} = e^0_{(a)} e^0_b R_{00} = \alpha^2 R_{00} \tag{6.11}$$

where $a, b = 0, 1$; here

$$R_{00} = -(H_{xx} + H_{yy}) = -\left(\frac{F''}{F} + \frac{G''}{G}\right). \tag{6.12}$$

Therefore $R_{(ab)} = \alpha^2 R_{00} I_{(ab)}$, where

$$I_{(ab)} = \begin{pmatrix} 1 & 1 \\ 1 & 1 \end{pmatrix}. \tag{6.13}$$

The PPON-frame stress-energy tensor takes the form

$$T_{(ab)} = \left(\frac{\alpha^2}{8\pi}\right) R_{00} I_{(ab)} = \alpha^2 T_{00} I_{(ab)}, \tag{6.14}$$

where T_{00} is the coordinate-frame energy density. All curvature scalars vanish, so no scalar curvature singularities can occur. Nonscalar curvature singularities are possible, however; these can be Ricci or Weyl singularities depending upon the properties of the Ricci tensor $R_{(ab)}$ and Weyl tensor $C_{(abcd)}$ in PPON frames. The only nonzero components of the Weyl tensor are

$$C_{(0202)} = -C_{(0303)} = -\frac{\alpha^2}{2}(H_{xx} - H_{yy}) = -\frac{\alpha^2}{2}\left(\frac{F''}{F} - \frac{G''}{G}\right) \qquad (6.15)$$

and those obtained by symmetry operations. Note that $R_{(ab)}$, $T_{(ab)}$, and $C_{(abcd)}$ differ from their coordinate counterparts by at most a constant. In the following sections we will use this fact and often refer only to the coordinate components.

We consider three null fields moving parallel to the gravitational wave: massless scalar fields, electromagnetic fields, and null dust. A massless scalar field Φ is a solution of $\Box\Phi = 0$. In Brinkmann coordinates any field of the form $\Phi = \Phi(u)$ is a solution. In this case the only nonzero element of the stress-energy tensor

$$T_{\mu\nu} = \frac{1}{4\pi}\left[\Phi_{,\mu}\Phi_{,\nu} - g_{\mu\nu}g^{\alpha\beta}\Phi_{,\alpha}\Phi_{,\beta}\right] \qquad (6.16)$$

is $T_{00} = \frac{1}{4\pi}(\Phi')^2$, where $\Phi' = d\Phi/du$. A corresponding source-free electromagnetic field A^μ is a solution of $A^{\mu;\nu}{}_{;\nu} = 0$. Wave solutions moving parallel to the gravitational wave have the form

$$A^\mu(u) = (0, f_1(u), f_2(u), f_3(u)), \qquad (6.17)$$

where f_1 , f_2 , f_3 are arbitrary differentiable functions of u. The only possibly nonzero elements of the electromagnetic field tensor are $F_{02} = -F_{20} = f_2'(u)$ and $F_{03} = -F_{30} = f_3'(u)$. The only nonzero element of the stress-energy tensor

$$T_{\mu\nu} = \frac{1}{4\pi}\left[F_{\mu\alpha}F_\nu{}^\alpha - \frac{1}{4}g_{\mu\nu}F_{\alpha\beta}F^{\alpha\beta}\right] \qquad (6.18)$$

is $T_{00} = \frac{1}{4\pi}\left[(f_2'(u))^2 + (f_3'(u))^2\right]$.
Null dust moving parallel to the wave has density $\rho(u)$ and four-velocity $u^\mu = (0,1,0,0)$. The only nonzero element of the stress-energy tensor $T_{\mu\nu} = \rho(u)u_\mu u_\nu$ is $T_{00} = \rho(u)$.

In each of these cases, if the field $\Phi(u)$, $A^\mu(u)$, or $\rho(u)$ has the same form in the test-field case as in the back-reaction case, the stress-energy tensors have the same form as well. In the PPON frame with unit vectors as given in Eq. (6.4) , the stress-energy tensor in each case is $T_{(ab)} = \alpha^2 T_{00} I_{(ab)}$, where the matrix $I_{(ab)}$ is given in Eq. (6.14). So also the PPON stress-energy tensors have the same form whether they are constructed from test fields or from exact back-reaction fields. That is because the stress-energy tensor of a field can only depend upon the field, its derivatives, and the metric tensor. But in a PPON frame the metric tensor becomes

$\eta_{(ab)} = diag(-1,1,1,1)$, the same for the test-field case as for the back-reaction case. Therefore the PPON frame stress-energy tensors must have the same form in the test-field and back-reaction cases if the fields themselves have the same form.

6.5 A plane-wave singularity theorem

In this section we compare the results of test fields and exact solutions. It is important to realize that because of linearity [1], if a parallel-propagating wave is added to a background gravitational plane wave, whatever properties the metric has on the initial hypersurface are retained for all future times. So if no singularity is introduced, none will develop, whereas if a singularity is introduced it will remain for all time. If a finite stress-energy tensor is introduced on the initial surface, the stability conjecture predicts that no singularity will occur in the backreaction solution, and that is true provided no Weyl singularity is introduced on the initial hypersurface.

We consider the several possible classes of metrics based upon their behavior near the Cauchy horizon of the background gravitational plane wave. Since $R_{(ab)}$, $T_{(ab)}$, and $C_{(abcd)}$ are related to their coordinate counterparts by only a constant multiple, we simplify our discussion by referring only to the coordinate tensor components. We take the Cauchy horizon to be located at $u = 1$, and consider the singular or nonsingular behavior of $H^f_{xx}(u)$ and $H^f_{yy}(u)$ as $u \to 1$, where $H^f_{xx}(u)$ and $H^f_{yy}(u)$ are the Brinkmann-form metric coefficients when the field has been added to the spacetime. There are several possibilities:

1. $H^f_{xx}(u)$ and $H^f_{yy}(u)$ are both nonsingular as $u \to 1$. Then the Ricci tensor $R_{00} = -(H^f_{xx} + H^f_{yy})$, the Weyl tensor $C_{0202} = -\frac{1}{2}(H^f_{xx} - H^f_{yy})$, and the PPON-frame tensors $R_{(ab)}$ and $C_{(abcd)}$, are all nonsingular. The back-reaction stress-energy tensor $T_{(ab)} = \frac{1}{8\pi} R_{(ab)}$ is also nonsingular, as is the test-field stress-energy tensor. The conjecture therefore correctly predicts that the CH remains nonsingular. A simple example is the case where the Rosen-form metric coefficients are $F = 1 + u$ and $G(u) = 1 - u$, so that

$$H^f_{xx} = \frac{F''}{F} = 0 \text{ and } H^f_{yy} = \frac{G''}{G} = 0, \tag{6.19}$$

   so that the corresponding stress-energy tensor $T_{00} = -\frac{1}{8\pi}(H^f_{xx} + H^f_{yy}) = 0$.

   For a massless scalar field Φ for which $T_{00} = \frac{1}{4\pi}\left[\Phi'(u)\right]^2$, the added scalar field is a constant, independent of u. For an electromagnetic field, all elements of the field tensor are similarly constant. The conjecture predicts that no singularity will form if such constant fields are added, and none is formed.

2. Exactly one of $H^f_{xx}(u)$ and $H^f_{yy}(u)$ is singular as $u \to 1$. In this case both R_{00} and C_{0202} are singular, as are the back-reaction and test-field stress-energy tensors. The conjecture predicts a nonscalar curvature singularity, and there is

one, both in the Ricci and Weyl tensors. An example of this case is the choice of Rosen-form metric coefficients $F = (1 - u^2)$ and $G = 1$, so that $H^f_{xx} = \frac{-2}{1-u^2}$ and $H^f_{yy} = 0$. The stress-energy tensor

$$T_{00} = -\frac{1}{8\pi}(H^f_{xx} + H^f_{yy}) = \frac{1}{4\pi}\left(\frac{1}{1-u^2}\right) \tag{6.20}$$

is singular at $u = 1$.

If the field is a scalar field, then $\Phi(u) = \sin^{-1}(u)$; an electromagnetic field with this stress-energy has (for example) $f_2(u) = \sin^{-1}(u)$, $f_2(u) = 0$. In either case the field is regular at $u = 1$ although the stress-energy tensor diverges. If only null dust is present, its density $\rho(u) = \frac{1}{4\pi(1-u^2)}$ diverges at $u = 1$.

3. Both $H^f_{xx} + H^f_{yy}$ and $H^f_{xx} - H^f_{yy}$ are singular as $u \to 1$. In this case there are three possibilities:

   (a) Both $H^f_{xx} + H^f_{yy}$ and $H^f_{xx} - H^f_{yy}$ are singular, so both the Ricci and Weyl tensors are singular, in a PPON frame as well as in the coordinate frame. The back-reaction and test-field stress-energy tensors in a PPON frame are both singular as well, so the conjecture correctly predicts the singularity (NSCS). An example is the choice $F = (1 - u^2)$, $G = (1 - u^3)$, giving $H^f_{xx} = \frac{-1}{1-u^2}$ and $H^f_{yy} = \frac{-6u}{1-u^3}$.

   (b) $H^f_{xx} + H^f_{yy}$ is singular, but $H^f_{xx} - H^f_{yy}$ is nonsingular. The Ricci tensor is singular, but not the Weyl tensor. The conjecture correctly predicts the singularity (NSCS). An example is $F = G = (1 - u^2)$, for which $H^f_{xx} = H^f_{yy} = \frac{-2}{1-u^2}$. The Ricci tensor and stress-energy tensor diverge, but the Weyl tensor is zero everywhere.

   (c) $H^f_{xx} + H^f_{yy}$ is nonsingular, but $H^f_{xx} - H^f_{yy}$ is singular. It follows that the Ricci tensor is nonsingular, as is the stress-energy tensor is a PPON frame. The conjecture therefore predicts that the Cauchy horizon will remain nonsingular. However, the Weyl tensor is singular, so the stability conjecture fails in this case. The reason for the failure is that the stability conjecture only has the ability to predict Ricci-tensor singularities, since the conjecture relies upon the behavior of the stress-energy tensor. An example is provided by the Rosen-form metric coefficients

$$F = 1 + \beta(1-u)^{3/2} \text{ and } G = 1 - \beta(1-u)^{3/2}. \tag{6.21}$$

   The corresponding Brinkmann-form coefficients are

$$H^f_{xx} = \frac{(3\beta/4)(1-u)^{-1/2}}{1+\beta(1-u)^{3/2}} \text{ and } H^f_{yy} = \frac{-(3\beta/4)(1-u)^{-1/2}}{1-\beta(1-u)^{3/2}}, \tag{6.22}$$

each of which is singular as $u \to 1$. The Ricci tensor

$$R_{00} = \frac{3\beta^2}{2} \frac{(1-u)}{1-\beta^2(1-u)^3} \tag{6.23}$$

is nonsingular, but the Weyl tensor

$$C_{0202} = -\frac{3\beta}{4} \frac{(1-u)^{-1/2}}{1-\beta^2(1-u)^3} \tag{6.24}$$

is singular as $u \to 1$. The non-zero stress-energy tensor element is $T_{00} = \frac{1}{8\pi} R_{00} = \frac{1}{4\pi} [\Phi'(u)]^2$ in the case of a scalar field, from which the field $\Phi(u)$can be found by integration. For an electromagnetic field the stress-energy tensor is $T_{00} = \frac{1}{4\pi}[(f_2')^2 + (f_3')^2]$, from which possible functions $f^2(u)$ and $f^3(u)$ can also be found. In the case of null dust, the dust density is $\rho(u) = \frac{3\beta^2}{16\pi} \frac{(1-u)}{1-\beta^2(1-u)^3}$. In any of these cases T_{00} is nonsingular at the Cauchy horizon, so the conjecture by itself fails to indicate the existence of a nonscalar curvature singularity of the Weyl tensor. As stated before, the Weyl tensor singularity already exists on the initial spacelike surface and so could be eliminated by simply outlawing it at the beginning.

It is clear that the stability conjecture can reliably indicate singularities in the Ricci tensor, because of the field-equation connection between the Ricci tensor and the stress-energy tensor. The Weyl tensor is not connected in this direct way, so a singularity in the Weyl tensor cannot be detected by inspection of the stress-energy tensor.

Using the above results, we can prove a theorem [10] for plane waves.

Theorem

Let $T_{(ab)}$ be a PPON-frame test field stress-energy tensor.

1. If $T_(ab)$ diverges at the CH, there is a NSCS in the exact back reaction solution, in agreement with the stability conjecture.
2. If $T_(ab)$ is nonsingular at the CH, then in the back-reaction spacetime either
   (a) there is no NSCS at the CH, or
   (b) there is a NSCS in only the Weyl portion of the Riemann tensor, with no divergence of the Ricci tensor.

6.6 Conclusions

Plane-wave spacetimes are the largest class of exact back-reaction solutions we have used so far as tests of the Cauchy-horizon stability conjecture. Plane waves of massless scalar fields, electromagnetic fields, and null dust can be added to a gravitational plane wave. The exact back reaction problem is easily solved if the added fields propagate in the same direction as the background gravitational wave. The results show that the behavior of the test-field stress-energy tensor unerringly predicts the appearance or nonappearance of Ricci-tensor nonscalar curvature singularities in the exact back reaction solution for plane waves. However, the results show also that test-field stress-energy tensors cannot predict the appearance or nonappearance of Weyl tensor singularities. If a plane-wave spacetime has a Weyl-tensor singularity, it must have been imposed by the fields placed on an initial Cauchy surface. Therefore in these spacetimes one need only check for initial Weyl singularities. If there are none, the conjecture works for all applied fields.

Acknowledgements

We thank the organizers of the Haifa conference for inviting us to a very stimulating meeting. One of us (D.A.K.) was partially funded by NSF grants Nos. PHY-9312448 and PHY-9602068 .

Bibliography

[1] J. B. Griffiths, *Colliding Plane Waves in General Relativity* (Oxford, Clarendon Press, 1991), p.p. 17-25.

[2] R. Penrose, Rev. Mod. Phys. **37**, 215 (1965).

[3] G. F. R. Ellis and B. G. Schmidt, Gen. Relativ. Gravit. **8**, 915 (1977).

[4] U. Yurtsever, Phys. Rev. D **36**, 1662 (1987); Class. Quantum. Grav. **10**, L17 (1993).

[5] T. M. Helliwell and D. A. Konkowski, Phys. Rev. D **51**, 5517 (1995).

[6] T. M. Helliwell and D. A. Konkowski, Phys. Rev. D **47**, 4322 (1993).

[7] D. Kramer, H. Stephani, M. A. H. MacCallum, and E. Herlt, *Exact Solutions of Einstein's Field Equations* (Cambridge University Press, Cambridge, 1980).

[8] M. W. Brinkmann, Proc. Natl. Acad. Sci. USA **9**, 1 (1923).

[9] N. Rosen, Phys. Z. Sowjet. **12**, 366 (1937).

[10] T. M. Helliwell and D. A. Konkowski, "Stability of Plane Wave Spacetimes", in preparation.

INTERNAL STRUCTURE OF NON–ABELIAN BLACK HOLES

D. V. Gal'tsov [a], E. E. Donets [b] and M. Yu. Zotov [c]

a) Department of Theoretical Physics, Moscow State University, 119899 Moscow, Russia
b) Laboratory of High Energies, JINR, 141980 Dubna, Russia
c) Skobeltsyn Institute of Nuclear Physics, Moscow State University, 119899 Moscow, Russia

Abstract

Recent results concerning an internal structure of static spherically-symmetric non-Abelian black holes in the Einstein–Yang–Mills (EYM) theory and its generalizations including scalar fields are reviewed and discussed with an emphasis on the problem of a generic singularity in black holes. It is argued that in the theories admitting a violation of the naive no-hair conjecture the structure of singularity is essentially affected by the "hair roots". This invalidates an image of a non-Abelian black hole as a Schwarzschild black hole sitting inside the soliton. We give an analytic description of the generic oscillatory approach to the singularity in the pure EYM theory in terms of a divergent discrete sequence and show that the mass function is exponentially growing "in average". The second type of a generic approach to the singularity in hairy black holes is a "power-law mass inflation" which is realized in the theories including scalar fields. Both singularities are spacelike and no Cauchy horizons are met in the full interior region in conformity with the Strong Cosmic Censorship conjecture. Black holes violating this conjecture exist only for certain discrete values of the event horizon radius thus forming a subset of zero measure.

7.1 Introduction

The no-hair conjecture [1] has played an important role in understanding the nature of black holes. As a result of the early investigation of various field theories coupled to gravity it was generally believed that the only fields which may extend outside the event horizon (apart from external ones not related to the black hole itself) are those associated with the conserved or topological charges carried by the hole.

Later a notable counter example to the naive no-hair conjecture was found in the framework of the Einstein–Yang–Mills (EYM) theory. Although no classical glueballs may exist in the flat spacetime because of purely repulsive nature of the constituent vector fields, such particle-like objects become possible when gravitational attraction is taken into account (Bartnik–McKinnon (BK) solutions [2]). It was shown [3] that the lower mass BK particle topologically is similar to the sphaleron of the Weinberg–Salam theory, with a substantial difference, however, due to existence in the BK case of the gravitational negative mode. Historically, just the gravitational instability of the BK solutions was discovered first [4] and later invoked as an argument favoring the sphaleron interpretation of the BK solutions [5]. More detailed analysis showed, however, that the *gravitational* (even parity) negative modes of the BK particles had nothing to do with an expected sphaleron picture. Genuine sphaleronic features of BK solutions are manifest via the *odd-parity* negative modes along the vacuum-to-vacuum direction in the functional space [6]. Physically the lower mass BK solution may be seen as interpolating between the neighboring topologically distinct YM vacua in the EYM theory. It carries neither electric nor magnetic charge, while the Chern–Simons number $1/2$ is naturally associated with it [3, 7].

It was soon realized that the EYM theory also admits the black hole counterparts to the BK solutions [8, 9]. The $SU(2)$ EYM black holes form a two-parameter family labeled by a continuously varying radius of the event horizon r_h and the number n of oscillations of the YM field in the strip $[-1,1]$ bounded by the neighboring YM vacua values. Since no asymptotic charges are present, these black holes apparently violate the no-hair principle. Other solutions sharing the same property were found in the non-Abelian gauge theories including scalar fields [10, 11, 12] as well as in the gravity coupled Skyrme model. Meanwhile, a common feeling reasonably persisted that this kind of hair should be distinguished from the "allowed" hair associated with the global charges. It was suggested [13] to interpret a non-Abelian field structure outside the event horizon as a "wig" rather than a genuine hair. Indeed, the EYM black hole can lose its YM hair as a result of the gravitational instability leaving a bare Schwarzschild black hole. As far as the black holes inside the magnetic monopoles are concerned, they were often thought of (at least for the horizon radius much smaller than the monopole size) as tiny Schwarzschild black holes sitting inside the solitons [14]. If so, the inside singularity should remain unaffected by the wig.

However, little was known until recently about the realistic internal structure of non-Abelian black holes (apart from some qualitative discussion in [9]). Closer investigation [15] has shown that the idea of a Schwarzschild black hole sitting inside regular solitons is essentially incorrect. The interior structure and the character of the singularity inside the EYM black hole with an arbitrary (continuously varying) radius of the horizon are strikingly different from those described by the Schwarzschild geometry. Both the Schwarzschild and Reissner–Nordström type singularities (predicted in [9]) are encountered indeed, but only for some discrete values of the horizon radius. This "second quantization" in non-Abelian black holes follows mathematically from the same kind of a non-linear boundary value problem (now in the interior region) as the "first quantization" of the BK particle mass. In this sense the picture of a

Schwarzschild black hole inside the BK particle covers only the set of solutions of zero measure in the parameter space. Although it could be anticipated by continuity that the non-Abelian hair should penetrate inside the event horizon, an essential role of hair in the formation of the singularity seems not to have been clearly realized before. A generic singularity in the spherically symmetric t-independent EYM coupled system turns out to be spacelike and of the oscillatory nature as was often suggested in view of the Bianchi IX analysis by Belinskii, Khalatnikov and Lifshitz (BKL) [16]. In the black hole context, however, the cosmological counterpart is given by the (non-Bianchi) closed Kantowski–Sachs cosmology, and the nature of oscillations observed is rather different from that in the BKL case.

Our results [15] concerning the EYM generic black hole interior solutions were confirmed by Breitenlohner et al. [17] who also attempted to extend the analysis to monopole black holes in the EYM-Higgs (EYMH) theory with the triplet Higgs. In the latter case no oscillations were observed numerically, and the behavior of the mass function near the singularity was found to be monotonous. This behavior was interpreted in [17] as exhibiting an exponential "mass inflation". Soon after we have found an analytical solution for the mass function [18] in the $SU(2)$ EYMH theories both with the (real) triplet and (complex) doublet Higgs using a consistent truncation of the system of equations near the singularity. Such a truncation is possible also in the EYM–dilaton (EYMD) theory [19], in both cases the behavior of the metric near the singularity is dominated by the kinetic (gradient) contribution of the scalar field. In terms of the radial dependence of the mass function one finds a *power-law* divergence towards the singularity. Therefore no mass-inflation (in the usual sense [20]) is detected inside the EYMH black holes where the singularity is dominated by the scalar field. A similar scalar dominated asymptotic behavior was found earlier in the framework of the Kantowski–Sachs cosmology [21] driven by a non-linear scalar field (in the Abelian model). In this latter case no black hole counterpart may exist because of the no-hair theorem for a non-linear one-component scalar field, but locally the solution is the same. Our formula for the power-law mass function in the scalar-dominated regime was reproduced later in [22] (see this volume) though without necessary restrictions on the power index following from the consistency of the truncation. It is worth noting that the domains of variation of the power index (depending on the horizon radius of the black hole) are different in the EYMH and EYMD cases, so in the Kantowski–Sachs interpretation the corresponding singularities look differently. The scalar-dominated singularity of the "power-law mass-inflationary" type inside spherical black holes in various field models including scalar fields seems to be a rather general phenomenon. This singularity is spacelike in conformity with the Strong Cosmic Censorship. Both oscillatory and the power-law singularities in hairy black holes therefore support this principle by the genericity argument.

The plan of the talk is as follows. First we discuss the local solutions near the singularity which may be seen as dressed Schwarzschild and Reissner–Nordström singularities. Then (Sec.7.3) we give an analytic description of the oscillatory approach to the singularity in the generic EYM black holes. In Sec. 7.4 it is shown that no exponential mass inflation develops inside the EYMH and EYMD black holes where the typical power-law divergence of the mass function is met. We conclude with a brief discussion of a new role of the black hole hair which was revealed in the investigation of black hole interiors.

7.2 Hairy Schwarzschild and Reissner–Nordström singularities

We start with the pure $SU(2)$ EYM system

$$S = \frac{1}{16\pi}\int \{-R + F^2\}\sqrt{-g}d^4x\,, \tag{7.1}$$

where F is the $SU(2)$ field, assuming the metric to be invariant under the time translations and the $SO(3)$ spatial rotations:

$$ds^2 = \frac{\Delta\sigma^2}{r^2}dt^2 - \frac{r^2}{\Delta}dr^2 - r^2(d\theta^2 + \sin^2\theta d\phi^2)\,. \tag{7.2}$$

Here the metric functions Δ, σ depend only on r. This parameterization of spacetime is suitable until a turning point of r is reached. It happens that for asymptotically flat solutions there are no such points neither in exterior, nor in interior regions up to the curvature singularity at $r = 0$. Such points exist, however, for asymptotically non-flat solutions, in which case another chart should be used [17]. Here we will not be dealing with *a priori* asymptotically non-flat solutions, so the coordinates (7.2) are good both in the exterior and interior regions. Of course one should keep in mind that in the region where $\Delta < 0$ the radial coordinate r is timelike, while t is spacelike, so that t-independence means spatial homogeneity of the spacetime rather than staticity.

As usual, we choose the t-independent spherically symmetric magnetic ansatz for the YM potential

$$A = (W(r) - 1)\,(T_\varphi d\theta - T_\theta \sin\theta d\varphi)\,, \tag{7.3}$$

where $T_{\varphi,\theta}$ are spherical projections of the $SU(2)$ generators. The value $W = 1$ is a trivial YM vacuum, while $W = -1$ corresponds to the vacuum of the next topological sector (this state is related to the trivial vacuum by a "large" gauge transformation).

The field equations consist of a coupled system for W, Δ

$$\Delta U' + \left(1 - \frac{V^2}{r^2}\right)W' = \frac{WV}{r}\,, \tag{7.4}$$

$$\left(\frac{\Delta}{r}\right)' + 2\Delta U^2 = 1 - \frac{V^2}{r^2}\,, \tag{7.5}$$

where $V = (W^2 - 1)$, $U = W'/r$, and a decoupled equation for σ:

$$\frac{d}{dr^2} \ln \sigma = U^2 . \tag{7.6}$$

These equations admit black hole solutions in the domain $r \geq r_h$ for any radius of the event horizon r_h [8]. The solutions for W outside the horizon lie within the strip $-1 < W < 1$ approaching ± 1 asymptotically. They are specified by the number n of nodes of W thus forming a discrete set for each r_h. Local solutions in the vicinity of the regular event horizon of a given radius contain one free parameter $W(r_h)$ [9]. A "quantization" of $W(r_h)$ results from an imposition of the boundary conditions at infinity. Each asymptotically flat exterior solution starts with some uniquely determined value $W^n(r_h)$ at the horizon. All other $W(r_h) \neq W^n(r_h)$ generate asymptotically non-flat solutions. This quantization does not lead to the quantization of the black hole mass since the radius of the horizon still remains an arbitrary continuously varying parameter.

To start the analysis of the internal structure of non-Abelian black holes, which is an essentially numerical problem, one has to explore first local solutions of the field equations near the singularity. Two such local branches were found earlier [9]. The first one is the Schwarzschild–like (S) solution, it corresponds to the vacuum value of the YM field $|W(0)| = 1$. Introducing the mass function $m(r)$, $\Delta = r^2 - 2mr$, one has

$$\begin{aligned} W &= -1 + br^2 + b^2(3 - 8b)r^5/(30m_0) + O(r^6) , \\ m &= m_0(1 - 4b^2r^2 + 8b^4r^4) + 2b^2r^3 + O(r^5) , \end{aligned} \tag{7.7}$$

where m_0, b are (the only) free parameters. This local branch is non-generic already by counting free parameters (a generic solution of the system (7.4, 7.5) should have three free parameters).

The second is the Reissner–Nordström type branch which can be found assuming the leading term of Δ to be a positive constant (related to the charge parameter $P^2 = \Delta(0)$). This requires $W(0) = W_0 \neq \pm 1$ and gives [9]

$$\begin{aligned} W &= W_0 - W_0 r^2/(2V_0) + cr^3/(2V_0) + O(r^4) , \\ \Delta &= V_0^2 - 2m_0 r + r^2 + 2W_0(c + m_0 W_0/V_0^2)r^3 + O(r^4) , \end{aligned} \tag{7.8}$$

what corresponds to the RN metric of the mass m_0 and the (magnetic) charge $P^2 = V_0^2$, $V_0 = V(W_0)$. The expansion contains three free parameters W_0, m_0, c (i.e., is locally generic).

These two local solutions may be regarded as describing "hairy" Schwarzschild and Reissner–Nordström singularities. Their essential distinction from the usual Schwarzschild and Reissner–Nordström singularities consists in presence of the additional free parameters (b in the first case and c in the second) responsible for hair degrees of freedom.

We have also found the third local power series solution [15] assuming a *negative* value for $\Delta(0)$ (i.e., *imaginary* P):

$$\begin{aligned} W &= W_0 \pm r - W_0 r^2/(2V_0) + O(r^3)\,, \\ \Delta &= -V_0^2 \mp 4 W_0 V_0 r + O(r^2)\,, \\ \sigma &= \sigma_1 (r^2 \mp 4 W_0 r^3/V_0) + O(r^3)\,. \end{aligned} \tag{7.9}$$

Here there is only one free parameter (W_0) for W, Δ. The corresponding space-time near the singularity is conformal to $R^2 \times S^2$: after a time rescaling one obtains

$$ds^2 = r^2 (dr^2 - dt^2 - d\theta^2 - \sin^2\theta d\varphi^2)\,. \tag{7.10}$$

This geometry was encountered in the previous study of black hole interiors in the framework of the perturbed Einstein–Maxwell theory [23, 24] and called homogeneous mass-inflation model (HMI).

It is easy to realize that neither of these asymptotics may correspond to a *generic* black hole. Imposing boundary conditions in the singularity, we obtain the same kind of the singular boundary value problem as one encountered in the exterior problem where a similar role is played by the asymptotic flatness condition. This interior boundary value problem leads to the *second quantization* condition, now for the event horizon radius r_h. Therefore, the EYM black holes with the S and RN type interiors may constitute only the zero measure set in the whole EYM black hole solution space.

The system (7.4–7.5) was integrated numerically in the region $0 < r < r_h$ using an adaptive step size Runge–Kutta method for various $r_h = 10^{-8}, ..., 10^6$. The integration started at the left vicinity of the event horizon r_h where the local power series solution contains one free parameter $W_h = W(r_h)$ satisfying the inequalities $|W_h| < 1$ and $1 - W_h^2 < r_h$ which are the necessary conditions for the asymptotic flatness [9]. For given r_h, the interior solutions meeting the expansions (7.7–7.9) may exist only for some discrete W_h. A numerical strategy used to find such $W(r_h)$ consisted in detecting the change of sign of the derivative W'. In the S-case we found the curve $W(r_h)$ which starts at -1 as $r_h \to 0$ and approaches -0.1424125 for large r_h (Fig. 7.1) (without loss of generality we choose $W_h < 0$). Our S-curve intersects the $n = 1$ branch of the family of trajectories $W^n(r_h)$ corresponding to the set of external asymptotically flat solutions. Parameters of this black hole are shown in Tab. 7.1, its global behavior is depicted in Fig. 7.2 (for higher n S-solutions see [17]).

Interior solutions of the RN-type, meeting the expansions (7.8) in the singularity, were found for $r_h > r_h^* = 0.990288617$. The corresponding curve $W(r_h)$ (also shown in Fig. 7.1) intersects the trajectories $W^n(r_h)$ for all $n \geq 2$. These solutions possess an inner Cauchy horizon at some $r_- < r_h$ with $|W(r_-)| > 1$ (Fig. 7.3).

Solutions of the third type (7.9) were studied numerically starting from the vicinity of the origin. The unique solution has been found for the horizon data subject to the necessary conditions for an asymptotic flatness $|W_h| < 1$, $1 - W_h^2 < r_h$ for the upper sign in (7.9) and $W(0) = -0.9330656$, corresponding to $r_h = 1.889088$. This solution,

Table 7.1: S– and RN–type solutions.

	S–type, $n=1$	RN–type, $n=2$	RN–type, $n=3$
r_h	0.613861419	1.273791	1.0318420
$W(r_h)$	−0.8478649145	−0.113763994	−0.10185163
r_-	—	0.02171654	0.08948446
$W(0)$	−1	−1.212296124	−1.3566052
$\sigma(0)$	0.2263801	5.991210×10^{-3}	1.751928×10^{-3}
Mass	0.8807931	1.018002	1.000277

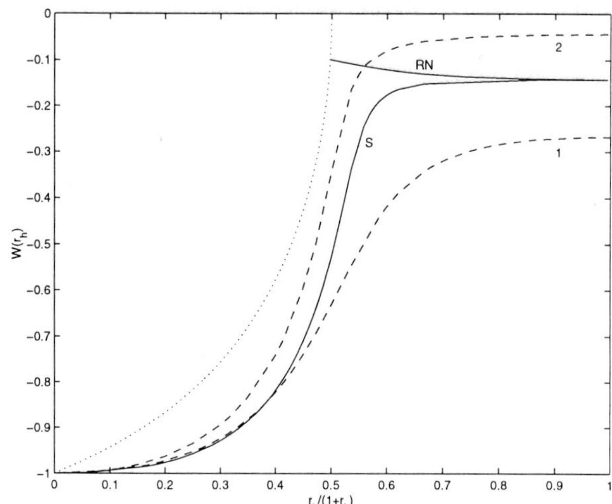

Figure 7.1: $W(r_h)$ for the S– and RN–type interior solutions. Dashed lines — $W^n(r_h)$ for $n=1,2$ (higher–n curves lie between the $n=2$ one and the boundary $r_h = 1 - {W_h}^2$, dotted line). Note that S and RN curves $W(r_h)$ do not merge.

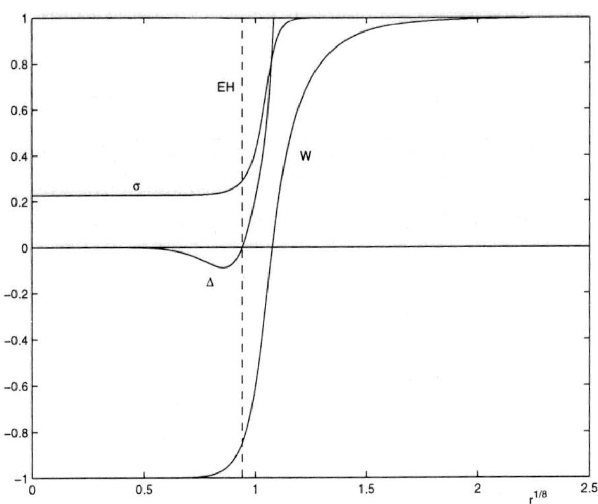

Figure 7.2: The $n = 1$ EYM black hole (S–type). EH — events horizon.

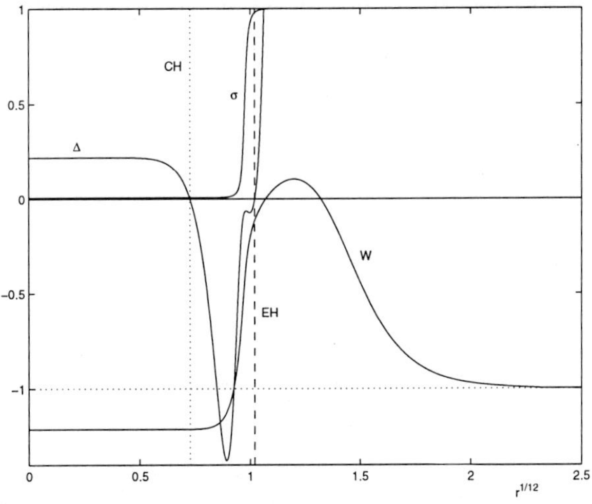

Figure 7.3: The $n = 2$ EYM black hole (RN–type). EH — events horizon, CH — Cauchy horizon.

however, does not meet any value $W^n(r_h)$ and thus does not represent a black hole. Thus a pure HMI interior can not be attributed to the EYM black holes. But it turns out to be an unstable fixed point of an asymptotic (truncated) dynamical system [15]. Deviations from this fixed point correspond to generic EYM interior solutions sharing the same property to possess no Cauchy horizons.

More general families of internal solutions not restricted by the asymptotical flatness condition were found in [17] (in particular, other internal solutions of the third type). They do not have, however, a direct significance for the EYM black holes.

7.3 Oscillatory approach to singularity

Since both hairy Schwarzschild and hairy Reissner–Nordström singularities are encountered only for certain discrete values of r_h (and hence the black hole masses), while the external solutions exist for continuously varying r_h, we must look for an alternative regime of approach to the singularity for generic r_h. It is generically observed during numerical integration from the events horizon towards the origin that sooner or later (depending on r_h and W_h) the metric function Δ starts to oscillate in the negative region with a very fast growing amplitude [15]. Because of huge numbers encountered in these oscillations the accuracy of the computation can hardly be maintained, so one has to seek appropriate truncations of the system of equations to describe the asymptotic regime analytically. Numerically it is observed that, when oscillation progress, the right hand side of the Eq. (7.4) becomes small with respect to terms at the left hand side. Neglecting it one obtains the following approximate first integral of the system:

$$Z = \Delta U \sigma / r = \text{const}\,, \tag{7.11}$$

which relates oscillations of the mass function to the evolution of σ. Numerical experiments also show that while the YM function W remains almost constant up to $r = 0$, its derivative is still non-zero and is rapidly changing on some very small intervals of r. The function U exhibits a step-like behavior being constant with high accuracy during almost all the chosen oscillation cycle (Fig. 7.4) and then jumping to a greater absolute value corresponding to the next cycle. It is clear from (7.6) that σ is exponentially falling down with decreasing r while $U \approx$ const, whereas in the tiny intervals of U-jumps σ remains almost unchanged. So σ tends to zero through an infinite sequence of exponential falls with increasing powers in the exponentially decreasing intervals. Combining this with (7.11) and the above mentioned properties of U one can deduce rather detailed description of the metric behavior.

Let us denote by r_k the value of radial coordinate where Δ has k-th local maximum. Soon after passing this point, the function U stabilizes at some value U_k approximately equal to the doubled value at the point of local maximum (similarly, U increases by about a factor of two when approaching the local maximum, whereas Δ is almost stationary). Then, according to (7.6), σ is equal to

$$\sigma(r) = \sigma(r_k) \exp\left[U_k^2(r^2 - r_k^2)\right] , \tag{7.12}$$

From (7.11) one finds that, while $U_k \approx \text{const}$,

$$\Delta(r) = \frac{\Delta(r_k)}{r_k} r \ \exp\left[U_k^2(r_k^2 - r^2)\right] . \tag{7.13}$$

This function falls down with decreasing r until it reaches a local minimum at

$$R_k = \frac{1}{\sqrt{2}\,|U_k|} \approx \frac{\sqrt{|\Delta(r_k)|}}{2|V(r_k)|} r_k . \tag{7.14}$$

In what follows, in view of the observed fact that in the course of oscillations of Δ the YM function W changes insignificantly, we will put $V = \text{const}$.

Therefore, the mass function is inflating exponentially while r decreases from r_k to R_k. After passing R_k, an exponential in (7.13) becomes of the order of unity, hence Δ starts to grow linearly, and the mass function $m(r)$ stabilizes at the value $M_k = m(R_k)$. Such a behavior holds until the point of local maximum of Δ/r^2 is reached; this takes place when $\Delta \approx -V^2$ at the point

$$r_k^* \approx \frac{V^2}{|\Delta(r_k)|} r_k \exp\left[-(U_k r_k)^2\right] . \tag{7.15}$$

After this a rapid fall of $|\Delta|$ is observed causing a violent rise of $|U|$. Then the term $2\Delta U^2$ in the Eq. (7.5) becomes negligible and consequently at this stage

$$U\Delta \approx -V^2 U_k , \tag{7.16}$$

while r practically stops. This implies that very soon Δ reaches the next local maximum at the point $r_{k+1} \approx r_k^*$, while $m(r)$ rapidly falls down to m_{k+1}. At the point of local maximum of Δ one has in the Eq. (7.6) $|\Delta| \ll V^2$, then in view of the smallness of r we find

$$|U(r_k)| \approx \frac{|V|}{\sqrt{2|\Delta(r_k)|} r_k} . \tag{7.17}$$

To obtain the estimates by the order of magnitude we will neglect all numerical coefficients elsewhere except for the power indices in exponentials, in particular putting $U(r_k) = U_k$, and omitting also (quasi-constant) factors V. With this accuracy one obtains from (7.13)–(7.17):

$$r_{k+1} = M_k^{-1} , \quad r_{k+1}^2 = R_k R_{k+1} , \quad M_k = \frac{R_k^2}{r_k^3} \exp\left(\frac{r_k^2}{2R_k^2}\right) , \tag{7.18}$$

$$|\Delta(r_k)| = \left(\frac{R_k}{r_k}\right)^2 , \quad \frac{r_{k+1}}{r_k} = \frac{r_k^2}{R_k^2} \exp\left[-\left(\frac{r_k^2}{2R_k^2}\right)\right] . \tag{7.19}$$

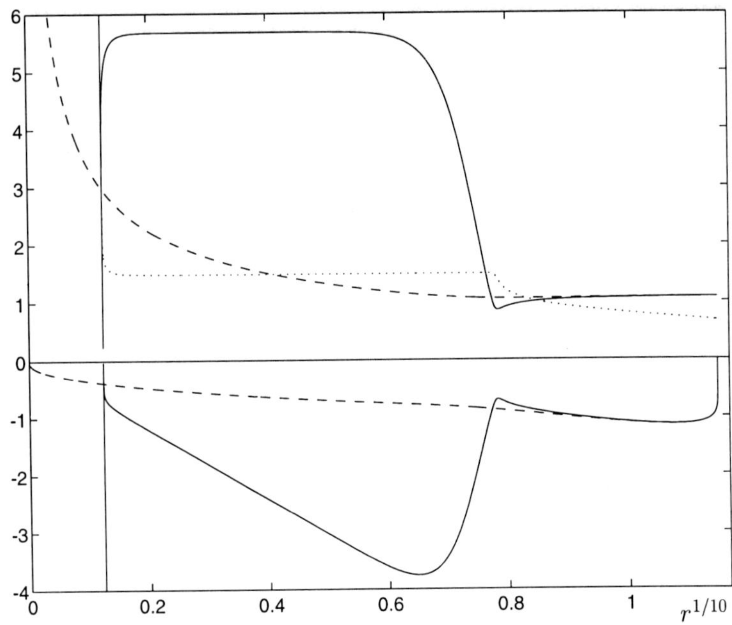

Figure 7.4: The first oscillation cycle for an EYM solution. The functions Δ for EYM (solid line) and EYMD (dashed) are shown in lower half-plane. In upper half-plane — mass functions $m(r)$ (analogously) and the function U for EYM (dotted). All functions are power rescaled with the power index 1/10. Here $r_h = 4$; $W_h = -0.283993$ for EYM, $W_h = -0.298357$, $\phi_h = 0.05623$ for EYMD (asymptotically flat solutions with one node of W.)

Thus, introducing a variable $x_k = (r_k/R_k)^2\,(\gg 1)$, we can derive the following recurrent formula

$$x_{k+1} = x_k^{-3} \mathrm{e}^{x_k}\,, \tag{7.20}$$

which shows that x_k is an exponentially diverging sequence. In terms of x_k one has

$$\frac{r_{k+1}}{r_k} = x_k \mathrm{e}^{-x_k/2}\,, \tag{7.21}$$

this relation can also be understood as a ratio of the neighboring oscillation periods since $r_k \gg r_{k+1}$. Values of the function $|\Delta|$ at the points r_k rapidly tend to zero:

$$|\Delta(r_k)| = x_k^{-1}\,, \tag{7.22}$$

so we deal with an infinite sequence of "almost" Cauchy horizons as $r \to 0$. At the same time the values of $|\Delta|$ at the points R_k grow rapidly:

$$|\Delta(R_k)| = x_k^{-3/2} \mathrm{e}^{x_k/2}\,, \tag{7.23}$$

and the values of the mass function grow correspondingly as

$$\frac{M_k}{M_{k-1}} = x_k^{-1} \mathrm{e}^{x_k/2}\,. \tag{7.24}$$

While r decreases form r_k to R_k, the function σ rapidly falls down to the value $\sigma_k = \sigma(R_k)$, which then remains unchanged until the point r_{k+1}. As the singularity is approached, the sequence σ_k decreases according to

$$\frac{\sigma_{k+1}}{\sigma_k} = \mathrm{e}^{-x_k/2}\,. \tag{7.25}$$

Therefore for a generic (continuously varying) r_h one observes a rather unexpected approach to the singularity through an infinite sequence of oscillations with an exponentially growing amplitude.

Apart from the discrete picture given above one can find a truncated two-dimensional dynamical system clearly showing an infinitely oscillating nature of the generic solution. As we have already noted, the oscillation region starts with an exponential fall of Δ which typically occurs after passing a local maximum r^{max} (Fig. 7.5), and the right hand side of (7.4) becomes comparatively small with respect to other terms. Another important feature is that the YM function W (contrary to Δ) possesses a *finite* limit W_0 in the singularity. Omitting the right hand side of (7.4), replacing W by its limiting value W_0, and neglecting 1 as compared with V^2/r^2 one can derive from the system (7.4–7.5) the following two-dimensional dynamical system

$$\begin{aligned} \dot{q} &= p\,, \\ \dot{p} &= (3e^{-q} - 1)p + 2e^{-2q} - 1/2\,, \end{aligned} \tag{7.26}$$

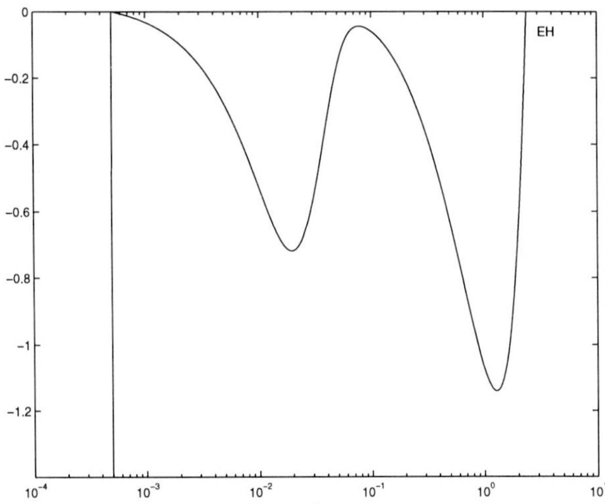

Figure 7.5: The beginning of Δ–oscillations for $n = 1$ EYM BH solution, $r_h = 2.4$, $W(r_h) = -0.31652531$. EH — events horizon.

where $\Delta = -(V_0^2/2)\exp(q)$, and a dot stands for derivatives with respect to $\tau = 2\ln(r_h/r)$. This system has one (focal) fixed point ($p = 0$, $q = \ln 2$), corresponding to some imaginary charge RN-like local solution of the type (7.9), with eigenvalues $\lambda = (1 \pm i\sqrt{15})/4$. Its phase portrait is shown in Fig. 7.6 together with an invariant set $p = -e^{-q} - 1/2$ corresponding to the RN–type local solution (7.8). The generic oscillating solutions lie above this curve. The phase motion in this region is unbounded, and there are no limiting circles. One can easily show that the rotation in the phase plane never stops and the limit $q = -\infty$ ($\Delta = 0$) can not be reached. The metric function Δ remains negative valued as $r \to 0$ and passes an infinite sequence of local maxima and minima. The class of oscillating solutions is stable in a sense that a small deviation from one such solution moves us to another oscillating solution. Thus the existence of the consistent two-dimensional truncation of the full set of equations proves that the oscillating approach to the singularity is generic for the EYM black holes. The dynamical system (7.26) was derived in our paper [15] and reproduced in a slightly different notation in [17].

7.4 Power-law "mass-inflationary" singularity

In non-Abelian theories including scalar fields, such as EYMH and EYMD, another regime of the generic approach to the singularity is realized. In this case the scalar kinetic term becomes dominant soon after entering an asymptotic regime what results

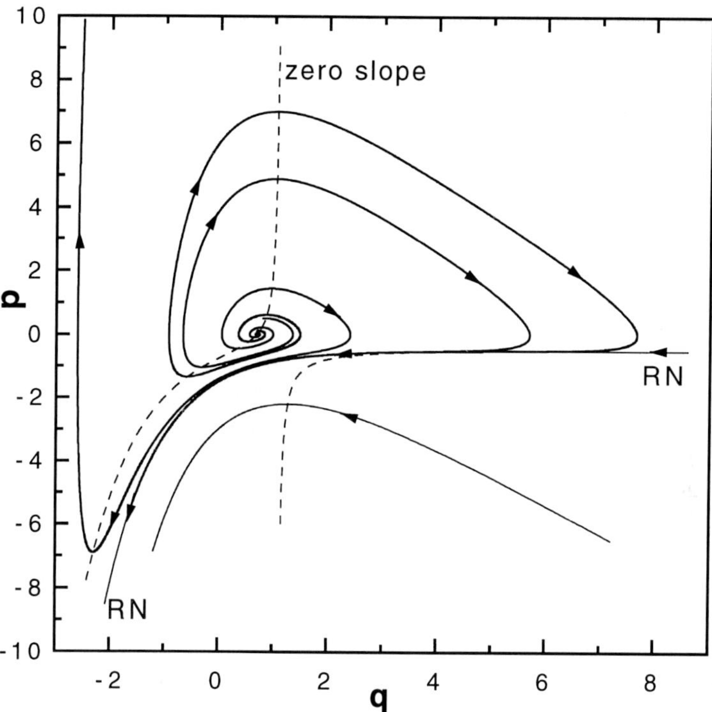

Figure 7.6: Phase portrait of the dynamical system (3.16), RN — an invariant set, corresponding to the RN–type solution (dashed — zero slope lines).

in a monotonous behavior of the metric. Consider first the $SU(2)$ EYMH theory

$$S = \frac{1}{16\pi}\int\left\{-R - F^2 + 2|D\Phi|^2 - \frac{\lambda}{2}(|\Phi|^2-\eta^2)^2\right\}\sqrt{-g}\,d^4x\,, \tag{7.27}$$

where Φ is a Higgs field in either vector (real triplet) or fundamental (complex doublet) representations, $D\Phi$ is the corresponding YM covariant derivative (in the doublet case $|D\Phi|^2 = (D\Phi)^\dagger(D\Phi)$, $|\Phi|^2 = \Phi^\dagger\Phi$) and without loss of generality both the Planck mass and the gauge coupling constant are set to unity. In the flat space-time the triplet version of the theory gives rise to regular magnetic monopoles, the doublet version — to sphalerons. New physically interesting configurations emerge when gravity is coupled in a self-consistent way, in particular, static spherical black holes exist in both cases: monopole [11] and sphaleron [12].

Static spherically symmetric configurations of the YM fields are still given by the ansatz (7.3), while the Higgs field is $\Phi^a T_a = \phi(r)\,T_r$ in the triplet case, and $\Phi = \phi(r)v$ in the doublet one, where v is some (here irrelevant) spinor depending only on the angle variables. In both cases $\phi(r)$ is the only real scalar function of the radial variable.

The system of equations following from (7.27) with this ansatz may be presented as a set of three coupled equations for W, ϕ, and $\Delta = r^2 - 2mr$:

$$\left(\frac{\Delta}{r^2}W'\right)' + \frac{\Delta}{r}W'\phi'^2 = \frac{1}{2}\frac{\partial\mathcal{V}}{\partial W} - Q\frac{W'}{r}\,, \tag{7.28}$$

$$\left(\frac{\Delta}{r}\right)' + \Delta\phi'^2 = 1 - 2\mathcal{V} - Q\,, \tag{7.29}$$

$$(\Delta\phi')' + \Delta r\phi'^3 = \frac{\partial\mathcal{V}}{\partial\phi} - Qr\phi'\,, \tag{7.30}$$

where $Q = 2\Delta W'^2/r^2$, and

$$\mathcal{V} \equiv \mathcal{V}(W,\phi,r) = \frac{V^2}{2r^2} + \frac{\lambda r^2}{8}(\phi^2-\eta^2)^2 + P^2, \tag{7.31}$$

with $P = W\phi$ in the triplet case and $P = (W+1)\phi$ in the doublet one. An equation for σ now becomes

$$(\ln\sigma)' = \frac{2}{r}W'^2 + r\phi'^2, \tag{7.32}$$

and can be easily integrated once W, ϕ are found.

Local expansions of three types listed in Sec. 7.2 can be easily generalized to include Higgs [17], and the same arguments can be used to prove that neither of the corresponding global solutions is generic. Meanwhile, the forth local branch can be found analytically in this case [18]. It can be derived using the consistent truncation of the non-linear systems of equations. An appropriate truncation consists in omitting from the equations all matter terms except for those related to the gradient of the scalar field. This reduces to dropping the right hand sides in Eqs. (7.28–7.30).

The resulting system can be easily disentangled leading to the following decoupled equations:

$$W'' - \frac{W'}{r} = 0\,, \tag{7.33}$$

$$\phi'' + \frac{\phi'}{r} = 0\,, \tag{7.34}$$

which can be solved as follows:

$$W = W_0 + br^2\,, \tag{7.35}$$

$$\phi = \phi_0 + k \ln r\,. \tag{7.36}$$

Here W_0, ϕ_0, b, k are free parameters. Note that contrary to the expansions discussed in Sec. 7.2, this is not a power series solution. Higgs field is logarithmically divergent, so that its derivative diverges as r^{-1}, this is why the corresponding terms become dominant for sufficiently small r. Once W and ϕ are found, the metric function Δ can be obtained by integrating the simple equation

$$\frac{\Delta'}{\Delta} = \frac{1}{r} - r\phi'^2\,, \tag{7.37}$$

what gives

$$\Delta = -2m_0 r^{(1-k^2)} \tag{7.38}$$

with the fifth (positive) constant m_0. Hence, by counting free parameters, this is a generic solution with non-positive Δ. Now, to find whether our truncation, one has to substitute the solution into the Eqs. (7.28–7.30) and to check whether the right hand side terms are indeed comparatively small. One finds the following condition: $k^2 > 1$. This means that the metric function Δ is divergent at the singularity. The corresponding mass-function is also divergent according to the power-law

$$m = \frac{m_0}{r^{k^2}}\,. \tag{7.39}$$

The asymptotic behavior of σ dominated by the scalar term then reads

$$\sigma = \sigma_0 r^{k^2}\,, \tag{7.40}$$

with (positive) constant σ_0.

This local solution can be interpreted as exhibiting a "power-law mass inflation". As a matter of fact this regime has nothing to do with the usual (exponential) mass-inflation which presumably takes place once a Cauchy horizon is approached.

The second example of the scalar-dominated singularity is given by the EYMD theory:

$$S = \frac{1}{16\pi} \int \left\{-R + 2(\nabla\phi)^2 - e^{-2\phi} F^2\right\} \sqrt{-g}\, d^4x\,, \tag{7.41}$$

The equations of motion for W, Δ, ϕ now take the form

$$\Delta U' - 2\Delta U\phi' = WV/r - \mathcal{F}W' , \tag{7.42}$$
$$(\Delta/r)' + \Delta\phi'^2 = \mathcal{F} - 2\Delta U^2 e^{-2\phi} , \tag{7.43}$$
$$(\Delta\phi')' + \Delta r\phi'^3 = \mathcal{F} - 2\Delta(\phi' r + 1)U^2 e^{-2\phi} - 1 , \tag{7.44}$$

where $\mathcal{F} = 1 - V^2 e^{-2\phi} r^{-2}$, $V = W^2 - 1$. The remaining equation for σ reads

$$(\ln\sigma)' = r\left(\phi'^2 + 2U^2 e^{-2\phi}\right) . \tag{7.45}$$

Similarly to the EYMH case for sufficiently small r the right sides of the Eqs. (7.42)–(7.44) become small in comparison with the left hand side terms and one gets the following truncated system

$$\begin{aligned} &(\ln U)' - 2\phi' = 0 , \\ &[\ln(\Delta/r)]' = [\ln(\Delta\phi')]' = -r\phi'^2 . \end{aligned} \tag{7.46}$$

Its integration gives the following five-parameter (i.e., generic) family of solutions

$$\begin{aligned} W &= W_0 + br^{2(1-\lambda)} , \\ \Delta &= -2\mu r^{(1-\lambda^2)} , \\ \phi &= c + \ln\left(r^{-\lambda}\right) , \end{aligned} \tag{7.47}$$

with constant W_0, b, c, μ, λ. The validity of the truncated equations (7.46) now can be checked by substituting the asymptotic solution (7.47) into the full system (7.42)–(7.44). For consistency it is sufficient that the following inequalities hold:

$$\sqrt{2} - 1 < \lambda < 1 , \tag{7.48}$$

which is in agreement with the numerical data.

From (7.47) it follows that the mass function diverges as $r \to 0$ according to the power law:

$$m(r) = \frac{\mu}{r^{\lambda^2}} . \tag{7.49}$$

The corresponding σ tends to zero as

$$\sigma(r) = \sigma_1 r^{\lambda^2} , \tag{7.50}$$

where $\sigma_1 = \text{const}$. A typical EYMD solution is shown in Fig. 7.4.

The only difference with the EYMH case is that the region of the power index (7.48) is different. This leads to different picture of the singularity in the Kantowski-Sachs interpretation: a point like singularity in the EYMH case, but a cigar singularity in the EYMD case [18].

7.5 Discussion

Non-Abelian black holes turned out to be a useful laboratory to explore the nature of the singularity inside "realistic" black holes. Unlike the gravity coupled Abelian field models such as Maxwell, Maxwell–dilaton–axion, or more general string-inspired systems, non-Abelian models have an advantage to give rise to several qualitatively different possibilities as far as the singularity is concerned. Moreover, in spite of the absence of exact analytic solutions, one can find a correspondence between the singularity structure and external parameters such as the black hole mass. Thus the relative weight of different type interior structures in the parameter space may be found. This opens a new way to probe the Strong Cosmic Censorship using the genericity argument. An advantage of this approach is that a non-trivial information may be extracted already at the level of static ("eternal") black hole solutions. One can speculate that non-Abelian electroweak theory should replace the Maxwell electrodynamics at high energies which are reached in the course of the mass-inflation inside a (perturbed) Reissner–Nordström black hole due to the phase transition similar to that in cosmology. Therefore a further fate of the black hole should be in the scope of a non-Abelian theory. Although the full time-dependent picture is likely to be quite complicated, it is reassuring to feel that the static spherical non-Abelian black holes choose the spacelike singularity in conformity with the Strong Cosmic Censorship.

The new features observed in the interiors of non-Abelian black holes are due to the field components which are not connected with the conserved charges and are usually termed as a hair. Violation of the no-hair conjecture in black holes traditionally was attributed to their external appearance. New results clearly show that the hair is equally important for the *internal* structure of black holes. Hair "roots" penetrate up to the singularity supplying it with additional degrees of freedom. A tiny black hole inside a large magnetic monopole generically has a very different internal structure than the Schwarzschild black hole. Hairy black holes have hairy singularities.

Qualitatively speaking, the interiors of the EYM black holes look like the hair-perturbed Schwarzschild or Reissner–Nordström interiors. So it is not surprising that one encounters an exponential growth of mass whenever the metric reaches an "almost" Cauchy horizon. This phenomenon, first noted in [15] and termed as "mass inflation" in [17], is, however, only a half of the story. In the oscillating regime one observes two qualitatively different "mass-inflations": the first is the local inflation at some very short interval of each oscillation cycle, the second is associated with the exponential growth of mass from cycle to cycle [19] while the singularity is approached.

If scalar fields are present, the dynamics gets a new dimension. Once the scalar component is excited, it soon becomes a dominant factor of the evolution which changes drastically the approach to the singularity. In the scalar-dominated regime no mass-inflation is manifest, instead one observes a behavior which we call a "power-law mass-inflation", i.e., a power-fashion divergence of the mass-function near the singularity. This behavior has a different origin as compared with the usual mass-inflation, in particular, it is not related to an approach to the Cauchy horizon. Rather it follows from the coupled Einstein–scalar field dynamics which is essentially Abelian. Indeed, the same behavior near the singularity was earlier observed in the Kantowski–Sachs cosmology with a non-linear (one-component) scalar field source. Still the non-Abelian nature of the model is substantial, otherwise the (asymptotically flat) static black holes are prohibited by the no-hair theorems.

The analysis given here is purely classical, a few words are in order about the relevance of the results to the full quantum theory. Vacuum polarization of the conformal scalar field on the HMI background was considered in [24]. It was found that the correction to the mass function diverges more strongly than in the classical case. Hence the singularity is not smoothened but rather intensified. In the oscillating regime there is no hope to compute quantum effects quasi-classically. Moreover, huge values of the mass function (in Planck's units) encountered soon after entering such a regime indicate that the quantum behavior of the model should be considered nonperturbatively and may well be qualitatively different from the classical picture. However, the conclusion about the spacelike nature of the generic singularity is unlikely to be changed.

D.V.G. thanks the Organizing Committee for hospitality and support during the Workshop. The research was supported in part by the RFBR grants 96–02–18899, 18126.

Bibliography

[1] R. Ruffini and J. A. Wheeler, Physics Today **24**, 30 (1970).

[2] R. Bartnik, J. McKinnon, Phys. Rev. Lett. **61**, 141 (1988).

[3] D. V. Gal'tsov, M. S. Volkov, Phys. Lett. **B 273**, 255 (1991).

[4] N. Straumann, Z. H. Zhou, Phys. Lett. **B 237**, 353 (1990).

[5] D. Sudarsky and R. M. Wald, Phys. Rev. D **46**, 1453 (1992).

[6] M. S. Volkov, D. V. Gal'tsov, Phys. Lett. **B 341**, 279 (1995).

[7] I. Moss and A. Wray, Phys. Rev. D **46**, 1215 (1992).

[8] M. S. Volkov and D. V. Gal'tsov, Pis'ma Zh. Eksp. Teor. Fiz. **50**, 312 (1989) [JETP Lett. **50**, 345 (1990)]; H. P. Kunzle, A. K. M. Masood–ul–Alam, J. Math. Phys. **31**, 928 (1990); P. Bizon, Phys. Rev. Lett. **64**, 2844 (1990).

[9] M. S. Volkov and D. V. Gal'tsov, Sov. J. Nucl. Phys. **51**, 747 (1990).

[10] D. V. Gal'tsov and E. E. Donets, Int. J. Mod. Phys. **D 3**, 755 (1994) and refences therein.

[11] K. Lee, V. P. Nair, and E. Weinberg, Phys. Rev. D **45**, 2751 (1992); M. Ortiz, Phys. Rev. D **45**, 2586 (1992); P. Breitenlohner, P. Forgács, and D. Maison, Nucl. Phys. **B 383**, 357 (1992); **B 442**, 126 (1995); T. Tachizawa, K. Maeda, and T. Torii, Phys. Rev. D **51**, 4051 (1995).

[12] B. R. Greene, S. D. Mathur, and C. M. O'Neill, Phys. Rev. D **47**, 2242 (1993).

[13] T. Torii, K. Maeda, Phys. Rev. D **48**, 1643 (1993).

[14] D. Maison, *Solitons of the Einstein–Yang–Mills Theory*, gr-qc/9605053.

[15] E. E. Donets, D. V. Gal'tsov, and M. Yu. Zotov, *Internal structure of Einstein–Yang–Mills black holes*, gr-qc/9612067, Phys. Rev. D **56**, 3459 (1997).

[16] V. A. Belinskii, I. M. Khalatnikov, and E. M. Lifshitz, Adv. Phys. **19**, 525 (1970).

[17] P. Breitenlohner, G. Lavrelashvili, and D. Maison, *Mass inflation and chaotic behavior inside hairy black holes*, MPI-PhT/97-20, BUTP-97/08, gr-qc/9703047.

[18] D. V. Gal'tsov and E. E. Donets, *Power-law mass inflation in Einstein–Yang–Mills–Higgs black holes* gr-qc/9706067.

[19] D. V. Gal'tsov, E. E. Donets, and M. Yu. Zotov, Pis'ma Zh. Eksp. Teor. Fiz. **65**, 855 (1997), [JETP Lett. **65**, 895 (1997)]; (gr-qc/9706063).

[20] E. Poisson and W. Israel, Phys. Rev. Lett. **63**, 1663 (1989); Phys. Rev. D **41**, 1976 (1990); A. Ori, Phys. Rev. Lett. **67**, 789 (1991); A. Bonnano, S. Droz, W. Israel, S. M. Morsink, Phys. Rev. D **50**, 7372 (1994); A. Ori, Phys. Rev. D **55**, 3575 (1997).

[21] B. C. Paul, D. P. Datta, and S. Mukherjee, Modern Physics Letters **A 1**, No. 2, 149 (1986).

[22] P. Breitenlohner, G. Lavrelashvili, and D. Maison, *Non-Abelian black holes: The inside story*, BUTP-97/23, gr-qc/9708036; these proceedings.

[23] D. N. Page, in Proc. NATO Advanced Summer Institute on Black Hole Physics (Erice 1991), ed. V. De Sabbata and Z. Zhang (Kluwer, 1992), p. 185.

[24] W. G. Anderson, P. R. Brady, and R. Camporesi, Class. Quant. Grav. **10**, 497 (1993).

[25] J. A. Smoller and A. G. Wasserman *Reissner–Nordström–like solution of the SU(2) Einstein–Yang/Mills equations*, gr-qc/9703062.

INTERNAL STRUCTURE OF EINSTEIN–YANG–MILLS–DILATON BLACK HOLES

O. Sarbach, N. Straumann and M. S. Volkov

Institute for Theoretical Physics, University of Zurich-Irchel, Winterthurerstrasse 190, CH–8057 Zurich, Switzerland

Abstract

We study the interior structure of the Einstein-Yang-Mills-Dilaton black holes as a function of the dilaton coupling constant $\gamma \in [0, 1]$. For $\gamma \neq 0$ the solutions have no internal Cauchy horizons and the field amplitudes follow a power law behavior near the singularity. As γ decreases, the solutions develop more and more oscillation cycles in the interior region, whose number becomes infinite in the limit $\gamma \to 0$.

8.1 Introduction

The much discussed mass inflation scenario [1] provides a picture of what happens to the unstable Cauchy horizons of the Kerr-Newman family when the back reaction of the blue-shifted perturbations is taken into account. In this scenario the blue-shifted influx, due to the inavoidable gravitational wave tail of a realistic gravitational collapse, causes the local mass function to blow up exponentially near the Cauchy horizon. Thereby, a singularity is produced along the Cauchy horizon, which leads to infinite tidal forces on a free-falling object. This singularity is, however, lightlike and mild, accompanied with a divergence of the conformal curvature, but with bounded shear and expansion. The genericity of the mass inflation picture has been established in a series of analytical and numerical studies of a class of vacuum and electrovacuum solutions [2, 3, 4, 5, 6].

What really happens very close to the singularity along the horizon cannot be decided by classical physics. This is presumably not even a physically relevant question, because the black hole will have evaporated or merged with other black holes in a cosmological big crunch. It may, nevertheless, be of some interest to study the interior black hole solutions for classical field–theoretical matter models which are important in high-energy particle physics.

Such investigations were initiated in ref. [7], where it was shown that for pure Yang-Mills fields a new type of infinitely oscillating behavior with exponentially growing amplitude is developed. It is natural to ask, whether this cyclic repetition with associated mass inflation is generic when other matter fields are included. In [8], [9] it was shown that the interior solution changes qualitatively when a Higgs field is included.

In the present contribution we discuss the interior structure of black holes for the Einstein-Yang-Mills-Dilaton (EYMD) system, considering the dilaton coupling constant, γ, as a bifurcation parameter of the associated dynamical system. It turns out that for $\gamma = 1$ the behavior is drastically different from that of the EYM system ($\gamma = 0$), in that there are no oscillations; this phenomenon has also been observed in [10]. All fields evolve quite monotonically when the radial Schwarzschild coordinate r (internal time) approaches 0. This has prompted us to study the change of the phase portrait as γ decreases from 1 to 0. Our analytical and numerical results show quite convincingly, that with decreasing γ more and more cycles are developing, and that the limit $\gamma \to 0$ is quite singular. For $\gamma \neq 0$ the interior solutions have no Cauchy horizons. We believe that this is true quite generically for a large class of nonlinear field–theoretical matter models. Further work on this issue is in progress.

8.2 The EYMD black holes

The black hole solutions we shall be considering arise within the context of the SU(2) EYMD theory, whose action reads in standard notations

$$\mathcal{S} = \int \left(-\frac{1}{4} R + \frac{1}{2}\, \partial_\mu \Phi\, \partial^\mu \Phi + \frac{1}{2} e^{2\gamma\Phi} \mathrm{tr}\, F_{\mu\nu} F^{\mu\nu} \right) \sqrt{-g}\, d^4x. \tag{8.1}$$

The dilaton coupling γ is assumed to have values in the interval $[0, 1]$. We are especially interested in the limiting behavior for $\gamma \to 0$. When γ vanishes, the dilaton decouples, in which case one can put $\Phi = 0$. In the static spherically symmetric case the metric is given by

$$ds^2 = S^2 N dt^2 - \frac{dr^2}{N} - r^2 (d\vartheta^2 + \sin^2 d\varphi^2) \tag{8.2}$$

and the gauge potential A for a purely magnetic YM field can be parametrized as

$$A = w(r)(-\mathrm{T}_2\, d\theta + \mathrm{T}_1 \sin\theta\, d\varphi) + \mathrm{T}_3 \cos\theta\, d\varphi, \tag{8.3}$$

where the group generators T_a are chosen as $\tau^a/2i$ (τ^a=Pauli matrices). The functions w, S, N and the dilaton Φ depend only on r. It will be sometimes convenient to express N in terms of the mass function m as $N = 1 - 2m/r$. The field equations, corresponding to (8.1) read

$$(rN)' + r^2 N \Phi'^{\,2} + U = 1,$$

$$(SNr^2\Phi')' = \gamma S U,$$

$$r^2\left(NSe^{2\gamma\Phi}\,w'\right)' = Se^{2\gamma\Phi}\,w(w^2-1),$$
$$S' = S\left(r\phi'^{\,2} + 2e^{2\gamma\Phi}\,w'^{\,2}/r\right), \tag{8.4}$$

where $U = 2e^{2\gamma\Phi}\left(Nw'^{\,2} + (w^2-1)^2/2r^2\right)$. These equations are invariant under the scale transformations

$$r \mapsto e^{\lambda}r, \quad N \mapsto N, \quad S \mapsto e^{-\lambda}S, \quad \Phi \mapsto \Phi + \frac{\lambda}{\gamma}, \quad w \mapsto w. \tag{8.5}$$

The function S can be eliminated from the system (8.4). When N, Φ and w are known, S can be expressed as

$$S = \exp\left(-\int_r^\infty (r\Phi'^2 + \frac{2}{r}e^{2\gamma\Phi}w'^2)dr\right). \tag{8.6}$$

For any $\gamma \in [0,1]$, equations (8.4) are known to possess a family of black hole solutions [11], [12]. Close to the event horizon they behave as follows:

$$\begin{aligned} N &= (1-V_h)x + O(x^2), \\ \Phi &= \Phi_h + \frac{\gamma V_h}{(1-V_h)}x + O(x^2), \\ w &= w_h + \frac{w_h(w_h^2-1)}{(1-V_h)}x + O(x^2), \end{aligned} \tag{8.7}$$

where $x = (r-r_h)/r_h$, $V_h = e^{2\gamma\Phi_h}(w_h^2-1)^2/r_h^2$, and the scaling symmetry (8.5) can be used to set $\Phi_h = 0$. Note that at the event horizon one has $N'(r_h) > 0$. The asymptotic behavior at infinity is described by

$$\begin{aligned} N &= 1 - \frac{2M}{r} + \frac{D^2}{r^2} + O(\frac{1}{r^3}), \\ \Phi &= \Phi_\infty - \frac{D}{r} + O(\frac{1}{r^2}), \\ w &= \pm\left(1 - \frac{c}{r}\right) + O(\frac{1}{r^2}), \end{aligned} \tag{8.8}$$

where M is the ADM mass and D is the dilaton charge. The black hole solutions with these asymptotics are numerically known in the whole interval $r_h \le r < \infty$ [11], [12]. Their behavior is qualitatively similar for all values of γ: the functions N and Φ are monotone in the exterior region, interpolating between the boundary values given by Eqs. (8.7), (8.8), whereas w oscillates within finite bounds. The solutions form a 2-parameter family labeled by r_h, and n, the number of nodes of w in $[r_h, \infty)$. In what follows we shall consider the extension of these solutions into the interior region of the black hole, $0 \le r < r_h$. It will turn out that the behavior of the solutions in the interior region changes dramatically with varying γ.

8.3 The interior black hole solutions

When extending the black hole solutions into the interior region, the following phenomenon is observed: For $\gamma = 0$ (the dilaton is absent) the interior solution is characterized by violent oscillations of the metric coefficients and the gauge field strength [7]. The amplitude of these oscillations tends to infinity as the system approaches the singularity. For $\gamma = 1$ the behavior is rather different, without any peculiar variations (see Fig.1). It is interesting to study more closely what happens when γ decreases form 1 to 0. Our analytical and numerical investigations lead to the following qualitative picture: For small values of γ the solution in the vicinity of the horizon follows closely that for $\gamma = 0$. The smaller the γ, the more oscillation cycles the solution exibits. However, as long as $\gamma \neq 0$, the functions assume a power law behavior in the vicinity of $r = 0$:

$$N \propto -N_1 r^{-(1+p^2)}, \quad \Phi \propto \Phi_1 + p\ln(r), \quad w \propto w_0 - br^{2(1-\gamma p)}, \tag{8.9}$$

with $b, N_1, p > 0$, and the values of p are restricted by

$$\sqrt{\gamma^2+1} - \gamma < p < \frac{1}{\gamma}. \tag{8.10}$$

In order to gain some qualitative understanding of this phenomenon, it is helpful to introduce the new variables

$$x = w' e^{\gamma\Phi}, \quad y = -\frac{(w^2-1)^2}{r^2 N} e^{2\gamma\Phi}, \quad z = r\Phi'. \tag{8.11}$$

In addition, the following approximations in the differential equations are numerically justified for the interior solutions:

- $w \simeq$ const. (but w' is kept),
- $e^{2\gamma\Phi}(w^2-1)^2 \gg r^2$,
- the term $w(w^2-1)/r^2$ in Eqs.(8.4) can be neglected.

This yields, as an approximate first integral of the field equations,

$$SNe^{2\gamma\Phi} w' \simeq \text{const.} \tag{8.12}$$

Eqs. (8.4) can then be truncated to the following dynamical system:

$$\begin{aligned} \dot{x} &= x(y + \gamma z - 1), \\ \dot{y} &= -y(2x^2 - y - 1 + z^2 + 2\gamma z), \\ \dot{z} &= -\gamma(2x^2 - y) + yz, \end{aligned} \tag{8.13}$$

where a dot stands for differentiation with respect to $t = -\ln(r)$. We are interested in the behavior of the solutions as $t \to \infty$. First of all, we note that the planes $\{x = 0\}$ and $\{y = 0\}$ are invariant sets and hence divide the phase space into four regions. Since N is negative and the equations are symmetric under $x \mapsto -x$, we restrict ourselves to the invariant region $x > 0$, $y > 0$. The critical points in this region are:

1. $x = \alpha\beta$, $y = \alpha\beta$, $z = -\alpha\gamma$, where $\alpha = 1/(1 + 2\gamma^2)$ and $\beta = 1 + \gamma^2$.

2. $x = 0$, $y = 0$, $z = p$, where p is arbitrary.

The eigenvalues corresponding to the first critical point are $\lambda_1 = \alpha\beta$, $\lambda_{2,3} = \alpha\beta(1 \pm i\sqrt{15 + 32\gamma})/2$, hence it is an attractive center in the limit $t \to -\infty$. The solutions starting at this point spiral outwards when t increases. For $\gamma = 0$ one has $z \equiv 0$ and Eqs.(8.13) reduce to the 2-dimensional system

$$\dot{x} = x(y - 1),$$

$$\dot{y} = y(-2x^2 + y + 1). \tag{8.14}$$

Now, for this system one can show that the spiraling motion around the focal point can never stop. The trajectories are bound to spiral around the center $(1, 1)$ for all t. At the same time, they have to remain in the region $x > 0$ and $y > 0$. As a result, after each revolution the trajectories come closer and closer to the second critical point, $(0, 0)$, which corresponds to the spacetime singularity, but can never be reached. This explains qualitatively the unbounded oscillations for the pure EYM case [7]. It is interesting to observe that the critical point around which the spiralling occures has no physical meaning by itself. Indeed, it can only be reached for $t \to -\infty$, which corresponds to $r \to +\infty$, contradicting the assumption $r < r_h$.

On the other hand, for $\gamma \neq 0$ the solutions can no longer stay in the plane $\{z = 0\}$ for all t. It turns out that the appearance of the additional degree of freedom completely changes the behavior of the system. When γ is small the trajectory can stay for a long time in the vicinity of the plane spiralling outwards. The smaller γ, the smaller $\dot{z}$ and more revolutions are executed by the trajectory. As long as $p = z$ is small, the second critical point is repulsive, since the corresponding eigenvalues are $\lambda_1 = 1 - \gamma p$, $\lambda_2 = p^2 + 2\gamma p - 1$, and $\lambda_3 = 0$. For small p one has $\lambda_1 > 0 > \lambda_2$. However, when $p = z$ becomes large enough to fulfill the condition (8.10), one will have $\lambda_1 > 0$, $\lambda_2 > 0$, and so the critical point $(0, 0, p)$ will become attractive. One can also show that it is stable, despite the zero eigenvalue. As a result, having performed only a finite number of oscillations, the trajectory will eventually be attracted and end up at the singularity. The linearized solution near the critical point reproduces the power law behavior given by Eq. (8.9).

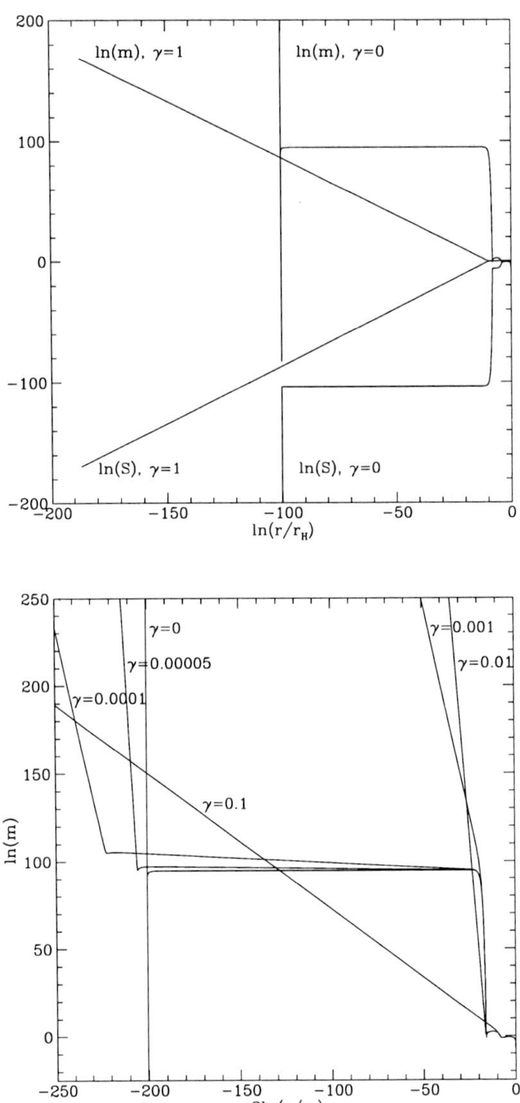

Figure 8.1: *The interior solutions for the $n = 1$, $r_h = 2$ EYMD black holes. Top: for $\gamma = 0$ the metric functions m and S can be approximated by step-functions whose amplitudes grow exponentially as $r \to 0$. For $\gamma = 1$, m and S are described asymptotically by linear functions – on a logarithmic scale. The picture on the bottom shows how the steps transform into straight lines as γ increases.*

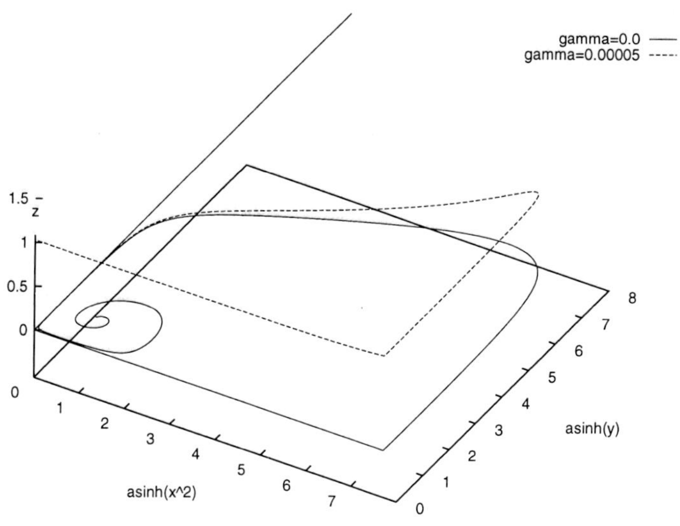

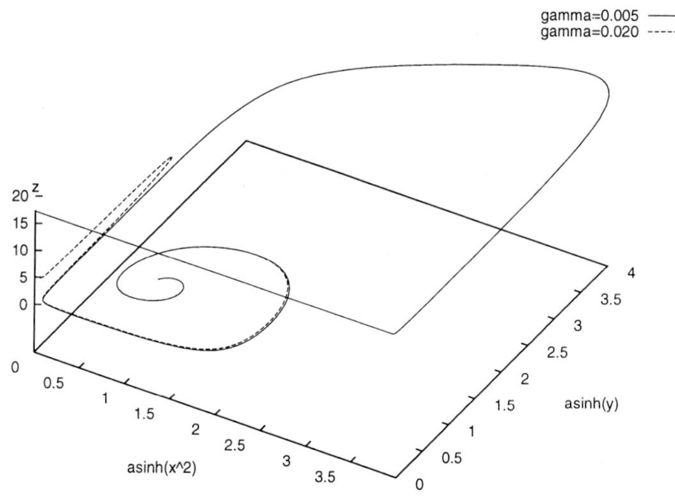

Figure 8.2: *Top: solutions of the dynamical system (13) for* $\gamma = 0$ *and* $\gamma = 0.00005$. *For* $\gamma = 0$ *the trajectory is bound to stay in the* $z = 0$ *plane always spiralling around the center, whereas for* $\gamma = 0.00005$ *it leaves the plane and converges to the* z*-axis. Bottom: the solutions for* $\gamma = 0.005$ *and* $\gamma = 0.020$. *For* $\gamma = 0.005$, *one additional oscillation cycle is performed.*

8.4 Absence of Cauchy horizons for the EYMD black holes

The interior solutions decribed above do not have inner horizons. It is natural to ask whether this property persists for all values of n and r_h. For $\gamma = 0$ some of the the solutions are known to display Cauchy horizons. Specifically, for each $n > 1$ there is one distinguished black hole solution with radius $r_h(n)$, which has an inner horizon at some $r = r_-(n)$ [7]. It turns out that for $\gamma \neq 0$ this can never happen. The proof of this statement is fairly simple and goes as follows: When $\gamma \neq 0$, the scaling symmetry (8.5) implies the existence of the conserved Noether current

$$J = r^2 \left(\frac{1}{2} S N' + N S' - N S \frac{\Phi'}{\gamma} \right) . \tag{8.15}$$

The conservation condition $J' = 0$ can be straightforwardly verified with the help of Eqs. (8.4). Taking the limit $r \to r_h$, we obtain

$$J = \frac{r_h^2}{2} S_h N'(r_h) > 0, \tag{8.16}$$

since $S(r)$ is everywhere positive, and $N' > 0$ at the event horizon. Now, assuming that an inner horizon exists at some $r_- < r_h$, such that $N(r_-) = 0$, $N'(r_-) < 0$, and taking the limit $r \to r_-$, we arrive at

$$J = \frac{r_-^2}{2} S(r_-) N'(r_-) < 0, \tag{8.17}$$

which contradicts (8.16). The addition of a dilaton field thus eliminates the Cauchy horizons. We believe that the same is true for a large class of non-linear matter models.

A more detailed account of this work is given in the diploma thesis by OS [13].

Acknowledgements

MSV thanks the organizers of "The Internal Structure of Black Holes and Space-time Singularities" workshop at Haifa for an interesting meeting and the generous hospitality.

Bibliography

[1] E. Poisson and W. Israel, Phys. Rev. D **41** 1796 (1990).

[2] A. Ori, Phys. Rev. Lett. **67**, 789 (1991).

[3] A. Ori, Phys. Rev. Lett. **68**, 2117 (1992).

[4] A. Bonanno, S. Droz, W. Israel, and S. M. Morsnik, Phys. Rev. D **50**, 7372 (1994).

[5] P. R. Brady and J. D. Smith, Phys. Rev. Lett. **75**, 1256 (1995).

[6] L. M. Burko, Phys. Rev. Lett **79**, 4958 (1997).

[7] E. E. Donets, D. V. Gal'tsov, and M. Yu. Zotov, *Internal structure of Einstein-Yang-Mills black Holes*, gr-qc/9612067.

[8] P. Breitenlohner, G. Lavrelashvili, and D. Maison, *Mass inflation and chaotic behaviour inside hairy black holes*, gr-qc/9703047.

[9] E. E. Donets, D. V. Gal'tsov, *Power-law mass inflation in Einstein-Yang-Mills-Higgs black holes*, gr-qc/9706067.

[10] E. E. Donets, D. V. Gal'tsov, and M. Yu. Zotov, *Singularities inside nonAbelian black holes*, gr-qc/9612063.

[11] E. E. Donets, D. V. Gal'tsov, Phys. Lett. **B 302**, 411 (1993).

[12] G. Lavrelashvili, and D. Maison, Nucl. Phys. **B 410**, 407 (1993).

[13] O. Sarbach, *Diploma Thesis*, Univerity of Zürich, Fall 1997 (unpublished).

NON–ABELIAN BLACK HOLES: THE INSIDE STORY

Peter Breitenlohner [a], George Lavrelashvili [b,c] and Dieter Maison [a]

a) Max-Planck-Institut für Physik, Werner Heisenberg Institut, Föhringer Ring 6, 80805 Munich, Fed. Rep. Germany
b) Institute for Theoretical Physics, University of Bern, Sidlerstrasse 5, CH-3012 Bern, Switzerland
c) On leave of absence from Tbilisi Mathematical Institute, 380093 Tbilisi, Georgia

Abstract

Recent progress in understanding the internal structure of non-Abelian black holes is discussed. The interior geometry of static, spherically symmetric black holes of the Einstein-Yang-Mills-Higgs theory is analyzed. It is found that in contrast to the Abelian case no inner (Cauchy) horizon is formed inside non-Abelian black holes in the generic case. The solutions approach an 'almost' Cauchy horizon, but close to it they undergo an enormous growth of the mass function, a phenomenon also observed for perturbations of the Reissner-Nordström and Kerr black holes (called 'mass inflation'). A significant difference between the theories with and without Higgs field is found. Without Higgs field the YM field induces a kind of cyclic behaviour leading to repeated cycles of mass inflation – taking the form of violent explosions – interrupted by quiescent periods and subsequent approaches to a next 'almost' Cauchy horizon. With the Higgs field no such cycles occur. Besides the generic solutions there exist non-generic families with a Schwarzschild, Reissner-Nordström resp. a pseudo Reissner-Nordström type singularity at $r = 0$.

9.1 Introduction

Let us start recalling what is known about the non-Abelian black holes in general. The non-Abelian story begun in 1988 when Bartnik and McKinnon (BK) unexpectedly found [1] numerically a discrete sequence of globally regular, particle like solutions of the Einstein–Yang–Mills (EYM) theory. Soon the same model was solved with the different boundary conditions corresponding to black holes [2]. Numerical findings were confirmed by mathematically rigorous existence proofs [3, 4] of both regular and black hole solutions. It turned out that all the above solutions of the EYM theory are classically unstable against small perturbations. In addition to the genuine gravitational instabilities [6] there also instabilities of topological origin [7, 8] related to the sphaleron nature of the solitons [9, 10].

A few related systems were investigated. It was shown [11, 12] that the gravitational field can be replaced by a dilaton, so that the YM-dilaton theory in flat space has a tower of solutions similar to the BK sequence. The combined EYMD theory [13, 14, 15, 16, 17] was shown to possess both regular and black hole solutions for any value of the dilaton coupling constant. For the EYMH theory with a Higgs doublet [18] it was shown that the theory in addition to the gravitating sphaleron solution contains its BK type excitations.

Furthermore the EYMH theory with a triplet Higgs [19, 20] was studied. This theory is interesting since in the flat limit it contains t'Hooft-Polyakov monopoles, which are known to be stable. It was shown that the basic monopoles continue to exist, when gravity is switched on, at least as long as the gravitational self-interaction is not too strong. In addition the monopole admits unstable BK type excitations.

Adding a cosmological constant to the EYM theory one obtains non-asymptotically flat analogues of the BK solutions [21, 22].

Whereas in the above study the globally regular solutions were completely analyzed, the investigation of the black hole solutions was not complete since their internal structure was unknown. Recently this problem was investigated independently by us [24] and by Donets, Gal'tsov and Zotov [23]. Our main results on the EYM case essentially agree with theirs, although we differ in some details.

We found it adequate to describe the generic behavior as a kind of mass inflation closely related to the "usual" mass inflation (see e.g. [25, 26, 27, 28, 30] and many references in the present proceedings).

In addition to the generic behavior there are three different types of special solutions, which are obtained by fine tuning of the initial data at the horizon. There are solutions with Reissner-Nordström (RN), Schwarzschild and pseudo-RN type behavior.

The present contribution is essentially based on our paper [24].

9.2 Field Equations

The action of the EYMH theory is

$$S = \frac{1}{4\pi}\int\Big(-\frac{1}{4G}R - \frac{1}{4g^2}F^2 + \frac{1}{2}|D_\mu\Phi|^2 - V(\Phi)\Big)\sqrt{-g}\, d^4x\,, \tag{9.1}$$

where g denote the gauge coupling constant, G is Newton's constant, F is the field strength of the *SU(2)* Yang-Mills field and $V(\Phi)$ is the usual quartic Higgs potential. The pure EYM action and corresponding equations can be trivially obtained from the EYMH ones by putting the Higgs field Φ and its potential $V(\Phi)$ to zero.

For the static, spherically symmetric metric we use the parametrization

$$ds^2 = A^2 B dt^2 - \frac{dR^2}{B} - r^2(R) d\Omega^2 , \tag{9.2}$$

with $d\Omega^2 = d\theta^2 + \sin^2\theta d\varphi^2$ and three independent functions A, B, r of a radial coordinate R, which has, in contrast to r, no geometrical significance. As long as $dr/dR \neq 0$ the simplest choice for r is $R = r$, i.e. Schwarzschild (S) coordinates. In this case it is common to express B through the "mass function" m defined by $B = 1 - 2m/r$.

For the *SU(2)* Yang-Mills field W^a_μ we use the standard minimal spherically symmetric (purely 'magnetic') ansatz

$$W^a_\mu T_a dx^\mu = W(R)(T_1 d\theta + T_2 \sin\theta d\varphi) + T_3 \cos\theta d\varphi , \tag{9.3}$$

and for the Higgs (triplet) field we assume the form

$$\Phi^a T_a = H(R) n^a T_a , \tag{9.4}$$

where T_a denote the generators of $SU(2)$ in the adjoint representation. One might also consider other representations for the Higgs , e.g. doublet, but we believe that the behavior near $r = 0$ is the same. Plugging these ansätze into the EYMH action results in

$$S = -\int dRA\Big[\frac{1}{2}\Big(1 + B((r')^2 + \frac{(A^2B)'}{2A^2B}(r^2)')\Big) - Br^2V_1 - V_2\Big] , \tag{9.5}$$

with

$$V_1 = \frac{(W')^2}{r^2} + \frac{1}{2}(H')^2 , \tag{9.6}$$

and

$$V_2 = \frac{(1-W^2)^2}{2r^2} + \frac{\beta^2 r^2}{8}(H^2 - \alpha^2)^2 + W^2H^2 . \tag{9.7}$$

Through a suitable rescaling we have achieved that the action depends only on the dimensionless parameters α and β representing the mass ratios $\alpha = M_W\sqrt{G}/g = M_W/gM_{\rm Pl}$ and $\beta = M_H/M_W$ (M_H and M_W denoting the Higgs resp. gauge boson mass).

Using S coordinates the field equations obtained from (9.5) are

$$(BW')' = W(\frac{W^2-1}{r^2} + H^2) - 2rBW'V_1 , \tag{9.8a}$$

$$(r^2BH')' = (2W^2 + \frac{\beta^2 r^2}{2}(H^2 - \alpha^2))H - 2r^3BH'V_1 , \tag{9.8b}$$

$$(rB)' = 1 - 2r^2BV_1 - 2V_2 , \tag{9.8c}$$

$$A' = 2rV_1A . \tag{9.8d}$$

If $dr/dR = 0$ (equator!) S coordinates become singular and one has to use a different choice (gauge) for R. A convenient possibility is given by $B \equiv r^{-2}$ for $B > 0$ resp. $B \equiv -r^{-2}$ for $B < 0$. We denote this radial coordinate by τ in order to distinguish it from the S coordinate r. With this choice the metric takes the form

$$ds^2 = -\frac{A(\tau)}{r^2(\tau)}dt^2 + r^2(\tau)(d\tau^2 - d\Omega^2) . \tag{9.9}$$

We will refer to this system of coordinates as isotropic coordinates. The equations obtained with isotropic coordinates are given in [24].

9.3 Singular points

Obviously the field Eqs. (9.8) are singular at $r = 0$, $r = \infty$ and for points where B vanishes. Note that a generic solution should have five free parameters for the EYMH case resp. three for the EYM case. Here we have ignored the variable A, which decouples (compare Eq. (9.8d)).

At the singular points the generic solution is singular and only a non-generic subset of solutions stays regular.

In the vicinity of $r = \infty$ one finds a 3-parameter family of asymptotically flat solutions for the EYMH theory (2-parameter family for the EYM case).

At $r = r_h$ for any given $r_h > 0$ one finds a 3-parameter family characterized by the value of a gauge and Higgs fields at the horizon, W_h and H_h respectively.

In case of the EYM theory it was shown [2, 3, 4] that for any given $r_h > 0$ there is a discrete set of solutions interpolating between the horizon and infinity. They can be characterized by an integer n, the number of nodes of the gauge amplitude W.

The EYMH case is more complicated [19, 20]. Let us here only recall that the previous research was concentrated on the investigation in the interval $[r_h, \infty]$.

Now we turn to the classification of the singular behavior at $r = 0$, which is of particular relevance for the internal structure of black hole solutions. We have to distinguish two cases, $B > 0$ and $B < 0$.

1. $B > 0$: For black holes this case is only possible, if there is a second, inner horizon. One finds a 5-parameter, i.e. generic, family of solutions.

$$W(r) = W_0 + \frac{W_0}{2(1 - W_0^2)}r^2 + W_3 r^3 + O(r^4) , \tag{9.10a}$$

$$H(r) = H_0 + H_1 r + O(r^2) , \tag{9.10b}$$

$$B(r) = \frac{(W_0^2 - 1)^2}{r^2} - \frac{2M_0}{r} + O(1) , \tag{9.10c}$$

where $W_0^2 \neq 1$.

According to the asymptotics of $B(r)$ we may call the singular behavior to be of RN-type. The special case $W_0^2 = 1, M_0 < 0$ leads to S-type behavior with a naked singularity. On the other hand $W_0^2 = 1, H_0 = 0, M_0 = 0$ gives regular solutions.

2. $B < 0$: This case is more involved, with two disjoint families of singular solutions.

2.1 There is a 3-parameter family of solutions with a S-type singularity, characterized by the asymptotics

$$W(r) = 1 + W_2 r^2 + O(r^3) , \tag{9.11a}$$

$$H(r) = H_0 + O(r) , \tag{9.11b}$$

$$B(r) = -\frac{2M_0}{r} + O(1) . \tag{9.11c}$$

where $M_0 > 0$.

Obviously the condition $B < 0$ ($M_0 > 0$) prevents the existence of regular solutions in this case.

2.2 There is an additional 2-parameter family of solutions with a pseudo-RN singularity (pseudo because $B < 0$).

$$W(r) = W_0 \pm r + O(r^2) , \tag{9.12a}$$

$$H(r) = H_0 + O(r^2) , \tag{9.12b}$$

$$B(r) = -\frac{(W_0^2 - 1)^2}{r^2} \pm \frac{4W_0(1 - W_0^2)}{r} + O(1) . \tag{9.12c}$$

with $W_0^2 \neq 1$. The eigenvalues of the linearized equations are $\lambda_{1,2} = -1/2(3 \pm i\sqrt{15})$ and $\lambda_3 = -1$. This is a repulsive focal point which will turn out to be important for the cyclic behavior in the EYM case.

Note that the corresponding Taylor series for the singular points in case of the EYM theory were listed (with minor mistakes) in [23], whereas in our work [24] the more general EYMH theory was studied and local existence proof was given. In other words we have shown that the expressions above are in fact the beginning of a **convergent** Taylor series for W, H and $r^2 B$.

In the case $B < 0$ we obtained no singular class that has enough parameters (three for EYM and five for EYMH) to describe the generic behavior. Since, also the appearance of a second, inner horizon is a non-generic phenomenon, one may wonder, what the generic behavior inside the horizon near $r = 0$ looks like. This situation is shown schematically on the Fig. 9.1 for the case of the EYM theory.

9.4 Numerical results

In order to investigate the generic behavior of non-Abelian black holes inside the event horizon, we integrate the field Eqs. (9.8) from the horizon assuming $B < 0$, ignoring the constraints on the initial data at the horizon required for asymptotic flatness.

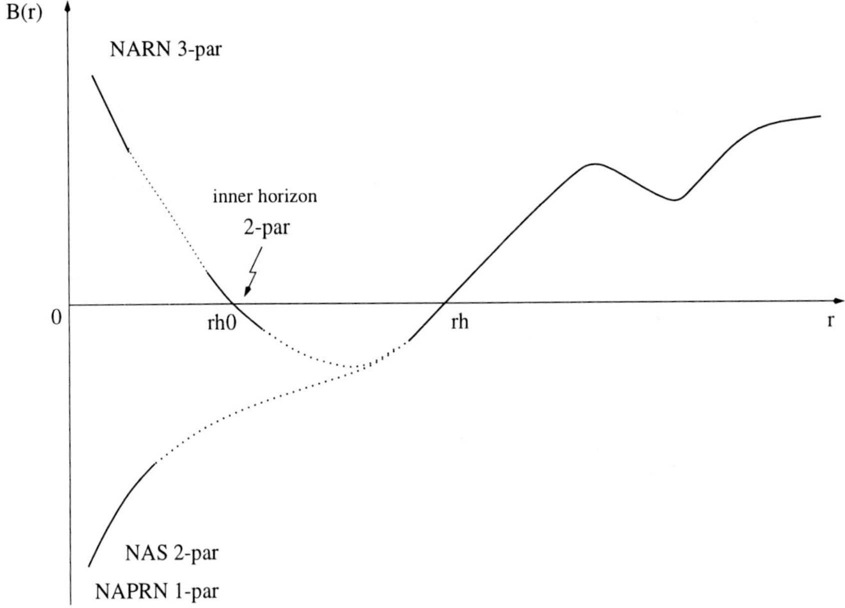

Figure 9.1: Schematic view of the different special solutions in the EYM theory.

Using the Killing time t the horizon is a singular point of the equations. Consequently one has to desingularize the equations in order to be able to start the integration right there. How this can be done, was described in [24, 19].

As one performs the numerical integration one quickly runs into problems due to the occurrence of a quasi-singularity, initiated by a sudden steep raise of W' and subsequent exponential growth of B resp. m (compare Figs. 9.6, and 9.8 for some examples). This inflationary behavior of the mass function is similar to the one observed for perturbations of the Abelian black hole solution at the Cauchy horizon [25, 26, 27, 28]. While this fast growth continues indefinitely for the EYMH system, it comes to a stop without the Higgs field. The mass function reaches a plateau and stays constant for a while until it starts to decrease again. When B has become small enough, i.e. the solution comes close to an inner horizon, the same inflationary process repeats itself. Generically this second "explosion" is so violent (we will give estimates on the increase of m in chapter 9.5) that the numerical integration procedure breaks down.

Besides these generic solutions there are certain families of special solutions obtained through suitable fine-tuning of the initial data at the horizon. There are two classes of such special solutions. The first class are black holes with a second, inner horizon, the second are solutions with one of the singular behaviors at $r = 0$ for $B < 0$ described in chapter 9.3. The numerical construction of such solutions is complicated by the fact that both boundary points are singular points of the equations. The strategies employed to solve such problems are well described in the paper on gravitating monopoles [19]. Actually, in order to control the numerical uncertainties we used two different methods, which may be called "matching" and "shooting and aiming". For matching we integrate independently from both boundary points with regular initial data, tuning these data at both ends until the two branches of the solution match. For shooting and aiming we integrate only from one end and try to suppress the singular part of the solution at the other end by suitably tuning the initial data at the starting point.

Our results concerning special solutions are shown in Fig. 9.2 and discussed in [24]. As already said, the first class of special solutions consists of black holes with a second, inner horizon; we call them non-Abelian RN-type (NARN) solutions. We have determined two such 1-parameter families for the EYM system, shown in Fig. 9.2. As may be inferred from Fig. 9.2, the (dotted) curve 2 corresponding to one such family intersects all (solid) curves describing asymptotically flat solutions except the one for $n = 1$. Curve continues straight through the parabola $r_h = 1 - W_h^2$ and runs all the way to $r_h = W_h = 0$. The branch to the left of the parabola cannot be obtained using S coordinates since solutions develop local maximum of r between the horizons.

Our second NARN family (curve 5 of Fig. 9.2) stays completely to the left of the parabola and ends at $r_h \approx 0.9$ close to the curve 3, whose significance will be explained below.

The second class are solutions without a second horizon (i.e. B stays negative) approaching the center $r = 0$ with one of the two singular behaviors described in chapter 9.3, i.e. those with a S-type singularity resp. with a pseudo-RN-type singularity; we denote them NAS resp. NAPRN solutions. We have determined several NAS families represented by the dashed-dotted curves of Fig. 9.2. The curve 1 staying to the right of the parabola coincides with the corresponding one found in [23], whereas the others, staying essentially to the left of the parabola are new [24]. The basic NAS curve 1 intersects (once) only $n = 1$ (solid) curve for asymptotically flat black holes. As will be explained in chapter 9.5, the two NAS curves 6 and 7 accompanying the (dotted) NARN curve 5 are expected to merge with the NAS curve 3 close to $r_h = 0.9$. Some of the NAS curves (e.g., 3 and 4) are expected to extend indefinitely to the right, but numerical difficulties (too violent "explosions") prevented us from continuing them further to larger values of r_h. They will intersect the (solid) curves for asymptotically flat solutions with $n = 2, 3, \ldots$ zeros of W and therefore yield additional asymptotically flat NAS black holes.

Asymptotically (for big r_h) the basic NARN curve 2 approaches the basic NAS curve 1. Why this happens can be "understood" from Fig. 9.4.

Finally there are the NAPRN solutions, which constitute a discrete set according to the number of available free parameters at $r = 0$. We found several such solutions [24]. Few NAPRN solutions are shown in Fig. 9.4 and Fig. 9.5. Only one of them has no maximum of r and was found in [23] as well. Let us stress that although

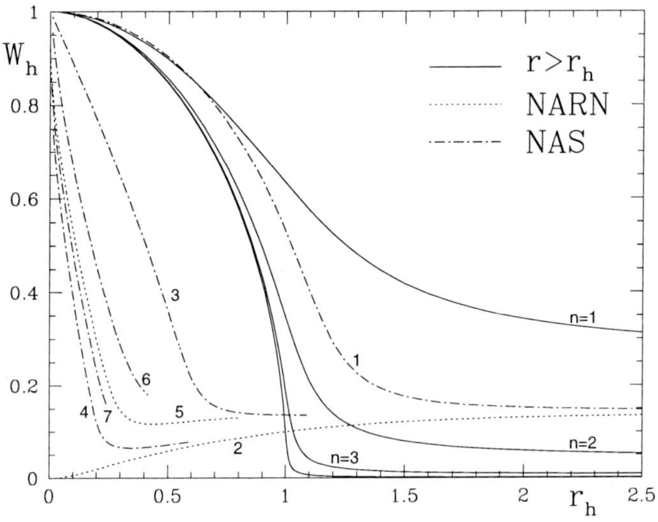

Figure 9.2: Initial data for special solutions. The solid curves represent asymptotically flat solutions with n zeros of W. The other curves represent various NARN and NAS families.

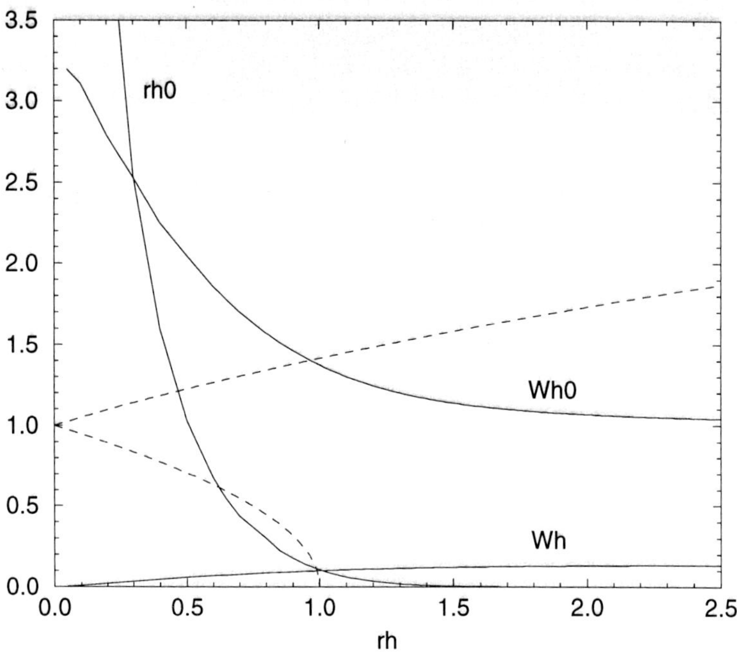

Figure 9.3: Parameters of the basic (lowest) NARN solutions. One clearly sees that with increasing r_h the value of the second (inner) horizon $r_{h0} \to 0$ and value of the gauge field there $W_{h0} \to 1$. This values correspond to NAS solutions.

the NAPRN solutions do not correspond to asymptotically flat black holes they play essential role in the explanation on the cyclic behavior of a generic solution in the EYM case (compare discussion in the next chapter 9.5).

9.5 Qualitative Discussion

We shall now give a qualitative picture of the solutions and try to explain our numerical results. Let us briefly summarize what can be done (and what in fact has been done [23, 24])

- get a qualitative understanding of the solutions [23, 24]

- obtain a plateau – to – plateau formula in the EYM case relating quantities at one plateau (before "explosion") to those on the next plateau (after "explosion") [24]

- describe a simplified dynamical system, which reflects the main properties of the generic solutions [23, 24]

Since the generic behavior of the solutions is rather different in the cases with and without Higgs field, we shall treat the two cases separately. Let us first concentrate on the case without Higgs field.

9.5.1 EYM theory

For a "naive" understanding of the cyclic behavior one can use a mechanical analogy. Introducing the "time" variable $\sigma = -\ln(r)$ the EYM equations can be written in the form

$$\ddot{W} = \frac{W(W^2-1)}{B} - \left[2 + \frac{1}{B}\left(\frac{(W^2-1)^2}{r^2} - 1\right)\right]\dot{W}\,, \tag{9.13a}$$

$$\dot{B} = \left(\frac{(W^2-1)^2}{r^2} - 1\right) + \left(1 + \frac{2\dot{W}^2}{r^2}\right)B\,, \tag{9.13b}$$

where $\dot{} \equiv d/d\sigma$. The first equation Eq. (9.13a)) resembles the motion of a fictitious particle in a potential with velocity dependent friction.

Note that the sign of the friction coefficient (term in square brackets in the Eq. (9.13a)) can be positive as well as negative. Close to the horizon $(W^2-1)^2/r^2 - 1 < 0$ and friction coefficient is positive, corresponding to a deceleration of the "particle". As the time σ increases (r decreases) this term changes sign and the friction turns into anti-friction. The particle starts to accelerate quickly. This leads to a domination of the second term (kinetic energy) in Eq. (9.13b), which in turn leads to a fast growth of the function B (respectively m). But growth of B stops the anti-friction in Eq. (9.13a) and the particle is again in the slow roll regime until the next "explosion".

For more detailed discussion of the generic behavior we introduce the notation $\bar{U} \equiv BW'$ and $\bar{B} \equiv rB$ and use again $\sigma \equiv -\ln(r)$ as a radial coordinate [24]. Note that $\bar{B} \approx -2m$ for small r. With these variables the field Eqs. (9.8)

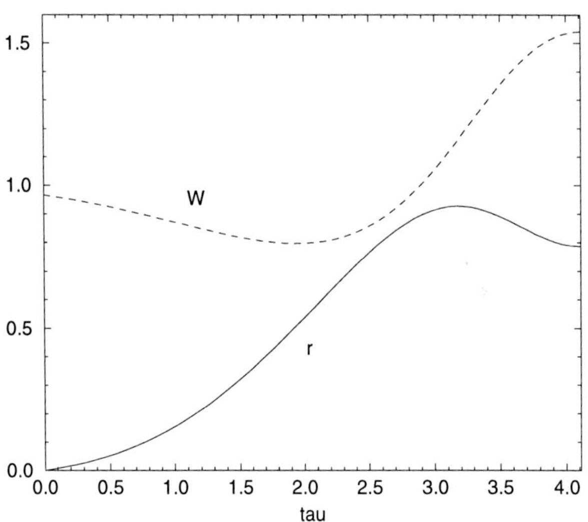

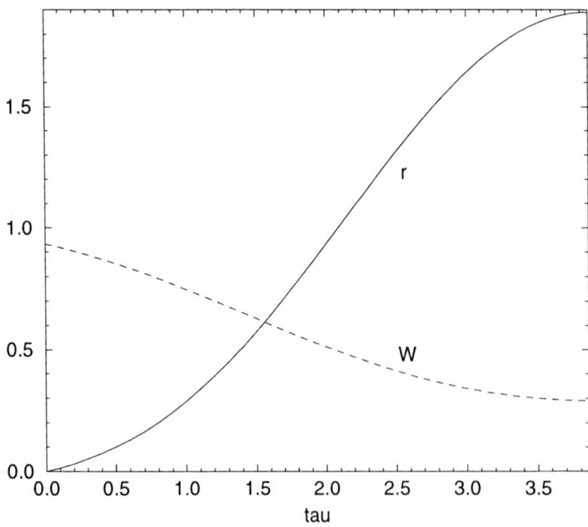

Figure 9.4: Two different NAPRN solutions with no zero of W between $r = 0$ and $r = r_h$.

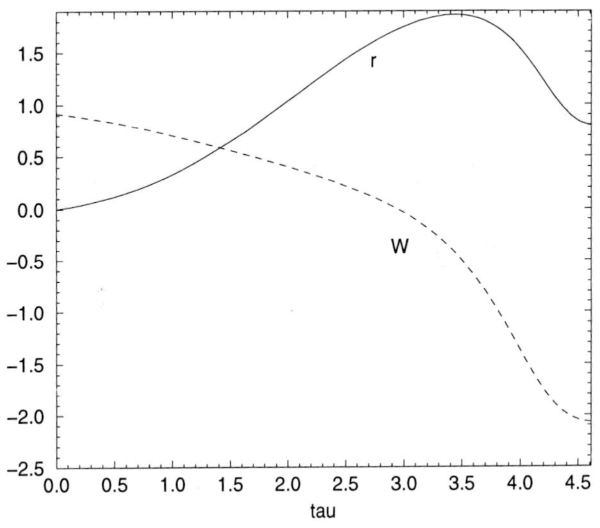

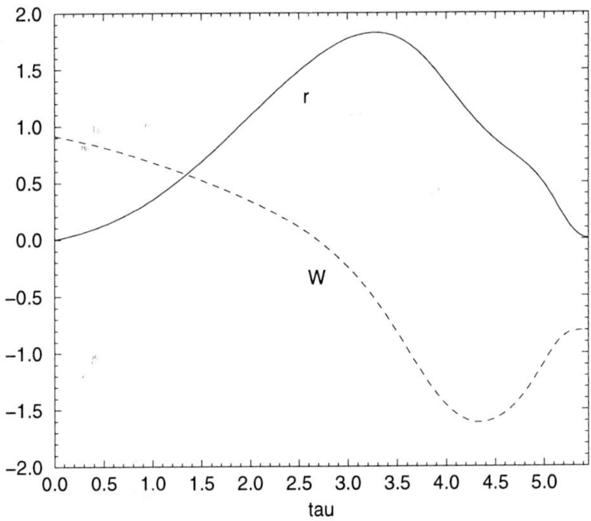

Figure 9.5: Two different NAPRN solutions with one zero of W between $r = 0$ and $r = r_h$.

$$\dot{W} = -r^2 \frac{\bar{U}}{\bar{B}} , \tag{9.14a}$$

$$\dot{\bar{U}} = -W \frac{W^2 - 1}{r} + 2r^2 \frac{\bar{U}^3}{\bar{B}^2} , \tag{9.14b}$$

$$\dot{\bar{B}} = r \left(\frac{(1 - W^2)^2}{r^2} - 1 \right) + 2r^2 \frac{\bar{U}^2}{\bar{B}} . \tag{9.14c}$$

Close to the horizon the first term in the equation for $\bar{B}$ dominates (since $\bar{U}$ vanishes at $r = r_h$) and thus $\bar{B}$ becomes negative. Provided W^2 does not tend to 1, this term will, however, change sign as r decreases and $\bar{B}$ will turn back to zero. Assuming further that $\bar{U}$ does not tend to zero simultaneously, the second term in the equation for $\bar{U}$ will grow very rapidly as $\bar{B}$ approaches zero, leading to a rapid increase of $\bar{U}$. This in turn induces a rapid growth of $\bar{B}$ (compare Fig. 9.6). Once the second terms in Eqs. (9.14b,c) dominate one gets $(\bar{U}/\bar{B})^{\cdot} \approx 0$ and thus $\bar{U}/\bar{B} = W'/r$ tends to a constant c. As long as $(rc)^2$ is sizable $\bar{U}$ and $\bar{B}$ increase exponentially, giving rise to the phenomenon of mass inflation. Eventually this growth comes to a stop when $(rc)^2$ has become small enough. Then $\bar{U}$ and $\bar{B}$ stay constant until the first terms in Eqs. (9.14b,c) become sizable again. As before $\bar{B}$ tends to zero inducing another "explosion" resp. cycle of mass inflation (compare Fig. 9.6).

In the discussion above we made two provisions – that W^2 stays away from 1 and that $\bar{U}$ does not tend zo zero simultaneously with $\bar{B}$. If the first condition is violated, i.e. $W^2 \to 1$ we get a NAS solution. If on the other hand $\bar{U}$ and $\bar{B}$ develop a common zero we get a NARN solution, i.e. a solution with a second horizon. Both these phenomena can occur after any finite number of cycles, giving rise to several NAS resp. NARN curves as in Fig. 9.2. Generically W changes very little during an inflationary cycle, with the exception of solutions that come very close to a second horizon, i.e. close to a NARN solution. In this case W may change by any amount, depending on how small $\bar{U}$ becomes at the start of the explosion. By suitably fine-tuning the initial data at the horizon one can then obtain new NAS solutions with $W \to \pm 1$ or a new NARN solution. In this way each NARN solution is the 'parent' of two NAS and one NARN solution. This schematically shown on the Fig. 9.7. Fig. 9.2 shows two such generations: the NARN solutions labeled 2 have the NAS children 3 and 4 and the NARN child 5; the curves labeled 6 and 7 are the NAS children of 5. Whenever the value of W at the second horizon of a NARN solution approaches ± 1 this NARN curve and its NAS children merge with the corresponding sibling NAS curve having one cycle less. This hierarchy of special solutions gives rise to a kind of chaotic structure in this region of "phase space".

Neglecting irrelevant terms one can integrate Eqs. (9.14) and obtain a plateau – to – plateau relation [24], which connects the quantities $W_0, \bar{U}_0$ and $\bar{B}_0$ before an explosion with $W_1, \bar{U}_1, \bar{B}_1$ after it [31]

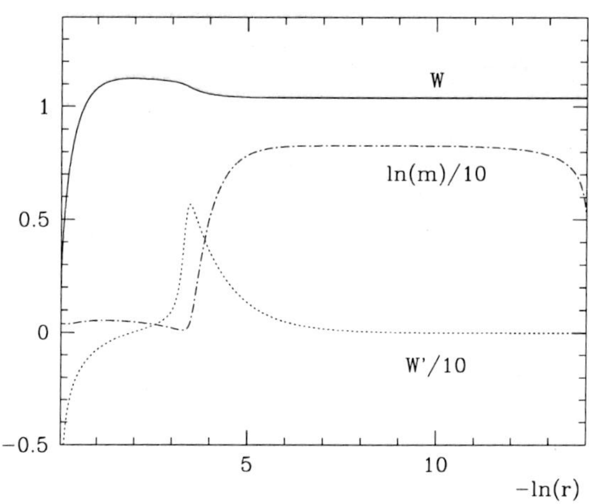

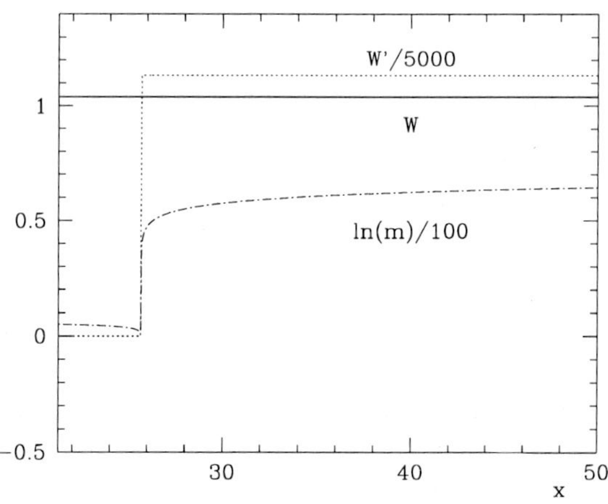

Figure 9.6: First two cycles of the solution with r_h=0.97 and W_h = 0.2. For the second cycle a suitably stretched coordinate x is used.

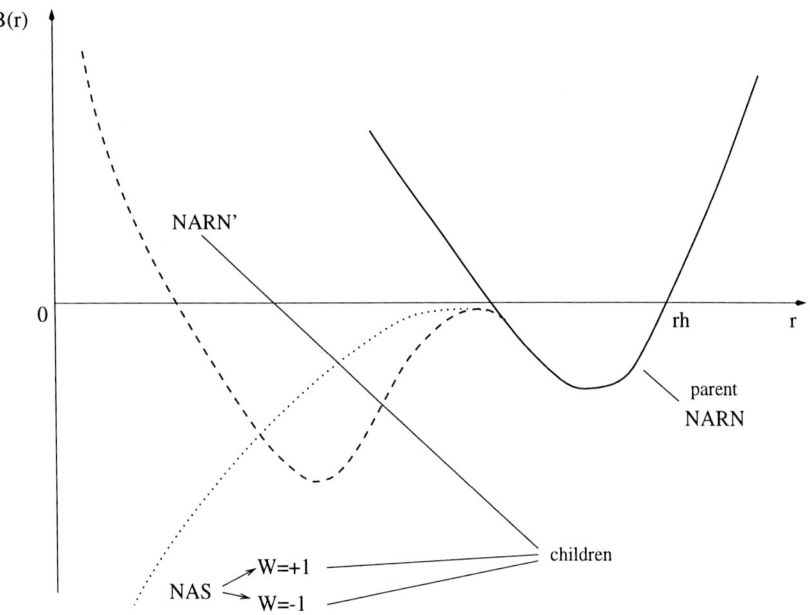

Figure 9.7: Schematic view of the NARN solutions and their children.

$$\bar{U}_1 = \bar{U}_0 e^{(cr_0)^2} , \quad \bar{B}_1 = \frac{\bar{U}_0}{c} e^{(cr_0)^2} , \quad W_1 = W_0 - \frac{c}{2} r_0^2 , \tag{9.15}$$

with

$$r_0 = -\frac{(W_0^2 - 1)^2}{\bar{B}_0} , \quad c = \frac{(W_0^2 - 1)^2}{2\bar{U}_0 r_0^3} . \tag{9.16}$$

It is instructive to illustrate these relations on an example. We take the fundamental black hole solution with $r_h = 1$ and $W_h = 0.6322$ shown in Fig. 1 in [24]. For the first explosion one finds the parameters $r_0 \approx 2.7 \cdot 10^{-4}$ and $c \approx 1.1 \cdot 10^5$ yielding $cr_0 \approx 30$ and thus $\bar{B}_1 \sim e^{900}$ and $W_1 - W_0 \approx 4 \cdot 10^{-3}$. The subsequent explosion will then take place at the fantastically small value $r_0 \sim e^{-900} \approx 10^{-330}$.

Since the change of W in one inflationary cycle has an extra factor r_0 the function W stays practically constant. If we furthermore concentrate on cases, where the first term in Eq. (9.14b) can be neglected we may use the simplified system [23, 24]

$$\dot{W} = 0 , \quad \dot{\bar{U}} = 2r^2 \frac{\bar{U}^3}{\bar{B}^2} , \quad \dot{\bar{B}} = \frac{(1 - W^2)^2}{r} + 2r^2 \frac{\bar{U}^2}{\bar{B}} . \tag{9.17}$$

Introducing the variables $x \equiv r\bar{U}/\bar{B} = W'$ and $y \equiv -(1 - W^2)^2/r\bar{B}$ one obtains the autonomous system

$$\dot{W} = 0 , \quad \dot{x} = (y - 1)x , \quad \dot{y} = y(y + 1 - 2x^2) . \tag{9.18}$$

Since the first of these equations may be ignored, we can concentrate on the x, y part. As usual for 2-dimensional dynamical systems the global behavior of the solutions can be analyzed determining its fixed points. Since the "large time" behavior $\sigma \to \infty$ corresponds to the limit $r \to 0$ these fixed points are related to the singular solutions at $r = 0$ discussed in chapter 9.3. There are essentially three different fixed points.

1. For $y < 0$ there is the fixed point $x = 0$, $y = -1$ giving the RN type singularity. Its eigenvalues are -1 and -2, hence it acts as an attracting center for $\sigma \to \infty$.
2. Then there is the point $x = y = 0$, a saddle with eigenvalues ± 1.
3. In addition there are the points $x = \pm 1$, $y = 1$ with the eigenvalues $1/2(1 \pm i\sqrt{15})$, related to the pseudo-RN type singularity. This fixed point acts as a repulsive focal point, from which the trajectories spiral outwards. Since solutions of the approximate system given by the Eqs. (9.18) cannot cross the coordinate axes, solutions in the quadrants $y > 0$, $x > 0$ resp. $x < 0$ stay there performing larger and larger turns around the focal point coming closer and closer to the saddle point $x = y = 0$ without ever meeting it. As observed in [23] this nicely explains the cyclic inflationary behavior of the solutions in the generic case.

It is interesting to note that the similar results were obtained in the Abelian case [27, 28] in the homogeneous mass inflation model [29].

9.5.2 EYMH theory

Finally let us discuss the black holes with Higgs field. Apart from the generic solutions there are the special ones approaching $r = 0$ with a singular behavior described in chapter 9.3. On the other hand, the generic behavior is much simpler than in the previously discussed situation without Higgs field. An easy way to understand this difference is to derive the analogue of the simplified system Eqs. (9.18). Introducing the additional variable $z \equiv -\dot{H}$ and ignoring again irrelevant terms one finds [24]

$$\dot{W} = 0 \, , \quad \dot{H} = -z \, , \tag{9.19a}$$

$$\dot{x} = (y-1)x \, , \quad \dot{z} = yz \tag{9.19b}$$

$$\dot{y} = y(y + 1 - 2x^2 - z^2) \, . \tag{9.19c}$$

Leaving aside the decoupled equations for W and H one may study the fixed points of the (x, y, z) system. For $z = 0$ one clearly finds the previous fixed points of the (x, y) system. However, for $z \neq 0$ the focal point disappears and the only fixed point for $y \geq 0$ is $x = y = 0$, $z = z_0$ with some constant z_0. For $z_0^2 < 1$ this point is a saddle with one unstable mode, whereas for $z_0^2 > 1$ it is a stable attractor. The latter describes the simple inflationary behavior described in chapter 9.4 and shown in the left part of Fig. 9.8. Solutions approaching a fixed point with $z_0^2 < 1$ eventually run away from it again and ultimately tend to one with $z_0^2 > 1$ as shown in the right part of Fig. 9.8.

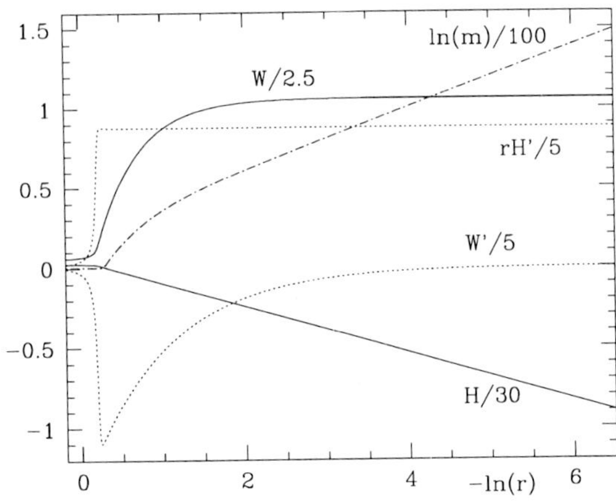

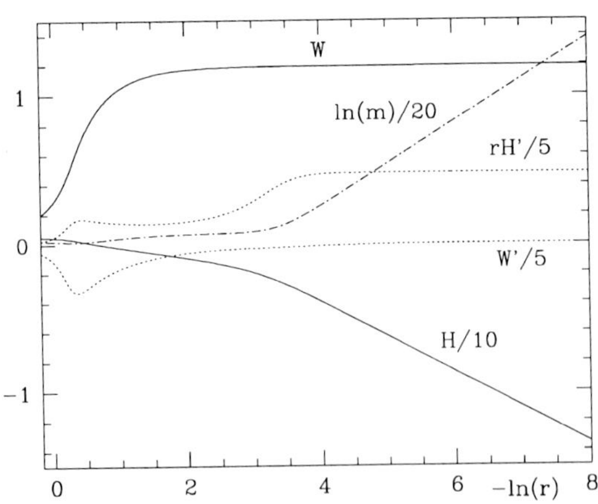

Figure 9.8: Two characteristic types of inflationary solutions with Higgs fields. Note that in both cases asymptotically $\ln(m)$ is linear in $\ln(r)$.

In the limit the equations can be trivially integrated with the result

$$y = y_0 e^{(1-z_0^2)\sigma} = y_0 r^{z_0^2 - 1}. \tag{9.20}$$

Thus the mass function grows exponentially in "time" σ or as a power in terms of r

$$m = m_0 e^{z_0^2 \sigma} = \frac{m_0}{r^{z_0^2}} \tag{9.21}$$

in perfect agreement with our numerical results Fig. 9.8. Note that the fixed point Eq. (9.20) differs from the ones listed in the chapter 9.3 since the nature (strength) of the singularity depends on the solution itself. This fixed point was "rediscovered" in [32].

Note that the inclusion of a scalar field in the homogeneous mass inflation model [28] has the same effect as the addition of the Higgs field to the EYM theory, namely that the mass inflation cycles disappear.

9.6 Concluding remarks

To summarize the interior geometry of non-Abelian black holes exhibits a very interesting and complicated structure. Besides the generic solutions there are special NAS, NARN and NAPRN solutions, which can be obtained by a fine tuning initial data at the horizon. The main conclusion is that no inner (Cauchy) horizon is formed inside non–Abelian black holes in generic case, instead one obtains a kind of mass inflation. Without a Higgs field, i.e. for the EYM theory, this mass inflation repeats itself in cycles of ever more violent growth.

A natural question to ask is what might be potential outcome of this investigation. A short answer would be

– illustration of singularity theorems [33]

– possible cosmological applications (see e.g. [32])

– interesting laboratory for non-perturbative study of (homogeneous) mass inflation phenomenon

Naturally one should bare in mind limitations which come from quantum corrections and instability. Our consideration was purely classical, but when r tends to zero the curvature diverges and quantum corrections will become important. As it was discussed in the Introduction the EYM and most of the EYMH black holes are classically unstable. What will be a fate of a non-static perturbations inside non-Abelian black holes is an interesting open question. Definitely these subjects require further study.

9.7 Acknowledgments

G.L wishes to thank the organizers of the workshop for the invitation and for the generosity extended during the workshop.

The work of G.L. was supported in part by the Tomalla Foundation and by the Swiss National Science Foundation.

Bibliography

[1] R. Bartnik and J. McKinnon, Phys. Rev. Lett. **61**, 141 (1988).

[2] H. P. Künzle and A. K. M. Masood-ul-Alam, J. Math. Phys. **31**, 928 (1990);
M. S. Volkov and D. V. Gal'tsov, JETP Lett. **50**, 345 (1990);
P. Bizon, Phys. Rev. Lett. **61**, 2844 (1990).

[3] J. A. Smoller, A. G. Wasserman, and S. T. Yau, Comm. Math. Phys. **154**, 377 (1993);
J. A. Smoller and A. G. Wasserman, Comm. Math. Phys. **151**, 303 (1993).

[4] P. Breitenlohner, P. Forgacs, and D. Maison, Commun. Math. Phys. **163**, 141 (1994).

[5] At least part of the solutions possess an equator, i.e. a local maximum of r. For those the use of S coordinates is excluded and we integrated the equations in the isotropic coordinates.

[6] N. Straumann and Z. H. Zhou, Phys. Lett. **B 237**, 353 (1990); **B 243** (1990) 33–35.

[7] G. Lavrelashvili and D. Maison, Phys. Lett. **B 343**, 214 (1995).

[8] M.S. Volkov, O. Brodbeck, G. Lavrelashvili and N. Straumann, Phys. Lett. **B 349**, 438 (1995).

[9] D. V. Gal'tsov and M. S. Volkov, Phys. Lett. **B 273**, 255 (1991) 255;
D. Sudarsky and R. M. Wald, Phys. Rev. D. **46**, 1453 (1992);
I. Moss and A. Wray, Phys. Rev. D. **46**, 1215 (1992);
G. W. Gibbons and A. R. Steif, Phys. Lett. **B 320**, 245 (1994) 245.

[10] G. Lavrelashvili, Mod. Phys. Lett. **A 9**, 3731 (1994).

[11] G. Lavrelashvili and D. Maison, Phys. Lett. **B 295**, 67 (1992).

[12] P. Bizon, Phys. Rev. D. **47**, 1656 (1993).

[13] G. Lavrelashvili and D. Maison, Nucl. Phys. **B 410**, 407, (1993);
G. Lavrelashvili, *Black holes and sphalerons in low–energy effective string theory*, hep-th/9410183.

[14] P. Bizon, Acta Physica Polonica **B 24**, 1209 (1993).

[15] E. E. Donets and D. V. Gal'tsov, Phys. Lett. **B 302**, 411 (1993).

[16] T. Torii and Kei-ichi Maeda, Phys. Rev. D. **48**, 1643 (1993).

[17] B. Kleihaus, J. Kunz, and A. Sood, Phys. Lett. **B 372**, 204 (1996).

[18] B. R. Greene, S. D. Mathur, and C. M. O'Neill, Phys. Rev. D. **47**, 2242 (1993).

[19] P. Breitenlohner, P. Forgacs, and D. Maison, Nucl. Phys. **B 383**, 357 (1992); **B 442**, 126 (1995).

[20] K. Lee, V. P. Nair, and E. J. Weinberg, Phys. Rev. D. **45**, 2751 (1992) 2751.

[21] T. Torii, Kei-ichi Maeda, and T. Tachizawa, Phys. Rev. D. **52**, 4272 (1995).

[22] M. S. Volkov, N. Straumann, G. Lavrelashvili, M. Heusler, and O. Brodbeck, Phys. Rev. D. **54**, 7243 (1996).

[23] E. E. Donets, D. V. Gal'tsov, M. Yu. Zotov, *Internal Structure of Einstein-Yang-Mills Black Holes*, gr-qc/9612067, versions 1–4.

[24] P. Breitenlohner, G. Lavrelashvili, and D. Maison, *Mass inflation and chaotic behavior inside hairy black holes*, gr-qc/9703047.

[25] E. Poisson and W. Israel, Phys. Rev. D. **41**, 1796 (1990).

[26] A. Ori, Phys. Rev. Lett. **67**, 789 (1991) 789; **68**, 2117 (1992).

[27] D.N. Page, in V. De Sabbata and Z. Zhang (eds.), *Black Hole Physics*, (Kluwer, Dordrecht, 1992), p. 185–224.

[28] A.Ori, 1991, (unpublished).

[29] We are thankful to A. Ori for bringing the ref. [27] to our attention and for communicating his unpublished results [28].

[30] A. Bonanno, S. Droz, W. Israel, and S. M. Morsink, Phys. Rev. D. **50**, 7372 (1994).

[31] Note that the similar relations are obtained recently in D. V. Gal'tsov, E. E. Donets, M. Yu. Zotov, JETP Lett. **65**, 855 (1997).

[32] D. V. Gal'tsov, and E. E. Donets, *Power-law mass inflation in Einstein–Yang–Mills–Higgs black holes*, gr-qc/9706067.

[33] S. W. Hawking, and G. F. R. Ellis, *The large scale structure of space-time*, (Cambridge University Press, Cambridge, 1973).

THE VALIDITY OF THE BACKGROUND FIELD APPROXIMATION

R. Parentani

Lab. de Mathématiques et Physique Théorique, UPRES A 6083 CNRS, Faculté des Sciences, Université de Tours, 37200 Tours, France

Abstract

In the absence of a tractable theory of quantum gravity, quantum matter field effects have been so far computed by treating gravity at the Background Field Approximation. The principle aim of this paper is to investigate the validity of this approximation which is not specific to gravity. To this end, for reasons of simplicity and clarity, we shall compare the descriptions of thermal processes induced by constant acceleration (i.e. the Unruh effect) in four dynamical frameworks. In this problem, the position of the "heavy" accelerated system plays the role of gravity. In the first framework, the trajectory is treated at the BFA: it is given from the outset and unaffected by radiative processes. In the second one, recoil effects induced by these emission processes are taken into account by describing the system's position by WKB wave functions. In the third one, the accelerated system is described by second quantized fields and in the fourth one, gravity is turned on. It is most interesting to see when and why transitions amplitudes evaluated in different frameworks but describing the same process do agree. It is indeed this comparison that determines the validity of the BFA. It is also interesting to notice that the abandonment of the BFA delivers new physical insights concerning the processes. For instance, in the fourth framework, the "recoils" of gravity show that the acceleration horizon area acts as an entropy in delivering heat to accelerated systems.

10.1 Introduction

In order to analyze the stability of a given configuration, one must choose a framework in which the dynamical consequences of certain class of fluctuations may be investigated. It is quite obvious that the stability conditions might drastically depend on the class of fluctuations considered. Moreover, in view of the complexity of most physically interesting situations, certain simplifying approximations must be made. These in turn specify the particular dynamical framework and may therefore restrict the class of fluctuations one is effectively taken into account. Because of this, stability conditions might also depend on the nature of the approximations used. In particular, upon dealing with gravitational systems, the absence of any tractable quantum theory of gravity imposes to treat gravity at the Background Field Approximation (BFA) even when one considers quantum effects induced by quantum matter fields. This implies that one can only compute the gravitational response induced by the mean (i.e. quantum average) energy-momentum tensor. One must thus inquire onto the validity of the results obtained by using the BFA for describing gravity. In particular, when one abandons the BFA, does one obtain a behavior more or less singular?

These considerations certainly apply to unstable or near unstable situations. Examples are provided by matter and gravity dynamics near inner horizons of charged black holes [1, 2] as well as the dynamics near the event horizon of extremal black holes [3, 4]. However, they might also apply to (apparently) more stable situations, such as, for instance, the evaporation process of Schwarzschild holes [5]. This is because the whole dynamical evolution might drastically depend on the nature of the approximations. Indeed, when working in the semi-classical approximation, one only deals with regular mean energy densities [6, 7, 8]. However, Hawking quanta issue, in usual quantum field theory, from vacuum fluctuations characterized by ultra-high frequencies (trans-Planckian), see [9]-[12]. These fluctuations give rise to trans-Planckian matrix elements of the stress energy-tensor [13, 14]. Therefore, since these matrix elements should determine (yet unknown) quantum gravitational back-reaction effects, the evolution in quantum gravity might strongly differ from the semi-classical one.

In this article, we shall investigate the relationships between the choice of the dynamical framework and the description of processes. To this end, we shall compare the descriptions of thermal effects induced by uniform acceleration, i.e. the Unruh [15] effect, in four dynamical frameworks. This comparative analysis clearly reveals the *generic* properties of the use of the BFA. (By generic we mean that these properties should also apply to quantum gravity.) In particular, it shows precisely when and why the description of processes obtained by using the BFA coincides to the description of the same processes in new frameworks in which the dynamics has been enlarged.

In order to perform this analysis, one should carefully choose the dynamical objects that shall be compared. We shall see that the comparison is the most instructive when focused on specific matrix elements controlling transition amplitudes [16]. Indeed these are the basic elements out of which all physical quantities can be computed. Furthermore they are very sensitive to the approximations used. Instead, integrated or averaged quantities, such as transition rates or the total energy emitted, are much less sensitive and therefore not appropriate to perform the comparison. Notice that the (relative) insensitivity of the transition rates is related to the thermodynamical nature of the Unruh effect.

The dynamical arena of the four frameworks we shall consider will be progressively enlarged and will incorporate the former one. Correspondingly, the description of the transitions will become more sophisticated. However, the enlargement of the dynamics shall provide new physical insights. In particular, the energy conservation law changes drastically when one abandons the BFA. In addition to this change which is not specific to the Unruh effect, there are also specific new informations obtained by the enlargement of the dynamics. The most interesting ones concern the strong relations between the Unruh effect and both pair creation in a constant electric field (in the third framework) and gravitational entropy (in the fourth one).

The first framework corresponds to the Unruh's original formulation of the problem [15]. In this case, one deals with a uniformly accelerated two-level atom coupled to a quantum massless field in Minkowski space time. The trajectory of the atom is treated at the BFA, i.e. it is once for all given and insensitive to the radiative events. Thus one can neither analyze the stability of the trajectory nor answer the question of the origin of the energy emitted by the accelerated system since momentum conservation is violated during the emission processes.

In the second framework, the position of the accelerated system is quantized. The particular model we shall consider consists of a "heavy" two-level ion immersed in a constant electric field and coupled, as before, to the massless field [17, 16]. In this second framework, the momentum and energy conservation laws are fully respected and one verifies that it is the electric field that provides the energy of the radiated quanta. We shall then establish under which conditions and for which reason the Unruh thermalization process is recovered in this new framework. The conditions are that both (i) the WKB approximation of the wave functions describing the system's position and (ii) a first order expansion in the energy changes induced by the transition should be valid. Then, Hamilton-Jacobi equations imply that the transition *rates* computed from both frameworks agree. (This agreement is absolutely generic in character: it also applies to quantum gravity when gravity plays the role of the heavy system [18].) This does not mean however that the transition *amplitudes* themselves coincide: there are indeed trans-Planckian phase shifts induced by exponentially large Doppler shifts. As a consequence, the mean flux radiated the accelerated system is smooth out by these recoil effects [16]. One thus realizes that the singular energy density obtained by treating the trajectory at the BFA [19, 20] was an *artifact* directly induced by the use of this approximation which ignored recoils.

However this second framework is still an approximation since pair creation effects have not been taken into account. Indeed, since one is working with quantized relativistic fields, it is mandatory to work in a second quantization. In this third framework, the new enlargement of the dynamics makes therefore contact with a an priori completely different phenomenon, namely pair creation in a constant electric field, i.e. the Schwinger effect [21]. It is remarkable that the amplitudes describing this latter effect are directly related to the radiative transitions by "crossing symmetry" [22], an intrinsic analytical property of amplitudes in QFT. Therefore, the Unruh effect and the Schwinger effect should be conceived as two aspects of a single dynamical theory, namely QFT in a constant electric field [23, 25]. In addition, this connection between pair creation amplitudes and radiative processes sheds light on the recently analyzed pair creation processes of charged black holes [26]-[30]. In particular, the role of the (gravitational) instanton which relates by tunneling process the Melvin geometry (in which the black holes are absent) to the Ernst geometry (present) is clarified [31].

This last remark naturally introduces the dynamical role of gravity that has been ignored so far. Indeed, upon turning on gravity, i.e. $G \neq 0$, the thermodynamical role of the acceleration horizon emerges. Then, the Unruh effect is included into the more general framework of thermodynamics of horizons. In particular, the role of the horizon area acting as an entropy in delivering heat to accelerated systems (which are at rest with respect to the Killing horizon) is established. These relations between Euclidean gravity and thermal phenomena can also be conceived as particular examples of the thermodynamical approach to gravity [32] presented by Ted Jacobson.

In resume, by abandoning the BFA for describing the system's trajectory, we shall successively

1. restore energy-momentum conservation,
2. suppress singular behavior of the emitted fluxes,
3. relate radiative transitions to pair creation amplitudes and
4. connect thermal effects induced by constant acceleration to horizon entropy.

Throughout the text, we shall also mention what are the properties revealed by the analysis of Unruh's effect that are generic in character.

10.2 The transition amplitudes in the absence of recoils

In the original Unruh framework [15], the two level atom is maintained, for all times, on a single uniformly accelerated trajectory

$$t_a(\tau) = a^{-1}\mathrm{sinh}a\tau \ , \ z_a(\tau) = a^{-1}\mathrm{cosh}a\tau \tag{10.1}$$

a is the acceleration and τ is the proper time of the accelerated atom. We work for simplicity in Minkowski space time in $1+1$ dimensions. The two levels of the atom are designated by $|-\rangle$ and $|+\rangle$ for the ground state and the excited state respectively. The transitions from one state to another are induced by the operators $A, A^\dagger$

$$\begin{aligned} A|-\rangle = 0, \quad A|+\rangle = |-\rangle \\ A^\dagger|-\rangle = |+\rangle, \quad A^\dagger|+\rangle = 0 \end{aligned} \tag{10.2}$$

The atom is coupled to a massless scalar field $\phi(t, z)$. The Klein-Gordon equation is $(\partial_t^2 - \partial_z^2)\phi = 0$ and the general solution is thus

$$\phi(U, V) = \phi(U) + \phi(V) \tag{10.3}$$

where U, V are the light like coordinates given by $U = t - z, V = t + z$. The right moving part may be decomposed into plane waves:

$$\varphi_\omega(U) = \frac{e^{-i\omega U}}{\sqrt{4\pi\omega}} \tag{10.4}$$

where ω is the Minkowski energy. Using this basis, the Heisenberg field operator ϕ reads

$$\phi(U) = \int_0^\infty d\omega \left(a_\omega \varphi_\omega(U) + a_\omega^\dagger \varphi_\omega^*(U) \right) \tag{10.5}$$

where $a_\omega, a_\omega^\dagger$ are operators of destruction and creation of a Minkowski quantum of energy ω. The coupling between the atom and the field is taken to be, see [16],

$$\begin{aligned} H_{\text{int}}(\tau) &= ga \left(Ae^{-i\Delta m \tau} + A^\dagger e^{i\Delta m \tau} \right) \phi(U_a(\tau)) \\ &= ga \int_0^\infty d\omega \left[\left(Ae^{-i\Delta m \tau} + A^\dagger e^{i\Delta m \tau} \right) \right. \\ &\qquad \left. \times \left(\frac{a_\omega e^{i\omega e^{-a\tau}/a}}{\sqrt{4\pi\omega}} + \frac{a_\omega^\dagger e^{-i\omega e^{-a\tau}/a}}{\sqrt{4\pi\omega}} \right) \right] \end{aligned} \tag{10.6}$$

where $U_a(\tau) = t_a(\tau) - z_a(\tau) = -e^{-a\tau}/a$. Because of the locality of the coupling, only $\phi(U_a(\tau))$ evaluated along the classical trajectory enters into the interaction.

The classically forbidden transition amplitude (spontaneous excitation) from the ground state $|-\rangle|0\rangle$, where $|0\rangle$ is Minkowski vacuum, to the excited state $|+\rangle|1_\omega\rangle$ containing one quantum of energy ω (where $|1_\omega\rangle = a_\omega^\dagger|0\rangle$) is

$$B(\omega, \Delta m, a) = \langle 1_\omega|\langle +|e^{-i\int d\tau H_{\text{int}}}|-\rangle|0\rangle \tag{10.7}$$

To first order in g, one finds

$$\begin{aligned} B(\omega, \Delta m, a) &= -iga \int_{-\infty}^{+\infty} d\tau \; e^{i\Delta m \tau} \frac{e^{-i\omega e^{-a\tau}/a}}{\sqrt{4\pi\omega}} \\ &= ig \; \Gamma(-i\Delta m/a) \; \frac{(\omega/a)^{i\Delta m/a}}{\sqrt{4\pi\omega}} \; e^{-\pi\Delta m/2a} \end{aligned} \tag{10.8}$$

where $\Gamma(x)$ is the Euler function. This transition amplitude is closely related to the β coefficient of the Bogoliubov transformation [33] which relates the Minkowski operators a_ω to the Rindler operators associated with the eigenmodes of $-i\partial_\tau$ (hence given by $\varphi_{Rindler}(\tau) = e^{-i\lambda\tau}$). This provides the dynamical justification of studying Bogoliubov coefficients which can be defined in the absence of any coupling to accelerated system. Their study is thus preparatory in character, exactly like the analysis of Green functions in free field theory. This point of view has been developed in [18] to incorporate gravitational "recoil" effects in quantum cosmology along the same lines that recoil effects of accelerated systems shall be treated in the next Section.

The transition amplitude for the inverse process, i.e. disintegration from $|+\rangle|0\rangle$ to $|-\rangle|1_\omega\rangle$, is given by $A(\omega, \Delta m, a) = B(\omega, -\Delta m, a)$. From eq. (10.8), one easily finds

$$A(\omega, \Delta m, a) = -B^*(\omega, \Delta m, a)\, e^{\pi \Delta m/a} \tag{10.9}$$

Thus for all ω one has

$$|\frac{B(\omega, \Delta m, a)}{A(\omega, \Delta m, a)}|^2 = e^{-2\pi\Delta m/a} \tag{10.10}$$

Since this ratio is independent of ω, the ratio of the probabilities of transitions (excitation and disintegration) is also given by eq. (10.10). Hence at equilibrium, the ratio of the probabilities P_-, P_+ to find the atom in the ground or excited state satisfy

$$\frac{P_+}{P_-} = |\frac{B(\omega, \Delta m, a)}{A(\omega, \Delta m, a)}|^2 = e^{-2\pi\Delta m/a} \tag{10.11}$$

This is the Unruh effect [15]: at equilibrium, the probabilities of occupation are thermally distributed with temperature $a/2\pi$.

Using the amplitudes $B(\omega, \Delta m, a)$ and $A(\omega, \Delta m, a)$, one can compute, to order g^2, the mean value of the flux emitted by the two level atom, see [34]-[37]. The point we wish to emphasize is that all the emitted Minkowski quanta, whatever is their energy ω, interfere so as to deliver a negative mean flux, for $U < 0$, whose interpretation is that some Rindler quantum has been absorbed [35]. However, the energy density is positive and singular on the horizon $U = 0$ [19]. This singular and physically pathological behavior arises from the extremely well tuned phases of $B(\omega, \Delta m, a)$ and $A(\omega, \Delta m, a)$ for $\omega \to \infty$. These phases will be inevitably washed out after a finite proper time when recoils will be taken into account [16]. Then, the new value of the flux will be will be rapidly positive and perfectly regular around and on $U = 0$.

10.3 The amplitudes in first quantization

In this section, we first introduce the model of [17] which is similar to the one used by Bell and Leinaas [38]. This will allow us to take into account the momentum transfers to the accelerated system which are caused by the radiative emission processes.

We shall then establish when and why the thermalization of the inner degrees of freedom of the accelerated system is recovered. In particular, we shall see that the thermalization does not require a well defined classical trajectory; therefore it neither requires well defined "Rindler" energies nor a well defined location of the horizon. Indeed, even when one deals with delocalized waves, the thermal equilibrium ratio eq. (10.11) is obtained.

The model consist on two scalar charged fields (ψ_M and ψ_m) of slightly different masses (M and m) which will play the role of the former states of the atom: $|+\rangle$

and $|-\rangle$. The quanta of these fields are accelerated by an external classical constant electric field E with a given by

$$\frac{E}{M} = a \simeq \frac{E}{m} \tag{10.12}$$

because one imposes

$$\Delta m = M - m \ll M \tag{10.13}$$

to have the "light" mass gap Δm well separated from the "heavy" rest mass of the ion.

We work in the homogeneous gauge ($A_t = 0,\ A_z = -Et$). In that gauge, the momentum k is a conserved quantity and the energy p of a relativistic particle of mass M is given by the mass shell constraint $(p_\mu - A_\mu)^2 = M^2$:

$$p^2(M,k,t) = M^2 + (k + Et)^2 \tag{10.14}$$

The classical equations of motion are easily obtained from this equation and are given in terms of the proper time τ by

$$\begin{aligned} p(M,k,t) &= M\cosh a\tau \\ t + k/E &= (1/a)\sinh a\tau \\ z - z_0 &= (1/a)\cosh a\tau \end{aligned} \tag{10.15}$$

Thus at fixed k, the time of the turning point (i.e. $dt/d\tau = 1$) is fixed whereas its position is arbitrary and given by $z_0 + 1/a$.

¿From eq. (10.14), the Klein Gordon equation for a mode $\psi_{k,M}(t,z) = e^{ikz}\chi_{k,M}(t)$ is

$$\left[\partial_t^2 + M^2 + (k + Et)^2\right]\chi_{k,M}(t) = 0 \tag{10.16}$$

When $\Delta m \simeq a$ and when eq. (10.13) is satisfied, one has $M^2/E \gg 1$. Then pair production amplitudes [21] may be safely ignored since the mean density of produced pairs scales like $e^{-\pi M^2/E}$, see next Section. Furthermore, the WKB approximation for the modes $\chi_{k,M}(t)$ is valid for all t. Indeed, the corrections to this approximation are smaller than $(M^2/E)^{-1}$. The modes $\psi_{k,M}(t,z)$ can be thus correctly approximated by

$$\psi^{WKB}_{k,M}(t,z) = \frac{e^{ikz}}{\sqrt{2\pi}} \frac{e^{-i\int^t p(M,k,t')dt'}}{\sqrt{p(M,k,t)}} \tag{10.17}$$

where $p(M,k,t')$ is the classical energy at fixed k given in eq. (10.14).

As emphasized in refs. [17, 39], the gaussian wave packets in k do not spread if their width is of the order of $E^{1/2}$. In that case, the spread in z at fixed t is of the order of $E^{-1/2} = (Ma)^{-1/2}$ and thus much smaller than the acceleration length $1/a$. Since the stationary phase condition of the $\psi^{WKB}_{k,M}(t,z)$ modes gives back the accelerated trajectory and since the wave packets do not spread, one has constructed a quantized version of the accelerated system which tends uniformly to the BFA model when $M \to \infty, E \to \infty$ with $E/M = a$ fixed.

The interacting Hamiltonian which induces transitions between the quanta of mass M and m by the emission or absorption of a massless neutral quantum of the ϕ field is simply

$$H_{\psi\phi} = \tilde{g} M^2 \int dz \left[\psi_M^\dagger(t,z)\psi_m(t,z) + \psi_M(t,z)\psi_m^\dagger(t,z)\right] \phi(t,z) \tag{10.18}$$

where $\tilde{g}$ is dimensionless. In momentum representation, by limiting ourselves to the right moving modes of the ϕ field, one obtains

$$\begin{aligned} H_{\psi\phi} &= \tilde{g} M^2 \int_{-\infty}^{+\infty} dk \int_0^\infty \frac{d\omega}{\sqrt{4\pi\omega}} \\ &\times \left\{ \left[b_{M,k-\omega}\chi_{M,k-\omega}(t)\; b_{m,k}^\dagger \chi_{m,k}^*(t) + \text{h.c.}\right] a_\omega e^{-i\omega t} \right. \\ &\left. + \left[b_{M,k+\omega}\chi_{M,k+\omega}(t)\; b_{m,k}^\dagger \chi_{m,k}^*(t) + \text{h.c.}\right] a_\omega^\dagger e^{+i\omega t} \right\} \end{aligned} \tag{10.19}$$

where the operator $b_{M,k-\omega}$ detroys a quantum of mass M and momentum $k-\omega$ and the operator $b_{m,k}^\dagger$ creates a quantum of mass m and momentum k. Therefore the product $b_{M,k-\omega}b_{m,k}^\dagger$ plays the role of the operator A in eq. (10.6). However, the interaction between the radiation field ϕ and the two levels of the accelerated system is no longer a priori restricted to a classical trajectory. In the present context of homogeneous electric field, this leads to an exact conservation of momentum. Notice that we have not introduced anti-ion creation operators in $H_{\psi\phi}$. This is a legitimate truncation when working with WKB waves. This condition shall be relaxed in the next Section.

We first establish when and why the behavior of the $\chi_M(t)$ modes re-delivers the Unruh effect [16]. To this end, we shall not consider well localized wave packets. We emphasize this point. An alternative route, a priori equally valid, would consist in *first* making well localized wave packets and only *then* computing transition amplitudes, see [17]. It turns out that the averaging over k in constructing wave packets is both unnecessary and a nuisance in that it erases the detailed mechanisms that ensure thermalization in this new framework.

Thus, we focus on the transition amplitude at fixed momentum k that replaces $B(\Delta m, \omega, a)$, eq. (10.8). In the present context, it corresponds to the amplitude to jump from the state $|1_k, m\rangle|0, M\rangle|0\rangle$ to the state $|0, m\rangle|1_{k'}, M\rangle|1_\omega\rangle$. $|0, M\rangle$ designates the vacuum state for the ψ_M field and $|1_k, m\rangle = b_{m,k}^\dagger|0, m\rangle$ is the one particle state of the ψ_m field of momentum k. Due to momentum conservation, this amplitude can be expressed as

$$\begin{aligned} \delta(k &- k' - \omega)\; \tilde{B}(M, m, k, \omega) \\ &= \langle 1_\omega|\langle 1_{k'}, M|\langle 0, m|e^{-i\int dt H_{\psi\phi}}|1_k, m\rangle|0, M\rangle|0\rangle \end{aligned} \tag{10.20}$$

compare this expression with eq. (10.7).
To first order in $\tilde{g}$, one finds

$$\begin{aligned}\tilde{B}(M,m,k,\omega) &= -i\tilde{g}M^2 \int_{-\infty}^{+\infty} dt\ \chi^*_{M,k-\omega}(t)\ \chi_{m,k}(t)\frac{e^{i\omega t}}{\sqrt{4\pi\omega}} \\ &= -i\tilde{g}M^2 \int_{-\infty}^{+\infty} dt \frac{e^{i\int^t dt'\left[p(M,k-\omega,t')-p(m,k,t')\right]}}{\sqrt{p(M,k-\omega,t)p(m,k,t)}}\frac{e^{i\omega t}}{\sqrt{4\pi\omega}} \end{aligned} \tag{10.21}$$

In the second line, we have used the WKB approximation eq. (10.17) for the χ modes.

As such, $\tilde{B}(M,m,k,\omega)$ seems very different from the BFA amplitude $B(\Delta m,\omega,a)$ given in eq. (10.8). Indeed, this latter expression was based on a well defined classical trajectory parametrized by τ whereas in the present case, it is momentum conservation that has been exactly taken into account.

However, it is precisely this conservation law that re-introduces the notion of classical trajectory (exactly like in quantum gravity the Wheeler-DeWitt constraint re-introduces the notion of time [18]). To explicitize this, we develop the phase and the norm of the *integrand* of eq. (10.21) in powers of ω and Δm. To first order in Δm and ω, the phase $\varphi(t)$ is *equal*, up to a constant, to the phase of the integrand of eq. (10.8). Indeed, one has

$$\begin{aligned}\varphi(t) &= \int^t dt' \left[p(M,k-\omega,t') - p(m,k,t')\right] + \omega t \\ &\simeq \Delta m\ \partial_M \int^t dt' p(M,k,t') - \omega\ \partial_k \int^t dt' p(M,k,t') + \omega t \\ &\simeq \Delta m \Delta\tau(t) - \omega(\Delta z_k(t) - t) \end{aligned} \tag{10.22}$$

$\Delta\tau(t)$ and $\Delta z_k(t)$ are the lapses of proper time and of space evaluated along a uniformly accelerated trajectory characterized by k. These relations may be checked explicitly by computing the integrals using eq. (10.14).

However, for establishing their universal validity, it is appropriate to realize that those relations are nothing but Hamilton-Jacobi relations. Indeed, these are

$$\partial_M S_{cl.}(M,k,t) = \partial_M \int^t dt' p(M,k,t') = \Delta\tau(t) \tag{10.23}$$

$$\partial_k S_{cl.}(M,k,t) = \partial_k \int^t dt' p(M,k,t') = \Delta z_k(t) \tag{10.24}$$

Thus, whatever is the nature of the external field which brings the system into constant acceleration, the first two terms of eq. (10.22) will always be found. More importantly and more generally, whatever is the problem one is considering, upon using WKB wave functions for describing "heavy" degrees of freedom, the transition amplitudes among "light" degrees will agree with the BFA expressions upon developing the amplitudes to first order in the light changes. Indeed, in that approximation, one must also neglect the dependence in ω and Δm in the denominator of eq. (10.21). Then the measure is $dt/p(M,k,t) = d\tau/M$. Thus, one has, as announced,

$$\begin{aligned}\tilde{B}(M,m,k,\omega) &= -i\tilde{g}M \int_{-\infty}^{+\infty} d\tau e^{i\Delta m\tau} \frac{e^{-i\omega e^{-a\tau}/a}}{\sqrt{4\pi\omega}} \times \left(e^{-i\omega k/E}\right) \\ &= \left[\frac{\tilde{g}M}{ga}\right] B(\Delta m, \omega, a) \times \left(e^{-i\omega k/E}\right) \qquad (10.25)\end{aligned}$$

Very important is the fact that the momentum k introduces only a phase shift with respect to the BFA amplitude $B(\Delta m, \omega, a)$. Thus any normalized superposition of modes $\psi_{m,k}$ specifying the initial wave function of the "heavy" system will give rise to the same probability to emit of photon of energy ω. Moreover the ratio of the square of $\tilde{B}(M,m,k,\omega)$ and $\tilde{A}(M,m,k,\omega)$ will necessarily satisfy eq. (10.10). Therefore, under these approximations, the two level ion thermalizes exactly as in the no-recoil case. Moreover, the total energy emitted by the "recoiling" ion also equals the corresponding energy evaluated at the BFA. The reason is simply that the total energy is a function of the norm of $B(M,m,\omega,k)$ only [16].

The lesson of this comparison of amplitudes is the following. When both the WKB approximation for the "heavy" system and a first order expansion in the light changes are valid, Hamilton-Jacobi equations guarantee that the norm of the transition amplitudes agree. Therefore all physical quantities that are functions of these norms only will automatically agree as well.

However, there are both conceptual and numerical differences between the amplitudes computed in the two frameworks. For instance the phase of the amplitudes do not agree even to first order in ω and Δm. We shall illustrate these differences by first considering the stationary phase condition in both cases and then by analyzing the consequences of the phase shifts in the determination of the flux emitted by the atom.

In this present framework, the stationary point t^* of the integrand of eq. (10.21) is at

$$p(M, k-\omega, t^*) - p(m,k,t^*) + \omega = 0 \qquad (10.26)$$

This is the conservation law of the Minkowski energy. Instead, in the Unruh framework, from the first line of eq. (10.8), one finds

$$\Delta m + \omega e^{-a\tau} = 0 \qquad (10.27)$$

which is the resonance condition in the accelerated frame, i.e. conservation of Rindler energy. This difference also arise in gravitational situations, see [18]: Upon dealing with gravity described at the BFA, one finds that matter processes satisfy energy conservation in the given background; this is the equivalent of eq. (10.27). And upon abandoning the BFA and solving twice Einstein's equations, one verifies that the resonance condition involves the energy of gravity, exactly like eq. (10.26) contains the difference of two heavy energy $p(M, k, t)$.

Of course, the two versions of energy conservation must coincide in the limit $\Delta m/M \to 0$. Indeed, by taking the square of eq. (10.26) and using eq. (10.14) one gets

$$\frac{M^2 - m^2}{2} = \omega\left[(k - \omega + Et^*) - p(M, k - \omega, t^*)\right] \tag{10.28}$$

Introducing once more the proper time of the heavy ion M, eq. (10.15), one finds

$$\Delta m(1 - \Delta m/2M) = -\omega e^{-a\tau^*} \qquad \text{QED} \tag{10.29}$$

The second point we wish to make is the following. There is a strict relation between conservation of momentum, eq. (10.20), and energy, eq. (10.26), and the modifications of the properties of the emitted fluxes, see [16] for more details. Indeed, the two level ion constantly loses energy and momentum in accordance with these conservation laws. Then the trajectory of the ion drifts from orbits to neighboring ones in t and z corresponding to laters times and greater z. The total change in the time of the turning point is $E\Delta t_{t.p.} = \sum_i \omega_i$, i.e. it is proportional to the total momentum lost. One also verifies that the total change in position is $\sum_i \omega_i/E$. These successive changes of hyperbolae lead to the decoherence of the emissions causing these changes since the very peculiar phases obtained in the BFA framework are washed out by the recoils. Then, both "unphysical" properties obtained in that framework, namely negativity of the energy density during arbitrary large proper times [35]-[37] and singular behavior on the horizon [19, 20], are eliminated.

There is a third point that can be addressed before considering corrections to the WKB approximation. It concerns the possibility in computing corrections of the equilibrium ratio, eq. (10.11), induced by higher order terms in $\Delta m/M$. This problem shall be discussed elsewhere.

10.4 The amplitudes in second quantization

In the former Section, we used WKB wave functions which are valid approximate solutions of eq. (10.16) when $M^2/E \gg 1$. Then, positive and negative energy solutions completely decouple. This is no longer true when one deals with the exact solutions of eq. (10.16). In this new case indeed, one must introduce two sets of modes $(\chi_M^{in}, \chi_M^{out})$ which only asymptotically, i.e. for $t \to \pm\infty$, define particle states. That the two states do not coincide (they are related by a linear (Bogoliubov) transformation) indicates that the initial vacuum $|0, M, in\rangle$ associated with the initial set χ_M^{in} spontaneously decays and that particles will be find at $t \to \infty$. In the present case of a constant electric field, one finds that the mean number of quanta of momentum k is given by

$$N_M = |\beta_M|^2 = e^{-\pi M^2/E} \tag{10.30}$$

where β_M is the Bogoliubov coefficient, i.e. the amount of negative energy *out* solution in a purely positive *in* solution, see *e.g.* [40].

Using Feynman rules, one can re-calculate the ratio of the transition rates to emit a photon starting from the ground state (m) and from the excited state (M) by taken into account vacuum instability with respect to pair creation of both ion fields, see Nikishov [23]. This ratio turns out to be intimately related [17, 22] to eq. (10.30) in that it is equal to

$$|\frac{\mathcal{B}(M,m,p,\omega)}{\mathcal{A}(M,m,p,\omega)}|^2 = e^{-\pi(M^2-m^2)/E} = \frac{N_M}{N_m} \tag{10.31}$$

To re-calculate this ratio, we analyze the amplitude $\mathcal{B}(M,m,k,\omega)$ to emit a massless quantum of energy ω starting from the state m. This amplitude strictly corresponds to the amplitude $\tilde{B}(M,m,k,\omega)$ of eq. (10.20). To first order in $\tilde{g}$, it is given by

$$\begin{aligned}
&\delta(k-k'-\omega)\mathcal{B}(M,m,k,\omega) \\
&\quad = \langle 1_\omega|\langle 1_{k'},M,out|\langle 0,m,out|e^{-i\int dtH_{\psi\phi}}|1_k,m,in\rangle|0,M,in\rangle|0\rangle \\
&\quad = -\delta(k-k'-\omega)\; i\tilde{g}M^2\left(Z_M Z_m \alpha_M^{-1}\alpha_m^{-1}\right) \\
&\quad \times \int_{-\infty}^{\infty} dt\; \chi^{in*}_{M,k-\omega}(t)\; \chi^{out}_{m,k}(t)\; \frac{e^{i\omega t}}{\sqrt{4\pi\omega}}
\end{aligned} \tag{10.32}$$

The factor $Z_M Z_m$ is the product of the overlaps of the *in* and *out* vacuum states of both charged fields. $\alpha_{M(m)}$ is a coefficient whose norm is given by $\sqrt{1+|\beta_{M(m)}|^2}$. These factors take into account the fact that the scattering process happens in the presence of pair production of charged quanta. They all reduce to one in the WKB limit, $M^2/E \to \infty$.

This amplitude can be exactly evaluated in terms of the integral representations of the χ modes, see [22]. One obtains,

$$\begin{aligned}
\mathcal{B}(M,m,k,\omega) &= -i\frac{\tilde{g}M^2}{2E}\;\Gamma(-i(M^2-m^2)/2E) \\
&\times \frac{(\omega)^{i(M^2-m^2)/2E}}{\sqrt{4\pi\omega}}\; e^{-\pi(M^2-m^2)/2E} \\
&\times e^{i(\omega p-\omega^2/2)/E}\left[\Gamma(i\frac{M^2}{2E}+\frac{1}{2})\frac{e^{\pi M^2/2E}(E/2)^{-iM^2/2E}}{\sqrt{2\pi}}\right]
\end{aligned} \tag{10.33}$$

This expression should be compared with eq. (10.8). As in that case, the determination of the equilibrium population requires only to know the ratio of the transition rates. And as in that case, this ratio can be obtained effortless from the analytical behavior of the amplitude in ω, since $\mathcal{A}^*(M,m,k,\omega) = \mathcal{B}(M,m,k,e^{-i\pi}\omega)$. (Notice that this analytical behavior in ω encodes the stability condition of the Minkowski vacuum of the radiation field. Notice also that Hawking radiation can be derived in similar terms [15].) From this relation and eq. (10.33), one immediately deduces that the ratio of the transition rates satisfy eq. (10.31).

Therefore we have proven that the equilibrium probabilities defined by radiative processes are equal to those defined by the Schwinger process. However, the direct proof that both processes are intimately related follows from the fact that their respective amplitudes are interchanged by crossing symmetry, see [22]. Indeed what corresponds to a pair creation diagram in the direct channel, describes a radiative transition in the "crossed" channel.

To conclude this Section, we analyze the relationships between this formulation of radiative processes with the former ones studied in Section 2 and 3. First notice that in the limit $M^2/E \to \infty$ at fixed $M - m = \Delta m$ and $M/E = 1/a$, the *integrand* of eq. (10.32) tends uniformly to the *integrand* based on WKB expressions in eq. (10.21). This indicates once more that the rates agree because matrix elements defining transition amplitudes can be put in strict correspondence, at fixed quantum number.

Secondly, it should not have escaped the reader that eq. (10.31) differs from the Boltzmanian ratio found using the BFA, i.e. eq. (10.10). Indeed, in order to make contact with this thermal *canonical* distribution, one must consider the limit $\Delta m <
< M$. Then, in the present second quantized framework, the concepts of acceleration and temperature are brought to bear for the first time. They both appear through a first order change in Δm. This emergence of classical concepts bears many similarities with statistical mechanics since it is also through a first order change in the energy that the concept of temperature arises from *microcanonical* ensembles.

To bring about this contact with statistical mechanics, it is most instructive to use again Hamilton-Jacobi equations but applied this time to Euclidean dynamics. Indeed relationships with both eq. (10.23) and black hole thermodynamics will become clear. We remind the reader that the Schwinger pair creation amplitudes may be understood from the Euclideanized version of the problem. In fact, it is $S_{euclid} = \pi M^2/E$, the classical action to complete a Euclidean (closed) orbit, which determines their rate, see eq. (10.30). From this action, one computes the Euclidean proper time necessary to complete this orbit. It is given by

$$\tau_{euclid} = \partial_M S_{euclid} = \partial_M \left(\frac{\pi M^2}{E}\right) = \left(\frac{a}{2\pi}\right)^{-1} \tag{10.34}$$

It equals the inverse Unruh temperature. Thus, to first order in Δm, it is meaningful to write eq. (10.31) as

$$|\frac{\mathcal{B}(M,m,k,\omega)}{\mathcal{A}(M,m,k,\omega)}|^2 \simeq e^{-\Delta m \partial_M S_{euclid}} = e^{-\Delta m 2\pi/a} = |\frac{B(\Delta m,a,\omega)}{A(\Delta m,a,\omega)}|^2 \tag{10.35}$$

This shows that the Unruh process can be viewed as an infinitesimal ratio of two Schwinger processes. Moreover the instanton action S_{euclid} acts as an *entropy* in delivering the Unruh temperature. Indeed,

$$\tau_{euclid}\Delta m = \Delta S_{euclid} \tag{10.36}$$

is the first law of thermodynamics [24].

In view of the similarities between Unruh effect and black hole radiation, it is inviting to inquire about the relationship between this "instantonic" entropy and the gravitational entropy of black holes whose variation determines Hawking temperature. This is the subject of next Section.

10.5 The amplitudes in the presence of gravity

This short Section is of a heuristic character and the level of mathematical rigor will be lowered simply because Quantum Gravity does not exist. Indeed, to describe quantum transitions in which gravity does play an active role, i.e. in which the BFA for gravity has been abandoned, new approximations should be adopted. The type of approximations we shall need and use are quite similar to those we used in Section 3 for describing the system's trajectory: Gravitational quantum effects will approximatively be taken into account like momentum recoils were taken into account in eq. (10.21), i.e. in matrix elements describing transition amplitudes of light (matter) degrees of freedom, the initial and final wave functions of the heavy system (gravity) bring their own classical action, see [18].

To illustrate these aspects, we start the discussion by recalling the results of [26]-[30]. In these references, the probability of creation charged black holes in a constant electric field was estimated by making use of two hypothesis. First the authors required that the Euclidean manifold describing the instanton responsible for the creation be regular. Secondly they assumed that the probability of creation depends on the instanton action computed from the Hilbert-Einstein like S_{euclid}, the classical action to complete a Euclidean orbit, determined the production rate in eq. (10.30). Using these hypothesis, they found that the probability to produce a pair of black holes characterized by the mass M and charge Q is

$$P_{M,Q} = e^{\delta \mathcal{A}_{acc}/4G} \times e^{\mathcal{A}_{BH}/4G} \tag{10.37}$$

$\mathcal{A}_{BH}(M,Q)$ is the area of the black hole horizon. In this expression, $e^{\mathcal{A}_{BH}/4G}$ furnishes the density of black hole states with mass M and charge Q thereby confirming the Bekenstein interpretation of $\mathcal{A}_{BH}/4G = S_{BH}$ as being the black hole entropy.

$\delta\mathcal{A}_{acc}(M,Q,E)$ is the *change* of the area of the acceleration horizon induced by the creation of the black hole pair [41].

The domain of the one parameter family (i.e. the values of M and Q which satisfy the regularity condition) which can be compared with the Schwinger mechanism is the one in which the black holes radius is much smaller than the inverse acceleration, i.e. in the point particle limit. This limit corresponds to the limit $G \to 0$. In that case, one finds

$$\frac{\delta \mathcal{A}_{acc}}{4G} = -\pi M^2/QE = -S_{euclid} \tag{10.38}$$

This is the usual instanton action to create charged particles.

First notice that it is independent of G, as it should be. Indeed, the limit $G \to 0$ in the present gravitational case strictly corresponds to the limit $\Delta m/M \to 0$ in Sections 3 and 4. In that case, we saw that the transition probabilities coincide with the BFA probabilities since the integrands of the amplitudes leading to these probabilities agree up to a phase. Similarly, by turning on gravity and then taking the limit $G \to 0$, one must recover the BFA probabilities. That this is the case, justify a posteriori the assumptions used in [30].

Secondly, eq. (10.38) and eq. (10.36) show that S_{euclid} acts as a gravitational entropy. Indeed, eq. (10.36) can be rewritten as

$$\frac{2\pi}{\kappa}\Delta m = \frac{\Delta \mathcal{A}_{acc}}{4G} \tag{10.39}$$

where $\kappa = a$ is the surface gravity of the accelerating horizon measured along the trajectory $z^2 - t^2 = 1/a^2$. $\Delta \mathcal{A}_{acc}$ is the change of the accelerating horizon area associated with the replacement of the heavy mass M by the lighter one m [31]. This rewriting strongly suggests that changes in area divided by $4G$ determine equilibrium distributions of accelerated systems according to the first law of horizon thermodynamics.

More details on these aspects can be found in [31, 25].

Acknowledgments

I am grateful to Serge Massar and Ted Jacobson for the helpful discussions we had during the "MG8" and "The Internal Structure of Black Holes and Space-time Singularities" meetings. I wish also to thank the organizers of these meetings for inviting me.

Bibliography

[1] W. G. Anderson, P. R. Brady, Werner Israel, and S. M. Morsink, Phys. Rev. Lett. **70**, 1041 (1993).

[2] L. M. Burko and A. Ori, Phys. Rev. Lett. **74**, 1064 (1995).

[3] S. P. Trivedi, "Semiclassical extremal black holes", hep-th/9211011.

[4] T. Jacobson, "Semiclassical decay of near extremal black holes", hep-th/9705017.

[5] S. W. Hawking, Commun. Math. Phys. **43**, 199 (1975).

[6] J. M. Bardeen, Phys. Rev. Lett. **46**, 382 (1981).

[7] R. Parentani and T. Piran, Phys. Rev. Lett. **73**, 2805 (1994).

[8] S. Massar, Phys. Rev. D **52**, 5861 (1995).

[9] G. 't Hooft, Nucl. Phys. **B 256**, 727 (1985).

[10] T. Jacobson, Phys. Rev. D **44**, 1731 (1991); D **48**, 728 (1993).

[11] W. G. Unruh, Phys. Rev. D **51**, 2827 (1995).

[12] R. Brout, S. Massar, R. Parentani and Ph. Spindel, Phys. Rev. D **52**, 4559 (1995).

[13] S. Massar and R. Parentani, Phys. Rev. D **54**, 7431 (1996).

[14] Y. Kiem, H. Verlinde, E. Verlinde *Quantum Horizons and Complementarity.* CERN-TH-7469-94, hep-th/9502074.

[15] W. G. Unruh, Phys. Rev. D **14**, 870 (1976).

[16] R. Parentani, Nucl. Phys. **B 454**, 227 (1995).

[17] R. Brout, R. Parentani and Ph. Spindel, Nucl. Phys. **B 353** , 209 (1991).

[18] R. Parentani, Nucl. Phys. **B 492**, 475 (1997); **B 492**, 501 (1997).

[19] W. G. Unruh, Phys. Rev. D **46**, 3271 (1992).

[20] S. Massar and R. Parentani, Phys. Rev. D **54**, 7426 (1996).

[21] J. Schwinger, Phys. Rev. **82**, 664 (1951).

[22] S. Massar and R. Parentani, Phys. Rev. D **55**, 3603 (1997).

[23] A. I. Nikishov, Sov. Phys. JETP **30**, 660 (1970).

[24] For the skeptical reader, we add that this analysis has been enlarged [25] by replacing the massless neutral field by a charged field. Then the new equilibrium ratio is determined by an extended thermodynamical relation in which the work done by the electric field contributes as well. The analogy with the thermodynamics of charged black holes is manifest.

[25] C. Gabriel, Ph. Spindel, S. Massar, R. Parentani, "Interacting Charged Particles in an Electric Field and the Unruh Effect", hep-th/9706030.

[26] G. W. Gibbons, in *Fields and Geometry*, Proceedings of the 22nd Karpacz Winter School of Theoretical Physics, Karpacz, Poland, 1986, edited by A. Jadczyk (World Scientific, Singapore, 1986)

[27] D. Garfinkle and A. Strominger, Phys. Lett. **B 256**, 146 (1991).

[28] S. W. Hawking and S. Ross, Phys. Rev. D **52**, 5865 (1995).

[29] F. Dowker, J. P. Gauntlett, S. B. Giddings and G. T. Horowitz, Phys. Rev. D **50**, 2662 (1994).

[30] S. W. Hawking, G. T. Horowitz and S. Ross, Phys. Rev. D **51**, 4302 (1995).

[31] S. Massar and R. Parentani, Phys. Rev. Lett. **78**, 3810 (1997).

[32] T. Jacobson, Phys. Rev. Lett. **75**, 1260 (1995).

[33] S. A. Fulling, Phys. Rev. D **7**, 2850 (1973).

[34] W. G. Unruh and R. M. Wald, Phys. Rev. D **29**, 1047 (1984).

[35] P. Grove, Class. Quant. Grav. **3**, 801 (1986).

[36] D. Raine, D. Sciama and P. Grove, Proc. R. Soc. **A 435**, 205 (1991).

[37] S. Massar, R. Parentani and R. Brout, Class. Quant. Grav. **10** , 385 (1993).

[38] J. S. Bell and J. M. Leinaas, Nucl. Phys. **B 212**, 131 (1983).

[39] R. Brout, S. Massar, R. Parentani, S. Popescu and Ph. Spindel, Phys. Rev. D **52**, 1119 (1995).

[40] R. Brout, S. Massar, R. Parentani and Ph. Spindel, Phys. Rep. **260**, 329 (1995).

[41] S. W. Hawking and G. T. Horowitz, Class. Q. Grav. **13**, 1487 (1996).

HOMOGENEOUS SPACELIKE SINGULARITIES INSIDE SPHERICAL BLACK HOLES

Lior M. Burko

Department of Physics, Technion—Israel Institute of Technology, 32000 Haifa, Israel

Abstract

Recent numerical simulations have found that the Cauchy horizon inside spherical charged black holes, when perturbed nonlinearly by a self-gravitating, minimally-coupled, massless, spherically-symmetric scalar field, turns into a null weak singularity which focuses monotonically to $r = 0$ at late times, where the singularity becomes spacelike.

Our main objective is to study this spacelike singularity. We study analytically the spherically-symmetric Einstein-Maxwell-scalar equations asymptotically near the singularity. We obtain a series-expansion solution for the metric functions and for the scalar field near $r = 0$ under the simplifying assumption of homogeneity. Namely, we neglect spatial derivatives and keep only temporal derivatives. We find that there indeed exists a *generic* solution of a spacelike singularity for these equations (in the sense that the solution depends on the right number of free parameters), with similar properties to those found in the numerical simulations. This singularity is strong in the Tipler sense, namely, every extended object would inevitably be crushed to zero volume. In this sense this is a similar singularity to the spacelike singularity inside uncharged spherical black holes. On the other hand, there are some important differences between the two cases. Our model can also be extended to the more general inhomogeneous case.

The question of whether the same kind of singularity evolves in more realistic models (of a spinning black hole coupled to non-spherical vacuum perturbations) is still an open question.

11.1 Introduction

The singularity theorems of Hawking and Penrose [1] predict the occurrence of spacetime singularities inside black holes under very plausible assumptions. However, they tell us nothing about the geometrical and physical nature and properties of these singularities. The Kerr black hole, which describes the exterior of realistic black holes, fails to describe adequately their interiors. In the Kerr solution the singularity has the shape of a ring, and it is timelike. Consequently, it violates the strong cosmic censorship hypothesis of Penrose [2]. However, the Kerr solution is unstable, in the sense that the Cauchy horizon is converted into a spacetime singularity. This singularity is null and weak [3]. Namely, despite the divergence of curvature, infalling extended observers would experience only finite tidal distortions at the singularity. Because of the complexities involved with the analysis of spinning black holes, a frequently used toy model is the spherical charged black hole. Although realistic black holes are not expected to be significantly charged, the spherical charged black hole shares many common properties and similar causal structure with the more realistic spinning black holes. With spherical symmetry, non-trivial dynamics can be modeled by a scalar field, which has a spherically symmetric radiative mode. In recent numerical analyses with this toy model (namely, a spherical charged black hole perturbed non-linearly by a self-gravitating scalar field) it has been shown that the generators of the Cauchy horizon are focused at the late parts of the Cauchy horizon [4, 5]. Namely, the value of the area coordinate r decreases slowly and monotonically along the Cauchy horizon, until the Cauchy horizon is completely focused, i.e., the area coordinate shrinks to zero value at a finite value of the affine parameter along the Cauchy horizon. Then, the singularity becomes spacelike. (The spacelike singularity was first found numerically by Gnedin and Gnedin [6].) It is still an open question whether for a realistic spinning black hole, perturbed non-linearly by a realistic physical field (such as gravitational waves), the generators of the Cauchy horizon would also focus completely, and create a strong spacelike singularity. In this paper we shall focus on the spacelike singularities inside spherical black holes (both uncharged and charged), and study their properties under the simplifying assumptions of spherical symmetry and a minimally-coupled massless scalar field. We shall also assume homogeneity and power-law behavior of the metric functions, which are justified by recent numerical simulations.

The spacelike singularity inside a spherically-symmetric uncharged black hole perturbed by a scalar field was studied by Doroshkevich and Novikov [7], who treated the scalar field as a linear perturbation and obtained the general solution for the linear scalar field, but did not study the backreaction of the scalar field on the geometry. Doroshkevich and Novikov found that the linear scalar field diverges logarithmically with r at the singularity. Therefore, the energy density of the scalar field diverges like an inverse power of r, and therefore, acting as a source term for the Einstein-Klein-Gordon equations, the scalar field might be expected to change the metric functions. However, the very divergence of the linear perturbation analysis indicates that non-linear effects are expected to be important. More recently, Krori, Goswami and Das Purkaystha [8] considered the same problem as Doroshkevich and Novikov [7], but failed to obtain a general solution. The regular solution obtained by these authors depends on just one arbitrary parameter, while a general solution clearly depends on two independent arbitrary parameters (see below). Consequently, the correct conclusions to be drawn from a linear analysis are the conclusions of Doroshkevich and Novikov [7] and not the conclusions of Krori, Goswami and Das Purkaystha [8]. However, with the presence of an electric charge a linear perturbation analysis cannot investigate a generic spacelike singularity, as this singularity is created by nonlinear effects, i.e., by the nonlinear focusing of the generators of the Cauchy horizon. Under linear perturbations there would not exist inside a spherical charged black hole a spacelike singularity at all. Therefore, to study the spacelike singularity inside a spherical charged black hole with a scalar field one has to consider the fully nonlinear case. Nonlinear analysis of spacetime singularities with a scalar field in the cosmological case was done by Belinskii and Khalatnikov [9], who found that the scalar field destroys the BKL oscillations and that in that case the singularity is monotonic. (Interestingly, Belinskii and Khalatnikov showed that if, in addition to the scalar field, there were also a vector field, the singularity would again be oscillatory. The analysis by Belinskii and Khalatnikov [9] is markedly different from the analysis we shall present below, as we are interested here in spherical symmetry. It turns out that the Kasner-metric based Belinskii–Khalatnikov analysis cannot be reconciled with spherical symmetry.)

The organization of the paper is as follows. In Section 11.2 we describe the physical model we employ. In Section 11.3 we discuss the properties of power-law spacelike singularities. In sections 11.4 and 11.5 we study the cases of uncharged and charged spherical black holes, correspondingly. We show that indeed there exists a generic solution of the spherically-symmetric coupled Einstein-Maxwell-Klein-Gordon equations with a spacelike singularity, with properties which one actually finds in numerical simulations. This result strengthens our confidence in the picture of black hole interiors as described above. In Section 11.6 we analyze the strength of the singularity we find. We find that it is strong in the sense of Tipler (whereas the null Cauchy horizon singularity is weak). We summarize and give some concluding remarks in Section 11.7.

11.2 Physical model

We performed numerical simulations of the collapse of a spherical, self-gravitating, minimally coupled, massless scalar field over a pre-existing Reissner-Nordström black hole with charge q. Our code is based on free evolution and on double-null coordinates and is described at length in Ref. [10]. The code is stable and converges with second order. The numerical set-up is described in Figure 11.1. Prior to the initial hypersurface the geometry is Reissner-Nordström. Then, at some advanced time v_0 the spacetime is perturbed by a high-amplitude self-gravitating, spherical, massless scalar field of squared-sine shape on the outgoing section of the initial hypersurface. The null coordinates we use in the numerical simulations are linear with the area coordinate r on the initial hypersurface.

This high amplitude scalar field increases the external mass of the black hole at

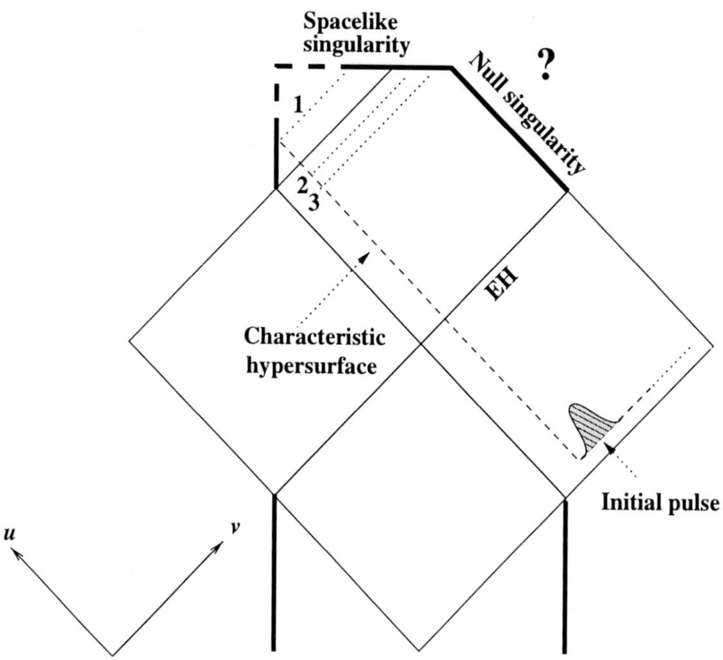

Figure 11.1: Spacetime diagram for a spherical charged black hole with a self-gravitating scalar field. Singularities are denoted by thick lines. Prior to the initial hypersurface (dashed) the geometry is Reissner-Nordström. The spacetime is then perturbed by the scalar field. The Cauchy horizon is converted into a null singularity. It is still an open question whether there is a continuation of the spacetime manifold beyond the null singularity, and if so, what the topology and the geometry are. The Cauchy horizon is focused to $r = 0$, where the singularity is spacelike. Our numerical setup does not allow us to investigate the late parts of the spacelike singularity, which are denoted by a thick dashed line, as they are located beyond the domain of influence of the characteristic hypersurface. We investigate the fields along outgoing null curves, denoted by 1, 2, and 3.

late times by over 10%. We then probe the metric functions and the scalar field in the black hole's interior along outgoing (and ingoing) null rays, approaching the spacelike singularity, which is located at some finite value of the ingoing null coordinate $v_*(u_p)$. Here, u_p is the value of the outgoing null ray u at which our outgoing null ray is located. Figure 11.1 shows three such rays, denoted by 1,2, and 3. The metric we use for the numerics is

$$ds^2 = -F(u,v)\,du\,dv + r^2(u,v)\,d\Omega^2. \tag{11.1}$$

The results we find numerically are as follow: First, the metric function $g_{uv} \equiv -F/2$ vanishes at the singularity, and F decays as a power-law with respect to the area coordinate r as one gets closer to the singularity. In addition, we find that the area coordinate r decays like a power of $v_* - v$ near the singularity. Numerically, we find in all our simulations this power to be very close to $\frac{1}{2}$, with a deviation of 1%. Therefore, we would like to show analytically, that in a generic solution indeed $r \propto (v_* - v)^{1/2}$ near the singularity. Our Numerical results are described in Figures 11.2 and 11.3. These results are independent of the precise definition of the null coordinates we were using, as any regular gauge transformation which preserves the metric-form (11.1) introduces just a multiplicative factor which does not change the power-law indices. In addition, we find similar results for all u_p's. That is, the power with which F decays to zero at the singularity depends on u_p, but only weakly so. The power $\frac{1}{2}$ we found for $r(v_* - v)$ is retained for all u_p's.

Therefore, we make the simplifying assumption of quasi homogeneity. Namely, we assume that only derivatives of the metric functions and of the scalar field normal to the singularity are non-vanishing, while derivatives tangent to the singularity are exactly zero on it. We thus write the line-element as

$$ds^2 = h(r)\,dt^2 + f(r)\,dr^2 + r^2\,d\Omega^2, \tag{11.2}$$

where $d\Omega^2 = d\theta^2 + \sin^2\theta\,d\phi^2$ is the usual metric on the unit two-sphere. As the singularity is spacelike, and t is the spacelike coordinate near the singularity and r is timelike, we are going to neglect all derivatives with respect to t, and keep only derivatives with respect to r. With these assumptions the t–t, r–r and θ–θ components of the Einstein-Maxwell-scalar equations in spherical symmetry are, correspondingly,

$$\frac{h}{r^2 f^2}\left(f'r + f^2 - f\right) = \frac{h}{f}\Phi'^2 + h\frac{q^2}{r^4} \tag{11.3}$$

$$\frac{1}{hr^2}\left(h'r - hf + h\right) = \Phi'^2 - f\frac{q^2}{r^4} \tag{11.4}$$

$$\frac{r}{4f^2h^2}\left(2h'fh - 2f'h^2 + 2rh''hf - rh'^2 f - rh'f'h\right) = \frac{q^2}{r^2} - \frac{r^2}{f}\Phi'^2. \tag{11.5}$$

[The ϕ–ϕ component of the field equations gives again Eq. (11.5)]. Here, a prime denotes differentiation with respect to r. In addition we also have the Klein-Gordon

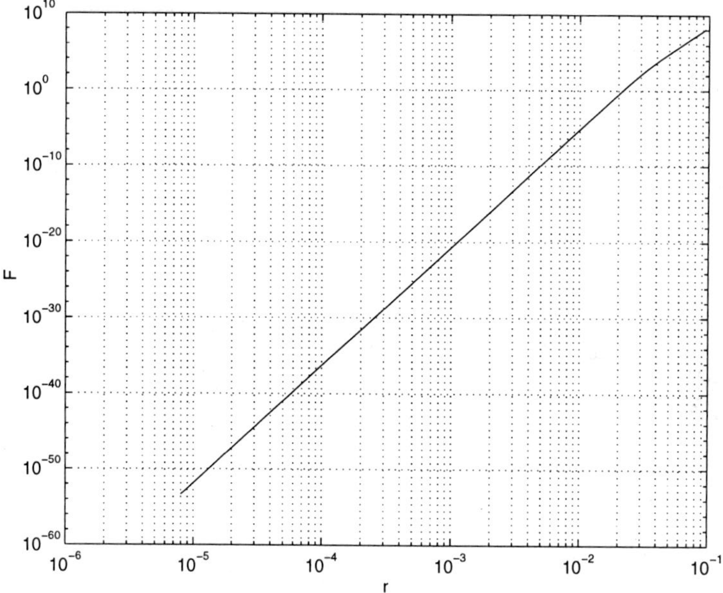

Figure 11.2: Metric function $F = -2g_{uv}$ as a function of r near the spacelike singularity along an outgoing null ray.

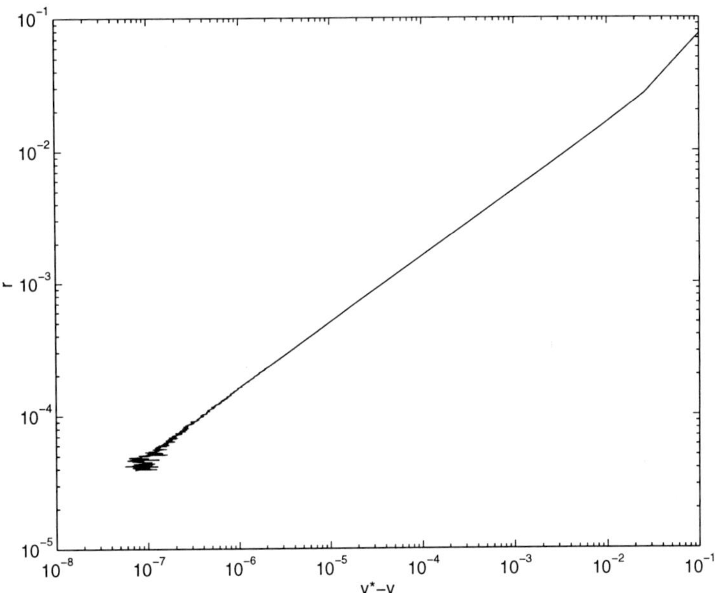

Figure 11.3: Area coordinate r as a function of $v_* - v$ near the spacelike singularity along an outgoing null ray. At small values of r there is a growing numerical noise. However, even before the noise becomes significant we have over two decades in r and four decades in $v_* - v$ of asymptotic behavior. At small r the numerical noise becomes significant. This numerical noise originates from free evolution of numerical errors, which dominate at small values of r.

equation for the scalar field $\Phi^{;\alpha}{}_{;\alpha} = 0$, whose first integral reads

$$\Phi'(r) = \frac{d}{\sqrt{-g}} f(r)\,\sin\theta, \tag{11.6}$$

where d is an integration constant and g is the metric determinant. (Note that because the metric determinant has a factor $\sin^2\theta$, $\Phi'(r)$ does not depend on θ.)

We now use Eq. (11.6) to eliminate the scalar field from Eqs. (11.3)–(11.5), and find that

$$f'r + f^2 - f + \frac{f^2}{hr^2}\left(d^2 - q^2 h\right) = 0 \tag{11.7}$$

$$h'r - hf + h + \frac{f}{r^2}\left(d^2 + q^2 h\right) = 0 \tag{11.8}$$

$$h'' - \frac{1}{2}\frac{h'^2}{h} + \frac{1}{r}\left(h' - \frac{f'}{f}h\right) - \frac{1}{2}\frac{h'f'}{f} - 2\frac{f}{r^4}\left(d^2 + q^2 h\right) = 0. \tag{11.9}$$

In what follows, we shall solve Eqs. (11.7) and (11.8), subject to our numerically-motivated homogeneous model. We then use Eq. (11.9) as a consistency check for our solution. Namely, we seek a generic solution for which both metric functions h, f vanish at the spacelike singularity. This is motivated by the vanishing of g_{uv} in the double-null metric, from which the vanishing of both h and f follows, by merit of the relations

$$g_{tt} \equiv h = 2\frac{\partial u}{\partial t}\frac{\partial v}{\partial t}g_{uv} = -2g_{uv}$$

$$g_{rr} \equiv f = g_{r_* r_*}\left(\frac{dr_*}{dr}\right)^2 = -g_{tt}\left(\frac{dr_*}{dr}\right)^2 = 2g_{uv}\left(\frac{dr_*}{dr}\right)^2 .$$

Here, the 'tortoise' coordinate r_* is defined by $g_{r_* r_*} = -g_{tt}$, and the null coordinates are defined by $u = r_* - t$ and $v = r_* + t$.

11.3 Properties of power-law spacelike singularities

In this section, we shall study several properties of spacelike singularities, which are common for any spatially homogeneous metric of the form (11.2) with $h(r), f(r)$ being powers of r. (Note, we still do not assume that $f(r), h(r)$ vanish at the singularity. We shall specialize to this case below.) First, we shall give a few examples of known spacelike singularities. Then, we shall study the general homogeneous power-law metric (11.2).

11.3.1 Examples for spacelike singularities

Schwarzschild singularity

The Schwarzschild solution assumes near the singularity the form

$$ds^2 \approx \left(\frac{r}{2M}\right)^{-1} dt^2 - \left(\frac{r}{2M}\right) dr^2 + r^2\, d\Omega^2, \tag{11.10}$$

where t is a spacelike coordinate tangent to the singularity and r is timelike and normal to the singularity. This metric obviously satisfies the power-law form of (11.2).

The Kretschmann scalar along radial $t =$ const curves is

$$R_{\alpha\beta\gamma\delta}R^{\alpha\beta\gamma\delta} = \frac{64}{27}\frac{1}{(\tau_* - \tau)^4}, \tag{11.11}$$

where τ is the proper time of the infalling object, and τ_* is the proper time at which the object arrives at the singularity at $r = 0$. We now denote by x^4 the coordinate tanget to the world line of an infalling observer. Because the Schwarzschild spacetime is vacuous, the projection of the Ricci tensor on the world line is identically zero, i.e., $R_{(4)(4)} \equiv R_{\mu\nu}e^{\mu}_{(4)}e^{\nu}_{(4)} = 0$, where $R_{(\alpha)(\beta)}$ is the $\alpha\beta$ tetrad component of the Ricci tensor. These results are independent of the mass M of the Schwarzschild black hole.

Friedmann-Robertson-Walker cosmology

The metric for the FRW cosmological model is given by

$$ds^2 = -dt^2 + R^2(t)\left(\frac{1}{1-kr^2}\,dr^2 + r^2\,d\Omega^2\right). \tag{11.12}$$

Let us consider here radiation-dominated perfect fluid with energy momentum tensor $T_{\mu\nu} = p\,g_{\mu\nu} + (p+\rho)u_\mu u_\nu$, u_μ being the four-velocity. We take $p = \rho/3$.

Then, in a co-moving frame near the singularity

$$R_{\alpha\beta\gamma\delta}R^{\alpha\beta\gamma\delta} = \frac{3}{2}\,\frac{1}{(\tau_*-\tau)^4}, \tag{11.13}$$

and

$$R_{(4)(4)}(\tau) = \frac{3}{4}\,\frac{1}{\tau^2}. \tag{11.14}$$

These results are independent of the density ρ (or the pressure p), and are also independent of k, i.e., independent of the spatial curvature.

Kasner solution

The Kasner solution with a spacelike parameter (similar results are obtained also for the Kasner solution with a timelike parameter) is given by the metric

$$ds^2 = x^{2p_1}\,dt^2 - dx^2 - x^{2p_2}\,dy^2 - x^{2p_3}\,dz^2, \tag{11.15}$$

with $p_1+p_2+p_3 = 1 = p_1^2+p_2^2+p_3^2$. (This solution is anisotropic and is not spherically symmetric like the other solutions we consider in this paper.)

For the Kasner solution one finds

$$R_{\alpha\beta\gamma\delta}R^{\alpha\beta\gamma\delta} = 16p_1^2(1-p_1)\,\frac{1}{(\tau_*-\tau)^4}, \tag{11.16}$$

and

$$R_{(4)(4)}(\tau) = 0. \tag{11.17}$$

This last result is obvious from the fact that the Kasner solution represents a vacuum spacetime.

11.3.2 General properties

In all the above examples, the Kretschmann scalar diverges near the spacelike singularity like $1/(\tau_*-\tau)^4$, and the Ricci tensor, if spacetime is non-vacuum, diverges like $1/\tau^2$. In this section we shall show that these results are general for any power-law metric of the form (11.2). Writing $h(r) = Br^m$ and $f(r) = -Ar^n$, with A, B being constants, a straightforward (but lengthy) calculation for the Kretschmann scalar yields

$$R_{\alpha\beta\gamma\delta}R^{\alpha\beta\gamma\delta} = 4\,\frac{m^4 - 4m^3 - 2m^3n + 12m^2 + 4m^2n + m^2n^2 + 8n^2 + 16}{(2+n)^4\,(\tau_* - \tau)^4}. \qquad (11.18)$$

In Schwarzschild, $m = -1$ and $n = 1$. Substitution recovers the known result for Schwarzschild (11.11). This result (11.18) is independent of A, B. This is the analogue of the independence of the Schwarzschild Kretschmann scalar of the mass M. One also finds that

$$R_{(4)(4)}(\tau) = \frac{2m + mn + 4n - m^2}{(n+2)^2}\,\frac{1}{\tau^2}. \qquad (11.19)$$

This result is of great importance, because one can show that if $2m+mn+4n-m^2 \neq 0$ than a sufficient condition for the singularity to be strong in the sense of Tipler is satisfied.

A strong singularity in the sense of Tipler can be defined as follows [11]: Let us consider a non-spacelike geodesic running into the singularity. Then, if the limit of the greatest lower bound of the volume element defined by the spatial metric induced by any three independent Jacobi fields vanishes as one approaches the singularity along every such geodesic, then the singularity is strong in the sense of Tipler. The physical content of this definition is that the volume of every extended physical object is compressed infinitely, such that every extended object will inevitably be crushed to zero volume. We note, that this is a 'strong' definition for destructive singularities, as even weaker singularities would be strong enough to destroy any extended physical object. Tipler [11] gives an example for such a singularity, where the spatial metric induced by the Jacobi fields behaves near the singularity like diag$(\tau,\ \tau,\ \tau^{-2})$. The volume element defined by the spatial metric equals unity, but any extended object would be infinitely stretched in one direction, and infinitely squeezed in two directions, in an infinite spaghettification.

A theorem by Clarke and Królak [12] relates $R_{(4)(4)}$ to the strength of the singularity: A sufficient condition for a singularity to be strong in the Tipler sense is that the integral $\int_0^{\tau_*} d\tau' \int_0^{\tau'} d\tau'' R_{(4)(4)}(\tau'')$ diverges at τ_*. For example, the FRW cosmological model has a Tipler strong singularity, as for this case this integral diverges logarithmically. Note that this is a sufficient condition, but not a necessary condition: the Schwarzschild and Kasner singularities are Tipler strong, but the integral vanishes identically by merit of their vacuous spacetimes. The important conclusion to be made here is as follows: For any non-vacuum homogeneous spacelike singularity near which the metric has a power-law metric (11.2) the singularity is strong in the Tipler sense by virtue of the theorem by Clarke and Królak and by Eq. (11.19).

11.4 Uncharged case

In this section we study the spacelike singularity in the presence of a self-gravitating scalar field with spherical symmetry and with vanishing electric charge ($q = 0$). This is the non-linear generalization in the homogeneous case of the linear analysis of Doroshkevich and Novikov [7]. In the next section we shall also study the charged case. We assume a series expansion for the metric functions of the general form $f^{(n)}(r) = \sum_{i=1}^{n} f_i(r)$ and $h^{(n)}(r) = \sum_{i=1}^{n} h_i(r)$, where at $r = 0$ $f^{(n)} = h^{(n)} = 0$. Here, n denotes the expansion order of the series, which, as will be shown below, are assumed to have a finite radius of absolute convergence. That is, $f^{(n=\infty)} \equiv f$, $h^{(n=\infty)} \equiv h$ for some finite interval $0 \leq r < r_0$. We next assume a power-law series behavior of the metric functions. That is, we assume that $h_i = h_{i0}\, r^{n_i}$ and $f_i = f_{i0}\, r^{m_i}$. Note that for the time being we do not restrict the parameter range of n_i, m_i.

11.4.1 Leading order approximation

To the leading order all nonlinear terms in the rhs of Eqs. (11.7) and (11.8) are negligible, and therefore Eqs. (11.7) and (11.8) assume the form

$$f_1' r - f_1 + \frac{d^2 f_1^2}{h_1 r^2} = 0 \tag{11.20}$$

$$h_1' r + h_1 + \frac{d^2 f_1}{r^2} = 0 \tag{11.21}$$

Substitution of the Ansatz $h_1 = B\, r^\beta$, $f_1 = -A\, r^\alpha$, with $A, B > 0$ (as we are looking for a spacelike singularity) into these equations yields $\alpha = \beta + 2$ and $A = (\beta+1)B/d^2$. From the positivity of A and B it then follows that $\beta > -1$. In Schwarzschild, $\beta = -1$, and the exponents are uniquely determined. Here, however, the scalar field endows the field equations with a freedom in the value of β. For small amplitudes of the scalar field on the initial hypersurface we would expect the deviation of β from -1 to be small, such that β would still be negative. However, there would also be situations with positive values for β, and even $\beta = 0$ as a special case. We thus find that

$$h^{(1)} = B\, r^\beta \tag{11.22}$$

$$f^{(1)} = -(\beta + 1)\, \frac{B}{d^2}\, r^{\beta+2}. \tag{11.23}$$

Recall that d is a constant of integration for the Klein-Gordon equation (11.6). In Schwarzschild $d = 0$, and therefore one needs to take

$$\lim_{\substack{d \to 0 \\ \beta \to -1}} \frac{\beta + 1}{d^2} = \frac{1}{4M^2} \qquad \lim_{\substack{d \to 0 \\ \beta \to -1}} B = 2M,$$

where M is the mass of the black hole, in order to recover in the limit of vanishing scalar field the Schwarzschild solution. An important conclusion to be drawn from this solution is that $r \propto (v_* - v)^{1/2}$. This can be shown as follows: For a general homogeneous metric (11.2) we define a 'tortoise' coordinate r_* by $g_{r_* r_*} = -g_{tt}$. The metric then takes the form

$$ds^2 = g_{r_* r_*} \left(dr_*^2 - dt^2 \right) + r^2 \, d\Omega^2, \tag{11.24}$$

where $g_{r_* r_*} = -g_{tt} = g_{rr} \, \left(dr / \, dr_* \right)^2$.
Consequently, $dr / \, dr_* = -\sqrt{-g_{tt}/g_{rr}} = -(1/r)\sqrt{d^2/(\beta + 1)}$. Integration yields $r_* = -\frac{1}{2}\sqrt{(\beta + 1)/d^2} \; r^2 + \text{const}$. Defining now future-directed null coordinates by $t = \frac{1}{2}(v - u)$ and $r_* = \frac{1}{2}(v + u)$, we find $v = 2r_* + \text{const}$, or

$$r = \left(\frac{d^2}{\beta + 1} \right)^{\frac{1}{4}} (v_* - v)^{1/2} . \tag{11.25}$$

The power $\frac{1}{2}$ we find is independent of the value of β, and is a direct consequence of the relation $\alpha = \beta + 2$. We note that for Schwarzschild $\alpha = 1$ and $\beta = -1$, which clearly satisfies this relation.

We next show that our solution is generic. In our solution there are three arbitrary parameters, namely β, d^2, and B. Without a scalar field the solution is Schwarzschild, where there is just one arbitrary parameter, by virtue of Birkhoff's theorem. Therefore, with the scalar field, we would expect one additional arbitrary parameter, i.e., two arbitrary parameters. Apparently, we have here three parameters. However, the arbitrariness in the fixing of B is just a trivial gauge mode, related to the possibility to make an arbitrary transformation $t \to t' = f(t)$ (*cf.* Ref. [13] for the analogue in Schwarzschild). Therefore, with a scalar field there are only two non-trivial arbitrary parameters, as should be expected. Therefore, our solution has the right number of arbitrary parameters, and in this sense is generic. Now, for a system of nonlinear equations the notion of a general solution is not unambiguous, and there may be other solutions of non-zero measure in solutions space, but our solution above also has a non-zero measure, and is therefore generic.

We can now substitute our solution in the Klein-Gordon equation (11.6), and obtain for the scalar field, to the leading order in r,

$$\Phi^{(1)} = \sqrt{\beta + 1} \; \ln r \tag{11.26}$$

which diverges logarithmically like the linear scalar field studied by Doroshkevich and Novikov [7]. We note that the amplitude of the scalar field depends on β, which is determined by the nonlinear equations, whereas in the linear analysis the amplitude is arbitrary. Therefore, even the leading order in the expansion for the scalar field is determined by nonlinear effects.

We now assess the error involved with the consideration of the leading order only of the series expansion. For this aim, we consider Eqs. (11.7)–(11.9). Substituting our leading order expression, we find the expressions for the deviation from zero of the rhs of each equation, which we shall refer to below as the error associated with the truncated solution at a certain order. We find that Eq. (11.9) is satisfied exactly. To the leading order in r^β the error in Eq. (11.7) is $(B^2/d^4)\,(\beta+1)^2\,r^{2\beta+4}$, and the error in Eq. (11.8) is $(B^2/d^2)\,(\beta+1)\,r^{2\beta+2}$.

11.4.2 Second order approximation

We shall now find the second order in the series expansions for the metric functions and for the scalar field. To the second order the field equations Eq. (11.7) and Eq. (11.8) reduce to

$$f_2'r-\left(1-2\frac{d^2f_1}{h_1r^2}\right)\,f_2+\left(1-\frac{d^2h_2}{r^2h_1^2}\right)\,f_1^2=0 \tag{11.27}$$

$$h_2'r+h_1f_1+h_2+\frac{d^2f_2}{r^2}=0, \tag{11.28}$$

where h_1 and f_1 are already known from Eqs. (11.22) and (11.23). We again assume a power-law behavior for f_2 and h_2, and obtain

$$f_2=-\frac{(\beta+1)^2(3\beta+4)}{(\beta+2)^2}\,\frac{B^2}{d^4}\,r^{2\beta+4} \tag{11.29}$$

$$h_2=\frac{\beta(\beta+1)}{(\beta+2)^2}\,\frac{B^2}{d^2}\,r^{2\beta+2}. \tag{11.30}$$

We thus find that the second-order expansion for f and h is

$$f^{(2)}=-(\beta+1)\frac{B}{d^2}\,r^{\beta+2}-\frac{(\beta+1)^2(3\beta+4)}{(\beta+2)^2}\,\frac{B^2}{d^4}\,r^{2\beta+4} \tag{11.31}$$

$$h^{(2)}=B\,r^\beta+\frac{\beta(\beta+1)}{(\beta+2)^2}\,\frac{B^2}{d^2}\,r^{2\beta+2}. \tag{11.32}$$

We now evaluate the errors, to leading order in r^β, in the field equations (11.7)–(11.9) for the second-order expansion. For Eq. (11.7) the error is $2(B^3/d^6)\,[(\beta+1)^3(5\beta+6)/(\beta+2)^2]\,r^{3\beta+6}$, the error in Eq. (11.8) is $4(B^3/d^4)\,[(\beta+1)^3/(\beta+2)^2]\,r^{3\beta+4}$, and the error in Eq. (11.9) is $[(\beta+1)(\beta^2+8\beta+8)/(\beta+2)^2]/d^4\,r^{3\beta+2}$.

The second order expansion we find from the Klein-Gordon equation (11.6) for the scalar field is

$$\Phi^{(2)}=\sqrt{\beta+1}\,\ln r+\frac{\sqrt{\beta+1}\,(5\beta^2+12\beta+6)}{(\beta+2)^3}\,\frac{B}{d^2}\,r^{\beta+2}. \tag{11.33}$$

Namely, under nonlinear effects the powers of the second (and higher) order term is different from the power found in the linear analysis [7].

11.4.3 General expression for the series expansion

We now turn to the general form of the series expansion. Based on the above expressions, we seek series expansions of the forms

$$f = \sum_{n=1}^{\infty} a_n \, r^{(\beta+2)n} \qquad a_n = a_n(d^2 \, , \beta \, , B) \tag{11.34}$$

$$h = \sum_{n=1}^{\infty} b_n \, r^{(\beta+2)n-2} \qquad b_n = b_n(d^2 \, , \beta \, , B). \tag{11.35}$$

Equations (11.34) and (11.35) will solve the field equations (11.7)–(11.9) if the following two conditions are satisfied: First, the series converge absolutely [this is to ensure that when Eqs. (11.34) and (11.35) are substituted in Eqs. (11.7)–(11.9) the series multiplication theorem will be applicable], and second, that the expansion coefficients a_n and b_n can be found uniquely for any n. If these two conditions are satisfied, than Eqs. (11.34) and (11.35) represent a generic solution of the field equations (as they contain the right number of arbitrary parameters). The first condition is hard to prove, without knowledge of the values of the arbitrary functions d^2, β and B. We thus assume that the first condition is satisfied for small enough values of r. We next prove the satisfaction of the second condition. For $n \geq 2$, Eqs. (11.34) and (11.35) yield the following algebraic equations for the expansion coefficients a_n and b_n:

$$d^2 \, a_n + [(\beta+2)(n+1) - 1] \, b_n = \sum_{k=0}^{n-1} a_{n-k-1} b_k \tag{11.36}$$

$$\left\{ [n(\beta+2) + \beta + 1] \, b_1 + 2d^2 \, a_1 \right\} \, a_n + (\beta+1) a_1 \, b_n = c, \tag{11.37}$$

where

$$c = - \sum_{k=1}^{n-1} \left\{ \left[(n-k)(\beta+2) \, b_k + d^2 \, a_k \right] \, a_{n-k} + b_k \sum_{l=0}^{n-k-1} a_l a_{n-k-l-1} \right\}$$
$$- \; b_1 \sum_{l=0}^{n-1} a_l a_{n-l-1}.$$

These equation will have a unique solution if the determinant of the homogeneous part does not vanish for all n. A straightforward calculation yields for this determinant $\Delta_n = n^2 B(\beta+2)^2 \neq 0$ (recall that $\beta > -1$), which proves that the second condition is satisfied.

The general series expansion for the scalar field is

$$\Phi = \sqrt{\beta + 1} \left(\ln r + \sum_{n=2}^{\infty} c_n \, r^{(\beta+2)(n-1)} \right), \tag{11.38}$$

where c_n are the expansion coefficient which can be found uniquely for each order n from the Klein-Gordon equation (11.6). We note that in Schwarzschild $\beta = -1$, and indeed for this choice of β the scalar field vanishes identically.

11.5 Charged case

In this section we shall assume that $q \neq 0$, and show that there exists a generic solution for a spacelike singularity, of similar properties with the singularity one finds numerically [4, 5]. We shall show that this singularity has many similarities to the singularity in the uncharged case we studied in the previous section, but also some important differences. We again make the Ansatz that the metric functions are functions of r only, and that r is a timelike coordinate. We thus neglect all derivatives with respect to spacelike coordinates. We first consider the leading order expression, then the second order correction, and finally the general expression for the series expansion.

11.5.1 Leading order approximation

The source term for the Einstein-Maxwell-Klein-Gordon equations Eqs. (11.3)–(11.5) contains two contributions: the contribution of the electric field and the contribution of the scalar field. In order to find the leading order of the series expansion for the metric functions, one can consider the following three possibilities for the relative contributions of the two sources near the spacelike singularity:
(a) The contribution of the scalar field near $r = 0$ is dominant, and the electric field's contribution is negligible.
(b) The contribution of the electric field near $r = 0$ is dominant, and the scalar field's contribution is negligible.
(c) The contributions of the scalar field and of the electric field near $r = 0$ are comparable.
It turns out that there is no consistent solution of the field equations with possibility (c). Let us now consider possibility (b): As near $r = 0$ the scalar field's contribution to the field equations is negligible compared with the electric field's contribution, the leading order expression is the same as the leading order expression without a scalar field at all. However, from the generalized Birkhoff theorem, the solution then is nothing but the Reissner-Nordström solution, written as a series expansion near $r = 0$. But the $r = 0$ singularity in Reissner-Nordström is timelike rather than spacelike, which is the type of singularity we are interested at here. Therefore, we do not expect possibility (b) to realize in our case. We note in passing that this case might be of some relevance for consideration of the self-gravitating scalar field solution for an hypothetical extension of the spacetime manifold *beyond* the weakly singular Cauchy horizon.

We study, then, possibility (a). Namely, we assume that near the singularity the contribution of the electric field to the energy-momentum tensor is negligible compared with the contribution of the scalar field. With this assumption, the equations which govern the first order expressions for the metric functions and for the scalar field are the same as their uncharged counterparts Eq. (11.20) and Eq. (11.21). Consequently, to the leading order, the spacelike singularities with an electric field or without an electric field have the same functional form [*cf.* Eqs. (11.22) and (11.23)]:

$$h^{(1)} = B\, r^{\beta} \tag{11.39}$$

$$f^{(1)} = -(\beta + 1)\, \frac{B}{d^2}\, r^{\beta+2}. \tag{11.40}$$

There is, however, a difference in the range of the parameters between the two cases. Whereas in the uncharged case β was free to assume negative, positive or zero values (subject to the restriction $\beta > -1$), we here find that in order to have a solution consistent with our conclusions based on numerical simulations β needs to be positive. Without a scalar field, the solution is Reissner-Nordström, where the singularity is timelike, and $\beta = -2$. Because there is no scalar field, the electric contribution to the energy-momentum tensor cannot be neglected, and the first order equations are dominated by the electric contribution. Therefore, there is no smooth analytic limit for a vanishing scalar field, and there is a discontinuity in the values that β can take. This can be understood in terms of the causal structure of the singularity, which with a scalar field is spacelike and without a scalar field is timelike. With Eqs. (11.39) and (11.40), the Klein-Gordon equation (11.6) yields for the scalar field again

$$\Phi^{(1)} = \sqrt{\beta + 1}\, \ln r \tag{11.41}$$

which is the same as $\Phi^{(1)}$ for the uncharged case [Eq. (11.26)]. Note that there is no value of β which nullifies the scalar field in this case, as with vanishing scalar field there is no spacelike singularity like the singularity we are studying.

A direct consequence of (11.39) and (11.40) is that near the singularity $r \propto (v_* - v)^{1/2}$ in the charged case too (see above). We note, that this square-root relation can be deduced directly from the full partial differential equations. From our numerical results $F(u,v)$ vanishes faster than $r(u,v)$ as functions of $v - v_*$ near the singularity. Consequently, from Eq. (6) of Ref. [10] it follows that near the singularity $(r^2)_{,uv} \approx 0$. This equation can be readily integrated to $r^2(u,v) = x_1(u) + x_2(v)$. As at the singularity r vanishes, one finds that along $u_p = \text{const}$ a series expansion yields

$$r^2(v)\big|_{u_p} = (v - v_*)\, \left.\frac{d(r^2)}{dv}\right|_{v_*} + O[(v - v_*)^2],$$

and consequently $r \propto (v - v_*)^{1/2}$.

The estimate for errors associated with the leading order terms in the charged case are, to the leading order in r^{β}: For Eq. (11.7) the error is $-(B^2/d^4)\ (\beta +$

$1)^2\ q^2\ r^{2\beta+2} + (B^2/d^4)\ (\beta+1)^2 r^{2\beta+4}$, for Eq. (11.8) the error is $-(B^2/d^2)\ (\beta+1)\ q^2\ r^{2\beta} + (B^2/d^2)\ (\beta+1)\ r^{2\beta+2}$, and for Eq. (11.9) the error is $2(B^2/d^2)\ (\beta+1)\ q^2\ r^{2\beta-2}$. Note the differences between these error estimates and their counterparts in the uncharged case. For each equation, the error estimate here has two terms, one of them does not depend on the charge, and the other does depend on it. The term which does not depend on the charge is the same as the error term in the uncharged case. However, the charge-dependent term has a smaller power index, and is therefore dominant. [In the error for Eq. (11.9) the term which does not depend on the charge vanishes.] The difference in the power indices between the charge-dependent and the charge-independent terms is 2. We shall show below that this is an indication for a more complicated series form for the metric functions than the form we encountered in the uncharged case.

11.5.2 Second order approximation

The electric field does not contribute then to the leading order terms in the series expansion for the metric functions or the scalar field. However, we find that it does contribute to the second order terms. The equations which govern the second order terms are:

$$f_2' r - \left(1 - 2\frac{d^2 f_1}{h_1 r^2}\right) f_2 - \left(\frac{q^2}{r^2} + \frac{d^2 h_2}{r^2 h_1^2}\right) f_1^2 = 0 \tag{11.42}$$

$$h_2' r + h_2 + \frac{d^2 f_2}{r^2} + q^2\ \frac{h_1 f_1}{r^2} = 0, \tag{11.43}$$

where again f_1 and f_2 are known from the first order terms (11.39) and (11.40), and where we again assumed a power-law behavior. These equations are different from their uncharged counterparts [Eqs. (11.27) and (11.28)] not only in the appearance of charge-dependent terms, but also in the absence of terms which appeared in the uncharged case. Substitution of the power-law behavior Ansatz yields

$$f_2 = \frac{(\beta+1)^2(3\beta+2)}{\beta^2}\ \frac{B^2}{d^4}\ q^2\ r^{2\beta+2} \tag{11.44}$$

$$h_2 = -\frac{(\beta+1)(\beta+2)}{\beta^2}\ \frac{B^2}{d^2}\ q^2\ r^{2\beta}, \tag{11.45}$$

such that

$$f^{(2)} = -(\beta+1)\ \frac{B}{d^2}\ r^{\beta+2} + \frac{(\beta+1)^2(3\beta+2)}{\beta^2}\ \frac{B^2}{d^4}\ q^2\ r^{2\beta+2} \tag{11.46}$$

$$h^{(2)} = B\ r^{\beta} - \frac{(\beta+1)(\beta+2)}{\beta^2}\ \frac{B^2}{d^2}\ q^2\ r^{2\beta} \tag{11.47}$$

and from the Klein-Gordon equation (11.6)

$$\Phi^{(2)} = \sqrt{\beta+1}\,\ln r - \frac{\sqrt{\beta+1}\,(\beta+1)(3\beta+2)}{2\beta^3}\,\frac{B}{d^2}\,q^2\,r^{\beta}. \tag{11.48}$$

The estimate for the leading terms in the errors associated with the second order terms are: For Eq. (11.7) the error is $(\beta+1)^2\,(B^2/d^4)\,r^{2\beta+4} + 2[(\beta+1)^3(5\beta+4)/\beta^2]\,(B^3/d^6)\,q^2\,r^{3\beta+2} + O(r^{3\beta+4})$, for Eq. (11.8) the error is $(\beta+1)\,(B^2/d^2)\,r^{2\beta+2} + 4[(\beta+1^3)/\beta^2]\,(B^3/d^4)\,q^4\,r^{3\beta} + O(r^{3\beta+4})$, and for Eq. (11.9) the error is $[(\beta+1)^2(\beta^2-6\beta-8)/\beta^2]\,(B^3/d^4)\,q^4\,r^{3m-2} + O(r^{4\beta-2})$.

Comparing these expressions for the errors associated with the second order expansions and the errors associated with the first-order expansions, we find that the effect of the second-order terms is to provide terms which balance the leading-order errors for the first-order expansions, and leave just the subsequent terms of the first-order expansions as the leading terms for the errors of the second-order expansions. Namely, the contribution to the solution of each order balances the leading-order terms in the errors associated with the lower-order solution, and thus reduces the errors.

11.5.3 General expression for the series expansion

One would be tempted to consider a series form similar to the series form we have for the uncharged case (11.34) and (11.35). However, when one considers the third order terms for the metric functions f_3 and h_3 in the charged case one finds that this is impossible, unless the exact value of β is known. A close inspection of our previous results and of the third order approximation shows that the increments in the power indices in the charged case are not equal. Namely, the increment changes from β to $\beta+2$. Therefore, it is more natural to seek a general solution in terms of a double series expansion than a single series. Therefore, the solution will be of the form

$$f = \sum_{m=1}^{\infty}\sum_{n=1}^{\infty} a_{mn}\, r^{\beta(m-1)+(\beta+2)n} \qquad a_{mn} = a_{mn}(d^2,\, q^2,\, B,\, \beta) \tag{11.49}$$

$$h = \sum_{m=1}^{\infty}\sum_{n=1}^{\infty} b_{mn}\, r^{\beta m+(\beta+2)(n-1)} \qquad b_{mn} = b_{mn}(d^2,\, q^2,\, B,\, \beta). \tag{11.50}$$

Again, in order that Eqs. (11.49) and (11.50) would be solutions of the field equations (11.7)–(11.9), the series should converge absolutely for some finite convergence interval, and the series expansion coefficients a_{mn} and b_{mn} should be found uniquely for each m, n. In fact, the terms we previously found explicitly are nothing but the a_{11}, a_{21} and b_{11}, b_{21} terms in this double series expansion. It is still an open question whether the satisfaction of these two conditions can be proved rigorously. A similar double series expansion is also obtained for the scalar field, with the leading term diverging logarithmically.

11.6 Singularity strength

From the expression for the metric functions we found for the charged case we can calculate the leading term for the Kretschmann scalar and for $R_{(4)(4)}$ in terms of the proper time of an observer who follows a t = const radial trajectory. We find that

$$R_{\alpha\beta\gamma\delta}R^{\alpha\beta\gamma\delta} = 64\,\frac{2\beta(\beta+1)+3}{(\beta+4)^4}\,\frac{1}{(\tau_*-\tau)^4} \tag{11.51}$$

$$R_{(4)(4)}(\tau) = 8\,\frac{\beta+1}{(\beta+4)^2}\,\frac{1}{\tau^2}. \tag{11.52}$$

From Eq. (11.52) and from the theorem by Clarke and Królak [12] it then follows that the spacelike singularity in spherical charged black holes with a self-gravitating scalar field is strong in the Tipler sense. Recall that the singularity we found is a generic one, in the sense that it relies on the correct number of arbitrary parameters. One indeed finds that these are the expressions which one obtains from the general expressions for power-law metrics (11.18) and (11.19) for this solution with $m = \beta$ and $n = \beta + 2$. In addition, as in the charged case $\beta > 0$, it turns out that there is no special case which nullifies either of the curvature invariant $R_{\alpha\beta\gamma\delta}R^{\alpha\beta\gamma\delta}$ or the second integral over $R_{(4)(4)}(\tau)$. Namely, for any choice of β the spacelike singularity is strong in the Tipler sense. (Note that because of the inevitable presence of the scalar field for the singularity to be spacelike with electric charge, there is no vacuum solution, and therefore the Ricci tensor does not vanish.)

11.7 Concluding remarks

We found a generic solution for a spacelike singularity for the spherically symmetric Einstein-Maxwell-scalar field equations, which was previously found numerically in scalar-field collapse simulations. The generic singularity we found has the same properties as the singularity which arises in the numerical simulations.

The singularity we found was obtained under the assumption of quasi-homogeneity, namely, that the metric functions do not depend on derivatives with respect to the spatial coordinates (in our case, because of the spherical symmetry, this means that there is no dependence on t). Namely, we sought a velocity-dominated singularity. The next obvious step in the analysis of the singularity would be to relax the homogeneity assumption. This could perhaps be done by allowing the arbitrary parameters in the solution to be functions of t. Namely, all we need to do is to make the transformation $\beta \to \beta(t)$, $B \to B(t)$, and $d^2 \to d^2(t)$ for the leading order terms of the solution. The higher-order terms will have to be corrected for the terms with dependence on derivatives with respect to t. (Because of the spherical symmetry there would still not be derivatives with respect to the angular coordinates θ and ϕ.) A preliminary check shows that these leading terms indeed satisfy the full spherically-symmetric

Einstein-Maxwell-scalar field equations, with error smaller than the leading terms. However, it is still needed to perform a fuller analysis of the higher-order contributions. Then, the solution would be inhomogeneous, and depend on three arbitrary functions, and in this sense would be a generic solution. The homogeneous solution is then nothing but the pointwise behavior of the inhomogeneous solution. However, it is yet to be shown that this inhomogeneous solution really solves the inhomogeneous Einstein-Maxwell-scalar field equations.

It is extremely important to test the predictions of the theoretical model against the numerical results. The numerics give us a chance to probe the inhomogeneous singularity, i.e., without the need to introduce the simplifying assumption of homogeneity. For instance, one could check the value of the parameter β in Eqs. (11.39) and (11.41), and test whether they are equal also for the inhomogeneous case. This in presently under investigation. However, preliminary results indicate that this is indeed the case, and hence we may conclude that the homogeneous model is able to describe the properties of the actual singularity adequately.

Special attention is required in order to analyze to behavior of the fields near the spacetime event corresponding to the change of the causal structure of the singularity, namely to the event where the null singularity becomes spacelike. There are two possibilities to approach this spacetime event numerically, using a double-null code such as ours. First, one could approach this event along an outgoing null rays. Then one would need to reduce the value of the outgoing null coordinate u, and compare the values of the fields approaching the singularity along various values of u. Second, one could also study the singularity along ingoing null rays. Late-time ingoing null rays actually probe the null mass-inflation singularity. However, one can also focus attention to the deep portions of these rays, and thus study the nearly-complete focusing domain, where the value of r is very small. We find numerically that the area coordinate r continues to decrease monotonically even at the nearly-complete focusing domain. This second approach would be more natural with a slight variation of the numerical code: The integration is normally carried out in our code along ingoing rays, namely, we intergate from the initial data on an ingoing ray and the data on the first grid-point on the next ingoing ray, the fields throughout this second ray, and so on (for more details see Ref. [10]). However, for this second approach to the special event where the causal structure of the singularity changes, one could benefit from changing the direction of integration, and integrate along *outgoing* null rays instead of ingoing rays. Then, for each value of advanced time v for a certain ingoing null ray one would eventually approach the spacelike singularity for large u, and the later v, the closer one would be the the spacetime event under consideration.

We again stress that it ramains an open question whether in a more realistic model our results would be preserved. Namely, a scalar field was introduced as a toy model for a physical field because it has a spherical radiative mode. However, it is possible that more realistic fields will not create a spacelike singularity. (Scalar fields are known to create unique phenomena [9].) In addition, a realistic black hole is not spherically symmetric. In fact, a spherical spacelike singularity is known to be unstable to non-spherical perturbations (the only known generic spacelike singularity is the BKL singularity). In contrast with the BKL singularity, the singularity we found here is monotonic rather than oscillatory, and might not be stable to non-spherical perturbations. Namely, allowing for non-sphericities, one might expect to find a BKL singularity instead of the monotonic singularity we found for the spherical case. Whether a spacelike singularity would be created inside spinning black holes is a question still awaiting investigations.

Acknowledgements

I thank Amos Ori for many stimulating discussions and Alexei Starobinsky for useful comments.

Bibliography

[1] S. W. Hawking and G. F. R. Ellis, *The large scale structure of space-time* (Cambridge University Press, Cambridge, 1973).

[2] R. Penrose, Rivista Nuovo Cimento **1**, 252 (1969).

[3] A. Ori, Phys. Rev. Lett. **68**, 2117 (1992).

[4] P. R. Brady and J. D. Smith, Phys. Rev. Lett. **75**, 1256 (1995).

[5] L. M. Burko, Phys. Rev. Lett. **79**, 4958 (1997).

[6] N. Y. Gnedin and M. L. Gnedin, Sov. Astron. **36**, 296 (1992); M. L. Gnedin and N. Y. Gnedin, Class. Quantum Grav. **10**, 1083 (1993).

[7] A. G. Doroshkevich and I. D. Novikov, Zh. Eksp. Teor. Fiz. **74**, 3 (1978) [Sov. Phys. JETP **47**, 1 (1978)].

[8] K. D. Krori, A. K. Goswami and A. Das Purkaystha, Class. Quantum Grav. **12**, 835 (1995).

[9] V. A. Belinskii and I. M. Khalatnikov, Zh. Eksp. Teor. Fiz. **63**, 1121 (1972) [Sov. Phys. JETP **36**, 591 (1973)].

[10] L. M. Burko and A. Ori, Phys. Rev. D **56**, 7820 (1997).

[11] F. J. Tipler, Phys. Lett. **A 64**, 8 (1977).

[12] C. J. S. Clarke and A. Królak, J. Phys. Geom. **2**, 127 (1985).

[13] L. D. Landau and E. M. Lifshiz, *The Classical Theory of Fields*, Fourth Edition (Pergamon, Oxford, 1975), §100.

TWO–DIMENSIONAL ACCELERATED BLACK HOLES AND THE COSMIC CENSORSHIP CONJECTURE

A. Fabbri

Laboratoire 'Gravitation et Cosmologie Relativistes',
Université Paris VI, CNRS/URA 769
Tour 22/12 4ème etage – Boite Courrier, 142–4,
Place Jussieu- 75252 Paris Cedex 05, France

Abtract

In this paper we review recent work on a 2d dilaton gravity theory that describes accelerated black holes. In particular, the stability analysis of the Cauchy horizons arising in its solutions shows that, although at the classical level the strong cosmic censorship appears to be violated, quantum effects suggest that it is not possible to extend the spacetime geometry across this surface.

12.1 Introduction

Theoretical physicists have always been fascinated by the possibility that an observer falling into a black hole would not be crushed at the spacetime singularity but, instead, reemerge into a (causally disconnected) "parallel" universe.

On one hand, it is a common belief that quantum gravity will smooth out any singular behavior existing in the classical theory (attempts to mimic this scenario have been made where the interior of the Schwarzschild black hole was replaced by a new expanding macroscopic universe [1]). However, already at the classical level there exist black hole solutions where the singularity, although present, can be avoided and timelike geodesics that cross the event horizon travel "safely" through a black-hole tunnel and emerge into another universe. The Reissner-Nordström and Kerr solutions are examples of this type. This last possibility has, recently and not, catalysed much interest and research.

There are at least three reasons for which this kind of trips shouldn't be possible in nature. Already in 1969 Penrose [2] noted that the Reissner-Nordström black hole possesses a surface, the inner or Cauchy horizon, which is highly unstable to small perturbations generated in the external universe. Due to an infinite (exponential) blueshift, in fact, free falling observers would see these perturbations growing without bound there. Moreover, the presence of Cauchy horizons is "dangerous" because it seems to threaten the predictability of Einstein field equations [3] : it turns out that in order to determine the form of the spacetime geometry at the future of such surface one needs to impose boundary conditions at the timelike singularities and these are, of course, completely arbitrary in the context of the classical theory. Finally, the presence of timelike singularities, where gravity is in some sense repulsive, is rather unphysical and their occurrence is in principle rejected by the strong cosmic censorship conjecture (see for instance [3]).

Poisson, Israel [4] and Ori [5] have constructed simple models where these 'contradictions' of the Einstein theory appear to be solved. They used the fact that in the gravitational collapse of massive stars small departures from spherical symmetry arise, which constitute the radiative tail determined by Price [6]. The backreaction of these perturbations on the spacetime geometry produces a curvature singularity at the Cauchy horizon in the form of an unbounded growth of the local mass function. This is the essence of the mass inflation phenomenon [7]. Whether or not quantum effects will be able to smooth out the mass inflation singularity and, if so, which is the right extension across the Cauchy horizon is still an open problem (see for example [8]).

In this paper we will consider the one parameter (n) family of (1+1)-dimensional dilaton gravity models introduced in [9]

$$S_n = \frac{1}{2\pi}\int d^2x\sqrt{-g}\left[e^{-\frac{2}{n}\phi}(R+\frac{4}{n}(\nabla\phi)^2)+4\lambda^2e^{-2\phi}\right]. \tag{12.1}$$

In the case $n = 1$ (12.1) reduces to the well known CGHS action [10]. We remarked in [9] and [11] that an interesting feature of this theory is that its static black hole solutions are, for $n \neq 1$, asymptotically Rindlerian. This suggests the idea that these black holes are seen as (uniformly) accelerated to distant inertial observers.

Due to their mathematical complexity, axisymmetric solutions describing accelerated black holes in four dimensions are not very easy to tract. In section 12.2 we show how our two-dimensional theory (12.1) can be viewed as describing "low energy excitations" along the throat of near extreme accelerated black holes (in the point particle limit) of a 4d theory introduced in [12].

In this simplified context, we will then introduce the accelerated (electrically) charged black hole solutions and perform the stability analysis of the Cauchy horizons arising in these spacetimes. The results of the calculations, performed in [13] and summed up in section 12.4, show that there exists a region of nonzero measure in the space of the parameters characterizing the solutions where the Cauchy horizon remains perfectly regular and the mass inflation behavior doesn't occur. Similarities of our solutions in the case $n > 1$ with the Reissner-Nordström-de Sitter spacetime are also remarked.

We then turn to evaluate quantum effects. It has been shown in [14] that, unlike the Reissner-Nordström case, the evaporation process is stopped before these black holes become extreme. The final (quantum mechanically) stable configuration, for which the Hawking temperature equals the Unruh temperature due to the acceleration, is regular at the event horizon, but a (strong) curvature singularity develops at the Cauchy horizon. This last result, while restoring the full predictive power of the theory that was apparently lost at the classical level, perfectly agrees with the strong cosmic censorship conjecture.

12.2 Motivation for studying our 1+1 dimensional dilaton gravity models

Following ref. [12] we consider the four dimensional theory

$$S_n^{(4)} = \frac{1}{2\pi}\int d^4x\sqrt{-g^{(4)}}\left\{e^{-2\phi}\left[R^{(4)} + \left(6-\frac{2}{n^2}\right)(\nabla\phi)^2\right] - \frac{1}{2}e^{-\frac{2\phi}{n}}F^2\right\}, \tag{12.2}$$

where $R^{(4)}$ is the Ricci scalar, ϕ is the dilaton field and $F_{\mu\nu}$ the e.m. field tensor. n is a real parameter that characterizes the theory; the case $n=1$ gives the well known low-energy effective string action.

The axisymmetric (magnetic) solutions of this theory are quite complicated. They are related to the dilaton Ernst metric derived in [15] through a conformal transformation involving the dilaton field ϕ and read

$$\begin{aligned} ds^2 &= \frac{e^{2\phi_0}\Lambda(x,y)^{1-n}}{A^2(x-y)^2}\left(\frac{F(y)}{F(x)}\right)^{-n}\left\{F(x)\left(G(y)\,dt^2 - \frac{dy^2}{G(y)}\right)\right.\\ &+ \left. F(y)\left(\frac{dx^2}{G(x)} + \frac{G(x)}{\Lambda^2(x,y)}\,d\varphi^2\right)\right\}, \end{aligned} \tag{12.3}$$

$$e^{-\frac{2\phi}{n}} = e^{-\frac{2\phi_0}{n}}\Lambda(x,y)\frac{F(y)}{F(x)}, \qquad A_\varphi = -\frac{e^{\frac{\phi_0}{n}}}{B\Lambda(x,y)}[1+Bqx]+k \tag{12.4}$$

where

$$\begin{aligned} \Lambda(x,y) &= [1+Bqx]^2 + \frac{B^2}{2A^2(x-y)^2}G(x)F(x), \\ F(\xi) &= (1+r_-A\xi), \\ G(\xi) &= (1-\xi^2-r_+A\xi^3). \end{aligned} \tag{12.5}$$

They describe magnetically charged black holes accelerated by an external magnetic field. r_+, r_- are related to the mass m and the magnetic charge q; A and B to the acceleration of the hole and to the strength of the external magnetic field. Choosing the parameters such that $r_-A \le r_+A < 2/(3\sqrt{3})$ the function $G(\xi)$ has three real

distinct roots, which we call ξ_2, ξ_3 and ξ_4 (with $\xi_2 < \xi_3 < \xi_4$). $y \equiv -1/(r_- A)$ denotes the boundary of the space-time in the region inside the black hole; $y = \xi_2$ represents the black hole horizon and $y = \xi_3$ the acceleration horizon.
x and φ are angular coordinates. x is restricted to the range $\xi_3 \leq x \leq \xi_4$ and by requiring the absence of nodal singularities at $x = \xi_3$ and $x = \xi_4$, i.e.

$$G'(\xi_3)\Lambda(\xi_4) = -G'(\xi_4)\Lambda(\xi_3), \tag{12.6}$$

we have that $0 \leq \varphi \leq \frac{4\pi\Lambda(\xi_3)}{G'(\xi_3)}$.

The extremal solutions are defined by the condition $\xi_1 = \xi_2$. To analyse near extremal accelerated black holes we introduce the small parameter

$$\epsilon \equiv \frac{\xi_1 - \xi_2}{\xi_2}, \tag{12.7}$$

which represents the deviation from extremality. We will also deal with another parameter

$$\tau = r_+ A. \tag{12.8}$$

For $\tau \ll 1$, i.e. in the point particle limit (or, equivalently, for small acceleration), we have (see [16])

$$\xi_2 \sim -\frac{1}{\tau} + \tau, \quad \xi_3 \sim -1 - \frac{\tau}{2}, \quad \xi_4 \sim 1 - \frac{\tau}{2}. \tag{12.9}$$

In this limit we have a better understanding of the various parameters entering the solutions. In fact eq. (12.6) gives, at lowest order in τ (and for $r_+ = 2m$)

$$qA = mB, \tag{12.10}$$

that is Newton's law.

In ref. [12] we have carried out the expansion of the solution in eqs. (12.3) and (12.4) in terms of the two small parameters ϵ and τ and in the limit $y \to \xi_2$. The calculations are a bit involved. Here we just mention that, apart from irrelevant constant terms that can be reabsorbed, at lowest order we get

$$\begin{aligned} ds^2 &= e^{2\phi_0}(1+\epsilon)\left[1 - \frac{\hat{r}_-}{\hat{r}}\right]^{1-n}\left\{-\frac{(1-\frac{\hat{r}_+}{r})}{(1-\frac{\hat{r}_-}{\hat{r}})}\,dt^2 \right. \\ &+ \left. \frac{dr^2}{(1-\frac{\hat{r}_+}{r})(1-\frac{\hat{r}_-}{r})} + \hat{r}_+^2\,d\Omega^2\right\}, \end{aligned} \tag{12.11}$$

$$e^{-\frac{2\phi}{n}} = e^{-\frac{2\phi_0}{n}}\left[1 - \frac{\hat{r}_-}{r}\right], \tag{12.12}$$

$$A_{\tilde{\varphi}} = \hat{q}(1 - \cos\theta). \tag{12.13}$$

Here the various constants appearing are suitable combinations of the old ones (with $\hat{r}_- \equiv \hat{r}_+(1-\epsilon)$); $y = \frac{\xi_2 \hat{r}_+}{r}$ and also t has been redefined through a multiplicative constant (see [14] for the details). The next-to-lowest order terms are $O(\epsilon\tau)$ and contain the x dependent perturbations. At this level, therefore, the solution down the throat of the near extremal holes is to a good approximation spherically symmetric in form with the radius of the unit two-sphere approximately constant $\simeq \hat{r}_+$.

We then parametrize our line element (12.11) as

$$ds^2 = g^{(2)}_{ab}\, dx^a\, dx^b + \tilde{R}^2\, d\Omega^2, \tag{12.14}$$

where $g^{(2)}_{ab}$ $(a, b = 1, 2)$ is the radial part and

$$\tilde{R} = exp[(\frac{n-1}{n})\phi + \frac{\phi_0}{n}]r \tag{12.15}$$

with $r \simeq \hat{r}_+$ (down the throat).
The dimensional reduction of the action (12.2) is now simply achieved and we get

$$S^{(2)}_n = \int d^2x\sqrt{-g^{(2)}}\left[e^{-\frac{2\phi}{n}}\left(R^{(2)} + \frac{4}{n}(\nabla\phi)^2\right) + 4\lambda^2 e^{-2\phi}\right], \tag{12.16}$$

where we have defined

$$4\lambda^2 \equiv \frac{2e^{-\frac{2\phi_0}{n}}}{\hat{r}_+^2} - \frac{Q^2 e^{-\frac{4\phi_0}{n}}}{\hat{r}_+^4}. \tag{12.17}$$

(12.16) is exactly the theory that was introduced in [9]. The correspondence between the two dimensional black holes of this action and the near extremal black holes of the theory (12.2) is better understood if one defines the spatial coordinate σ through

$$e^{2\lambda\sigma} = 2\lambda r. \tag{12.18}$$

We then have, concerning the radial part of the metric [see eq. (12.11)],

$$ds^2 = e^{2(1-n)\lambda\sigma}[-(1-\epsilon e^{-2\lambda\sigma})\,dt^2 + \frac{d\sigma^2}{(1-\epsilon e^{-2\lambda\sigma})}], \quad \phi = -n\lambda\sigma. \tag{12.19}$$

Provided one identifies the small parameter ϵ with the two dimensional ADM mass M via the relation

$$\epsilon = \frac{M}{\lambda} \tag{12.20}$$

the solution (12.19) represents the black holes of our theory (12.16) ([9], [11]).

12.3 Two dimensional accelerated black holes

12.3.1 Physical interpretation of the solutions and a brief look to the global causal structure

The main message of the previous section is that the two-dimensional theory (12.16) can be regarded as a toy model for describing four dimensional accelerated black holes. Let us now analyse the features of the solutions eq. (12.19) for $n \neq 1$. The curvature scalar is

$$R = 4M\lambda n e^{-2(2-n)\lambda\sigma}. \tag{12.21}$$

For the purposes of this paper it is enough to consider the range $0 < n < 2$. A complete analysis of the solutions for all values of n can be found in ref. [11].

The surface $\sigma = -\infty$ is a curvature singularity and $\sigma = -\frac{1}{2\lambda}\ln\frac{M}{\lambda}$, where g_{00} vanishes, is an event horizon.

To interpret physically the surface $\sigma = +\infty$ we note that in this limit the line element in (12.19) takes the Rindler form

$$ds^2 \sim e^{2(1-n)\lambda\sigma}(-dt^2 + d\sigma^2) \tag{12.22}$$

with acceleration parameter $a \equiv \lambda|1-n|$.

We can introduce Minkowskian coordinates X, T defined by

$$(1-n)\lambda(T+X) = e^{(1-n)\lambda(t+\sigma)}, \quad -(1-n)\lambda(T-X) = e^{-(1-n)\lambda(t-\sigma)} \tag{12.23}$$

but the solution is no more static. The fact that these black holes are static only when viewed by distant Rindler observers is an indication of the fact that they are accelerated. This is the reason why in the inertial frame (T, X) the gravitational field is changing with time.

Moreover, by looking at the X, T diagram of Fig. 12.1 we easily realize that in the case $n < 1$ $\sigma = +\infty$ (12.23) is the asymptotic part of the Rindler wedge R. On the other hand, when $n > 1$ it is rather an horizon, the acceleration horizon that divides the region L from F and P. The solution can then be extended across this surface and the results of this analysis are presented in ref. [11].

12.3.2 Charged black holes

We now add one more ingredient and consider the electrically charged solutions. In the full four dimensional theory (12.2) the e.m. field has dilaton coupling $e^{-\frac{2\phi}{n}}$. The dimensional reduction will change the coupling through the term (12.15). It is, in fact, $\sqrt{-g^{(4)}}e^{-\frac{2\phi}{n}} \sim \sqrt{-g^{(2)}}e^{-\frac{2(2-n)}{n}\phi}$ and therefore we add to our two-dimensional theory the term

$$S_{EM} = \frac{1}{2\pi}\int d^2x\sqrt{-g}e^{\frac{2n-4}{n}\phi}(-2F_{\mu\nu}F^{\mu\nu}). \tag{12.24}$$

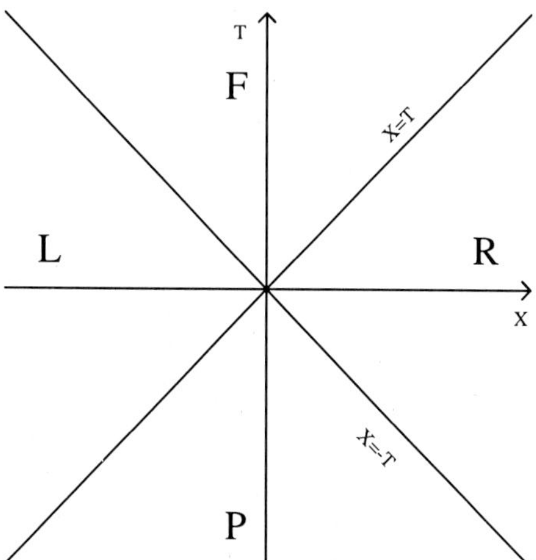

Figure 12.1: The full (X, T) plane is divided in four regions. The Rindler wedges are R and L. The other sectors are F (future) and P (past).

The solutions to the equations of motion of the theory defined by $S_n^{(2)} + S_{EM}$ can be expressed in the Schwarzschild-Rindler gauge (σ, t) (see [14] for the details)

$$ds^2 = e^{2(1-n)\lambda\sigma}\left[-f\,dt^2 + \frac{1}{f}\,d\sigma^2\right], \quad f = 1 - \frac{2m_0}{\lambda}e^{\frac{2}{n}\phi} + \frac{Q^2}{\lambda^2}e^{\frac{4}{n}\phi}, \tag{12.25}$$

$$\phi = -n\lambda\sigma, \tag{12.26}$$

$$F_{\mu\nu} = Qe^{\frac{4-2n}{n}\phi}e_{\mu\nu}, \quad e_{\mu\nu} = e_{[\mu\nu]}, \quad e_{01} = \sqrt{-g} \tag{12.27}$$

(note that M of the previous sections has been rewritten as $2m_0$ for convenience).

With respect to the neutral case the only addition in the metric (12.25) is the term proportional to Q^2, the electric charge, in complete analogy with the Reissner-Nordström solution. As a consequence, the inner structure of the black holes gets modified and an inner, or Cauchy, horizon appears. The black hole horizons are as usual defined by the equation $f = 0$. They are given by $\sigma_\pm$ such that (provided that $m_0 > |Q|$)

$$e^{2\lambda\sigma_\pm} = e^{-\frac{2}{n}\phi_\pm} = \frac{\left(m_0 \pm \sqrt{m_0^2 - Q^2}\right)}{\lambda}; \tag{12.28}$$

σ_+ is the outer and σ_- the inner horizon.

Concerning the causal structure, the singularity is still located at $\sigma = -\infty$ and the region $\sigma = +\infty$ is, as in the previous subsection, either asymptotic as $n < 1$ or an acceleration horizon when $n > 1$. In the first case the Penrose diagram is the same as for the Reissner-Nordström black hole (see Fig. 12.2); for $n > 1$ the asymptotic part

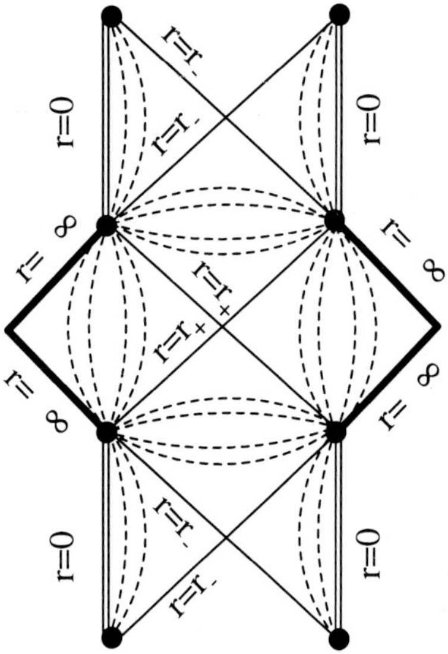

Figure 12.2: Penrose diagram of the Reissner-Nordström spacetime for $m_0 > |Q|$. Double lines represent the singularity, broken lines the lines r = const, regular lines the horizons and thick lines the asymptotic region.

of the diagram is replaced by an horizon and the geometry could be further extended. We have not studied the global causal structure in this case, but the analysis can be easily carried out on the basis of the uncharged case ref. [11] and following the simple rules given in [17].

A quantity that will be useful in the following is the surface gravity κ, defined by (see [18])

$$\kappa = \frac{1}{2}\left|\frac{\partial_\sigma g_{tt}}{\sqrt{-g_{\sigma\sigma}g_{tt}}}\right| = \left|\lambda(1-n)f + \frac{1}{2}f_{,\sigma}\right| . \tag{12.29}$$

At the black hole horizons it is

$$\kappa_\pm = \frac{1}{2}|f_{,\sigma_\pm}| = \lambda e^{-2\lambda\sigma_\pm}(e^{2\lambda\sigma_+} - e^{2\lambda\sigma_-}) \tag{12.30}$$

(note that $m_0 > |Q|$ implies $\kappa_- > \kappa_+$) and in the case $n > 1$ at the acceleration horizon

$$\kappa_{ah} \equiv \kappa(\sigma = \infty) = \lambda(n-1). \tag{12.31}$$

12.4 Cauchy horizon stability

In this section we will consider the stability analysis of the Cauchy horizons arising in the electrically charged solutions eqs. (12.25)–(12.27). We will then have to perturb these geometries with incoming and outgoing fluxes of radiation. These perturbations will be described by conformally coupled massless scalar fields f_i given by the action

$$S_M = -\frac{1}{4\pi}\sum_{i=1}^{N}\int d^2x\sqrt{-g}(\nabla f_i)^2. \tag{12.32}$$

N is the number of such fields. In the purely incoming case $f_i = f_i(v)$ (where $v = t+\frac{d\sigma}{f}$ is an advanced null time coordinate) the solutions of the theory $S_n^{(2)} + S_{EM} + S_M$ take the Vaidya-Rindler form

$$ds^2 = e^{2(1-n)\lambda\sigma}(-f\,dv^2 + 2\,dv\,d\sigma), \quad \phi = -n\lambda\sigma, \tag{12.33}$$

where

$$f = 1 - \frac{2m(v)}{\lambda}e^{\frac{2}{n}\phi} + \frac{Q^2}{\lambda^2}e^{\frac{4}{n}\phi} \tag{12.34}$$

and

$$\frac{dm}{dv} = \frac{1}{4}\sum_{i=1}^{N}(\partial_v f_i)^2. \tag{12.35}$$

Due to the different form of the perturbations, we treat separately the two cases $n < 1$ and $n > 1$. They are discussed in the next two subsections.

12.4.1 Asymptotically flat spacetimes ($n < 1$)

We noted in subsection 12.3.2 that in this range the solution eq. (12.25) is, at least for what concerns the causal structure, similar to the Reissner-Nordström spacetime. We will therefore follow the analysis of refs. [4], [5] for the introduction of the perturbations in the external region of the background spacetime. We consider these perturbations, as in ref. [6], to follow at late times ingoing null geodesics and decay as an inverse power law in the advanced time of asymptotic inertial observers. We note that in our case the "right" time is not v but $X^+ = X + T$, where X and T have been defined in eq. (12.23) (the retarded time is $X^- = T - X$). In terms of the scalar fields introduced in (12.32) we then write (see [13] and [19] in the case $n = 1$)

$$T_{++} = \frac{1}{2}\sum_{i=1}^{N}(\frac{df_i}{dX^+})^2 = 2\gamma(\lambda X^+)^{-q}, \tag{12.36}$$

where γ is a constant and q is a number that depends on the multipole order of the perturbing field (we will suppose that at least $q > 2$). Transforming to the coordinate $v = \frac{1}{\lambda(1-n)}\ln(\lambda(1-n)X^+)$ the solution takes the form given in (12.33), (12.34) with mass [see (12.35)]

$$m(v) = m_0 - \frac{\gamma(1-n)^{q-1}}{\lambda(q-2)}e^{\lambda(1-n)(-q+2)v}. \tag{12.37}$$

Now let us propagate these perturbations along the Cauchy horizon, defined by $v = +\infty$, $X^+ = +\infty$, and consider a free falling observer who is going to cross this surface. The energy density that he measures is given by $\rho_{obs} = T_{\mu\nu}u^\mu u^\nu$, where u^μ is his velocity. The blueshift factor is no more exponential as in the Reissner-Nordström and $n = 1$ cases (given by $e^{2k_- v}$), but power law and consequently we have that (see [13])

$$\rho_{obs} \sim (\lambda X^+)^{\frac{2\kappa_-}{\lambda(1-n)}+(2-q)}. \tag{12.38}$$

This quantity is finite when $(1-n)(-q+2)\lambda + 2k_- \leq 0$.

The additional ingredient that usually triggers the mass inflation mechanism is to allow also an outflux of null radiation which crosses the Cauchy horizon. It is striking that, no matter how T_{--} is as long as it is nonvanishing, the combination of both fluxes causes in [4] the mass function to diverge at the Cauchy horizon resulting in a scalar curvature singularity there.

We have performed a similar analysis in our theory in [13], to which we refer for the details of the calculations. The perturbing fields are described by T_{++} given in (12.36) and T_{--} of arbitrary strength. Here we just state that in the region where both fluxes are present and close to the Cauchy horizon it is possible to solve the field equations and calculate the scalar curvature R, which behaves as

$$R \sim (X^+)^{\frac{\kappa_-}{\lambda(1-n)}+(2-q)}. \tag{12.39}$$

The final result is therefore that when

$$\lambda(1-n)(-q+2) + k_- \leq 0 \tag{12.40}$$

the Cauchy horizon is completely stable and no mass inflation occurs. Using eq. (12.30) the inequality (12.40) can be rewritten more elegantly in terms of the parameters characterizing our solutions as

$$m_0^2 \leq Q^2 \frac{[2+(1-n)(q-2)]^2}{4+4(1-n)(q-2)}. \tag{12.41}$$

Note that in the case $n = 1$ (12.41) reduces to $m_0 = |Q|$, i.e. the extremal solution; for $n < 1$, on the other hand, we have a region of nonzero measure in the space (m_0, Q, λ, n) (in the plane (m_0^2, Q^2) it is the dotted region of Fig. 12.3).

12.4.2 The case $n > 1$

An important difference with respect to $n < 1$ is that for the solutions characterized by $n > 1$ the asymptotic region is replaced by an acceleration horizon. This is the basic feature of the spacetime that we need for our analysis. This case, provided the acceleration horizon is replaced with a cosmological horizon, is in some sense similar to the Reissner-Nordström de Sitter spacetime considered in refs. [20] and [21].

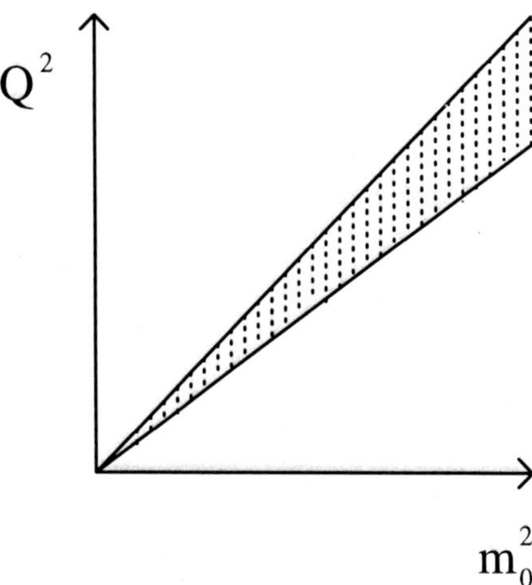

Figure 12.3: The dotted part represents the region of the (m_0^2, Q^2) plane where the Cauchy horizon is completely stable (here $n = \frac{1}{2}$ and $q = 12$).

An appropriate perturbation for these spacetimes which is regular at the acceleration horizon is described by a stress-energy tensor of the form (note that now the coordinates $X^{\pm} = X \pm T$ define the Kruskal frame regular there)

$$T_{vv} \sim e^{-2\kappa_{ah} v}, \tag{12.42}$$

where κ_{ah} has been given in eq. (12.31).
Following the same steps as in section 12.4.1, an infalling observer crossing the Cauchy horizon will measure an influx of energy

$$\rho_{obs} \sim e^{2(k_- - k_{ah})v}, \tag{12.43}$$

which is finite for $k_- \leq k_{ah}$. The addition of a generic outflux of null radiation on the spacetime is such that at the Cauchy horizon it is

$$R \sim e^{(k_- - 2k_{ah})v}. \tag{12.44}$$

This quantity diverges unless $k_- \leq 2k_{ah}$ or, using (12.30) and (12.31),

$$m_0^2 \leq Q^2 \frac{n^2}{2n-1}. \tag{12.45}$$

The analysis of this section has shown that at least at the classical level the possibility of having a regular Cauchy horizon is not completely ruled out. The maximal analytic extension of the spacetime suggests then the existence of infinite disconnected universes through which, at least in principle, one can travel.

12.5 Quantum theory and the cosmic censorship

In two dimensions conformal anomaly and conservation equations completely determine the $\langle T_{\mu\nu}\rangle$ of matter fields living in a given background spacetime. In the case of the static metric eq. (12.25) and for N massless scalar fields it is given by (see ref. [14])

$$\langle T_{uu}\rangle = \frac{\hbar N}{96\pi}[ff_{,\sigma\sigma} - \frac{1}{2}f_{,\sigma}^2 - 2(1-n)^2\lambda^2 f^2 + t_u], \tag{12.46}$$

$$\langle T_{vv}\rangle = \frac{\hbar N}{96\pi}[ff_{,\sigma\sigma} - \frac{1}{2}f_{,\sigma}^2 - 2(1-n)^2\lambda^2 f^2 + t_v], \tag{12.47}$$

$$\langle T^{\mu}_{\mu}\rangle = \frac{\hbar N R}{24\pi} = -\frac{\hbar N}{24\pi}e^{-2(1-n)\lambda\sigma}[f_{,\sigma\sigma} + 2(1-n)\lambda f_{,\sigma}]. \tag{12.48}$$

Here the null directions are given by $u = t - \int \frac{d\sigma}{f}$, $v = t + \int \frac{d\sigma}{f}$. We recall that $f = 1 - \frac{2m_0}{\lambda}e^{-2\lambda\sigma} + \frac{Q^2}{\lambda^2}e^{-4\lambda\sigma}$ and that t_u, t_v are just constants depending on physical conditions such as the quantum state in which the expectation values are taken.

In the $n = 1$ Reissner-Nordström like case static configurations with no radiation at infinity (i.e. the ones which are suitable for describing the final state of the evaporation) have $t_u = t_v = 0$. As a consequence, one finds divergent fluxes at the event horizon in a Kruskal frame

$$\langle T_{UU}\rangle \sim -\frac{k_+^2}{U^2}, \quad \langle T_{VV}\rangle \sim -\frac{k_+^2}{V^2}, \tag{12.49}$$

where U, V vanish there. In the case $\kappa_+ = 0$, which corresponds to the extremal black hole, there is still a mild divergence $\sim \frac{f'''}{f'}$ [22]. This quantity, however, doesn't affect the regularity of the "quantum corrected solution" at the event horizon. This configuration corresponds to the final stable state of the evaporation.

The situation is different in the case of the solutions with $n \neq 1$. Choosing $t_u = t_v = 2\lambda^2(1-n)^2$ we select a state which is both nonradiating 'asymptotically' [23] and regular at the event horizon provided

$$\kappa_+ = \lambda|1-n|. \tag{12.50}$$

The physical meaning of this last formula is quite clear: the end-point of Hawking evaporation in our models is reached when the Hawking temperature $T_H = \frac{\hbar\kappa_+}{2\pi}$ equals the Unruh temperature $T_U = \frac{\hbar\lambda|1-n|}{2\pi}$ due to the acceleration of these black holes. Using the expression for the surface gravity at the event horizon (12.30) we can rewrite (12.50) as

$$Q^2 = \frac{4n}{(1+n)^2}m_0^2. \tag{12.51}$$

We can check that for $n = 1$ this gives $m_0 = |Q|$, i.e. the extremal solution for which the two horizons coalesce. Our final configurations are instead non extremal

and exhibit two distinct horizons.
Despite the regularity at the event horizon, there appear divergent fluxes at the Cauchy horizon, namely

$$\langle T_{++} \rangle \sim -\frac{1}{x^{+2}}, \tag{12.52}$$

where $x^+ \sim e^{-k_- v}$ is the Kruskal advanced time there.
In [14] and [13] we have determined how these quantum effects influence the background geometry. The "quantum corrected solution" tends to the classical one as $\sigma \to \infty$ and is regular at the event horizon (the location of this latter receives corrections $\sim O(\hbar)$, see [14] for the details). Strong deviations from the classical behavior, as expected, arise close to the Cauchy horizon. The analysis performed in [13] shows that the scalar curvature R there behaves as

$$R \sim \frac{1}{x^+(-\ln(-k_- x^+))^{2-n}}, \tag{12.53}$$

i.e. it diverges for any value of n.
This result, analogous to that found in the $n = 1$ theory in [19], forbids further extension of the geometry across the Cauchy horizon.

Acknowledgements

It is a pleasure to thank R. Balbinot for many interesting and useful discussions.

Bibliography

[1] V. P. Frolov, M. A. Markov and V. F. Mukhanov, Phys. Let. **B 216**, 272 (1989).

[2] R. Penrose in *Battelle Rencontres*, ed. by C. M. De Witt and J. A. Wheeler (Benjamin, New York, 1968).

[3] S.W. Hawking and G.F.R. Ellis, *The large scale structure of space-time* (Cambridge University Press, Cambridge, England, 1973).

[4] E. Poisson and W. Israel, Phys. Rev. D. **41**, 1796 (1990).

[5] A. Ori, Phys. Rev. Lett. **67**, 789 (1991).

[6] R. H. Price, Phys. Rev. D. **5**, 2419 (1972).

[7] Ori, however, noted that the mass inflation singularity is a "weak" one, since tidal distortions are finite at the Cauchy horizon. In some sense, therefore, the problem of the extension of the geometry across this surface is not completely understood.

[8] R. Balbinot and E. Poisson, Phys. Rev. Lett. **70**, 13 (1993); W. G. Anderson, P. R. Brady, W. Israel and S. M. Morsink, Phys. Rev. Lett. **70**, 1041 (1993).

[9] A. Fabbri and J. G. Russo, Phys. Rev. D **53**, 6995 (1996).

[10] C. Callan, S. Giddings, J. Harvey and A. Strominger, Phys. Rev. D. **45**, R1005 (1992).

[11] R. Balbinot and A. Fabbri, Class. Quantum Grav. **13**, 2457 (1996).

[12] R. Balbinot, A. Fabbri and L. Mazzacurati, *Dilaton gravity black holes with regular interior*, gr-qc/9708017.

[13] A. Fabbri, *Cauchy horizon stability in 2d accelerated black holes*, preprint Laboratoire de Gravitation et Cosmologie Relativistes, Université Pierre et Marie Curie, Paris.

[14] R. Balbinot and A. Fabbri, Class. Quantum Grav. **14**, 463 (1997).

[15] F. Dowker, J. P. Gauntlett, D. A. Kastor and J. Traschen, Phys. Rev. D **49**, 2909 (1994).

[16] F. Dowker, J. P. Gauntlett, S. B. Giddings and G. T. Horowitz, Phys. Rev. D **50**, 2662 (1994).

[17] T. Klösch and T. Strobl, Class. Quantum Grav. **13**, 2395 (1996).

[18] R. M. Wald, *General Relativity* (University of Chicago Press, Chicago, 1984).

[19] R. Balbinot and P. R. Brady, Class. Quantum Grav. **11**, 1763 (1994).

[20] P. R. Brady and E. Poisson, Class. Quantum Grav. **9**, 121 (1992).

[21] P. R. Brady, D. Nunez and S. Sinha, Phys. Rev. D. **47**, 4239 (1993).

[22] S.P. Trivedi, Phys. Rev. D. **47** (1993) 4233.

[23] Remember that $\sigma = +\infty$ is a "true" asymptotic region only for $n < 1$.

GENERIC WORMHOLE THROATS

Matt Visser [a] and David Hochberg [b]

a) Physics Department, Washington University, Saint Louis, Missouri 63130-4899, USA
b) Laboratorio de Astrofísica Espacial y Física Fundamental, Apartado 50727, 28080 Madrid, Spain

Abstract

Wormholes and black holes have, apart from a few historical oddities, traditionally been treated a quite separate objects with relatively little overlap. The possibility of a connection arises in that wormholes, if they exist, might have profound influence on black holes, their event horizons, and their internal structure. For instance: (1) small wormhole-induced perturbations in the geometry can lead to massive non-perturbative shifts in the event horizon, (2) Planck-scale wormholes near any spacelike singularity might let information travel in effectively spacelike directions, (3) vacuum polarization effects near any singularity might conceivably lead to a "punch through" into another asymptotically flat region—effectively transforming a black hole into a wormhole. After discussing these connections between black hole physics and wormhole physics we embark on an overview of what can generally be said about traversable wormholes and their throats. We discuss the violations of the energy conditions that typically occur at and near the throat of any traversable wormhole and emphasize the generic nature of this result. We discuss the original Morris–Thorne wormhole and its generalization to a spherically symmetric time-dependent wormhole. We discuss spherically symmetric Brans–Dicke wormholes as examples of how to hide the energy condition violations in an inappropriate choice of definitions. We also discuss the relationship of these results to the topological censorship theorem. Finally we turn to a rather general class of wormholes that permit explicit analysis: generic static traversable wormholes (without any symmetry). We show that topology is too limited a tool to accurately characterize a generic traversable wormhole—in general one needs geometric information to detect the presence of a wormhole, or more precisely to locate the wormhole throat. For an arbitrary static spacetime we shall define the wormhole throat in terms of a 2–dimensional constant-time hypersurface of minimal area. (Zero trace for the extrinsic curvature plus a "flare–out" condition.) This enables us to severely constrain the geometry of spacetime at the wormhole throat and to derive generalized theorems regarding violations of the energy conditions—theorems that do not involve geodesic averaging but nevertheless apply to situations much more general than the spherically symmetric Morris–Thorne traversable wormhole. [For example: the null energy condition (NEC), when suitably weighted and integrated over the wormhole throat, must be violated.]

13.1 Introduction

Traversable wormholes [1, 2, 3] have traditionally been viewed as quite distinct from black holes, with essentially zero overlap in techniques and topics. (Historical oddities that might at first seem to be counter-examples to the above are the Schwarzschild wormhole, which is not traversable, and the Einstein–Rosen bridge, which is simply a bad choice of coordinates on Schwarzschild spacetime [3, pages 45–51].) Since this workshop is primarily directed toward the study of black holes we shall start by indicating some aspects of commonality and inter-linkage between these objects.

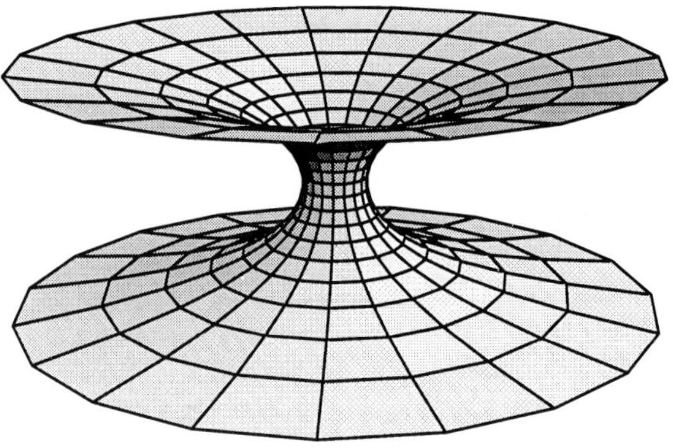

Figure 13.1: Schematic representation of an inter-universe wormhole connecting two asymptotically flat regions.

First: We point out that traversable wormholes, if they exist, can have violent non-perturbative influence on black hole event horizons. This is fundamentally due to the fact that the event horizon is defined in a global manner, so that small changes in the geometry, because they have all the time in the universe to propagate to future null infinity, can have large effects on the event horizon.

Second: We point out that the region near any curvature singularity is expected to be subject to large fluctuations in the metric. If Wheeler's spacetime foam picture is to be believed, the region near a curvature singularity should be infested with wormholes. If it's a spacelike singularity (meaning, it's spacelike before you take the wormholes into account) then the wormholes can connect regions that would otherwise be outside each other's lightcones, and so lead to information transport in an effectively spacelike direction. You still have to worry about how to transfer the information across the event horizon, but this is a start towards a lazy way out of the black hole information paradox. (The information paradox is only a paradox if you take the Schwarzschild singularity too seriously, in particular if you take the spacelike nature of the singularity too seriously.)

Third: In addition to the fluctuations that take place near a curvature singularity, we would also expect large effects on the expectation value of the metric due to gravitational vacuum polarization. Since gravitational vacuum polarization typically violates the energy conditions, this could lead us to expect a "punch through" to another asymptotically flat region. The possibility that singularities might heal themselves by automatic conversion into wormhole throats, though maybe unlikely, should at least be kept in mind.

After developing these linkages between black hole physics and wormhole physics, we embark on an overview of traversable wormholes, concentrating on the region near the throat of the wormhole. One of the most important features that characterizes traversable wormholes is the violation of the energy conditions, in particular the null energy condition, *at or near the throat*. The energy condition violations were first discovered in the static spherically symmetric Morris–Thorne wormholes, but the result is generic (modulo certain technical assumptions) as borne out by the topological censorship theorem, and also by the generic analysis of static wormholes developed later in this survey.

We shall discuss the case of time-dependent spherically symmetric wormholes, wherein the energy violation conditions can be isolated at particular regions in time (in the same way that static thin-shell wormholes permit one to isolate the energy condition violations at particular regions of space.) We show that these results are compatible with the topological censorship theorem. Perhaps surprisingly, we show that cosmological inflation is useless in terms of generating even temporary suspension of the energy condition violations.

We further discuss the case of spherically symmetric Brans–Dicke wormholes as an example of what happens in non-Einstein theories of gravity: In this case it is possible to mistakenly conclude that the energy conditions are not violated, but this would merely be a consequence of an inappropriate choice of definitions. For instance: If one works in the Einstein frame, defines the existence of a wormhole in terms of the Einstein metric, and calculates using the total stress energy tensor, then the null energy condition must be violated at or near the throat. If one artificially divides the total stress-energy into (Brans–Dicke stress-energy) plus (ordinary stress-energy), then since the Brans–Dicke stress energy (calculated in the Einstein frame) sometimes violates the energy conditions (if the Brans–Dicke parameter ω is less than $-3/2$), the "ordinary" part stress-energy can sometimes satisfy the null energy condition.

Similarly, suppose one works in the Jordan frame, and defines the existence of a wormhole using the Jordan metric. If one calculates the total stress energy tensor then it is still true that the null energy condition must be violated at or near the throat. If one now artificially divides the total stress-energy into (Brans–Dicke stress-energy) plus (ordinary stress-energy), then for suitable choices of the Brans–Dicke parameter ω ($\omega < -2$) one can hide all the energy condition violations in the Brans-Dicke field and permit the ordinary stress-energy to satisfy the energy conditions.

To further confuse the issue, one could define the existence (or nonexistence) of a wormhole using one frame, and then calculate the Einstein tensor and total stress-energy in the other frame. This is a dangerous and misleading procedure: We shall show that the definition of the existence and location of a wormhole throat is not frame independent (because it is not conformally invariant). Jumping from one frame to the other in the middle of the calculation can easily lead to meaningless results. It is critical to realize that the energy condition violations must still be there and in fact are still there: they have merely been hidden by sleight of hand. Qualitatively, these comments also apply to other non–Einstein theories of gravity such as the Einstein–Cartan theory, Dilaton gravity, Lovelock gravity, Gauss–Bonnet gravity, etc...

Finally, we wrap up by presenting a general analysis for static traversable wormholes that completely avoids all symmetry requirements and even avoids the need to assume the existence of asymptotically flat regions. This exercise is particularly useful in that it places the notion of wormhole in a much more general setting. Indeed, wormholes are often viewed as intrinsically topological objects, occurring in multiply connected spacetimes. The Morris–Thorne class of inter-universe traversable wormholes is even more restricted, requiring both exact spherical symmetry and the existence of two asymptotically flat regions in the spacetime. To deal with intra-universe traversable wormholes, the Morris–Thorne analysis must be subjected to an approximation procedure wherein the two ends of the wormhole are distorted and forced to reside in the same asymptotically flat region. The existence of one or more asymptotically flat regions is an essential ingredient of the Morris–Thorne approach [1].

However, there are many other classes of geometries that one might still quite reasonably want to classify as wormholes, that either have trivial topology [3], or do not possess any asymptotically flat region [4], or exhibit both these phenomena.

A simple example of a wormhole lacking an asymptotically flat region is two closed Friedman–Robertson–Walker spacetimes connected by a narrow neck (see figure 13.3), you might want to call this a "dumbbell wormhole". A simple example of a wormhole with trivial topology is a single closed Friedman–Robertson–Walker spacetime connected by a narrow neck to ordinary Minkowski space (see figure 13.4). A general taxonomy of wormhole exemplars may be found in [3, pages 89–93], and discussions of wormholes with trivial topology may also be found in [3, pages 53–74].

To set up the analysis for a generic static throat, we first have to define exactly what we mean by a wormhole—we find that there is a nice *geometrical* (not topological) characterization of the existence of, and location of, a wormhole "throat". This

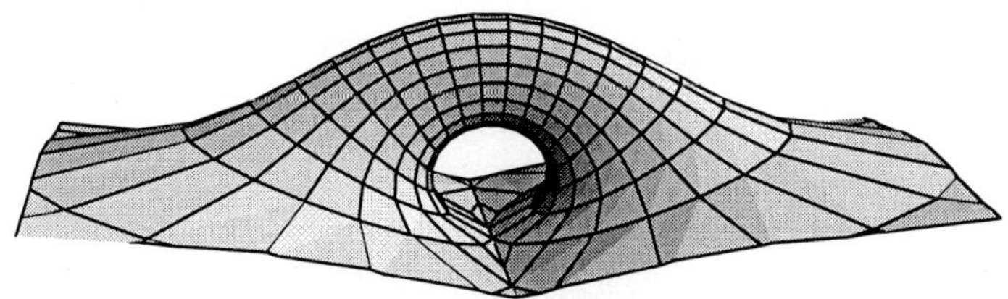

Figure 13.2: Schematic representation of an intra-universe wormhole formed by deforming an inter-universe wormhole and forcing the two asymptotically flat regions to merge.

characterization is developed in terms of a hypersurface of minimal area, subject to a "flare–out" condition that generalizes that of the Morris–Thorne analysis.

With this definition in place, we can develop a number of theorems about the existence of "exotic matter" at the wormhole throat. These theorems generalize the original Morris–Thorne result by showing that the null energy condition (NEC) is generically violated at some points on or near the two-dimensional surface comprising the wormhole throat. These results should be viewed as complementary to the topological censorship theorem [5]. The topological censorship theorem tells us that in a spacetime containing a traversable wormhole the averaged null energy condition must be violated along at least some (not all) null geodesics, but the theorem provides very limited information on where these violations occur. The analysis of this paper shows that some of these violations of the energy conditions are concentrated in the expected place: on or near the throat of the wormhole. The present analysis, because it is purely local, also does not need the many technical assumptions about asymptotic flatness, future and past null infinities, and global hyperbolicity that are needed as ingredients for the topological censorship theorem [5].

The key simplifying assumption in the present analysis is that of taking a static wormhole. While we believe that a generalization to dynamic wormholes is possible, the situation becomes technically much more complex and one is rapidly lost in an impenetrable thicket of definitional subtleties and formalism. (A suggestion, due to Page, whereby a wormhole throat is viewed as an *anti-trapped surface* in spacetime holds promise for suitable generalization to the fully dynamic case [6].)

In summary, the violations of the energy conditions at wormhole throats are unavoidable. Many of the attempts made at building a wormhole without violating the energy conditions do so only by hiding the energy condition violations in some subsidiary field, or by hiding the violations at late or early times.

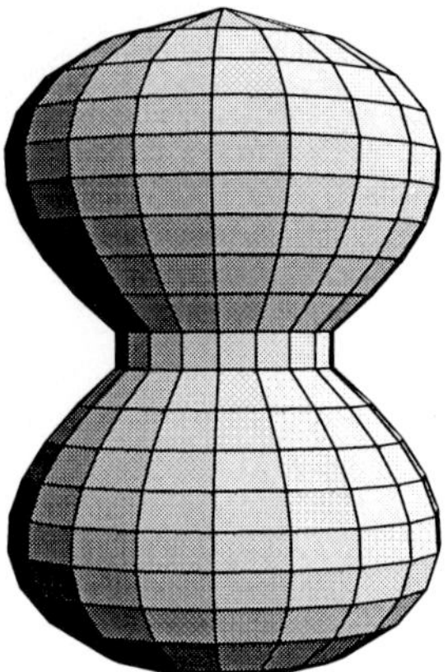

Figure 13.3: A "dumbbell wormhole": Formed (for example) by two closed Friedman–Robertson–Walker spacetimes connected by a narrow neck.

13.2 Connections

We start by describing a few scenarios whereby wormholes might prove of interest to black hole physics. Don't take any of these scenarios too seriously: they are presented more as suggestions for things to consider than as definite proposals for serious models.

13.2.1 Non-perturbative changes in the event horizon

Traversable wormholes, if they exist, can lead to massive nonperturbative changes in the event horizon of a black hole at the cost of relatively minor perturbations in the geometry. (This should not be too surprising: it is a general feature of Einstein gravity that small perturbations in the geometry can lead to massive perturbations in the event horizon.) One of the simplest examples of this effect is to take a wormhole (with two mouths), and a black hole, and then throw one wormhole mouth down into the black hole while keeping the second wormhole mouth outside [7]. Everything in the past lightcone of the mouth that falls into the wormhole, including the segment that is behind where the event horizon used to be before the wormhole mouth fell in, can now influence future null infinity by taking a shortcut through the wormhole.

This can best be seen visually; we idealize the wormhole to have a mouth with an extremely small radius, and model the wormhole by a pair of timelike lines in

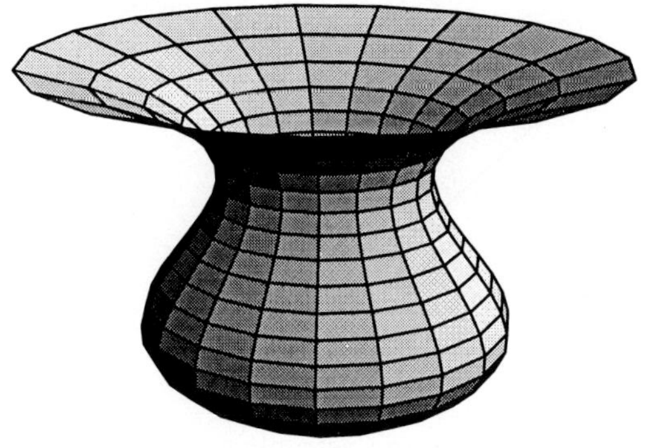

Figure 13.4: A wormhole with trivial topology: Formed (for example) by connecting a single closed Friedman–Robertson–Walker spacetime to Minkowski space by a narrow neck.

spacetime that are mathematically identified. The black hole will be idealized in the usual way with a Penrose diagram. The geometry of the spacetime is affected by the wormhole only in the immediate vicinity of the wormhole mouths, but the global properties (such as the event horizon) suffer drastic non-perturbative changes.

With a swarm of traversable wormholes one can chip away at the event horizon to the extent that it "almost" disappears. The event horizon is progressively eaten away as more wormhole mouths fall into the region behind the apparent horizon.

13.2.2 Effectively spacelike information transfer

Near any spacetime curvature singularity, in the region where curvatures are large, the spacetime foam picture developed by Wheeler [8, 9] strongly suggests that Lorentzian wormholes will be rapidly popping in and out of existence. (We ignore for the sake of present discussion potential difficulties associated with topology change in Lorentzian manifolds: certainly something peculiar has to happen in the region where the wormhole is created or destroyed. See [3, pages 61–73].)

We can think of at least two plausible models for Lorentzian wormhole creation. (We again take the approximation that the wormholes have infinitely small mouth radius, and so can be modeled by infinitely thin spacetime world-lines.) In model

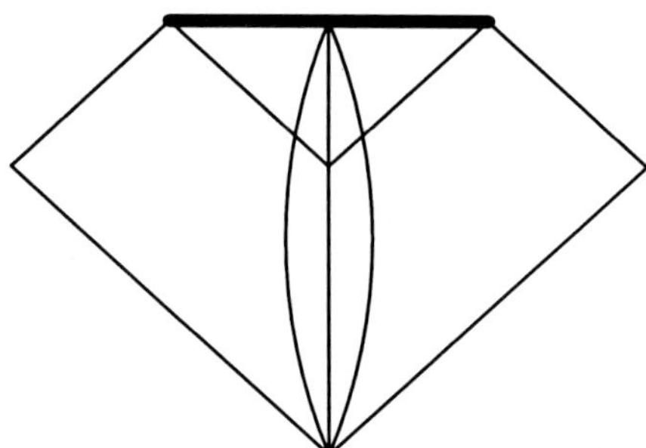

Figure 13.5: Penrose diagram for an astrophysical black hole formed by (say) stellar collapse.

(1) the two mouths are created (pulled out of the Planck slop?) at different places, and after creation one simply has two timelike world-lines that are to be identified. In model (2) the two mouths are created at the same spacetime point and the two mouths then subsequently move off along distinct timelike world-lines (which are again identified to produce the wormhole structure).

In model (1) the spacetime distribution of the creation points for the two mouths is something we have no idea how to calculate. we should hope that the two creation points are spacelike separated since otherwise the Chronology Protection Conjecture [10] has failed at the outset. Beyond that, very little can be said. In model (2) the Chronology Protection Conjecture is respected at the initial creation point, but one still has to worry about the subsequent evolution of the wormhole mouths. Since all of this is presumably taking place deep inside the Planck slop, issues of reliability of the entire semiclassical approximation also deserve attention [11].

Whichever of these models (1) or (2) you pick [and both models are perfectly compatible with both the typical noises generated in this field, and our current ignorance of quantum gravity] the wormhole mechanism takes some points that would be spacelike separated if the wormholes were not present and makes them timelike separated. This is quite sufficient to allow "effectively spacelike" information transfer in a thin region near the curvature singularity.

To actually get information out of the black hole, you will either have to live with an extension of model (1) wherein the second wormhole mouth pops into existence just outside the event horizon, or rely on an external wormhole pair one of whose mouths is permitted to fall into the horizon.

13.2.3 "Punch-through" near the singularity

Near a spacelike curvature singularity, the quantum vacuum expectation value of the stress-energy tensor is likely to be heading off to infinity due to gravitational vacuum polarization effects. But gravitational vacuum polarization quite typically leads to violations of the energy conditions [12, 13, 14, 15, 16]. And violations of the energy conditions are a generic feature of wormholes.

This suggests (hints) that curvature singularities might heal themselves by "punching through" to another asymptotically flat region in a manner similar to a wormhole. (See for example the minisuperspace model discussed in [3, pages 347–359] and [17, 18, 19].) For a concrete suggestion along these lines consider the metric

$$ds^2 = -\left(1 - \frac{2GM}{\sqrt{\ell^2 + r^2}}\right) dt^2 + \left(1 - \frac{2GM}{\sqrt{\ell^2 + r^2}}\right)^{-1} dr^2 + (\ell^2 + r^2)\left\{d\theta^2 + \sin^2\theta \, d\phi^2\right\}. \tag{13.1}$$

This is almost the Schwarzschild geometry, apart from the parameter ℓ. The radial variable r can now be extended all the way from $+\infty$ to $-\infty$, and there is a symmetry under interchange $r \to -r$. For $\ell \ll 2GM$ (but $\ell \neq 0$) and r not too close to zero (which is where the spacelike singularity is for $\ell = 0$), this is indistinguishable from the Schwarzschild solution. There is a thin region near $r = 0$ where the stress-energy is appreciably different from zero. In fact the Einstein tensor is

$$G_{\hat{t}\hat{t}} = -\frac{\ell^2\left(1 - \frac{4GM}{\sqrt{\ell^2+r^2}}\right)}{(\ell^2 + r^2)^2}, \tag{13.2}$$

$$G_{\hat{\theta}\hat{\theta}} = +\frac{\ell^2\left(1 - \frac{GM}{\sqrt{\ell^2+r^2}}\right)}{(\ell^2 + r^2)^2}, \tag{13.3}$$

$$G_{\hat{r}\hat{r}} = -\frac{\ell^2}{(\ell^2 + r^2)^2}, \tag{13.4}$$

Note that this is not a traversable wormhole in the usual sense, since the region near the "throat" ($r = 0$) it is the radial direction that is timelike. Thus travel is definitely one-way, and if anything one should call this a "timehole". The hope of course is that some sort of geometry similar to the above might emerge as a self-consistent solution to the semiclassical field equations. A timelike version of the Hochberg–Popov–Sushkov analysis [4] is what we have in mind. Note in particular that at $r = 0$ we have

$$G_{\hat{t}\hat{t}} + G_{\hat{r}\hat{r}} = +\frac{(4GM - 2\ell)}{\ell^3}. \tag{13.5}$$

Thus the null energy condition is *not* violated at $r = 0$, indicating that one-way "timeholes" of this type are qualitatively different from the two-way traversable wormholes considered in the rest of this survey. (There are some similarities here with the Aichelburg–Schein wormholes, which are not true traversable wormholes in that they are only one-way traversable [20]. In contrast the Aichelburg–Israel–Schein wormholes are true traversable wormholes which violate the energy conditions in the usual manner [21].)

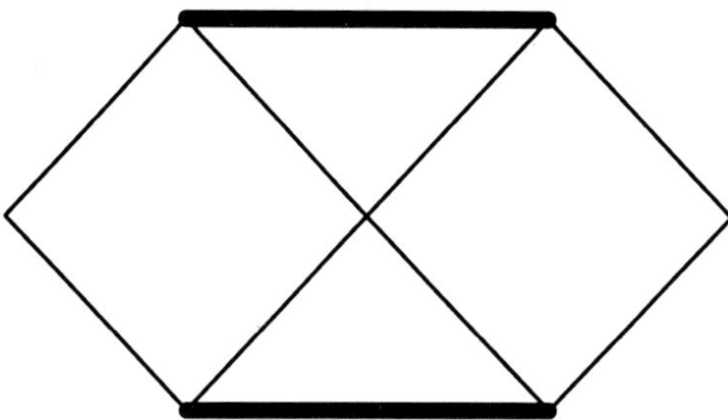

Figure 13.6: Penrose diagram for the maximally extended Schwarzschild geometry.

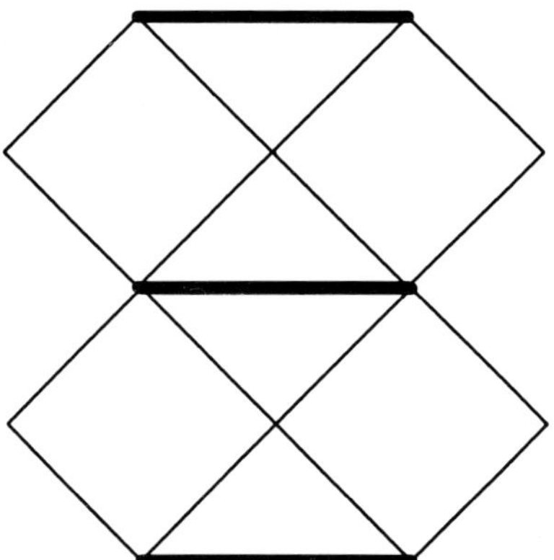

Figure 13.7: Penrose diagram for a "timehole". Gravitational vacuum polarization near where the spacelike singularity would have been could plausibly lead to "punch through" to another asymptotically flat region.

13.2.4 "Punch-through" near the horizon

A bolder proposal is that "punch through" might occur not near where the central singularity would have been, but instead that "punch through" might occur out where one would naively expect the event horizon to be. For instance, if one insists on using the Boulware vacuum, and insists on a spacetime containing an event horizon, then the renormalized stress-energy will diverge at the horizon. Therefore, if one insists on having a self-consistent solution to the semiclassical field equations in the Boulware vacuum, then no event horizon can survive. Given that test-field calculations on Schwarzschild spacetime show infinite gravitational vacuum polarization at the would-be event horizon, and that this gravitational vacuum polarization violates the null energy condition, one might suspect that the self-consistent solution contains a traversable wormhole.

Bolstering this suspicion is the fact that the Hochberg–Popov–Sushkov analysis [4] has numerically integrated the fourth-order differential equations derived from semiclassical gravity, and found wormhole throats in the Boulware vacuum. (There are a tangle of technical issues here to do with the size of the wormhole throat and whether or not it is possible to extend these numerical solutions out into truly asymptotically flat regions. We refer the interested reader to the literature.)

A simple and explicit toy model geometry along these lines is to start with the Schwarzschild solution written in isotropic coordinates

$$\begin{aligned} ds^2 &= -\left(\frac{1-GM/2r}{1+GM/2r}\right)^2 dt^2 \\ &\quad +(1+GM/2r)^4\left\{dr^2+r^2[d\theta^2+\sin^2 d\phi^2]\right\}. \end{aligned} \tag{13.6}$$

The radial coordinate r runs from $r=0$ to $r=+\infty$. The horizon is at $r=R\equiv GM/2$, and there is symmetry under radial inversion $r\to R^2/r$. Now perturb this geometry by distorting the g_{tt} component of the metric:

$$\begin{aligned} ds^2 &= -\left\{\frac{(R^2-r^2)^2+\ell^2 r^2}{(R+r)^4+\ell^2 r^2}\right\} dt^2 \\ &\quad +(1+R/r)^4\left\{dr^2+r^2[d\theta^2+\sin^2 d\phi^2]\right\}. \end{aligned} \tag{13.7}$$

The radial coordinate r still runs from $r=0$ to $r=+\infty$ and there is still a symmetry under radial inversion $r\to R^2/r$. The particular form of the distortion given above has been chosen so that the asymptotic form of the metric at spatial infinity agrees with the Schwarzschild metric up to order $O[r^{-3}]$. But $r=R\equiv GM/2$ is no longer an event horizon, it is now the location of a wormhole throat connecting two asymptotically flat regions (one at $r=+\infty$ and the other at $r=0$). The throat is at finite redshift $1+z=\sqrt{1+(16R^2/\ell^2)}$. For $\ell\ll R$ the region outside the would-be event horizon is arbitrarily close to Schwarzschild geometry. (This is an example of a "proximal Schwarzschild" wormhole as discussed in [3, pages 147–149].) The stress-energy tensor is a bit messy, but falls of as order $O[r^{-5}]$ for large r. Thus there is a thin layer of energy condition violating matter near the throat of the wormhole, but by the time you are a little distance away from the throat you cannot tell the difference between this geometry and the usual Schwarzschild solution.

13.2.5 Summary

The key point to be extracted from the above discussion is that small perturbations on the geometry of spacetime can make drastic perturbations to the notion of a black hole. Small perturbations of the geometry can eat away at the event horizon like cancer, can drastically affect the properties of spacelike singularities (spacelike information transfer, time holes), and in the wrong hands can even make event horizons vanish (proximal Schwarzschild wormholes). All of the specific examples discussed above use wormholes, or things that are almost wormholes, so we feel it is a good idea for black hole physicists to have some basic understanding of the wormhole system and its limitations.

13.3 Energy conditions: An overview

The Morris–Thorne analysis [1] revitalized interest in Lorentzian traversable wormholes. Morris and Thorne were able to show that traversable wormholes were compatible with our current understanding of general relativity and semiclassical quantum gravity — but that there was a definite price to be paid — one had to admit violations of the null energy condition.

More precisely, what Morris and Thorne showed was equivalent to the statement that for spherically symmetric traversable wormholes there must be an open region surrounding the throat over which the null energy condition is violated [1, 3]. The striking nature of this result has led numerous authors to try to find ways of evading or minimizing the energy condition violations, and on the other hand has led to a number of general theorems guaranteeing the existence of these violations.

For instance, static but not spherically symmetric thin-shell wormholes and their variants allow you to move the energy condition violations around in space, so that there are some routes through the wormhole that do not encounter energy condition violations. (This is most easily seen using "cut and paste" wormholes constructed using the thin shell formalism [22]. See also [3, pages 153–194].)

Time dependent but spherically symmetric wormholes allow you to move the energy condition violations around in time [23, 24, 25, 26, 27]. (Unfortunately radial null geodesics through the wormhole will still encounter energy condition violations, subject to suitable technical qualifications.)

The Friedman–Schleich–Witt topological censorship theorem [5], under suitable technical conditions, guarantees that spacetimes containing traversable wormholes must contain some null geodesics that violate the averaged null energy condition (but gives little information on where these null geodesics must be).

Attempts at eliminating the energy condition violations completely typically focus on alternative gravity theories (Brans–Dicke gravity, Dilaton gravity, gravity with torsion). We argue that such attempts are at best sleight of hand—it is sometimes possible to hide the energy condition violations in the Brans–Dicke field, or the dilaton field, or the torsion, but the energy condition violations are still always there.

13.3.1 Morris–Thorne wormhole

Take any spherically symmetric static spacetime geometry and use the radial proper distance as the radial coordinate. The metric can then without loss of generality be written as

$$ds^2 = -e^{2\phi(l)}dt^2 + dl^2 + r^2(l)\left[d\theta^2 + \sin^2\theta\, d\varphi^2\right]. \tag{13.8}$$

If we wish this geometry to represent a Morris–Thorne wormhole then we must impose conditions both on the throat and on asymptotic infinity [1, 3].

- Conditions at the throat:
  - The absence of event horizons implies that $\phi(l)$ is everywhere finite.
  - The radius of the wormhole throat is defined by
  $$r_0 = \min\{r(l)\}. \tag{13.9}$$
  - For simplicity one may assume that there is only one such minimum and that it is an isolated minimum. Generalizing this point is straightforward.
  - Without loss of generality, we can take this throat to occur at $l = 0$.
  - The metric components should be at least twice differentiable as functions of l.
- Conditions at asymptotic infinity:
  - The coordinate l covers the entire range $(-\infty, +\infty)$.
  - There are two asymptotically flat regions, at $l \approx \pm\infty$.
  - In order for the spatial geometry to tend to an appropriate asymptotically flat limit we impose
  $$\lim_{l\to\pm\infty}\{r(l)/|l|\} = 1. \tag{13.10}$$
  - In order for the spacetime geometry to tend to an appropriate asymptotically flat limit, we impose finite limits
  $$\lim_{l\to\pm\infty}\phi(l) = \phi_\pm. \tag{13.11}$$
- These are merely the minimal requirements to obtain a wormhole that is "traversabl in principle". For realistic models, "traversable in practice", one should address additional engineering issues such as tidal effects [3, pages 137–152].

- We shall argue, later in this survey, that the conditions imposed at asymptotic infinity can be relaxed (and for certain questions, asymptotic infinity can be ignored completely), and that it is the throat of the wormhole that is often more of direct interest.

It is an easy exercise to show that the Einstein tensor is [3, pages 149-150]

$$G_{\hat{t}\hat{t}} = -\frac{2r''}{r} + \frac{1-(r')^2}{r^2}, \tag{13.12}$$

$$G_{\hat{r}\hat{r}} = \frac{2\phi' r'}{r} - \frac{1-(r')^2}{r^2}, \tag{13.13}$$

$$G_{\hat{\theta}\hat{\theta}} = G_{\hat{\varphi}\hat{\varphi}} = \phi'' + (\phi')^2 + \frac{\phi' r' + r''}{r}. \tag{13.14}$$

Thus in particular

$$G_{\hat{t}\hat{t}} + G_{\hat{r}\hat{r}} = -\frac{2r''}{r} + \frac{2\phi' r'}{r}. \tag{13.15}$$

But by definition $r' = 0$ at the throat. We also know that $r'' \geq 0$ at the throat. The possibility that $r'' = 0$ at the throat forces us to invoke some technical complications. By definition the throat is a local minimum of $r(l)$. Thus there must be an open region $l \in (0, l_*^+)$ such that $r''(l) > 0$, and on the other side of the throat an open region $l \in (-l_*^-, 0)$ such that $r''(l) > 0$. This is the Morris–Thorne "flare–out" condition. So for the Einstein tensor we have

$$\exists\, l_*^-, l_*^+ > 0: \qquad \forall l \in (-l_*^-, 0) \cup (0, l_*^+), \qquad G_{\hat{t}\hat{t}} + G_{\hat{r}\hat{r}} < 0. \tag{13.16}$$

This constraint on the components of the Einstein tensor follows directly from the definition of a traversable wormhole and the definition of the Einstein tensor—it makes no reference to the dynamics of general relativity and automatically holds *by definition* regardless of whether one is dealing with Einstein gravity, Brans–Dicke gravity, or any other exotic form of gravity. As long as you have a spacetime metric that contains a wormhole, and calculate the Einstein tensor using that *same* spacetime metric, the above inequality holds by definition.

We can always define the total stress energy by enforcing the Einstein equations

$$G^{\mu\nu} = 8\pi G\, T^{\mu\nu}_{total}. \tag{13.17}$$

In terms of the total stress energy

$$\exists\, l_*^-, l_*^+ > 0: \qquad \forall l \in (-l_*^-, 0) \cup (0, l_*^+), \qquad T^{total}_{\hat{t}\hat{t}} + T^{total}_{\hat{r}\hat{r}} < 0. \tag{13.18}$$

Thus the null energy condition for the total stress energy must be violated on some open region surrounding the throat. The only requirements for this result are essentially matters of definition: use the same metric to define the wormhole and to calculate the Einstein tensor, and use that same Einstein tensor to identify the total stress-energy.

(This also shows where the maneuvering room is in exotic theories of gravity. If you have multiple metrics, multiple Einstein tensors, and multiple definitions of stress-energy then it becomes easy to hide the violations of the energy conditions in inappropriate definitions.)

13.3.2 Spherically symmetric time-dependent wormholes

After the initial treatment by Morris and Thorne it was quickly realized that by eschewing spherical symmetry it is quite possible to minimize the violations of the null energy condition [22]. In particular it is quite possible to move the regions subject to energy condition violations around in space so as to hide them in the woodwork and permit at least some travelers through the wormhole completely avoid any personal contact with exotic matter.

Somewhat later, it was realized that *time dependence* lets one move the energy condition violating regions around in time [25, 26], and so leads to a temporary suspension of the need for energy condition violations. (For related analyses see also [23, 24, 27].) The most direct presentation of the key results can best be exhibited by taking a spacetime metric that is conformally related to a zero-tidal force wormhole by a simple time-dependent but space-independent conformal factor. Thus

$$ds^2 = \Omega(t)^2 \left\{ -dt^2 + \frac{dr^2}{1 - b(r)/r} + r^2 \left(d\theta^2 + \sin^2\theta d\phi^2 \right) \right\}. \tag{13.19}$$

(This form of the metric has the advantage that null geodesics remain null geodesics as Ω is altered.) It is an easy exercise to see that

$$G_{\hat{t}\hat{t}} = +\Omega^{-2} \left(\frac{b'(r)}{r^2} + \frac{\dot{\Omega}^2}{\Omega^2} \right), \tag{13.20}$$

$$G_{\hat{\theta}\hat{\theta}} = +\Omega^{-2} \left(\frac{b(r)}{2r^3} - \frac{b'(r)}{2r^3} + \frac{\dot{\Omega}^2}{\Omega^2} - 2\frac{\ddot{\Omega}}{\Omega} \right), \tag{13.21}$$

$$G_{\hat{r}\hat{r}} = -\Omega^{-2} \left(\frac{b(r)}{r^3} + 2\frac{\ddot{\Omega}}{\Omega} - \frac{\dot{\Omega}^2}{\Omega^2} \right). \tag{13.22}$$

In particular, looking along the radial null direction

$$G_{\hat{t}\hat{t}} + G_{\hat{r}\hat{r}} = \Omega^{-2}\left(-\frac{b(r)}{r^3} + \frac{b'(r)}{r^3}2 - 2\frac{\ddot{\Omega}}{\Omega} + 4\frac{\dot{\Omega}^2}{\Omega^2}\right) \tag{13.23}$$

We can rewrite this in terms of the Hubble parameter H and deceleration parameter q, by first relating conformal time t to comoving time T via $\Omega\, dt = dT$. We then deduce

$$H = \Omega^{-1}\frac{d\Omega}{dT} = \Omega^{-2}\,\dot{\Omega}, \tag{13.24}$$

and

$$q = -\Omega\,\frac{d^2\Omega}{dT^2}\left(\frac{d\Omega}{dT}\right)^{-2} = 1 - \frac{\Omega\,\ddot{\Omega}}{\dot{\Omega}^2}. \tag{13.25}$$

Then

$$G_{\hat{t}\hat{t}} + G_{\hat{r}\hat{r}} = \Omega^{-2}\left(-\frac{b(r)}{r^3} + \frac{b'(r)}{r^3}\right) + 2H^2(1+q). \tag{13.26}$$

So if the universe expands quickly enough we can (temporarily) suspend the violations of the null energy condition. Note that the first term above is just the static $\Omega = 1$ result rescaled to the expanded universe. At the throat of the wormhole this is of order $-\xi/a^2$ where a is the physical size of the wormhole mouth and ξ is a dimensionless number typically of order unity unless some fine tuning is envisaged. This implies that the expansion of the universe can overcome the violations of the energy condition only if

$$a > \sqrt{\frac{\xi}{1+q}}\,\frac{1}{H}. \tag{13.27}$$

This requires a wormhole throat with radius of order the Hubble distance or larger (modulo possible fine tuning). This indicates that suspension of the energy condition violations may be of some interest during the very early universe when the Hubble parameter is large, but that in the present epoch it would be quite impractical to rely on the expansion of the universe to avoid the need for energy condition violations.

It is perhaps surprising to realize that the inflationary epoch, though it can be invoked to make wormholes larger by inflating them out of the Planck slop (Wheeler's spacetime foam) up to more manageable sizes [23, 24], cannot be relied upon to aid in the suspension of energy condition violations. This arises because $q = -1$ during the inflationary epoch so that for the metric considered above

$$G_{\hat{t}\hat{t}} + G_{\hat{r}\hat{r}} = \Omega^{-2}\left(-\frac{b(r)}{r^3} + \frac{b'(r)}{r^3}\right). \tag{13.28}$$

This should be compared to equation (3.8) of [23], which was obtained directly in terms of a comoving time formalism assuming inflationary expansion from the outset.

In the formalism developed above, this can be checked by noting that during the inflationary epoch

$$\Omega = \exp(HT) = \frac{1}{1 - Ht}, \tag{13.29}$$

where we have normalized $t = 0$ at $T = 0$. It is then easy to explicitly check that $q = -1$.

If we want to compare this result with that of the topological censorship theorem [5] it is important to realize that the topological censorship theorem applies to the current situation only if we switch off the expansion of the universe at sufficiently early and late times. This is because the topological censorship theorem uses asymptotic flatness, in both space and time, as an essential ingredient in setting up both the statement of the theorem and the proof. (Scri$^-$, past null infinity, simply makes no sense in a big bang spacetime.)

So, provided we switch off the expansion of the universe at sufficiently early and late times, we deduce first that null energy condition violations reappear at sufficiently early and late times, and more stringently, that radial null geodesics must violate the ANEC.

In brief, the suspension of the energy condition violations afforded by time dependence are either transitory or intimately linked to the existence of a big bang singularity, and in either case are most likely limited to small scale microscopic wormholes in the pre-inflationary epoch.

13.3.3 Brans–Dicke wormholes

Brans–Dicke wormholes [28, 29, 30] are particular examples of the effect that choosing a non-Einstein model for gravity has on the behavior of traversable wormholes. The Brans–Dicke theory of gravity is perhaps the least violent alteration to Einstein gravity that can be contemplated. In the Brans–Dicke theory the gravitational field is composed of two components: a spacetime metric plus a dynamical scalar field. (A word of caution: some papers dealing with Brans–Dicke wormholes are actually in Euclidean signature [31]. Euclidean wormholes are qualitatively different from Lorentzian wormholes and will not be discussed in this survey.)

The Jordan Frame

In the so-called Jordan frame the Action is [36, page 1070]

$$S = \int \sqrt{-g} \left\{ \phi R - \omega \frac{(\nabla\phi)^2}{\phi} + 16\pi G\, \mathcal{L}_{matter} \right\}. \tag{13.30}$$

The equations of motion are

$$G_{\alpha\beta} = \frac{8\pi G}{\phi} T_{\alpha\beta}^{matter} + \frac{\omega}{\phi^2}\left\{\phi_\alpha\phi_\beta - \frac{1}{2}g_{\alpha\beta}(\nabla\phi)^2\right\} + \frac{1}{\phi}\left\{\phi_{;\alpha\beta} - g_{\alpha\beta}\nabla^2\phi\right\}, \quad (13.31)$$

and

$$\nabla^2\phi = \frac{8\pi G}{3+2\omega} T. \quad (13.32)$$

If we wish the metric g to describe a spherically symmetric static wormhole, then the Morris-Thorne analysis implies that the total stress energy defined by $T_{\alpha\beta}^{total} = G_{\alpha\beta}/(8\pi G)$ must violate the null energy condition. Whether or not the "matter" part of the total stress-energy violates the null energy condition depends on how the Brans–Dicke scalar field behaves at and near the throat.

If we assume that the wormholes we are looking for have $\phi \neq 0$ and $\phi \neq \infty$ at the throat, then for any radial null vector k^α we have

$$G_{\alpha\beta}\, k^\alpha\, k^\beta = \frac{8\pi G}{\phi}\, T_{\alpha\beta}^{matter}\, k^\alpha\, k^\beta + \omega\frac{(k^\alpha\phi_\alpha)^2}{\phi^2} + \frac{\phi_{;\alpha\beta}\, k^\alpha\, k^\beta}{\phi} \leq 0. \quad (13.33)$$

This implies that the only way in which the "matter" part of the of the stress-energy can avoid violating the null energy condition is if either $\omega < 0$ and $\nabla\phi \neq 0$ at the throat, or if ϕ is convex at the throat. We now verify these general conclusions by looking at some specific exact solutions.

Vacuum Brans–Dicke wormholes [Jordan frame]

For vacuum Brans–Dicke gravity a suitably large class of solutions to the field equations is [28]

$$\begin{aligned} ds^2 \;=\; & -\left[\frac{1-R/r}{1+R/r}\right]^{2A} dt^2 + [1+R/r]^4 \left[\frac{1-R/r}{1+R/r}\right]^{2+2B} \\ & \times \left[dr^2 + r^2\left(d\theta^2 + \sin^2\theta d\phi^2\right)\right]. \end{aligned} \quad (13.34)$$

$$\phi = \phi_0 \left[\frac{1-B/r}{1+B/r}\right]^{-(A+B)}. \quad (13.35)$$

(Note there is a non-propagating typo in equation (8) of [28].) Here we have chosen to work in isotropic coordinates. We have

$$R = \sqrt{\frac{3+2\omega}{4+2\omega}}\,\frac{G\,M}{2}. \quad (13.36)$$

(This is necessary to get the correct asymptotic behavior as $r \to \infty$.) We also have the constraints that

$$\phi_0 = \frac{4+2\omega}{3+2\omega}. \tag{13.37}$$

$$A = \sqrt{\frac{4+2\omega}{3+2\omega}} > 0. \tag{13.38}$$

$$B = -\frac{1+\omega}{2+\omega}\sqrt{\frac{4+2\omega}{3+2\omega}}. \tag{13.39}$$

The metric is real only for $\omega > -3/2$ or $\omega < -2$, the square root always being taken to be positive when it is real, and the metric reduces to the Schwarzschild geometry for $\omega \to \pm\infty$. The geometry also has a symmetry under $r \to R^2/r$, but this is potentially misleading. The surface $r = R$ is *not* the throat of a wormhole. To see what is going on, consider the proper circumference of a circle at radius r circumscribing the geometry. we have

$$C(r) = 2\pi r \left[1 + R/r\right]^2 \left[\frac{1 - R/r}{1 + R/r}\right]^{1+B}. \tag{13.40}$$

Assume that R (and hence M) is positive, this can be generalized if desired.

- If $B > -1$ then $C(r) \to 0$ as $r \to R$. In this case the surface $r = R$ is a naked curvature singularity as may be verified by direct computation of the curvature invariants. The region $r \in (0, R)$ is a second asymptotically flat region, isomorphic to the first, that connects to the first only at the curvature singularity $r = R$. [$B > -1$ corresponds to $\omega \in (-3/2, +\infty)$.]

- If $B = -1$ then $C(r) \to 8\pi R$ as $r \to R$. In this case the surface $r = R$ is at least a surface of finite area. In fact $B = -1$ is achieved only for $\omega = \pm\infty$ in which case the geometry reduces to Schwarzschild. (In this case we also have $A = 1$ and $\phi = \phi_0$.)

- If $B < -1$ then $C(r) \to \infty$ as $r \to R$. This corresponds to $\omega \in (-\infty, -2)$. In this case the surface $r = R$ is the second asymptotic spatial infinity associated with a traversable wormhole. The location of the wormhole throat is specified by looking for the minimum value of $C(r)$ which occurs at

$$r_{throat} = R\left[-B + \sqrt{B^2 - 1}\right] > R. \tag{13.41}$$

The wormhole is in this case *asymmetric* under interchange of the two asymptotic regions ($r = \infty$ and $r = R$). Since $A + B < 0$ in this parameter regime, $\phi \to 0$ as $r \to R$, and so the effective Newton constant $G_{eff} = G/\phi$ tends to infinity on the other side of the wormhole. The region near $r = R$ is asymptotically large, but not asymptotically flat, as may be verified by direct computation of the curvature. (It would be interesting to know a little bit more about what this region actually looks like, and to develop a better understanding of the physics on the other side of this class of Brans–Dicke wormholes.) The region $r \in (0, R)$ is now a second completely independent universe, isomorphic to the first, with its own wormhole occurring at

$$r_{throat} = R\left[-B - \sqrt{B^2 - 1}\right] < R. \tag{13.42}$$

- There is even more parameter space to explore if we look at the extended class of Brans–Dicke solutions discussed in [29], or let the total mass go negative.

In summary, for suitable choices of the parameters, $\omega < -2$, there are wormholes in vacuum Brans–Dicke gravity expressed in the Jordan frame. The null energy condition is still violated, with the Brans–Dicke field providing the exotic matter.

The Einstein Frame

By making a conformal transformation it is possible to express the Brans–Dicke theory in the so-called Einstein frame. The action is now (see [31], modified for Lorentzian signature)

$$S = \int \sqrt{-\tilde{g}} \left\{ \frac{R(\tilde{g})}{2\omega + 3} + \frac{1}{2}(\nabla\sigma)^2 + 16\pi G \exp(2\sigma)\mathcal{L}_{matter} \right\}. \tag{13.43}$$

Here

$$\tilde{g}_{\mu\nu} = \exp(\sigma)\, g_{\mu\nu}. \tag{13.44}$$

$$\phi = \frac{1}{(2\omega + 3)} \exp(\sigma). \tag{13.45}$$

(And note that the matter Lagrangian, which implicitly depends on the Jordan metric, also has to be carefully rewritten in terms of σ and the Einstein metric.) Now provided ϕ is neither zero nor infinite the conformal transformation from the Jordan to the Einstein frame is globally well-defined. If in addition ϕ goes to a finite non-zero constant at in the asymptotically flat region then it is clear that wormholes, in the sense of topologically nontrivial curves from Scri$^-$ to Scri$^+$ (past null infinity to future null infinity) can neither be created no destroyed by a change of frame. (For background see [5] and [3, pages 195–199]). What can, and in general does, change however is the location and even the number of wormhole throats that are encountered in crossing from one asymptotically flat region to another. A wormhole throat as defined by the Einstein frame metric is not necessarily a wormhole throat as defined by the Jordan frame metric.

Vacuum Brans–Dicke wormholes [Einstein frame]

If we look at the vacuum Brans–Dicke field equations in the Einstein frame we get

$$G_{\alpha\beta} = (2\omega + 3)\left\{\sigma_\alpha\sigma_\beta - \frac{1}{2}g_{\alpha\beta}(\nabla\sigma)^2\right\}, \tag{13.46}$$

and

$$\nabla^2\sigma = 0. \tag{13.47}$$

In the Einstein frame, the Brans–Dicke field violates the null energy condition only for $\omega < -3/2$, and so vacuum Brans–Dicke wormholes can exist only for $\omega < -3/2$. (This is a necessary but not sufficient condition.) Note that the Jordan and Einstein frames are globally conformally equivalent only if ϕ is never zero. Since the $\omega < -2$ Jordan frame wormholes have $\phi = 0$ at $r = R$ there is a risk of pathological behavior.

If we take the class of vacuum Brans–Dicke solutions considered previously, and transform them to the Einstein frame we have

$$\begin{aligned}(2\omega + 4)d\tilde{s}^2 &= -\left[\frac{1 - R/r}{1 + R/r}\right]^{A-B} dt^2 + [1 + R/r]^4 \left[\frac{1 - R/r}{1 + R/r}\right]^{2+B-A} \\ &\quad \times \left[dr^2 + r^2\left(d\theta^2 + \sin^2\theta d\phi^2\right)\right].\end{aligned} \tag{13.48}$$

The proper circumference of a circle at radius r circumscribing the geometry is now

$$\tilde{C}(r) = \frac{2\pi r}{\sqrt{|2\omega + 4|}}[1 + R/r]^2 \left[\frac{1 - R/r}{1 + R/r}\right]^{1+(B-A)/2}. \tag{13.49}$$

Assume again that R (and hence M) is positive. It is now the combination $B - A$ that is of central interest. Indeed

$$A - B = 2\sqrt{\frac{3 + 2\omega}{4 + 2\omega}} > 0. \tag{13.50}$$

- If $B - A > -2$ then $\tilde{C}(r) \to 0$ as $r \to R$. This again corresponds to $\omega \in (-3/2, +\infty)$, and the geometry again contains a naked curvature singularity as may be verified by direct computation of the curvature invariants. The region $r \in (0, B)$ is a second asymptotically flat region, isomorphic to the first, that connects to the first only at the curvature singularity $r = B$.

- If $B - A = 2$ then $\tilde{C}(r) \to 8\pi R$ as $r \to R$. In this case the surface $r = R$ is at least a surface of finite area. It is easy to see that $B - A = -2$ implies $\omega = \pm\infty$ and thus $A = 1$, $B = -1$. This again reproduces the Schwarzschild solution.

- If $B - A < -2$ then $\tilde{C}(r) \to \infty$ as $r \to R$. The surface $r = R$ is the second asymptotic spatial infinity associated with a traversable wormhole. The location of the wormhole throat is specified by looking for the minimum value of $\tilde{C}(r)$ which occurs at

$$r_{throat} = R\left[\frac{(A-B)}{2} + \sqrt{\frac{(A-B)^2}{4} - 1}\right]. \tag{13.51}$$

The wormhole is again *asymmetric* under interchange of the two asymptotic regions ($r = \infty$ and $r = R$), and the throat is located at a different place.

- In some sense we have been lucky: The field equations have forced the conformal transform that relates the Jordan and Einstein frames to be sufficiently mild that the two frames agree as to the range of values of the ω parameter that lead to traversable wormhole geometries. There is no a priori necessity for this agreement. (In fact when considering $O(4)$ Euclidean Brans–Dicke wormholes these differences are rather severe [31].) The two frames do differ however in the precise location of the wormhole throat, and in the precise details of the energy condition violations. In the Einstein frame the energy condition violations are obvious from the relative minus sign (for $\omega < -3/2$) in front of the kinetic energy term for the σ field. In the Jordan frame the energy condition violations are still encoded in the Brans–Dicke field (now ϕ) but in a more subtle manner.

In summary, for suitable choices of the parameters ($\omega < -2$), there are wormholes in vacuum Brans–Dicke gravity expressed in the Einstein frame. The null energy condition is still violated, with the Brans–Dicke field providing the exotic matter.

Other exotic wormholes

Similar analyses can be performed for other more or less natural generalizations of Einstein gravity. Dilaton gravity is a particularly important example, inspired by string theory, that is very closely related to Brans-Dicke gravity. Other examples include Einstein–Cartan gravity, Lovelock gravity, Gauss–Bonnet gravity, higher-derivative gravity, etc...

13.4 Generic static wormholes

We now set aside the special cases we have been discussing, and seek to develop a general analysis of the energy condition violations in static wormholes. (This analysis is based largely on [32]. For an analysis using similar techniques applied to static vacuum and electrovac black holes see Israel [33, 34]. A related decomposition applied to the collapse problem is addressed in [35].) In view of the preceding discussion we want to get away from the notion that topology is the intrinsic defining feature of wormholes and instead focus on the geometry of the wormhole throat. Our strategy is straightforward

- Take any static spacetime, and use the natural time coordinate to slice it into space plus time.
- Use the Gauss–Codazzi and Gauss–Weingarten equations to decompose the $(3+1)$–dimensional spacetime curvature tensor in terms of the 3–dimensional spatial curvature tensor, the extrinsic curvature of the time slices [zero!], and the gravitational potential.
- Take any 3–dimensional spatial slice, and look for a 2-dimensional surface of strictly minimal area. Define such a surface, if it exists, to be the throat of a wormhole. This generalizes the Morris–Thorne flare out condition to arbitrary static wormholes.
- Use the Gauss–Codazzi and Gauss–Weingarten equations again, this time to decompose the 3–dimensional spatial curvature tensor in terms of the 2–dimensional curvature tensor of the throat and the extrinsic curvature of the throat as an embedded hypersurface in the 3–geometry.
- Reassemble the pieces: Write the spacetime curvature in terms of the 2 curvature of the throat, the extrinsic curvature of the throat in 3-space, and the gravitational potential.
- Use the generalized flare-out condition to place constraints on the stress-energy at and near the throat.

13.4.1 Static spacetimes

In any static spacetime one can decompose the spacetime metric into block diagonal form [36, 37, 38]:

$$
\begin{aligned}
ds^2 &= g_{\mu\nu}\, dx^\mu dx^\nu && (13.52)\\
&= -\exp(2\phi) dt^2 + g_{ij}\, dx^i dx^j. && (13.53)
\end{aligned}
$$

Notation: Greek indices run from 0–3 and refer to space-time; latin indices from the middle of the alphabet $(i,\ j,\ k,\ \ldots)$ run from 1–3 and refer to space; latin indices from the beginning of the alphabet $(a,\ b,\ c,\ \ldots)$ will run from 1–2 and will be used to refer to the wormhole throat and directions parallel to the wormhole throat.

Being static tightly constrains the space-time geometry in terms of the three-geometry of space on a constant time slice, and the manner in which this three-geometry is embedded into the spacetime. For example, from [36, page 518] we have

$$
\begin{aligned}
{}^{(3+1)}R_{ijkl} &= {}^{(3)}R_{ijkl}. && (13.54)\\
{}^{(3+1)}R_{\hat{t}abc} &= 0. && (13.55)\\
{}^{(3+1)}R_{\hat{t}i\hat{t}j} &= \phi_{|ij} + \phi_{|i}\,\phi_{|j}. && (13.56)
\end{aligned}
$$

The hat on the t index indicates that we are looking at components in the normalized t direction

$$
X_{\hat{t}} = X_t\sqrt{-g^{tt}} = X_t\ \exp(-\phi). \tag{13.57}
$$

This means we are using an orthonormal basis attached to the fiducial observers (FIDOS). We use $X_{;\alpha}$ to denote a space-time covariant derivative; $X_{|i}$ to denote a three-space covariant derivative, and will shortly use $X_{:a}$ to denote two-space covariant derivatives taken on the wormhole throat itself.

Now taking suitable contractions,

$$
\begin{aligned}
{}^{(3+1)}R_{ij} &= {}^{(3)}R_{ij} - \phi_{|ij} - \phi_{|i}\,\phi_{|j}. && (13.58)\\
{}^{(3+1)}R_{\hat{t}i} &= 0. && (13.59)\\
{}^{(3+1)}R_{\hat{t}\hat{t}} &= g^{ij}\left[\phi_{|ij} + \phi_{|i}\phi_{|j}\right]. && (13.60)
\end{aligned}
$$

So

$$
{}^{(3+1)}R = {}^{(3)}R - 2g^{ij}\left[\phi_{|ij} + \phi_{|i}\phi_{|j}\right]. \tag{13.61}
$$

To effect these contractions, we make use of the decomposition of the spacetime metric in terms of the spatial three-metric, the set of vectors e_i^μ tangent to the time-slice, and the vector $V^\mu = \exp[\phi]\,(\partial/\partial t)^\mu$ normal to the time slice:

$$
{}^{(3+1)}g^{\mu\nu} = e_i^\mu e_j^\nu\, g^{ij} - V^\mu V^\nu. \tag{13.62}
$$

Finally, for the spacetime Einstein tensor (see [36, page 552])

$$
\begin{aligned}
{}^{(3+1)}G_{ij} &= {}^{(3)}G_{ij} - \phi_{|ij} - \phi_{|i}\,\phi_{|j} + g_{ij}\,g^{kl}\left[\phi_{|kl} + \phi_{|k}\phi_{|l}\right]. && (13.63)\\
{}^{(3+1)}G_{\hat{t}i} &= 0. && (13.64)\\
{}^{(3+1)}G_{\hat{t}\hat{t}} &= +\frac{1}{2}{}^{(3)}R. && (13.65)
\end{aligned}
$$

This decomposition is generic to *any* static spacetime. (You can check this decomposition against various standard textbooks to make sure the coefficients are correct. For instance see Synge [39, page 339], Fock [40], or Adler–Bazin–Schiffer [41])

Observation: Suppose the strong energy condition (SEC) holds then [3]

$$\begin{aligned} SEC \quad &\Rightarrow \quad (\rho + g_{ij}T^{ij}) \geq 0 && (13.66)\\ &\Rightarrow \quad g^{ij}\left[\phi_{|ij} + \phi_{|i}\phi_{|j}\right] \geq 0 && (13.67)\\ &\Rightarrow \quad \phi \text{ has no isolated maxima.} && (13.68) \end{aligned}$$

This is a nice consistency check, and also helpful in understanding the physical import of the strong energy condition.

13.4.2 Definition of a generic static throat

We define a traversable wormhole throat, Σ, to be a 2–dimensional hypersurface of *minimal* area taken in one of the constant-time spatial slices. Compute the area by taking

$$A(\Sigma) = \int \sqrt{^{(2)}g}\; d^2x. \tag{13.69}$$

Now use Gaussian normal coordinates, $x^i = (x^a; n)$, wherein the hypersurface Σ is taken to lie at $n = 0$, so that

$$^{(3)}g_{ij}\; dx^i dx^j = {}^{(2)}g_{ab}\; dx^a dx^b + dn^2. \tag{13.70}$$

The variation in surface area, obtained by pushing the surface $n = 0$ out to $n = \delta n(x)$, is given by the standard computation

$$\delta A(\Sigma) = \int \frac{\partial\sqrt{^{(2)}g}}{\partial n}\; \delta n(x)\; d^2x. \tag{13.71}$$

Which implies

$$\delta A(\Sigma) = \int \sqrt{^{(2)}g}\; \frac{1}{2}\; g^{ab}\; \frac{\partial g_{ab}}{\partial n}\; \delta n(x)\; d^2x. \tag{13.72}$$

In Gaussian normal coordinates the extrinsic curvature is simply defined by

$$K_{ab} = -\frac{1}{2}\frac{\partial g_{ab}}{\partial n}. \tag{13.73}$$

(See [36, page 552]. We use MTW sign conventions. The convention in [3, page 156] is opposite.) Thus

$$\delta A(\Sigma) = -\int \sqrt{^{(2)}g}\; \mathrm{tr}(K)\; \delta n(x)\; d^2x. \tag{13.74}$$

[We use the notation $\mathrm{tr}(X)$ to denote $g^{ab}\; X_{ab}$.] Since this is to vanish for arbitrary $\delta n(x)$, the condition that the area be *extremal* is simply $\mathrm{tr}(K) = 0$. To force the area to be *minimal* requires (at the very least) the additional constraint $\delta^2 A(\Sigma) \geq 0$. (We shall also consider higher-order constraints below.) But by explicit calculation

$$\delta^2 A(\Sigma) = -\int \sqrt{^{(2)}g} \left(\frac{\partial \mathrm{tr}(K)}{\partial n} - \mathrm{tr}(K)^2 \right) \delta n(x)\, \delta n(x)\, d^2 x. \tag{13.75}$$

Extremality [$\mathrm{tr}(K) = 0$] reduces this minimality constraint to

$$\delta^2 A(\Sigma) = -\int \sqrt{^{(2)}g} \left(\frac{\partial \mathrm{tr}(K)}{\partial n} \right) \delta n(x)\, \delta n(x)\, d^2 x \geq 0. \tag{13.76}$$

Since this is to hold for arbitrary $\delta n(x)$ this implies that at the throat we certainly require

$$\frac{\partial \mathrm{tr}(K)}{\partial n} \leq 0. \tag{13.77}$$

This is the generalization of the Morris–Thorne "flare-out" condition to arbitrary static wormhole throats.

We now invoke some technical fiddles related to the fact that we eventually prefer to have a strong inequality ($<$) at or near the throat, in reference to a weak inequality ($\leq$) at the throat. Exactly the same type of technical fiddle is required when considering the Morris–Thorne spherically symmetric wormhole. In the following definitions, the two-surface referred to is understood to be embedded in a three-dimensional space, so that the concept of its extrinsic curvature (relative to that embedding space) makes sense.

Definition: Simple flare-out condition.
A two-surface satisfies the "simple flare-out" condition if and only if it is extremal, $\mathrm{tr}(K) = 0$, *and also satisfies* $\partial \mathrm{tr}(K)/\partial n \leq 0$.

This flare-out condition can be rephrased as follows: We have as an identity that

$$\frac{\partial \mathrm{tr}(K)}{\partial n} = \mathrm{tr}\left(\frac{\partial K}{\partial n} \right) + 2\mathrm{tr}(K^2). \tag{13.78}$$

So minimality implies

$$\mathrm{tr}\left(\frac{\partial K}{\partial n} \right) + 2\mathrm{tr}(K^2) \leq 0. \tag{13.79}$$

Generically we would expect the inequality to be strict, in the sense that $\partial \mathrm{tr}(K)/\partial n < 0$, for at least some points on the throat. (See figure 13.8.) This suggests the modified definition below.

Definition: Strong flare-out condition.
A two surface satisfies the "strong flare-out" condition at the point x if and only if it is extremal, $\mathrm{tr}(K) = 0$, *everywhere satisfies* $\partial \mathrm{tr}(K)/\partial n \leq 0$, *and if at the point x on the surface the inequality is strict:*

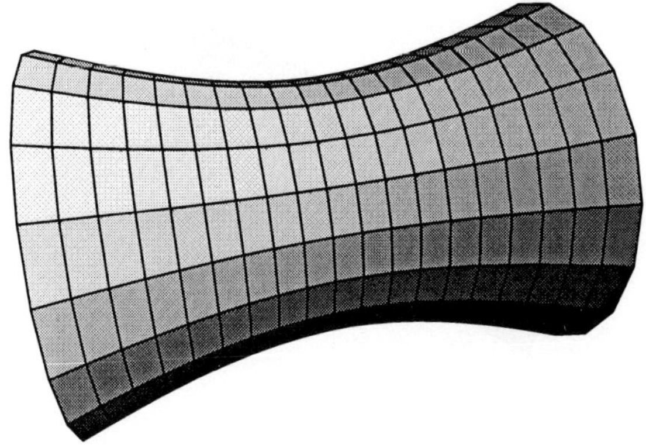

Figure 13.8: Generically, we define the throat to be located at a true minimum of the area. The geometry should flare-out on either side of the throat, but we make no commitment to the existence of any asymptotically flat region.

$$\frac{\partial \mathrm{tr}(K)}{\partial n} < 0. \tag{13.80}$$

It is sometimes sufficient to demand a weak integrated form of the flare-out condition.

Definition: Weak flare-out condition.
A two surface satisfies the "weak flare-out" condition if and only if it is extremal, $\mathrm{tr}(K) = 0$, *and*

$$\int \sqrt{^{(2)}g}\, \frac{\partial \mathrm{tr}(K)}{\partial n} d^2x < 0. \tag{13.81}$$

Note that the strong flare-out condition implies both the simple flare-out condition and the weak flare-out condition, but that the simple flare-out condition does not necessarily imply the weak flare-out condition. (The integral could be zero.) Whenever we do not specifically specify the type of flare-out condition being used we deem it to be the simple flare-out condition.

The conditions under which the weak definition of flare-out are appropriate arise, for instance, when one takes a Morris–Thorne traversable wormhole (which is symmetric under interchange of the two universes it connects) and distorts the geometry by placing a small bump on the original throat. (See figure 13.9.)

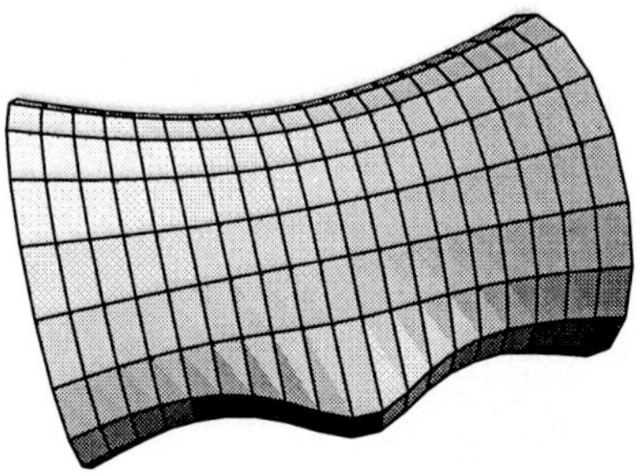

Figure 13.9: Strong versus weak throats: Placing a small bump on a strong throat typically causes it to tri-furcate into two strong throats plus a weak throat.

The presence of the bump causes the old throat to trifurcate into three extremal surfaces: Two minimal surfaces are formed, one on each side of the old throat, (these are minimal in the strong sense previously discussed), while the surface of symmetry between the two universes, though by construction still extremal, is no longer minimal in the strict sense. However, the surface of symmetry is often (but not always) minimal in the weak (integrated) sense indicated above.

A second situation in which the distinction between strong and weak throats is important is in the cut-and-paste construction for traversable wormholes [3, 22, 42]. In this construction one takes two (static) spacetimes ($\mathcal{M}_1$, $\mathcal{M}_2$) and excises two geometrically identical regions of the form $\Omega_i \times \mathcal{R}$, Ω_i being compact spacelike surfaces with boundary and $\mathcal{R}$ indicating the time direction. One then identifies the two boundaries $\partial\Omega_i \times \mathcal{R}$ thereby obtaining a single manifold ($\mathcal{M}_1 \# \mathcal{M}_2$) that contains a wormhole joining the two regions $\mathcal{M}_i - \Omega_i \times \mathcal{R}$. We would like to interpret the junction $\partial\Omega_{1=2} \times \mathcal{R}$ as the throat of the wormhole.

If the sets Ω_i are convex, then there is absolutely no problem: the junction $\partial\Omega_{1=2} \times \mathcal{R}$ is by construction a wormhole throat in the strong sense enunciated above.

On the other hand, if the Ω_i are concave, then it is straightforward to convince oneself that the junction $\partial\Omega_{1=2} \times \mathcal{R}$ is not a wormhole throat in the strong sense. If one denotes the *convex hull* of Ω_i by $\mathrm{conv}(\Omega_i)$ then the *two* regions $\partial[\mathrm{conv}(\Omega_i)] \times \mathcal{R}$ *are* wormhole throats in the strong sense. The junction $\partial\Omega_{1=2} \times \mathcal{R}$ is at best a wormhole throat in the weak sense.

For these reasons it is useful to have this notion of a weak throat available as an alternative definition. Whenever we do not qualify the notion of wormhole throat it will refer to a throat in the simple sense. Whenever we refer to a throat in the weak sense or strong senses we will explicitly say so. Finally, it is also useful to define

Definition: Weak f-weighted flare-out condition

A two surface satisfies the "weak f-weighted flare-out" condition if and only if it is extremal, $\mathrm{tr}(K) = 0$, *and*

$$\int \sqrt{^{(2)}g}\; f(x)\; \frac{\partial \mathrm{tr}(K)}{\partial n} d^2x < 0. \tag{13.82}$$

(We will only be interested in this condition for $f(x)$ some positive function defined over the wormhole throat.)

The constraints on the extrinsic curvature embodied in these various definitions lead to constraints on the spacetime geometry, and consequently constraints on the stress-energy.

Technical point I: Degenerate throats

A class of wormholes for which we have to extend these definitions arises when the wormhole throat possesses an accidental degeneracy in the extrinsic curvature at the throat. The previous discussion has tacitly been assuming that near the throat we can write

$$\begin{aligned} {}^{(2)}g_{ab}(x,n) &= {}^{(2)}g_{ab}(x,0) + n \left.\frac{\partial\left[{}^{(2)}g_{ab}(x,n)\right]}{\partial n}\right|_{n=0} \\ &\quad + \frac{n^2}{2} \left.\frac{\partial^2\left[{}^{(2)}g_{ab}(x,n)\right]}{(\partial n)^2}\right|_{n=0} + O[n^3]. \end{aligned} \tag{13.83}$$

with the linear term having trace zero (to satisfy extremality) and the quadratic term being constrained by the flare-out conditions.

Now if we have an accidental degeneracy with the quadratic (and possibly even higher order terms) vanishing identically, we would have to develop an expansion such as

$$\begin{aligned} {}^{(2)}g_{ab}(x,n) &= {}^{(2)}g_{ab}(x,0) + n \left.\frac{\partial\left[{}^{(2)}g_{ab}(x,n)\right]}{\partial n}\right|_{n=0} \\ &\quad + \frac{n^{2N}}{(2N)!} \left.\frac{\partial^{2N}\left[{}^{(2)}g_{ab}(x,n)\right]}{(\partial n)^{2N}}\right|_{n=0} + O[n^{2N+1}]. \end{aligned} \tag{13.84}$$

Applied to the metric determinant this implies an expansion such as

$$\sqrt{^{(2)}g(x,n)} = \sqrt{^{(2)}g(x,0)} \left(1 + \frac{n^{2N}\, k_N(x)}{(2N)!} + O[n^{2N+1}]\right). \tag{13.85}$$

Where $k_N(x)$ denotes the first non-zero sub-dominant term in the above expansion, and we know by explicit construction that

$$k_N(x) = +\frac{1}{2}\mathrm{tr}\left(\left. \frac{\partial^{2N}\left[{}^{(2)}g_{ab}(x,n)\right]}{(\partial n)^{2N}} \right|_{n=0} \right) \tag{13.86}$$

$$= -\mathrm{tr}\left(\left. \frac{\partial^{2N-1} K_{ab}(x,n)}{(\partial n)^{2N-1}} \right|_{n=0} \right) \tag{13.87}$$

$$= -\left(\left. \frac{\partial^{2N-1} K(x,n)}{(\partial n)^{2N-1}} \right|_{n=0} \right), \tag{13.88}$$

since the trace is taken with ${}^{(2)}g^{ab}(x,0)$ and this commutes with the normal derivative. We know that the first non-zero subdominant term in the expansion (13.85) must be of even order in n (*i.e.* n^{2N}), and cannot correspond to an odd power of n, since otherwise the throat would be a point of inflection of the area, not a minimum of the area. Furthermore, since $k_N(x)$ is by definition non-zero the flare-out condition must be phrased as the constraint $k_N(x) > 0$, with this now being a strict inequality. More formally, this leads to the definition below.

Definition: N-fold degenerate flare-out condition:
A two surface satisfies the "N-fold degenerate flare-out" condition at a point x if and only if it is extremal, $\mathrm{tr}(K) = 0$, if in addition the first $2N - 2$ normal derivatives of the trace of the extrinsic curvature vanish at x, and if finally at the point x one has

$$\frac{\partial^{2N-1}\mathrm{tr}(K)}{(\partial n)^{2N-1}} < 0, \tag{13.89}$$

where the inequality is strict. (In the previous notation this is equivalent to the statement that $k_N(x) > 0$.)

Physically, at an N-fold degenerate point, the wormhole throat is seen to be extremal up to order $2N - 1$ with respect to normal derivatives of the metric, i.e., the flare-out property is delayed spatially with respect to throats in which the flare-out occurs at second order in n. The way we have set things up, the 1-fold degenerate flare-out condition is completely equivalent to the strong flare-out condition.

If we now consider the extrinsic curvature directly we see, by differentiating (13.85), first that

$$K(x,n) = -\frac{n^{2N-1}\, k_N(x)}{(2N-1)!} + O[n^{2N}], \tag{13.90}$$

and secondly that

$$\frac{\partial K(x,n)}{\partial n} = -\frac{n^{2N-2}\, k_N(x)}{(2N-2)!} + O[n^{2N-1}]. \tag{13.91}$$

From the dominant $n \to 0$ behavior we see that if (at some point x) $2N$ happens to equal 2, then the flare-out condition implies that $\partial K(x,n)/\partial n$ must be negative *at and near the throat.* This can also be deduced directly from the equivalent strong flare-out condition: if $\partial K(x,n)/\partial n$ is negative and non-zero at the throat, then it must remain negative in some region surrounding the throat. On the other hand, if $2N$ is greater than 2 the flare-out condition only tells us that $\partial K(x,n)/\partial n$ must be negative *in some region surrounding the throat*, and does not necessarily imply that it is negative at the throat itself. (It could merely be zero at the throat.)

Thus for degenerate throats, the flare out conditions should be rephrased in terms of the first non-zero normal derivative beyond the linear term. Analogous issues arise even for Morris–Thorne wormholes [1, page 405, equation (56)], see also the discussion presented in [3, pages 104–105, 109]. Even if the throat is non-degenerate (1-fold degenerate) there are technical advantages to phrasing the flare-out conditions this way: It allows us to put constraints on the extrinsic curvature near but not on the throat.

Technical point II: Hyperspatial tubes

A second class of wormholes requiring even more technical fiddles arises when there is a central section which is completely uniform and independent of n. [So that $K_{ab} = 0$ over the whole throat for some finite range $n \in (-n_0, +n_0)$.] This central section might be called a "hyperspatial tube". The flare-out condition should then be rephrased as stating that whenever extrinsic curvature first deviates from zero [at some point $(x, \pm n_0)$] one must formulate constraints such as

$$\left.\frac{\partial \mathrm{tr}(K)}{\partial n}\right|_{\pm n_0^{\pm}} \leq 0. \tag{13.92}$$

In this case $\mathrm{tr}(K)$ is by definition not an analytic function of n at n_0, so the flare-out constraints have to be interpreted in terms of one-sided derivatives in the region outside the hyperspatial tube. [That is, we are concerned with the possibility that $\sqrt{g(x,n)}$ could be constant for $n < n_0$ but behave as $(n - n_0)^{2N}$ for $n > n_0$. In this case derivatives, at $n = n_0$, do not exist beyond order $2N$.]

13.4.3 Geometry of a generic static throat

Using Gaussian normal coordinates in the region surrounding the throat

$${}^{(3)}R_{abcd} = {}^{(2)}R_{abcd} - (K_{ac}K_{bd} - K_{ad}K_{bc}). \tag{13.93}$$

See [36, page 514, equation (21.75)]. Because two dimensions is special this reduces to:

$${}^{(3)}R_{abcd} = \frac{{}^{(2)}R}{2}\,(g_{ac}g_{bd} - g_{ad}g_{bc}) - (K_{ac}K_{bd} - K_{ad}K_{bc}). \tag{13.94}$$

Of course we still have the standard dimension-independent results that:

$$^{(3)}R_{nabc} = -(K_{ab:c} - K_{ac:b}). \tag{13.95}$$

$$^{(3)}R_{nanb} = \frac{\partial K_{ab}}{\partial n} + (K^2)_{ab}. \tag{13.96}$$

See [36, page 514, equation (21.76)] and [36, page 516 equation (21.82)]. Here the index n refers to the spatial direction normal to the two-dimensional throat.

Thus far, these results hold both on the throat and in the region surrounding the throat: these results hold as long as the Gaussian normal coordinate system does not break down. (Such breakdown being driven by the fact that the normal geodesics typically intersect after a certain distance.) In the interests of notational tractability we now particularize attention to the throat itself, but shall subsequently indicate that certain of our results can be extended off the throat itself into the entire region over which the Gaussian normal coordinate system holds sway.

Taking suitable contractions, *and using the extremality condition* $\mathrm{tr}(K) = 0$,

$$^{(3)}R_{ab} = \frac{^{(2)}R}{2} g_{ab} + \frac{\partial K_{ab}}{\partial n} + 2(K^2)_{ab}. \tag{13.97}$$

$$^{(3)}R_{na} = -K_{ab}{}^{:b}. \tag{13.98}$$

$$^{(3)}R_{nn} = \mathrm{tr}\left(\frac{\partial K}{\partial n}\right) + \mathrm{tr}(K^2) \tag{13.99}$$

$$= \frac{\partial \mathrm{tr}(K)}{\partial n} - \mathrm{tr}(K^2). \tag{13.100}$$

So that

$$^{(3)}R = {}^{(2)}R + 2\mathrm{tr}\left(\frac{\partial K}{\partial n}\right) + 3\mathrm{tr}(K^2) \tag{13.101}$$

$$= {}^{(2)}R + 2\frac{\partial \mathrm{tr}(K)}{\partial n} - \mathrm{tr}(K^2). \tag{13.102}$$

To effect these contractions, we make use of the decomposition of the three-space metric in terms of the throat two-metric and the set of two vectors e^i_a tangent to the throat and the three-vector n^i normal to the 2-surface

$$^{(2+1)}g^{ij} = e^i_a\, e^j_b\, g^{ab} + n^i n^j. \tag{13.103}$$

For the three-space Einstein tensor (*cf.* [36, page 552]) we see

$$^{(3)}G_{ab} = \frac{\partial K_{ab}}{\partial n} + 2(K^2)_{ab} - g_{ab}\frac{\partial \mathrm{tr}(K)}{\partial n} + \frac{1}{2} g_{ab}\, \mathrm{tr}(K^2). \tag{13.104}$$

$$
\begin{aligned}
{}^{(3)}G_{na} &= -K_{ab}{}^{:b}. && (13.105)\\
{}^{(3)}G_{nn} &= -\frac{1}{2}{}^{(2)}R - \frac{1}{2}\mathrm{tr}(K^2). && (13.106)
\end{aligned}
$$

Aside: Note in particular that by the flare-out condition ${}^{(3)}R_{nn} \leq 0$. This implies that the three-space Ricci tensor ${}^{(3)}R_{ij}$ has at least one negative semi-definite eigenvalue everywhere on the throat. If we adopt the strong flare-out condition then the three-space Ricci tensor has at least one negative definite eigenvalue somewhere on the throat. A similar result for Euclidean wormholes is quoted in [43] and the present analysis can of course be carried over to Euclidean signature with appropriate definitional changes.

This decomposition now allows us to write down the various components of the space-time Einstein tensor. For example

$$
\begin{aligned}
{}^{(3+1)}G_{ab} &= -\phi_{|ab} - \phi_{|a}\,\phi_{|b} + g_{ab}g^{kl}\left[\phi_{|kl} + \phi_{|k}\phi_{|l}\right]\\
&\quad + \frac{\partial K_{ab}}{\partial n} + 2(K^2)_{ab} - g_{ab}\frac{\partial \mathrm{tr}(K)}{\partial n} + \frac{1}{2}g_{ab}\,\mathrm{tr}(K^2)\\
&= 8\pi G\,T_{ab}. \qquad (13.107)
\end{aligned}
$$

But by the definition of the extrinsic curvature, and using the Gauss–Weingarten equations,

$$
\begin{aligned}
\phi_{|ab} &= \phi_{:ab} + K_{ab}\,\phi_{|n}. && (13.108)\\
\phi_{|na} &= K_a{}^b\,\phi_{:b}. && (13.109)
\end{aligned}
$$

[See, for example, equations (21.57) and (21.63) of [36].] Thus

$$
g^{kl}\phi_{|kl} = g^{ab}\phi_{:ab} + (g^{ab}K_{ab})\,\phi_{|n} + \phi_{|nn}. \qquad (13.110)
$$

But remember that $\mathrm{tr}(K) = 0$ at the throat, so

$$
g^{kl}\phi_{|kl} = g^{ab}\phi_{:ab} + \phi_{|nn}. \qquad (13.111)
$$

This finally allows us to write

$$
\begin{aligned}
{}^{(3+1)}G_{ab} &= -\phi_{:ab} - \phi_{:a}\,\phi_{:b} - K_{ab}\,\phi_{|n}\\
&\quad + g_{ab}\left[g^{cd}(\phi_{:cd} + \phi_{:c}\phi_{:d}) + \phi_{|nn} + \phi_{|n}\phi_{|n}\right]\\
&\quad + \frac{\partial K_{ab}}{\partial n} + 2(K^2)_{ab} - g_{ab}\frac{\partial \mathrm{tr}(K)}{\partial n} + \frac{1}{2}g_{ab}\,\mathrm{tr}(K^2)\\
&= 8\pi G\,T_{ab}. \qquad (13.112)
\end{aligned}
$$

$$
\begin{aligned}
{}^{(3+1)}G_{na} &= -K_a{}^b \phi_{:b} - \phi_{|n}\, \phi_{:a} - K_{ab}{}^{:b} \\
&= 8\pi G\, T_{na}. \qquad (13.113) \\
{}^{(3+1)}G_{nn} &= g^{cd}\left[\phi_{:cd} + \phi_{:c}\phi_{:d}\right] - \frac{1}{2}{}^{(2)}R - \frac{1}{2}\mathrm{tr}(K^2) \\
&= -8\pi G\, \tau. \qquad (13.114) \\
{}^{(3+1)}G_{\hat{t}a} &= 0. \qquad (13.115) \\
{}^{(3+1)}G_{\hat{t}n} &= 0. \qquad (13.116) \\
{}^{(3+1)}G_{\hat{t}\hat{t}} &= \frac{{}^{(2)}R}{2} + \frac{\partial \mathrm{tr}(K)}{\partial n} - \frac{1}{2}\mathrm{tr}(K^2) \\
&= +8\pi G\, \rho. \qquad (13.117)
\end{aligned}
$$

Here τ denotes the *tension* perpendicular to the wormhole throat, it is the natural generalization of the quantity considered by Morris and Thorne, while ρ is simply the energy density at the wormhole throat.

The calculation presented above is in its own way simply a matter of brute force index gymnastics—but we feel that there are times when explicit expressions of this type are useful.

13.4.4 Constraints on the stress-energy tensor

We can now derive several constraints on the stress-energy.

—**First constraint**—

$$
\tau = \frac{1}{16\pi G}\left[{}^{(2)}R + \mathrm{tr}(K^2) - 2g^{cd}(\phi_{:cd} + \phi_{:c}\phi_{:d})\right]. \qquad (13.118)
$$

(Unfortunately the signs as given are correct. Otherwise we would have a lovely lower bound on τ. We will need to be a little tricky when dealing with the ϕ terms.) The above is the generalization of the Morris–Thorne result that

$$
\tau = \frac{1}{8\pi G r_0^2} \qquad (13.119)
$$

at the throat of the special class of model wormholes they considered. (With MTW conventions ${}^{(2)}R = 2/r_0^2$ for a two-sphere.) If you now integrate over the surface of the wormhole

$$
\int \sqrt{{}^{(2)}g}\; \tau\; d^2x = \frac{1}{16\pi G}\left[4\pi\chi + \int \sqrt{{}^{(2)}g}\left\{\mathrm{tr}(K^2) - 2g^{cd}\phi_{:c}\phi_{:d}\right\} d^2x\right]. \qquad (13.120)
$$

Here χ is the Euler characteristic of the throat, while the $g^{cd}\phi_{:cd}$ term vanishes by partial integration, since the throat is a manifold without boundary.

—**Second constraint**—

$$\rho = \frac{1}{16\pi G}\left[{}^{(2)}R + 2\frac{\partial \mathrm{tr}(K)}{\partial n} - \mathrm{tr}(K^2)\right]. \tag{13.121}$$

The second term is negative semi-definite by the flare-out condition, while the third term is manifestly negative semi-definite. Thus

$$\rho \leq \frac{1}{16\pi G}\,{}^{(2)}R. \tag{13.122}$$

This is the generalization of the Morris–Thorne result that

$$\rho = \frac{b'(r_0)}{8\pi G r_0^2} \leq \frac{1}{8\pi G r_0^2} \tag{13.123}$$

at the throat of the special class of model wormholes they considered. (See [3, page 107].)

Note in particular that if the wormhole throat does not have the topology of a sphere or torus then there *must* be places on the throat such that ${}^{(2)}R < 0$ and thus such that $\rho < 0$. Thus wormhole throats of high genus will always have regions that violate the weak and dominant energy conditions. (The simple flare-out condition is sufficient for this result. For a general discussion of the energy conditions see [3] or [37].)

If the wormhole throat has the topology of a torus then it will generically violate the weak and dominant energy conditions; only for the very special case ${}^{(2)}R = 0$, $K_{ab} = 0$, $\partial \mathrm{tr}(K)/\partial n = 0$ will it possibly satisfy (but still be on the verge of violating) the weak and dominant energy conditions. This is a particular example of a degenerate throat in the sense discussed previously.

Wormhole throats with the topology of a sphere will, provided they are convex, at least have positive energy density, but we shall soon see that other energy conditions are typically violated.

If we now integrate over the surface of the wormhole

$$\int \sqrt{{}^{(2)}g}\;\rho\; d^2x = \frac{1}{16\pi G}\left[4\pi\chi + \int \sqrt{{}^{(2)}g}\left\{2\frac{\partial \mathrm{tr}(K)}{\partial n} - \mathrm{tr}(K^2)\right\} d^2x\right]. \tag{13.124}$$

So for a throat with the topology of a torus ($\chi = 0$) the simple flare-out condition yields

$$\int \sqrt{{}^{(2)}g}\;\rho\; d^2x \leq 0, \tag{13.125}$$

while the strong or weak flare-out conditions yield

$$\int \sqrt{{}^{(2)}g}\;\rho\; d^2x < 0, \tag{13.126}$$

guaranteeing violation of the weak and dominant energy conditions. For a throat with higher genus topology ($\chi = 2 - 2g$) the simple flare-out condition is sufficient to yield

$$\int \sqrt{^{(2)}g}\, \rho \, d^2x \le \frac{\chi}{4G} < 0. \tag{13.127}$$

—**Third constraint**—

$$\rho - \tau = \frac{1}{16\pi G}\left[+2\frac{\partial \mathrm{tr}(K)}{\partial n} - 2\mathrm{tr}(K^2) + 2g^{cd}(\phi_{:cd} + \phi_{:c}\phi_{:d})\right]. \tag{13.128}$$

Note that the two-curvature $^{(2)}R$ has conveniently dropped out of this equation. As given, this result is valid only on the throat itself, but we shall soon see that a generalization can be constructed that will also hold in the region surrounding the throat. The first term is negative semi-definite by the simple flare-out condition (at the very worst when integrated over the throat it is negative by the weak flare-out condition). The second term is negative semi-definite by inspection. The third term integrates to zero though it may have either sign locally on the throat. The fourth term is unfortunately positive semi-definite on the throat which prevents us from deriving a truly general energy condition violation theorem without additional information.

Now because the throat is by definition a compact two surface, we know that $\phi(x^a)$ must have a maximum somewhere on the throat. At the global maximum (or even at any local maximum) we have $\phi_{:a} = 0$ and $g^{ab}\phi_{:ab} \le 0$, so at the maxima of ϕ one has

$$\rho - \tau \le 0. \tag{13.129}$$

Generically, the inequality will be strict, and generically there will be points on the throat at which the null energy condition is violated.

Integrating over the throat we have

$$\begin{aligned}\int \sqrt{^{(2)}g}\, [\rho - \tau]\, d^2x \quad &= \quad \frac{1}{16\pi G}\int \sqrt{^{(2)}g}\left[+2\frac{\partial \mathrm{tr}(K)}{\partial n}\right. \\ &\qquad \left. -2\mathrm{tr}(K^2) + 2g^{cd}(\phi_{:c}\phi_{:d})\right] d^2x.\end{aligned} \tag{13.130}$$

Because of the last term we must be satisfied with the result

$$\int \sqrt{^{(2)}g}\, [\rho - \tau]\, d^2x \le \int \sqrt{^{(2)}g}\, \left[2g^{cd}(\phi_{:c}\phi_{:d})\right] d^2x. \tag{13.131}$$

—Fourth constraint—

We can rewrite the difference $\rho - \tau$ as

$$\rho - \tau = \frac{1}{16\pi G}\left[+2\frac{\partial \mathrm{tr}(K)}{\partial n} - 2\mathrm{tr}(K^2) + 2\exp(-\phi)\ {}^{(2)}\Delta \exp(+\phi)\right]. \tag{13.132}$$

So if we multiply by $\exp(+\phi)$ before integrating, the two-dimensional Laplacian ${}^{(2)}\Delta$ vanishes by partial integration and we have

$$\begin{aligned}\int \sqrt{{}^{(2)}g}\ \exp(+\phi)\ [\rho - \tau]\ d^2x &= \frac{1}{8\pi G}\int \sqrt{{}^{(2)}g}\ \exp(+\phi) \\ &\times \left[\frac{\partial \mathrm{tr}(K)}{\partial n} - \mathrm{tr}(K^2)\right] d^2x. \end{aligned} \tag{13.133}$$

Thus the strong flare-out condition (or less restrictively, the weak e^{ϕ}–weighted flare-out condition) implies the violation of this "transverse averaged null energy condition" (TANEC, the NEC averaged over the throat)

$$\int \sqrt{{}^{(2)}g}\ \exp(+\phi)\ [\rho - \tau]\ d^2x < 0. \tag{13.134}$$

This TANEC, and its off-throat generalization to be developed below, is perhaps the central result of this generic static wormhole analysis.

—Fifth constraint—

We can define an average transverse pressure on the throat by

$$\bar{p} \equiv \frac{1}{16\pi G}\ g^{ab}\ {}^{(3+1)}G_{ab} \tag{13.135}$$

$$= \frac{1}{16\pi G}\left[g^{cd}(\phi_{:cd} + \phi_{:c}\phi_{:d}) + 2\phi_{|nn} + 2\phi_{|n}\phi_{|n} - \frac{\partial \mathrm{tr}(K)}{\partial n} + \mathrm{tr}(K^2)\right]. \tag{13.136}$$

The last term is manifestly positive semi-definite, the penultimate term is positive semi-definite by the flare-out condition. The first and third terms are of indefinite sign while the second and fourth are also positive semi-definite. Integrating over the surface of the throat

$$\int \sqrt{{}^{(2)}g}\ \bar{p}\ d^2x \ \geq \ \frac{1}{8\pi G}\int \sqrt{{}^{(2)}g}\ \phi_{|nn}\ d^2x. \tag{13.137}$$

A slightly different constraint, also derivable from the above, is

$$\int \sqrt{^{(2)}g}\; e^{\phi}\; \bar{p}\, d^2x \;\; \geq \;\; \frac{1}{8\pi G} \int \sqrt{^{(2)}g}\; (e^{\phi})_{|nn}\; d^2x. \tag{13.138}$$

These inequalities relate transverse pressures to normal derivatives of the gravitational potential. In particular, if the throat lies at a minimum of the gravitational red-shift the second normal derivative will be positive, so the transverse pressure (averaged over the wormhole throat) must be positive.

—**Sixth constraint**—

Now look at the quantities $\rho - \tau + 2\bar{p}$ and $\rho - \tau - 2\bar{p}$. We have

$$\rho - \tau + 2\bar{p} \;\; = \;\; \frac{1}{4\pi G}\left\{g^{cd}(\phi_{:cd} + \phi_{:c}\phi_{:d}) + \phi_{|nn} + \phi_{|n}\phi_{|n}\right\} \tag{13.139}$$

$$\;\; = \;\; \frac{1}{4\pi G}\left\{g^{ij}(\phi_{|ij} + \phi_{|i}\phi_{|j})\right\}. \tag{13.140}$$

This serves as a nice consistency check. The combination of stress-energy components appearing above is equal to $\rho + g^{ij}T_{ij}$ and is exactly that relevant to the strong energy condition. See equations (13.66)—(13.68). See also equations (13.63)—(13.65). Multiplying by e^{ϕ} and integrating

$$\int \sqrt{^{(2)}g}\; e^{\phi}\; [\rho - \tau + 2\bar{p}]\; d^2x = \frac{1}{4\pi G} \int \sqrt{^{(2)}g}\; (e^{\phi})_{|nn}\; d^2x. \tag{13.141}$$

This relates this transverse integrated version of the strong energy condition to the normal derivatives of the gravitational potential.

On the other hand

$$\rho - \tau - 2\bar{p} = \frac{1}{4\pi G}\left\{-\phi_{|nn} - \phi_{|n}\phi_{|n} + \frac{\partial \mathrm{tr}(K)}{\partial n} - \mathrm{tr}(K^2)\right\}. \tag{13.142}$$

The second and fourth terms are negative semi-definite, while the third term is negative semi-definite by the flare-out condition.

—**Summary**—

There are a number of powerful constraints that can be placed on the stress-energy tensor at the wormhole throat simply by invoking the minimality properties of the wormhole throat. Depending on the precise form of the assumed flare-out condition, these constraints give the various energy condition violation theorems we are seeking. Even under the weakest assumptions (appropriate to a degenerate throat) they constrain the stress-energy to at best be on the verge of violating the various energy conditions.

13.4.5 Special case: The isopotential throat

Suppose we take $\phi_{:a} = 0$. This additional constraint corresponds to asserting that the throat is an *isopotential* of the gravitational red-shift. In other words, $\phi(n, x^a)$ is simply a constant on the throat. For instance, all the Morris–Thorne model wormholes [1] possess this symmetry. Under this assumption there are numerous simplifications.

We will not present anew all the results for the Riemann curvature tensor but instead content ourselves with the Einstein tensor

$$\begin{aligned} {}^{(3+1)}G_{ab} &= +g_{ab}\left(\phi_{|nn} + \phi_{|n}\phi_{|n}\right) - K_{ab}\,\phi_{|n} \\ &\quad + \frac{\partial K_{ab}}{\partial n} + 2(K^2)_{ab} - g_{ab}\frac{\partial \mathrm{tr}(K)}{\partial n} + \frac{1}{2} g_{ab}\,\mathrm{tr}(K^2) \\ &= 8\pi G\, T_{ab}. \end{aligned} \tag{13.143}$$

$${}^{(3+1)}G_{na} = -K_{ab}{}^{:b} = 8\pi G\, T_{na}. \tag{13.144}$$

$${}^{(3+1)}G_{nn} = -\frac{1}{2}{}^{(2)}R - \frac{1}{2}\mathrm{tr}(K^2) = -8\pi G\,\tau. \tag{13.145}$$

$${}^{(3+1)}G_{\hat{t}a} = 0. \tag{13.146}$$

$${}^{(3+1)}G_{\hat{t}n} = 0. \tag{13.147}$$

$${}^{(3+1)}G_{\hat{t}\hat{t}} = \frac{{}^{(2)}R}{2} + \frac{\partial \mathrm{tr}(K)}{\partial n} - \frac{1}{2}\mathrm{tr}(K^2) = +8\pi G\,\rho. \tag{13.148}$$

Thus for an isopotential throat

$$\tau = \frac{1}{16\pi G}\left[{}^{(2)}R + \mathrm{tr}(K^2)\right] \geq \frac{1}{16\pi G}{}^{(2)}R. \tag{13.149}$$

$$\begin{aligned} \rho &= \frac{1}{16\pi G}\left[{}^{(2)}R + 2\frac{\partial \mathrm{tr}(K)}{\partial n} - \mathrm{tr}(K^2)\right] \\ &\leq \frac{1}{16\pi G}{}^{(2)}R. \end{aligned} \tag{13.150}$$

$$\rho - \tau = \frac{1}{16\pi G}\left[+2\frac{\partial \mathrm{tr}(K)}{\partial n} - 2\mathrm{tr}(K^2)\right] \leq 0. \tag{13.151}$$

This gives us a very powerful result: using only the simple flare-out condition, the NEC is on the verge of being violated everywhere on an isopotential throat.

By invoking the strong flare-out condition the NEC is definitely violated somewhere on an isopotential throat.

Invoking the weak flare-out condition we can still say that the surface integrated NEC is definitely violated on an isopotential throat.

13.4.6 Special case: The extrinsically flat throat

Suppose now that we take $K_{ab} = 0$. This is a much stronger constraint than simple minimality of the area of the wormhole throat and corresponds to asserting that the three-geometry of the throat is (at least locally) symmetric under interchange of the two regions it connects. For instance, all the Morris–Thorne model wormholes [1] possess this symmetry and have throats that are extrinsically flat. Under this assumption there are also massive simplifications. (Note that we are not making the isopotential assumption at this stage.)

Again, we will not present all the results but content ourselves with the Einstein tensor

$$
\begin{aligned}
{}^{(3+1)}G_{ab} &= -\phi_{:ab} - \phi_{:a}\,\phi_{:b} + g_{ab}\left[g^{cd}(\phi_{:cd} + \phi_{:c}\phi_{:d}) + \phi_{|nn} + \phi_{|n}\phi_{|n}\right] \\
&\quad + \frac{\partial K_{ab}}{\partial n} - g_{ab}\frac{\partial \mathrm{tr}(K)}{\partial n} \\
&= 8\pi G\, T_{ab}. && (13.152) \\
{}^{(3+1)}G_{na} &= -\phi_{|n}\,\phi_{|a} = 8\pi G\, T_{na}. && (13.153) \\
{}^{(3+1)}G_{nn} &= g^{cd}\left[\phi_{:cd} + \phi_{:c}\phi_{:d}\right] - \frac{1}{2}{}^{(2)}R = -8\pi G\,\tau. && (13.154) \\
{}^{(3+1)}G_{\hat{t}a} &= 0. && (13.155) \\
{}^{(3+1)}G_{\hat{t}n} &= 0. && (13.156) \\
{}^{(3+1)}G_{\hat{t}\hat{t}} &= \frac{{}^{(2)}R}{2} + \frac{\partial \mathrm{tr}(K)}{\partial n} = +8\pi G\,\rho. && (13.157)
\end{aligned}
$$

Though the stress-energy tensor is now somewhat simpler than the general case, the presence of the $\phi_{:a}$ terms precludes the derivation of any truly new general theorems.

13.4.7 Special case: The extrinsically flat isopotential throat

Finally, suppose we take both $K_{ab} = 0$ and $\phi_{:a} = 0$. A wormhole throat that is both extrinsically flat and isopotential is particularly simple to deal with, even though it is still much more general than the Morris–Thorne wormhole. Once again, we will not present all the results but content ourselves with the Einstein tensor

$$
\begin{aligned}
{}^{(3+1)}G_{ab} &= +g_{ab}\left(\phi_{|nn} + \phi_{|n}\phi_{|n}\right) + \frac{\partial K_{ab}}{\partial n} - g_{ab}\frac{\partial \mathrm{tr}(K)}{\partial n} \\
&= 8\pi G\, T_{ab}. && (13.158) \\
{}^{(3+1)}G_{na} &= 0. && (13.159) \\
{}^{(3+1)}G_{nn} &= -\frac{1}{2}{}^{(2)}R = -8\pi G\,\tau. && (13.160)
\end{aligned}
$$

$$^{(3+1)}G_{\hat{t}a} = 0. \tag{13.161}$$

$$^{(3+1)}G_{\hat{t}n} = 0. \tag{13.162}$$

$$^{(3+1)}G_{\hat{t}\hat{t}} = \frac{^{(2)}R}{2} + \frac{\partial \text{tr}(K)}{\partial n} = +8\pi G\, \rho. \tag{13.163}$$

In this case $\rho - \tau$ is particularly simple:

$$\rho - \tau = \frac{1}{8\pi G}\frac{\partial \text{tr}(K)}{\partial n}. \tag{13.164}$$

This quantity is manifestly negative semi-definite by the simple flare-out condition.

- For the strong flare-out condition we deduce that the NEC must be violated somewhere on the wormhole throat.
- Even for the weak flare-out condition we have

$$\int \sqrt{^{(2)}g}\ [\rho - \tau]\ d^2x < 0. \tag{13.165}$$

- We again see that generic violations of the null energy condition are the rule.

13.4.8 The region surrounding the throat

Because the spacetime is static, one can unambiguously define the energy density everywhere in the spacetime by setting

$$\rho = \frac{^{(3+1)}G_{\hat{t}\hat{t}}}{8\pi G}. \tag{13.166}$$

The normal tension, which we have so far defined only on the wormhole throat itself, can meaningfully be extended to the entire region where the Gaussian normal coordinate system is well defined by setting

$$\tau = -\frac{^{(3+1)}G_{nn}}{8\pi G}. \tag{13.167}$$

Thus in particular

$$\rho - \tau = \frac{^{(3+1)}G_{\hat{t}\hat{t}} + {}^{(3+1)}G_{nn}}{8\pi G} = \frac{^{(3+1)}R_{\hat{t}\hat{t}} + {}^{(3+1)}R_{nn}}{8\pi G}, \tag{13.168}$$

with this quantity being well defined throughout the Gaussian normal coordinate patch. (The last equality uses the fact that $g_{\hat{t}\hat{t}} = -1$ while $g_{nn} = +1$.) But we have already seen how to evaluate these components of the Ricci tensor. Indeed

$$
\begin{aligned}
{}^{(3+1)}R_{\hat{t}\hat{t}} &= g^{ij}\left[\phi_{|ij} + \phi_{|i}\phi_{|j}\right]. && (13.169)\\
{}^{(3+1)}R_{nn} &= {}^{(3)}R_{nn} - \left[\phi_{|nn} + \phi_{|n}\phi_{|n}\right] && (13.170)\\
&= \frac{\partial \mathrm{tr}(K)}{\partial n} - \mathrm{tr}(K^2) - \left[\phi_{|nn} + \phi_{|n}\phi_{|n}\right], && (13.171)
\end{aligned}
$$

where we have been careful to *not* use the extremality condition $\mathrm{tr}(K) = 0$. Therefore

$$
\begin{aligned}
\rho - \tau &= \frac{1}{8\pi G}\left[\frac{\partial \mathrm{tr}(K)}{\partial n} - \mathrm{tr}(K^2) + g^{ab}\left(\phi_{|ab} + \phi_{|a}\phi_{|b}\right)\right] && (13.172)\\
&= \frac{1}{8\pi G}\left[\frac{\partial \mathrm{tr}(K)}{\partial n} - \mathrm{tr}(K^2) + \mathrm{tr}(K)\phi_{|n} + g^{ab}\left(\phi_{:ab} + \phi_{:a}\phi_{:b}\right)\right], && (13.173)
\end{aligned}
$$

where in the last line we have used the Gauss–Weingarten equations.

- If the throat is *isopotential*, where isopotential now means that near the throat the surfaces of constant gravitational potential coincide with the surfaces of fixed n, this simplifies to:

$$
\rho - \tau = \frac{1}{8\pi G}\left[\frac{\partial \mathrm{tr}(K)}{\partial n} - \mathrm{tr}(K^2) + \mathrm{tr}(K)\phi_{|n}\right]. \qquad (13.174)
$$

  - If the throat is non-degenerate and satisfies the simple flare-out condition, then at the throat the first and second terms are negative semi-definite, and the third is zero. Then the null energy condition is either violated or on the verge of being violated at the throat.
  - If the throat is non-degenerate and satisfies the strong flare-out condition at the point x, then the first term is negative definite, the second is negative semi-definite, and the third is zero. Then the null energy condition is violated at the point x on the throat.
  - If the throat satisfies the N-fold degenerate flare-out condition at the point x, then by the generalization of the flare-out conditions applied to degenerate throats the first term will be $O[n^{2N-2}]$ and negative definite in some region surrounding the throat. The second term is again negative semi-definite. The third term can have either sign but will be $O[n^{2N-1}]$. Thus there will be some region $n \in (0, n_*)$ in which the first term dominates. Therefore the null energy condition is violated along the line $\{x\} \times (0, n_*)$. If at every point x on the throat the N-fold degenerate flare-out condition is satisfied for some *finite* N, then there will be an open region surrounding the throat on which the null energy condition is everywhere violated.

- This is the closest one can get in generalizing to arbitrary wormhole shapes the discussion on page 405 [equation (56)] of Morris–Thorne [1]. Note carefully their use of the phrase “at or near the throat”. In our parlance, they are considering a spherically symmetric extrinsically flat isopotential throat that satisfies the N-fold degenerate flare-out condition for some finite but unspecified N. See also page 104, equation (11.12) and page 109, equation (11.54) of [3], and contrast this with equation (11.56).

- If the throat is not isopotential we multiply by $\exp(\phi)$ and integrate over surfaces of constant n. Then

$$\int \sqrt{{}^{(2)}g}\ \exp(\phi)\ [\rho - \tau]\ d^2x = \frac{1}{8\pi G} \int \sqrt{{}^{(2)}g}\ \exp(\phi) \left[\frac{\partial \mathrm{tr}(K)}{\partial n} - \mathrm{tr}(K^2) + \mathrm{tr}(K)\phi_{|n} \right] d^2x. \tag{13.175}$$

This generalizes the previous version (13.133) of the transverse averaged null energy condition to constant n hypersurfaces near the throat. For each point x on the throat, assuming the N-fold degenerate flare-out condition, we can by the previous argument find a range of values $[n \in (0, n_*(x))]$ that will make the integrand negative. Thus there will be a set of values of n for which the integral is negative. Again we deduce violations of the null energy condition.

13.5 Discussion

In this survey we have sought to give an overview of the energy condition violations that occur in traversable wormholes. We point out that in static spherically symmetric geometries these violations of the energy conditions follow unavoidably from the definition of a wormhole and the definition of the total stress-energy via the Einstein equations. In spherically symmetric time dependent situations limited temporary suspensions of the energy condition violations are possible. In non-Einstein theories of gravity it is often possible to push the energy condition violations into the nonstandard parts of the stress-energy tensor and let the ordinary part of the stress-energy satisfy the energy conditions. The total stress-energy tensor, however, must still violate the energy conditions.

To show the generality of the energy condition violations, we have developed an analysis that is capable of dealing with static traversable wormholes of arbitrary shape. We have presented a definition of a wormhole throat that is much more general than that of the Morris–Thorne wormhole [1]. The present definition works well in any static spacetime and nicely captures the essence of the idea of what we would want to call a wormhole throat.

We do not need to make any assumptions about the existence of any asymptotically flat region, nor do we need to assume that the manifold is topologically non-trivial. It is important to realize that the essence of the definition lies in the geometrical structure of the wormhole throat.

Starting from our definition we have used the theory of embedded hypersurfaces to place restrictions on the Riemann tensor and stress-energy tensor at the throat of the wormhole. We find, as expected, that the wormhole throat generically violates the null energy condition and we have provided several theorems regarding this matter. These theorems generalize the Morris–Thorne results on exotic matter [1], and are complementary to the topological censorship theorem [5].

Generalization to the time dependent situation is clearly of interest. Unfortunately we have encountered many subtleties of definition, notation, and formalism in this endeavor. (A formulation in terms of *anti-trapped surfaces* appears promising [6].) We defer the issue of time dependent wormhole throats to a future publication.

Acknowledgements

M.V. wishes to gratefully acknowledge the hospitality shown during his visits to the Laboratory for Space Astrophysics and Fundamental Physics (LAEFF, Madrid). This work was supported in part by the US Department of Energy (M.V.) and by the Spanish Ministry of Science and Education (D.H.).

Bibliography

[1] M.S. Morris and K.S. Thorne, Am. J. Phys. **56**, 395 (1988).

[2] M.S. Morris, K.S. Thorne and U. Yurtsever, Phys. Rev. Lett, **61**, 1446 (1988).

[3] M. Visser, *Lorentzian Wormholes: From Einstein to Hawking* (American Institute of Physics, Woodbury, N.Y., 1995).

[4] D. Hochberg, A. Popov and S.V. Sushkov, Phys. Rev. Lett. **78**, 2050 (1997).

[5] J.L Friedmann, K. Schleich and D.M. Witt, Phys. Rev. Lett. **71**, 1486 (1993).

[6] D. Page, cited as a note added in proof in [1].

[7] V. Frolov and I. D. Novikov Phys. Rev. D **48**, 1607 (1993).

[8] J. A. Wheeler, Phys. Rev. **97**, 511 (1955).

[9] J. A. Wheeler, Ann. Phys. (NY) **2**, 604 (1957).

[10] S. W. Hawking, Phys. Rev. D **46**, 603 (1992).

[11] M. Visser, *The reliability horizon for semiclassical quantum gravity: Metric fluctuations are often more important than back reaction.* gr-qc/9702041; Physics Letters B, in press.

[12] M. Visser, gr-qc/9409043, Phys. Lett. **B 349**, 443 (1995).

[13] M. Visser, gr-qc/9604007, Phys. Rev. D **54**, 5103 (1996).

[14] M. Visser gr-qc/9604008, Phys. Rev. D **54**, 5116 (1996).

[15] M. Visser, gr-qc/9604009, Phys. Rev. D **54**, 5123 (1996).

[16] M. Visser, gr-qc/9703001, Phys. Rev. D **56**, 936 (1997) 936.

[17] M. Visser, Phys. Rev. D **43**, 402 (1991).

[18] M. Visser, Phys. Lett. **B 242**, 24 (1990).

[19] M. Visser, *Quantum wormholes in Lorentzian signature*, in B. Bonner and H. Miettinen, editors, *Proceedings of the Rice meeting: 1990 meeting of the Division of Particles and Fields of the American Physical Society*, volume 2, pages 858–860. (World Scientific, Singapore, 1990).

[20] F. Schein and P. C. Aichelburg, gr-qc/9606069, Phys. Rev. Lett. **77**, 4130 (1996).

[21] F. Schein, P. C. Aichelburg, and W. Israel, gr-qc/9602053, Phys. Rev. D **54**, 3800 (1996).

[22] M. Visser, Phys. Rev. D **D39**, 3182 (1989).

[23] T. A. Roman, Phys. Rev. D **47**, 1370 (1993).

[24] D. Hochberg and T. W. Kephart, Phys. Rev. Lett **70**, 2665 (1993).

[25] S. Kar, Phys. Rev. D **49**, 862 (1994).

[26] S. Kar and D. Sahdev, Phys. Rev. D **53**, 722 (1996).

[27] S. W. Kim, Phys. Rev. D **53**, 6889 (1996).

[28] A. G. Agnese and M. La Camera, Phys. Rev. D **51**, 2011 (1995).

[29] K. K. Nandi, A. Islam, and J. Evans, Phys. Rev. D **55**, 2497 (1997).

[30] L. A. Anchordoqui, S. Perez Bergliaffa, and D. F. Torres, Phys. Rev. D **55**, 5226 (1997).

[31] H. H. Yang and Y. Z. Zhang, Phys. Lett. **A 212**, 39 (1996).

[32] D. Hochberg and M. Visser, *Geometric structure of the generic static traversable wormhole throat*, gr-qc/9704082; Phys. Rev. D (in press).

[33] W. Israel, Phys. Rev. **164**, 1776 (1967).

[34] W. Israel, Commun. Math. Phys. **8**, 254 (1968).

[35] W. Israel, Can. J. Phys. **64**, 120 (1986).

[36] C.W. Misner, K.S. Thorne and J.A. Wheeler *Gravitation* (W.H. Freeman, San Francisco, 1973).

[37] S.W. Hawking and G.F.R. Ellis, *The Large Scale Structure of Space-Time* (Cambridge University Press, Cambridge, England, 1973).

[38] R.M. Wald, *General Relativity* (University of Chicago Press, Chicago, 1984).

[39] J.L. Synge, *Relativity: the General Theory* (North-Holland, Amsterdam, 1964).

[40] V. Fock, *The Theory of Space, Time, and Gravitation* (Pergamon, New York, 1964).

[41] R. Adler, M. Bazin, and M. Schiffer, *Introduction to General Relativity* (McGraw–Hill, New York, 1965).

[42] M. Visser, Nucl. Phys. **B**328, 203 (1989).

[43] S. Giddings and A. Strominger, Nucl. Phys. **B 306**, 890 (1988), see esp. page 894.

BEHAVIOR OF SINGULARITIES OF KERR–NEWMAN AND KERR–SEN SOLUTIONS BY ARBITRARY BOOST

Alexander Burinskii [a] and Giulio Magli [b]

a) Gravity Research Group, IBRAE, Russian Academy of Sciences, B. Tulskaya 52, Moscow 113191, Russia
b) Dipartimento di Matematica del Politecnico di Milano, Piazza Leonardo Da Vinci 32, 20133 Milano, Italy

Abstract

The behavior of the singularities of rotating BH under an arbitrary boost is considered on the basis of a complex representation of the Kerr theorem. We give a simple algorithm allowing to get explicit expressions for the metric and the position of the singularities for arbitrary direction and magnitude of the boost, including the ultrarelativistic case. The non-smoothness of the ultrarelativistic limit is discussed. The Kerr-Sen black hole-solution to low energy string theory is also analyzed.

14.1 Introduction

Recently, the problem of finding the ultrarelativistic limit of exact particle-like solutions of the Einstein field equations received considerable attention, especially in connection with some non-trivial gravitational effects which are expected to occur in the interparticle interactions at extreme energies due to the appearance of gravitational shock waves [1, 2, 3, 4, 5, 6, 7, 8].

First results in this field were obtained by Aichelburg and Sexl [1], who considered the ultrarelativistic boost of the Schwarzschild solution to analyze the behavior of the gravitational field of a massless point particle in the light-like limit.

A similar treatment for the Kerr geometry, which can be considered as a model of a spinning particle in general relativity, has to take into account the orientation of the angular momentum with respect to the boost [4, 5, 6, 7, 8].

There are three different physical situations connected with boosted black hole (BH)-solutions. The first one is the original Aichelburg–Sexl problem of application of such solutions to describe the gravitational field of light–like particles with or without spin. The second application consists in modeling the gravitational field of elementary particles with finite rest mass under the boost, and it is connected with an analysis of possible effects generated in relativistic collisions. Finally, there are astrophysical applications, namely boosting black holes. In this case also the behavior of the horizon and of the ergosphere under the boost are of interest.

The analysis of the boosted Kerr solution [4, 5, 6, 7, 8] exhibits some difficulties in the interpretation of the results in the limiting, ultrarelativistic case. In particular, there are technical difficulties due to the absence of smooth ultrarelativistic limits, as well as an ambiguity in performing the limits when more than one parameter is involved in the limiting procedure simultaneously (for example, the parameters m and a in the non–charged Kerr case).

In any of the above cited approaches, the boosted Kerr solution is given by approximate expressions so that one cannot obtain an invariant description of the behavior of the singularity under the boost. We propose here a different method of description of the boosted Kerr solution based on the Debney, Kerr and Schild formalism [9] (DKS) and on the Kerr theorem [9, 10, 11, 12, 13].

The advantage of this approach relies in the possibility of obtaining exact, *explicit* expressions for the metric and its singularities in the case of an *arbitrary* boost, namely a boost with an arbitrary orientation with respect to the angular momentum. In fact, being represented in the Kerr-Schild form, the boosted Kerr metric can be linked with an auxiliary Minkowski space having a "rigid" coordinate system. This allows us to represent shock waves and singularities in asymptotically flat Cartesian coordinates.

14.2 The DKS-formalism and the Kerr theorem

In our notations we follow the work of Debney, Kerr and Schild [9]. All the BH-solutions in Einstein's gravity can be described by the simple Kerr-Schild metric

$$g_{\mu\nu} = \eta_{\mu\nu} + 2he^3_\mu e^3_\nu, \tag{14.1}$$

where $\eta_{\mu\nu}$ is the metric of an auxiliary Minkowski space M^4 with signature $(-+++)$ and Cartesian coordinates t, x, y, z. For a non-rotating BH the scalar function h has the form

$$h = m/r - e^2/2r^2,$$

where m and e are the mass and the charge of the BH. The vector $e^{3\mu} = (1, \vec{k})$ is a field of principal null directions which is spherically symmetric ($\vec{k} = (x, y, z)/r$ in the auxiliary Minkowski space with metric $\eta_{\mu\nu}$). In the case of rotating BH-solutions the metric is still of the Kerr-Schild form but its twisting structure is determined by a different null congruence e^3 and by a modification of the radial coordinate.

In null coordinates

$$2^{\frac{1}{2}}\zeta = x + iy, \quad 2^{\frac{1}{2}}\bar{\zeta} = x - iy, \quad 2^{\frac{1}{2}}u = z + t, \quad 2^{\frac{1}{2}}v = z - t, \tag{14.2}$$

the null vector e^3 can be expressed via a scalar function $Y(x)$ in the following way:

$$e^3 = du + \bar{Y}\, d\zeta + Y\, d\bar{\zeta} - Y\bar{Y}\, dv. \tag{14.3}$$

The determination of e^3 is possible since the principal null congruences of rotating BH solutions are geodesic and shear-free, and the *Kerr Theorem* [9, 10, 11, 12, 13] gives a rule to construct all such congruences: an arbitrary, geodesic shear-free null congruence in Minkowski space is defined by a function $Y(x)$ which is a solution of the equation

$$F = 0, \tag{14.4}$$

where $F(\lambda_1, \lambda_2, Y)$ is an arbitrary analytic function of the *projective twistor coordinates*

$$\lambda_1 = \zeta - Yv, \qquad \lambda_2 = u + Y\bar{\zeta}, \qquad Y\ . \tag{14.5}$$

A consequence of the Kerr Theorem is also the expression for the complex radial coordinate

$$\tilde{r} \equiv PZ^{-1} = dF/dY, \tag{14.6}$$

which characterizes "dilatation $+i$ twist " of the congruence. Correspondingly, the singular regions of the metrics are defined by the system of equations

$$F = 0, \qquad dF/dY = 0. \tag{14.7}$$

The BH-solutions belong to a class of metrics for which the singularities are contained in a bounded region of space. In this case the equation $F = 0$ can be solved in explicit form. Moreover, in this case there exists a complex representation of the function F in which the congruence is defined by an effective "source" moving in complex Minkowski space CM^4 along a complex world line. Such a complex representation was initially suggested by Lind and Newman [14, 15] in the Newman-Penrose formalism. The field e^3 can be used as one of the vectors of null tetrad e_1, e_2, e_3, e_4 satisfying

$$g_{ab} = e^{\mu}_{a} e_{b\mu} = \begin{pmatrix} 0 & 1 & 0 & 0 \\ 1 & 0 & 0 & 0 \\ 0 & 0 & 0 & 1 \\ 0 & 0 & 1 & 0 \end{pmatrix} = g^{ab}, \tag{14.8}$$

(e^3, e^4 are real null vectors, e^1, e^2 are complex conjugates). The null tetrad e^{μ}_{a} can be completed as follows:

$$e^1 = d\zeta - Y dv; \;\; e^2 = d\bar{\zeta} - \bar{Y} dv; \;\; e^4 = dv - he^3. \tag{14.9}$$

The field e^3, given by (14.3), can be used as one of the vectors of null tetrad e_1, e_2, e_3, e_4.

The inverse tetrad has the form

$$\partial_1 = \partial_\zeta - \bar{Y}\partial_u; \;\; \partial_2 = \partial_{\bar{\zeta}} - Y\partial_u; \;\; \partial_3 = \partial_u - h\partial_4; \;\; \partial_4 = \partial_v + Y\partial_\zeta + \bar{Y}\partial_{\bar{\zeta}} - Y\bar{Y}\partial_u. \tag{14.10}$$

The function h of the Kerr-Newman solution has the form

$$h = m(Z + \bar{Z})/P^3 - e^2/(Z\bar{Z}), \tag{14.11}$$

while the electromagnetic field can be obtained from the potential

$$A = -e(Z + \bar{Z})e^3/(2P^2). \tag{14.12}$$

14.3 Weak stationarity and congruences having singularities contained in a bounded region

The null congruence with tangent e^3 is stationary in M^4 if $\partial_t e^3 = 0$. However, in general one can also consider a "weak nonstationarity" corresponding to the fact that the stationarity can be restored by a Lorentz transformation. In this case there exists a real time-like vector field K such that

$$KY = K\bar{Y} = 0, \tag{14.13}$$

and consequently $Ke^3 = 0$. The congruences stationary in this weak sense and having singularities contained in a bounded region have been considered in [16, 17, 18]. In this case the function F must be at most quadratic in Y,

$$F \equiv a_0 + a_1 Y + a_2 Y^2 + (qY + c)\lambda_1 - (pY + \bar{q})\lambda_2, \tag{14.14}$$

where the coefficients c and p are real constants and $a_0, a_1, a_2, q, \bar{q}$, are complex constants. The solutions of the equation $F = 0$ and the equations for the singularities can be found in this case in explicit form. The solution $Y(x)$ of the equation $F = 0$ satisfies the weak stationarity condition (14.13) if

$$K = c\partial_u + \bar{q}\partial_\zeta + q\partial_{\bar{\zeta}} - p\partial_v. \tag{14.15}$$

In the papers [16, 17] another, equivalent form of F was suggested. This form allows to represent the parameters of the function F and the vector field K as retarded-time fields starting from an "effective" complex world line $x_0^\mu(\tau)$ depending from a complex time parameter τ. This form is the following

$$F \equiv (\lambda_1 - \lambda_1^0)K\lambda_2 - (\lambda_2 - \lambda_2^0)K\lambda_1. \tag{14.16}$$

Here the twistor components with zero indices

$$\lambda_1^0(\tau) = \zeta_0(\tau) - Yv_0(\tau), \qquad \lambda_2^0(\tau) = u_0(\tau) + Y\bar{\zeta}_0(\tau), \tag{14.17}$$

denote the values on the points of the complex world-line represented in null coordinates $\phi_0(\tau) = (\zeta_0, \bar{\zeta}_0, u_0, v_0)$ ($\bar{\zeta}_0$ and ζ_0 are not necessarily complex conjugates). The vector K can be expressed in the form

$$K(\tau) = \dot{x}_0^\mu(\tau)\partial_\mu, \tag{14.18}$$

where the dot denotes ∂_τ.

The Kerr congruences with weak nonstationarity are determined by straight analytic world lines with constant 3-velocity $\bar{v}$:

$$x_0^\mu(\tau) = x_0^\mu(0) + \xi^\mu \tau; \qquad \xi^\mu = (1, \bar{v}), \tag{14.19}$$

correspondingly, the vector $K = \xi^\mu \partial_\mu$ is a constant Killing vector of the solutions. The form (14.16) has the remarkable property that, in spite of an explicit dependence of the parameters of the function F in (14.15) on τ, this dependence is absent really, since in consequence of the relations

$$\lambda_1^0(x_0(\tau)) = \lambda_1^0(x_0(0)) + \tau K\lambda_1, \quad \lambda_2^0(x_0(\tau)) = \lambda_2^0(x_0(0)) + \tau K\lambda_2, \tag{14.20}$$

the terms proportional to τ cancel. Therefore the expressions (14.13) and (14.16) are equivalent.

The relation (14.16) is very convenient in order to obtain explicit representation of the congruences of the boosted Kerr solution. By writing the function F in the form

$$F = AY^2 + BY + C, \tag{14.21}$$

where

$$\begin{aligned} A &= (\bar{\zeta} - \bar{\zeta}_0)\dot{v}_0 - (v - v_0)\dot{\bar{\zeta}}_0; \\ B &= (u - u_0)\dot{v}_0 + (\zeta - \zeta_0)\dot{\bar{\zeta}}_0 - (\bar{\zeta} - \bar{\zeta}_0)\dot{\zeta}_0 - (v - v_0)\dot{u}_0; \\ C &= (\zeta - \zeta_0)\dot{u}_0 - (u - u_0)\dot{\zeta}_0, \end{aligned} \tag{14.22}$$

one can find two explicit solutions for the function $Y(x)$

$$Y_{1,2} = (-B \pm \Delta)/2A, \tag{14.23}$$

where $\Delta = (B^2 - 4AC)^{1/2}$.

On the other hand differentiating $F = 0$ and using (14.6) one finds

$$Y = -(B + PZ^{-1})/2A, \tag{14.24}$$

and consequently

$$PZ^{-1} = \mp\Delta. \tag{14.25}$$

One can find also

$$P = \dot{x}_o^\mu(\tau)e_\mu^3. \tag{14.26}$$

The field e^3 can be normalized by introducing $l^\mu = e^{3\mu}/P$ so that $\dot{x}_0^\mu l_\mu = 1$, that yields the following form of the Kerr-Newman metric

$$g_{\mu\nu} = \eta_{\mu\nu} + [m(\tilde{r}^{-1} + \bar{\tilde{r}}^{-1}) - e^2(\tilde{r}\bar{\tilde{r}})^{-1}]l_\mu l_\nu. \tag{14.27}$$

where the complex radial coordinate $\tilde{r} \equiv PZ^{-1}$ is given by the expression (14.25) or can be represented in the form

$$\tilde{r} = -dF/dY = -B - 2AY. \qquad (14.28)$$

It is convenient to represent $\tilde{r}$ as a sum of the real radial distance r and an angular coordinate $\tilde{r} = r + ia\cos\theta$. Then the equation (14.25) fixes the relation between the polar coordinates r, θ, ϕ and the null Cartesian coordinates through the expressions (14.22) for the coefficients A, B, C.

14.4 Behavior of singularities of the Kerr-Newman solution by the boost

In the "gauge" $x_0^0 = \tau$ the complex world line (14.19) can be represented as $x_0^\mu(\tau) = \{\tau, \vec{x}_0(0) + \vec{v}\tau\}$. The complex initial displacement can be decomposed as $\vec{x}_0(0) = \vec{c} + i\vec{d}$, where $\vec{c}$ and $\vec{d}$ are real 3-vectors with respect to the space O(3)-rotation. The real part $\vec{c}$ defines the initial shift of the solution, and the imaginary part $\vec{d}$ defines the size and the position of the singular ring as well as the corresponding angular momentum. It can be easily shown that in the rest frame, when $\vec{V} = 0$, $\vec{d} = \vec{d}_0$, the singular ring lies in the plane orthogonal to $\vec{d}$ and has a radius $a = |\vec{d}_0|$. The corresponding angular momentum is $\vec{J} = m\vec{d}_0$.

In the case of a boost orthogonal to the direction of $\vec{d}$, this vector is not altered by Lorentz contraction ($\vec{d} = \vec{d}_0$, $|\vec{d}| = a$), while if $\vec{d}$ and $\vec{V}$ are collinear we have

$$\vec{d}_0 = \vec{d}/\sqrt{1 - |\vec{V}|^2}\,. \qquad (14.29)$$

This shows that the parameter a coincides with its rest value a_0 if $\vec{d}$ and $\vec{V}$ are orthogonal, while

$$a_0 = a/\sqrt{1 - |\vec{V}|^2}\,, \qquad (14.30)$$

if $\vec{V}$ and $\vec{d}$ are collinear.

In order to calculate the parameters A, B, C it is convenient to express the complex world line in null coordinates

$$\begin{aligned} 2^{\frac{1}{2}}\zeta_0 &= x_0 + iy_0, \\ 2^{\frac{1}{2}}\bar{\zeta}_0 &= x_0 - iy_0, \\ 2^{\frac{1}{2}}u_0 &= z_0 + t_0, \\ 2^{\frac{1}{2}}v_0 &= z_0 - t_0\,. \end{aligned} \qquad (14.31)$$

The Killing vector of the solution will then be

$$\xi^\mu = 2^{-1/2}\{\dot{u}_0 - \dot{v}_0, \dot{\zeta}_0 + \dot{\bar{\zeta}}_0, -i(\dot{\zeta}_0 + \dot{\bar{\zeta}}_0), \dot{u}_0 + \dot{v}_0\},$$

while the functions P takes the form

$$P = e^3_\mu \dot{x}^\mu_0 = \dot{u}_0 + \bar{Y}\dot{\zeta}_0 + Y\dot{\bar{\zeta}}_0 - Y\bar{Y}\dot{v}_0. \quad (14.32)$$

The complex radial coordinate $\tilde{r} \equiv PZ^{-1}$ is given by the expression (14.16). As for the unboosted Kerr solution, one can represent $\tilde{r}$ as a "sum" of the real radial distance r and an angular coordinate. Then equation (14.16) can be used to fix the relation between the polar coordinates r, θ, ϕ and the null Cartesian coordinates (14.31) through the expressions (14.15) for the coefficients A, B, C. Due to the formula (14.28), the singular regions are defined by the zeros of the function $\tilde{r}$. In what follows, we present some examples of boosted Kerr solutions and then discuss the general features exhibited by them.

Example I.

Spinning particle moves with speed of the light in the positive direction of the z-axis, 3-vector $\vec{d} = (0, 0, a)$ is also directed along the z-axis. We have the following coordinates of complex world line
$x^0_0(\tau) \equiv \tau,\ z_0(\tau) = ia + \tau, \quad x_0(\tau) = y_0(\tau) = 0.$
In the null coordinates it gives

$$\sqrt{2}u_0 = z_0 + \tau = ia + 2\tau; \sqrt{2}v_0 = z_0 - \tau = ia, \zeta_0 = \bar{\zeta}_0 = 0, \quad (14.I.1)$$

that yields
$\dot{u}_0 = \sqrt{2}, \quad \dot{v}_0 = 0, \quad \dot{\zeta}_0 = \dot{\bar{\zeta}}_0 = 0$, and

$$\begin{aligned} u - u_0 &= (z - ia + t - 2\tau)/\sqrt{2}, \\ v - v_0 &= (z - ia - t)/\sqrt{2}, \\ \zeta - \zeta_0 &= \zeta, \\ \bar{\zeta} - \bar{\zeta}_0 &= \bar{\zeta}. \end{aligned} \quad (14.I.2)$$

The coefficients A, B, C calculated from (14.17) will be

$$A = 0; \quad B = t - z + ia; \quad C = x + iy. \quad (14.I.3)$$

As a result the function F acquires the form $F = x + iy - Y(z - ia - t)$, and solution of the equation $F = 0$ is

$$Y = (x + iy)/(z - ia - t). \quad (14.I.4)$$

The function

$$\tilde{r} = -dF/dY = z - ia - t. \quad (14.I.5)$$

One can see that there is no singularity if $a \neq 0$ since there is no real solutions to the system of equations $F = F_Y = 0$.

On the other hand, setting $a = 0$ we have got the case of spinless particle, and a moving singular plane which is placed at $z = t$. Therefore there is no smooth limit by $a \to 0$.

Example II.

The motion with speed of the light in the positive direction of the x- axis, orthogonal to the 3-vector $\vec{d}$ which defines the direction and the value of the angular momentum $\vec{J} = m(0,0,a)$, $a = |\vec{d}|$. We have the complex world line $x_0(\tau) = \tau$, $y_0(\tau) = 0$, $z_0(\tau) = ia$, $t_0 = \tau$. Correspondingly, the world line in the null coordinates is

$$\sqrt{2}u_0 = ia + \tau, \quad \sqrt{2}v_0 = ia - \tau, \quad \sqrt{2}\zeta_0 = \tau, \quad \sqrt{2}\bar{\zeta}_0 = \tau; \tag{14.II.1}$$

and the velocities are
$\sqrt{2}\dot{u}_0 = 1, \quad \sqrt{2}\dot{v}_0 = -1, \quad \sqrt{2}\dot{\zeta}_0 = 1, \quad \sqrt{2}\dot{\bar{\zeta}}_0 = 1.$
We have also

$$\begin{aligned} \sqrt{2}(u - u_0) &= z + t - ia - \tau, \\ \sqrt{2}(v - v_0) &= z - t - ia + \tau, \\ \sqrt{2}(\zeta - \zeta_0) &= x + iy - \tau, \\ \sqrt{2}(\bar{\zeta} - \bar{\zeta}_0) &= x - iy - \tau. \end{aligned} \tag{14.II.2}$$

The coefficients A, B, C take the form

$$A = (-x + iy - z + t + ia)/2; \qquad B = ia + iy - z; \qquad C = (x + iy - z - t + ia)/2. \tag{14.II.3}$$

The function $\tilde{r} \equiv PZ^{-1}$ takes the form

$$PZ^{-1} = -dF/dY = x - t. \tag{14.II.4}$$

There is therefore a moving singular plane placed at $x = t$. The function Y will be

$$Y = (dF/dY - B)/2A = (x + iy - z - t + ia)/(x - iy + z - t - ia). \tag{14.II.5}$$

Example III.

To understand better the absence of smooth limit in the example I we consider here an intermediate case with a boost with a speed $v = \alpha c$, $\alpha \leq 1$ in the positive direction of the z- axis, and then we will consider the limit $\alpha \to 1$. As in the example I, the 3-vector $\vec{d} = (0,0,a)$ is directed along the z-axis. We have the complex world line

$$x_0(\tau) = y_0(\tau) = 0, \quad z_0(\tau) = ia + \alpha\tau, \quad t_0 = \tau. \tag{14.III.1}$$

In null coordinates we have $\sqrt{2}\dot{u}_0 = 1 + \alpha$, $\sqrt{2}\dot{v}_0 = -1 + \alpha$, $\sqrt{2}\dot{\xi}_0 = 0$, $\sqrt{2}\dot{\bar{\xi}}_0 = 0$.
It yields

$$\begin{aligned} \sqrt{2}(u - u_0) &= z + t - ia - (\alpha + 1)\tau, \\ \sqrt{2}(v - v_0) &= z - t - ia + (1 - \alpha)\tau, \end{aligned}$$

$$\sqrt{2}(\xi - \xi_0) = x + iy,$$
$$\sqrt{2}(\bar{\xi} - \bar{\xi}_0) = x - iy. \qquad (14.III.2)$$

The coefficients A, B, C will be

$$A = -(x - iy)(1 - \alpha)/2; \quad B = ia - z + \alpha t; \quad C = (x + iy)(1 + \alpha)/2. \quad (14.III.3)$$

The expression for complex radial distance (3.9) yields

$$\tilde{r}^2 = B^2 - 4AC = (z - \alpha t)^2 + (1 - \alpha^2)(x^2 + y^2) - a^2 - 2ia(z - \alpha t). \quad (14.III.4)$$

Like to the standard Kerr solution one can represent the complex radial coordinate $\tilde{r}$ as a sum of the real radial distance r and an angular coordinate $\tilde{r} = r + ia\cos\theta$. Then, selecting the real and imagine parts of the expression (14.III.4) one obtains the following relations between the polar coordinates r, θ, ϕ and Cartesian coordinates of the auxiliary Minkowski space

$$x + iy = (r + ia)e^{i\phi}\sin\theta/\sqrt{1 - \alpha^2},$$
$$z - \alpha t = r\cos\theta. \qquad (14.III.4)$$

For the case $\alpha = 0$ this coincides with the coordinate relations of the standard Kerr solution. Setting $r = \cos\theta = 0$ we obtain the equation of singular ring

$$x^2 + y^2 = a^2/(1 - \alpha^2), \qquad z - \alpha t = 0. \qquad (14.III.5)$$

It may be seen that size of the ring grows by the increasing of α, and in the limiting case $\alpha = 1$ the singularity is placed on infinity. The cause of this effect is the above mentioned relation (14.30) $a_0 = a/\sqrt{1 - \alpha^2}$. The increasing of the singular ring by $\alpha = v/c \to 1$ is a seeming effect connected with using the parameter a instead of its rest value a_0. Being to expressed via the rest value singular region takes the form of moving ring of constant radius a_0, however, if we consider a light-like particle its rest mass is infinitely small and singularity is to be placed on infinity.

Example IV.

Intermediate case clarifying the limiting result of example II. The boost with a speed $v = \alpha c$, $\alpha \leq 1$ in the direction x, orthogonal to direction of angular momentum $\vec{d} = (0, 0, a)$.

We have the complex world line

$$x_0(\tau) = \alpha\tau, \quad y_0(\tau) = 0, \quad z_0(\tau) = ia, \quad t_0 = \tau; \qquad (14.IV.1)$$

In the null coordinates it yields

$$\sqrt{2}\dot{u}_0 = 1, \quad \sqrt{2}\dot{v}_0 = -1, \quad \sqrt{2}\dot{\xi}_0 = \alpha, \quad \sqrt{2}\dot{\bar{\xi}}_0 = \alpha.$$

$$\begin{aligned} \sqrt{2}(u - u_0) &= z + t - ia - \tau, \\ \sqrt{2}(v - v_0) &= z - t - ia + \tau, \\ \sqrt{2}(\xi - \xi_0) &= x + iy - \alpha\tau, \\ \sqrt{2}(\bar{\xi} - \bar{\xi}_0) &= x - iy - \alpha\tau. \end{aligned} \qquad (14.IV.2)$$

The coefficients A, B, C will be the following

$$\begin{aligned} A &= -(x - iy)/2 - (z - t - ia)\alpha/2 \\ B &= ia - z + iy\alpha \\ C &= (x + iy)/2 - \alpha(z + t - ia)/2. \end{aligned} \qquad (14.IV.3)$$

From the equation (4.4) we obtain

$$\tilde{r}^2 = B^2 - 4AC = (x - \alpha t)^2 + (1 - \alpha^2)[y^2 + (z - ia)^2]. \qquad (14.IV.4)$$

Representing $\tilde{r} = r + ia\sqrt{1 - \alpha^2}\cos\theta$ and selecting the real and imaginary parts of (14.IV.4) one obtains the following coordinate relations which generalize corresponding relations of the stationary Kerr solution

$$\begin{aligned} (x - \alpha t)/\sqrt{1 - \alpha^2} + iy &= (r/\sqrt{1 - \alpha^2} + ia)e^{i\phi}\sin\theta, \\ z &= r\cos\theta/\sqrt{1 - \alpha^2}. \end{aligned} \qquad (14.IV.5)$$

Singular region $r = \cos\theta = 0$ will be

$$z = 0; \qquad (x - \alpha t)^2 + (1 - \alpha^2)y^2 = a^2(1 - \alpha^2). \qquad (14.IV.6)$$

This is a moving ring placing in the $z = 0$ plane. It is oblate in x direction with the Lorentz factor $\sqrt{1 - \alpha^2}$.

Therefore, in the limit $\alpha = 1$ singular region will be moving segment of the line $z = 0, x = t; \quad -a \le y \le a$, which is parallel to the y-axis. One can mention that this limit is also non-smooth.

So, setting $\alpha = 1$ in the equations (14.IV.4) and (14.IV.5) we come to equation $x = t$, coordinate y is not restricted, however, the coordinate z is indefinite $\sim 0/0$. It will be equal to zero if we set first $r = \cos\theta = 0$ corresponding the singular region and then take the limit $\alpha \to 1$, however it is equal to infinity if the limit $\alpha = 1$ is taken first.

Example V.

We finally consider the general case in which the value of the velocity is arbitrary as well as its direction with respect to the angular momentum. Without lost of generality, we can consider the boost performed with a parameter α in the z-direction ($\alpha = v_z/c$), and a parameter β in the x-direction ($\beta = v_x/c$), while the angular momentum is defined by $\vec{a} = (0, 0, a)$. Denoting $w^2 = \alpha^2 + \beta^2$ the following general formula for the coordinate relations can be easily obtained:

$$(x - \beta t)\sqrt{1-\alpha^2} + iy\sqrt{1-w^2} = (r + ia\sqrt{1-\beta^2})e^{i\phi}\sin\theta,$$

$$z - \alpha t = -r\cos\theta/\sqrt{1-\beta^2}. \tag{14.V.1}$$

The singular region $r = 0, \cos\theta = 0$ is placed on the plane $z = \alpha t$ and is described by

$$(1-\alpha^2)(x-\beta t)^2 + (1-w^2)y^2 = a^2(1-\beta^2). \tag{14.V.2}$$

The singularity is a moving ring which is distorted in the x direction by a factor of $\sqrt{(1-\beta^2)/(1-\alpha^2)}$ and in the y direction by a factor of $\sqrt{(1-\beta^2)/(1-w^2)}$. The ultrarelativistic limit corresponds to $w = 1$ and the singular region is a couple of straight lines parallel to the y axis. Therefore, we can conclude that the non-smoothness and the non-commutativeness of the limiting procedure are a *general* feature of the boosted Kerr solutions. The main consequences of the considered examples are the non-smoothness and also the non-commutativeness of the limiting procedure as well as an unexpected behavior of the Kerr singular ring which is connected with the definition of the parameters of the solution after the boost and shows that such parameters must be "renormalized" by the boost.

It is interesting also to observe from the coordinate relations that the coordinate r is 'scaled' by the boost with respect to the asymptotically flat Cartesian coordinates x, y, z; as a consequence the region of small values of r (and big values of h) is stretched to big values of the x, y, z-coordinates, this is the origin of the shock waves in the ultrarelativistic limit since the region of big h can be very far from the "center" of the solution.

In astrophysical applications, the behavior of the horizon and of the ergosphere after the boost also has a physical interest. It can be easily shown that in the above suggested coordinates r and θ the horizon as well as the ergosphere are simply given by the known formulae for the Kerr case where the mass parameter m must be scaled by the Lorentz factor.

14.5 Boost of the Kerr-Sen solution

Recently, rotating BH-solutions received attention also in string theory. The Kerr-Sen BH-solution is a generalization of the Kerr solution to low energy string theory [19] (or to axion-dilaton gravity). We would like to show that the above formalism is also applicable to the Kerr-Sen solution. The metric of the Kerr-Sen BH may be written in the form [20]

$$ds^2_{dil} = 2e^{-2(\Phi-\Phi_0)}\tilde{e}^1\tilde{e}^2 + 2\tilde{e}^3\tilde{e}^4, \tag{14.33}$$

where

$$\tilde{e}^1 = (PZ)^{-1}dY, \qquad \tilde{e}^2 = (P\bar{Z})^{-1}d\bar{Y}, \tag{14.34}$$

$$\tilde{e}^3 = P^{-1}e^3, \tag{14.35}$$

$$\tilde{e}^4 = dr + iaP^{-2}(\bar{Y}dY - Yd\bar{Y}) + (H_{dil} - 1/2)e^3, \tag{14.36}$$

and

$$H_{dil} = Mr/\Sigma_{dil}; \quad \Sigma_{dil} = e^{-2(\Phi-\Phi_0)}(Z\bar{Z})^{-1}; \tag{14.37}$$

$$e^{-2(\Phi-\Phi_0)} = 1 + (Q^2/2M)(Z + \bar{Z}); \qquad Z^{-1} \equiv \tilde{r}. \tag{14.38}$$

The field of principal null directions is $\tilde{e}^3$. Following eq.(6.1) of [9] this tetrad is related to the Kerr-Schild tetrad (14.3),(14.9) as follows

$$\tilde{e}^1 = e^1 - P^{-1}P_{\bar{Y}}e^3, \qquad \tilde{e}^2 = e^2 - P^{-1}P_Y e^3, \tag{14.39}$$

$$\tilde{e}^3 = P^{-1}e^3, \tag{14.40}$$

$$\tilde{e}^4 = Pe^4_{dil} + P_Y e^1 + P_{\bar{Y}}e^2 - P_Y P_{\bar{Y}} P^{-1}e^3. \tag{14.41}$$

Therefore the Kerr-Sen metric (14.33) may be reexpressed in the form containing the Kerr-Schild tetrad e^a, the dilaton factor $e^{-2(\Phi-\Phi_0)}$, and a "deformed" function

$$H_{dil} = he^{2(\Phi-\Phi_0)}$$

instead of the function h.

It was shown in [20] that the Kerr-Sen metric is of type I contrary to the Kerr solution which is type D. However one of the principal null directions e^3 of the Kerr and the Kerr-Newman solutions survives in the Kerr-Sen solution and retains the property of being geodesic and shear free. It means that the Kerr theorem is applicable to this solution, since it has just the same principal null congruence and positions of caustics. Therefore the above analysis can be extended to the Kerr-Sen solution.

14.6 Conclusions

We discussed here a method allowing to describe in *explicit* form the metric and the behavior of the singular region of the Kerr solution under arbitrary boost and with arbitrary orientations of angular momentum. In particular, we have shown that the Kerr theorem automatically allows to obtain an asymptotically flat coordinate system and the equations describing the singularities in these coordinates. These results throw some light on the somewhat mysterious "standard" procedure commonly used to obtain shock waves metrics. In fact this procedure gives only approximate expressions before taking the ultrarelativistic limits [5, 6, 7, 8]. Of course, the tail of the shock wave (logarithmic term) cannot be obtained using the present method, because first of all we work always with vacuum, singular solutions of the Einstein field equations. To obtain the profile of the wave located on the delta–like singularity of the metric, one must first re–interpret the metric itself as being created by a singular distribution of matter on an extended manifold.

Our results are not very encouraging as far as the physical content of all such ultrarelativistic solutions is concerned. In fact, we obtained a quite general picture of non-smoothness and non-commutativeness of the limits $a \rightarrow 0$, $v \rightarrow 1$ and $r \rightarrow 0$. The absence of a smooth limit explains the well known fact that the limiting solution belongs to a completely different class with respect to the starting one: it is type N and not type D as the original Kerr solution, it is not asymptotically flat and has another group of symmetry.

Acknowledgements

The authors acknowledge Elisa Brinis Udeschini for many interesting discussions. One of us, A.B., is grateful to P. Aichelburg for useful discussions and hospitality at the Vienna University.

Bibliography

[1] P. C.Aichelburg and R. U. Sexl, Gen. Rel. Grav. **2**, 303 (1971).

[2] T. Dray and G. 't Hooft, Nucl. Phys. **B 253**,173 (1985).

[3] M. Fabbichesi, R. Pettorino, G. Veneziano and G. A. Vilkovisky, Nucl. Phys. **B 419**, 147 (1994).

[4] H. Balasin and H. Nachbagauer, Class. Quantum Grav. **12**, 707 (1995).

[5] H. Balasin and H. Nachbagauer, Class. Quantum Grav. **13**, 731 (1996).

[6] V. Ferrari and P. Pendenza, Gen. Rel. Grav. **22**, 1105 (1990).

[7] C. O. Luosto and N. Sanchez, Nucl. Phys. **B 355**, 231 (1991).

[8] C. O. Luosto and N. Sanchez, Nucl. Phys. **B 383**, 377 (1992).

[9] G. C. Debney, R. P. Kerr and A. Schild, J. Math. Phys. **10**, 1842 (1969).

[10] R. Penrose, J. Math. Phys. **8**, 345 (1967).

[11] D. Cox and E. J. Flaherty, Comm. Math. Phys. **47**, 75 (1976).

[12] D. Kramer, H. Stephani, E. Herlt, M. MacCallum, *Exact Solutions of Einstein's Field Equations,* (Cambridge University Press, Cambridge 1980).

[13] E. Brinis Udeschini and G. Magli, J. Math. Phys. **37**, 5695 (1996).

[14] E. T. Newman, J. Math. Phys. **14**, 102 (1973).

[15] R. W. Lind and E. T. Newman, J. Math. Phys. **15**, 1103 (1974).

[16] D. Ivanenko and A. Ya. Burinskii, Izvestiya Vuzov Fiz. N^0 7, 113 (1978) [Sov. Phys. J. (USA)].

[17] A. Burinsklii, R. P. Kerr, Z. Perjés, Report No. gr-qc/9501012 (unpublished).

[18] R. P.Kerr and W. B. Wilson, Gen. Rel. Grav. **10**, 273 (1979).

[19] A. Sen, Phys. Rev. Lett. **69**, 1006 (1992).

[20] A. Burinskii, Phys. Rev. D **52**, 5826 (1995).

TOPOLOGICAL BLACK HOLES — OUTSIDE LOOKING IN

R. B. Mann

Department of Physics
University of Waterloo, Waterloo, Ontario, Canada N2L 3G1

Abstract

I describe the general mathematical construction and physical picture of topological black holes, which are black holes whose event horizons are surfaces of non-trivial topology. The construction is carried out in an arbitrary number of dimensions, and includes all known special cases which have appeared before in the literature. I describe the basic features of massive charged topological black holes in $(3+1)$ dimensions, from both an exterior and interior point of view. To investigate their interiors, it is necessary to understand the radiative falloff behavior of a given massless field at late times in the background of a topological black hole. I describe the results of a numerical investigation of such behavior for a conformally coupled scalar field. Significant differences emerge between spherical and higher genus topologies.

15.1 Introduction

It has become clear over the last decade that black holes have an essential role to play in the development of a quantum theory of gravity. Each aspect of the physics of black holes – their formation due to gravitational collapse, their interior structure after formation, their thermodynamic properties, the singularities cloaked by their event horizons, and the endpoint of their evolution after thermally evaporating – present a set of interconnected puzzles whose ultimate resolution presumably entails the development of a full theory of quantum gravity. The most promising developments along this line have been in terms of the string-theoretic and Ashtekar proposals for quantum gravity, although a number of other candidate ideas exist, including non-commutative geometries, gauge-theoretic formulations, quantization of topologies, gravity as an induced phenomenon, and so on. These ideas are not mutually exclusive, and it is conceivable that the full theory of quantum gravity could contain some elements of each. Despite enormous effort, however, a full theory of quantum gravity remains elusive.

However the efforts expended towards achieving such a theory have yielded a large and still growing panoply of black objects, including black holes, black strings, and black branes. In general a spacetime containing a black object is one which has at least one event horizon: a surface which bounds a region of spacetime that is causally cut off from future timelike infinity. This surface is a null surface beyond which light (and hence all other forms of mass-energy) cannot escape. Each black object presents us with a theoretical laboratory in which we can test some of our basic ideas about the fundamental nature of gravity and its interactions with other forms of mass-energy.

One of the more interesting black objects to have entered the scene in recent times arose from the realization that certain identifications in anti-de Sitter spacetime in (2+1) dimensions yields a spacetime which can be interpreted as a black hole of definite mass M and angular momentum J in a spacetime with cosmological constant $\Lambda = -1/\ell^2$. The construction was originally given by Bañados, Teitelboim and Zanelli [1] and is generally referred to as the BTZ black hole. Although originally presented as a formal, mathematical construction [1, 2], it was quickly realized that these black holes can 'physically' form from the collapse of $(2+1)$-dimensional matter [3] and indeed have many features in common with their more conventional $(3+1)$-dimensional counterparts [4].

Since the exterior spacetime of a collapsing $(2+1)$ dimensional black hole is locally anti-de Sitter, its singularity structure differs considerably from that of its Schwarzschild anti-de Sitter counterpart in $(3+1)$ dimensions. The Kretschmann scalar $\mathcal{K} = R_{\mu\nu\alpha\beta}R^{\mu\nu\alpha\beta}$ has a power-law divergence at the origin in the latter case, whereas for the BTZ black hole it is everywhere finite. Instead the BTZ black hole has a causal singularity, which occurs where the identifications surfaces merge in $(2+1)$ dimensional anti-de Sitter spacetime. These properties suggest that a study of its interior structure along the lines of that carried out for mass inflation of $(3+1)$ dimensional Reissner-Nordstrom-Vaidya black holes [5] would be interesting. Such a study was carried out by Chan *et. al.* for the rotating BTZ-Vaidya black hole [6] and yielded several interesting results. Assuming the radiative falloff in the BTZ-Vaidya solution has a power-law tail, the mass function diverges at the Cauchy horizon in the interior of the black hole, but all the scalar curvature invariants remain finite, and tidal distortions remain bounded. However some higher-derivative curvature scalars diverge. This forms an interesting test [7] of the Konkowski-Helliwell conjecture, which predicts the stability of Cauchy horizons based upon the behavior of test fields, and in the case of instability also predicts the nature of the singularities produced [8]. More recently, it has been shown that the assumption of a power-law falloff for the BTZ-Vaidya black hole is not correct, and that the falloff rate is exponential [9]. This does not change any of the basic results mentioned above provided the falloff rate is sufficiently weak. However for $|\Lambda|J^2/M^2 > .64$, the exponential falloff rate is so large that mass inflation is cut off.

Intriguing as these effects are, their restriction to $(2+1)$ dimensions suggests at least that a considerable amount of caution be exercised in applying them to $(3+1)$ dimensions. One way of extending these results to $(3+1)$ dimensions is to embed the BTZ solution within a $(3+1)$-dimensional dilaton theory of gravity. The resultant solution may be interpreted as a spinning black string [10] and it exhibits mass inflation effects similar to its $(2+1)$ dimensional BTZ counterpart [11].

However within the last year it has been shown that higher dimensional generalizations of the $(2+1)$-dimensional BTZ construction exist. This yields an interesting new set of black holes with event horizons of non-trivial topologies. One can obtain these black holes by either constructing higher-dimensional counterparts of the identifications in anti-de Sitter spacetime [12, 13] or by systematic elimination of the conical singularities in the cosmological C-metric [14].

In this article I shall describe the basic properties of these topological black holes from both an exterior and an interior viewpoint. I shall first describe their basic construction in terms of identifications on anti-de Sitter spacetime. I shall then move on to outline some of their basic properties, and discuss how they can form from a distribution of collapsing dust. I shall then explore the behavior of the radiative falloff of waves outside these black hole, a necessary precursor to obtaining a more detailed understanding of their interior structure. Finally, I shall close with a discussion of the possible relevance of topological black holes to physics and of some recent research into this subject.

15.2 Describing Topological Black Holes

Topological black holes may be constructed by beginning with anti-de Sitter spacetime and then judiciously imposing certain identifications which have the effect of generating event horizons in the resultant quotient spacetime, thereby yielding a black hole. The treatment given here has not appeared before in the literature, although various aspects of certain details have [12, 13, 14].

Start with $(n+1)$ dimensional anti-de Sitter spacetime. This can be described by a hypersurface in an $(n+2)$-dimensional flat space of signature $(n-2,2)$

$$ds^2 = -dx_0^2 + dx_1^2 + \cdots + dx_{m-1}^2 + dx_m^2 + \cdots + dx_n^2 - dx_{n+1}^2 \qquad (15.1)$$

where the equation of constraint for the hypersurface is

$$-x_0^2 + x_1^2 + \cdots + x_{m-1}^2 + x_m^2 + \cdots + x_n^2 - x_{n+1}^2 = -\ell^2 \qquad (15.2)$$

where ℓ is constant.

Consider next the subspace described by the coordinates $(x_0, \ldots, x_m)$. This m-dimensional Minkowski subspace has the isometry group SO(m-1,1), with the associated Killing vectors

$$J_{\alpha\beta} = -J_{\beta\alpha} = (x_\alpha \partial_\beta \pm x_\beta \partial_\alpha) \qquad (15.3)$$

where $\alpha, \beta = 0, \ldots, m-1$, and the plus sign is chosen if one of α or β is 0. The idea now is to identify points in this subspace which are connected by some discrete subgroup Γ of the isometries which are generated by the $J_{\alpha\beta}$. This of course can be done in a variety of ways for the many different discrete subgroups. However most of these identifications will yield closed timelike curves (CTCs) because points in the x_0 direction will end up being identified.

Hence the identification subspaces must lie in a region where the Killing vectors are spacelike to avoid CTCs. This yields the constraint

$$x_0^2 - \left(x_1^2 + \cdots + x_{m-1}^2\right) = R^2 > 0 \tag{15.4}$$

on the Minkowskian subspace coordinates. Reparametrizing the coordinates so that

$$RX_\alpha = \frac{R_+}{\ell} x_\alpha \tag{15.5}$$

(where R_+ is an arbitrary constant) implies that the m-dimensional Minkowski subspace metric may be written as

$$\begin{aligned} ds_m^2 &= -dx_0^2 + dx_1^2 + \cdots + dx_{m-1}^2 \\ &= \frac{\ell^2}{R_+^2}\left[-dR^2 + R^2\left(-dX_0^2 + dX_1^2 + \cdots + dX_{m-1}^2\right)\right] \end{aligned} \tag{15.6}$$

where

$$-X_0^2 + X_1^2 + \cdots + X_{m-1}^2 = -\frac{R_+^2}{\ell^2} \quad . \tag{15.7}$$

This latter constraint means that we can write the quotient subspace metric as

$$ds_m^2 = -\frac{\ell^2}{R_+^2} dR^2 + R^2 d\sigma_{m-1}^2 \tag{15.8}$$

after identification. The metric

$$d\sigma_{m-1}^2 = \frac{\ell^2}{R_+^2}\left(-dX_0^2 + dX_1^2 + \cdots + dX_{m-1}^2\right) \tag{15.9}$$

with the constraint (15.7) describes, after identification, a compact $(m-1)$-dimensional space Σ of negative curvature, with $\Sigma = H^{m-1}/\Gamma$.

The coordinate R is timelike within the m-dimensional subspace, but within the full spacetime is actually timelike. The full metric may now be written as

$$ds^2 = -\frac{\ell^2}{R_+^2} dR^2 + R^2 d\sigma_{m-1}^2 + dx_m^2 + \cdots + dx_n^2 - dx_{n+1}^2 \tag{15.10}$$

where the constraint (15.2) becomes

$$x_m^2 + \cdots + x_n^2 - x_{n+1}^2 = \ell^2\left(\frac{R^2}{R_+^2} - 1\right) \quad . \tag{15.11}$$

From (15.11) is clear that the above identification procedure has broken up the original anti-de Sitter spacetime into various causal regions that are parametrized by the magnitude of R. When $R = R_+$ the coordinates $(x_m, \ldots, x_{n+1})$ describe a null hypersurface. This will later be seen to be the event horizon of the black hole. The region $R > R_+$ will correspond to the exterior of the black hole, and the region $R < R_+$ will correspond to the black hole interior, with $R = 0$ being a singular point where the identification procedure becomes degenerate, the condition (15.4) being violated.

To see that the metric (15.10) actually describes a black hole, it is helpful to reparametrize the remaining coordinates

$$\sqrt{R^2 - R_+^2}\, Y_I = \frac{R_+}{\ell} x_{n+1-I} \qquad I = m, \ldots, n+1 \tag{15.12}$$

in which case (15.11) becomes

$$Y_1^2 + \cdots + Y_{n-m+1}^2 - Y_0^2 = 1 \tag{15.13}$$

in the region $R > R_+$. The orthogonal subspace metric may be written as

$$\begin{aligned} ds_{n-m+1}^2 &= dx_m^2 + \cdots + dx_{n+1}^2 \\ &= \frac{\ell^2}{R_+^2}\left[\frac{R^2}{R^2 - R_+^2} dR^2\right. \\ &+ \left.(R^2 - R_+^2)\left(-dY_0^2 + dY_1^2 + \cdots + dY_{n-m+1}^2\right)\right] \end{aligned} \tag{15.14}$$

in terms of the Y-coordinates and R. The metric in the spatial Y coordinates may be rewritten in terms of $(n - m + 1)$ dimensional spherical coordinates

$$dY_1^2 + \cdots + dY_{n-m+1}^2 = d\rho^2 + \rho^2 d\Omega_{n-m}^2 \tag{15.15}$$

and the constraint (15.13) becomes

$$\rho^2 - Y_0^2 = 1 \quad . \tag{15.16}$$

This suggests the coordinate transformation

$$\rho = \cosh\left(\frac{R_+}{\ell^2} t\right) \qquad Y_0 = \sinh\left(\frac{R_+}{\ell^2} t\right) \tag{15.17}$$

which yields

$$\begin{aligned} ds_{n-m+1}^2 &= \frac{\ell^2}{R_+^2}\left[\frac{R^2}{R^2 - R_+^2} dR^2\right. \\ &+ \left.(R^2 - R_+^2)\left(-\frac{R_+^2}{\ell^4} dt^2 + \cosh^2\left(\frac{R_+}{\ell^2} t\right) d\Omega_{n-m}^2\right)\right] \end{aligned} \tag{15.18}$$

for the orthogonal subspace metric.

Inserting (15.18) into (15.10) yields the final result

$$ds^2 = \frac{\ell^4}{R_+^2} N(R) \left(-\frac{R_+^2}{\ell^4} dt^2 + \cosh^2 \left(\frac{R_+}{\ell^2} t \right) d\Omega_{n-m}^2 \right) + \frac{dR^2}{N(R)} + R^2 d\sigma_{m-1}^2 \quad (15.19)$$

which is the general metric for an $(n + 1)$-dimensional topological black hole. The function $N(R)$ is given by

$$N(R) = \frac{R^2 - R_+^2}{\ell^2} \quad (15.20)$$

and plays the role of a generalized lapse function in the metric (15.19).

The metric (15.19) is an exact solution to the Einstein equations with negative cosmological constant in $(n + 1)$ dimensions. The event horizon at $R = R_+$ is the direct product of null $(n - m)$-sphere with Σ, a generalization of the usual event horizon of a spherically symmetric black hole in $(3 + 1)$ dimensions which is a direct product of a pair of crossed null lines (*i.e.* a null 0-sphere) with a 2-sphere. Similarly, using (15.4), (15.10), and (15.11), one can see that there is a singularity at $R = 0$ which is the direct product of a hyperbolic surface with Σ, generalizing the usual singularity which is the direct product of a hyperbola with a 2-sphere.

It is easier to appreciate some of the features of this metric by considering a few special cases.

- $m = n = 2$ The metric is

$$ds^2 = -N(R)dt^2 + \frac{dR^2}{N(R)} + R^2 d\phi^2 \quad (15.21)$$

which is the static BTZ metric. The compact space Σ is a circle. The event horizon is the direct product of a pair of crossed null lines with this circle, and the singularity at $R = 0$ is the direct product of a hyperbola with this circle.

- $m = n = 3$ The metric is now

$$ds^2 = -N(R)dt^2 + \frac{dR^2}{N(R)} + R^2 \left(d\theta^2 + \sinh^2(\theta) d\phi^2 \right) \quad (15.22)$$

which are the $(3 + 1)$ dimensional topological black holes discussed in refs. [12, 14]. The space $\Sigma = \Sigma_g$ is a Riemann 2-surface of genus g. The event horizon is the direct product of a pair of crossed null lines with Σ_g and the singularity at $R = 0$ is the direct product of a hyperbola with Σ_g.

The compactness of Σ_g can be understood in the following way. The coordinates (θ, ϕ) are the coordinates of a hyperbolic space or pseudosphere. Geodesics on the pseudosphere are formed from intersections of the psuedosphere with planes through the origin, and are the analogs of great circles on a surface of constant

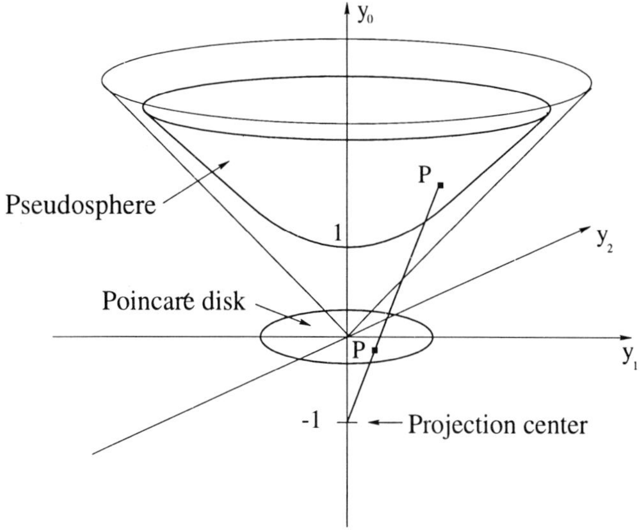

Figure 15.1: The pseudosphere is one half of the hyperboloid. Beneath the pseudosphere is the Poincaré disk, the center of which is the origin.

positive curvature (a sphere), which are intersections of the sphere and planes through the origin. A projection of the psuedosphere onto the (y_1, y_2) plane is known as a the Poincaré disk. On it, geodesics are segments of circles, orthogonal to the disk boundary at the edges. The pseudosphere, its associated Poincaré disk and the geodesics are shown in figure 15.1. Consider a polygon centered at the origin of the pseudosphere whose sides are geodesics. By identifying opposite sides of this polygon a compact surface on the pseudosphere can be obtained. The angles of the polygon must sum to 2π or more, and the number of sides must be a multiple of four in order to avoid conical singularities [15]. Since the geodesics on the pseudosphere meet at angles smaller than those for geodesics meeting on a flat plane, an octagon is the simplest solution, yielding a surface of genus 2. This construction is shown in figure 15.2. In general, a polygon of $4g$ sides yields a surface of genus g, where $g \geq 2$.

- $m = 2, n = 3$ The metric is

$$ds^2 = N(R)\left(-dt^2 + \frac{\ell^4}{R_+^2}\cosh^2\left(\frac{R_+}{\ell^2}t\right)d\theta^2\right) + \frac{dR^2}{N(R)} + R^2 d\phi^2 \qquad (15.23)$$

which describes the metric of an alternate generalization of the BTZ black hole discussed recently by Holst and Peldan [16]. The coordinate θ is periodically identified. The compact space Σ is again a circle with coordinate ϕ. The event

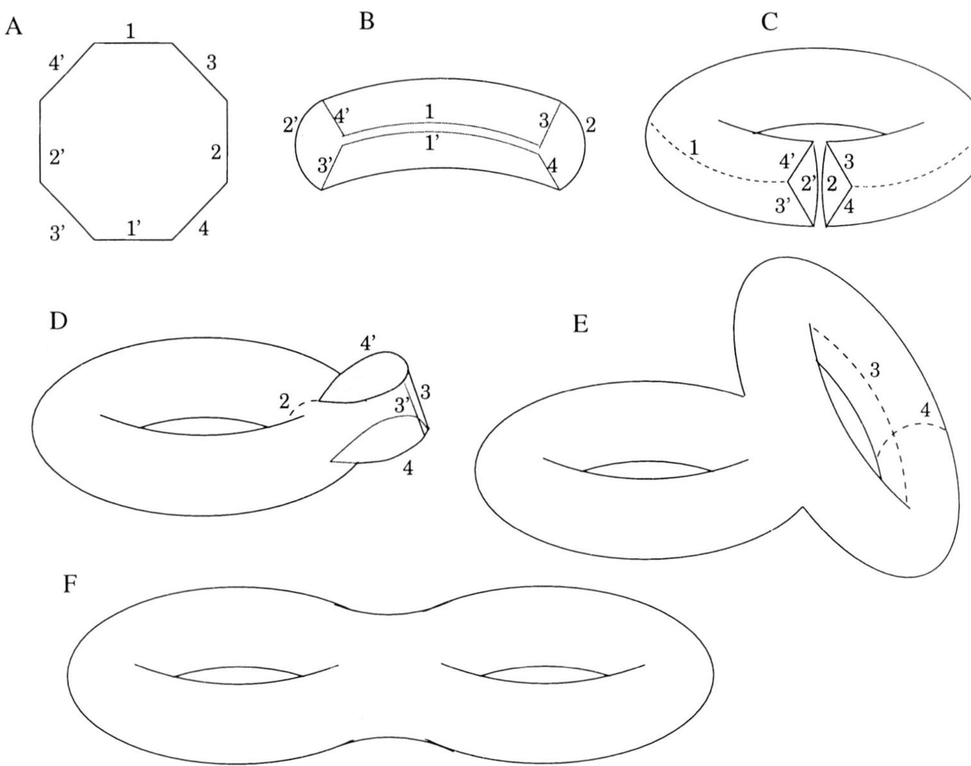

Figure 15.2: The identification of the octagon is shown. Opposite sides of the octagon are identified as in 2A, with sides drawn straight for clarity. Dashed lines indicate where sides have been sewn together. Sides 1 and $1'$ are first identified, folding the top and bottom of the octagon away from view (2B). The sides 2 and $2'$ are then brought together to form a torus with a diamond shaped hole, as in 2C. Next, sides 3 and $3'$ are stretched out and joined in 2D. The loop is lengthened along the direction of identification 3 and bent until 4 and $4'$ meet, forming a second torus. Finally the topology is deformed to the preferred shape, seen in 2F. Identification of a polygon of $4g$ sides will result in g attached tori or, equivalently, a g-holed pacifier.

horizon is the direct product of a null conoid (*i.e.* a null circle) with this circle, and the singularity is the direct product of a 2-dimensional hyperboloid with this circle.

Under the coordinate transformation

$$\sinh(\frac{R_+ t}{\ell^2}) = \cosh(T/\ell)\sqrt{X^2-1} \qquad \tan(\theta) = \sinh(T/\ell)\frac{\sqrt{X^2-1}}{X} \tag{15.24}$$

the (t,θ) section transforms as

$$-dt^2 + \frac{\ell^4}{R_+^2}\cosh^2\left(\frac{R_+}{\ell^2}t\right)d\theta^2 = \frac{\ell^4}{R_+^2}\left(-\frac{dX^2}{X^2-1} + (X^2-1)dT^2/\ell^2\right) \tag{15.25}$$

where the coordinate transformation is valid provided $|X| > 1$. It is clear from (15.25) that these coordinates can be extended to the $|X| \leq 1$ region, in which X becomes a spatial coordinate and T becomes a timelike coordinate. Writing $X = \cos\lambda$ yields

$$\frac{\ell^4}{R_+^2}\left(-\frac{dX^2}{X^2-1} + (X^2-1)dT^2/\ell^2\right) = \frac{\ell^4}{R_+^2}\left(d\lambda^2 - \sin^2\lambda\, dT^2/\ell^2\right) \tag{15.26}$$

which implies that

$$ds^2 = N(R)\frac{\ell^4}{R_+^2}\left(d\lambda^2 - \sin^2\lambda\, dT^2/\ell^2\right) + \frac{dR^2}{N(R)} + R^2 d\phi^2 \tag{15.27}$$

which is also a solution to the Einstein equations with negative cosmological constant.

The metric (15.27) is the constant curvature black hole (CCBH) discussed earlier [13]. Its global properties differ from those of the metric (15.23) in that the event horizon is the direct product of a pair of null conoids joined at their apexes with the circle Σ. Note that $\partial/\partial T$ is a Killing vector of the metric (15.27), but that $\partial/\partial t$ is not a Killing vector of the metric (15.23). In this sense the event horizon of (15.23) evolves with respect to the time coordinate t [16].

15.3 Properties of Topological Black Holes

I shall consider only (3 + 1) dimensional black holes throughout the sequel. For the solutions (15.22) it is possible to add mass and charge [14], yielding

$$\begin{aligned} ds^2 &= -\left(R^2/l^2 - 1 - 2m/R + q^2/R^2\right)dt^2 \\ &+ \frac{dR^2}{R^2/l^2 - 1 - 2m/R + q^2/R^2} \\ &+ R^2\left(d\theta^2 + \sinh^2(\theta)d\phi^2\right) \end{aligned} \tag{15.28}$$

which is an exact solution of the Einstein-Maxwell equations with negative cosmological constant $\Lambda = -3/\ell^2$. The metric for fixed (t, R) is assumed to be identified in the (θ, ϕ) coordinates as described previously, so that it describes a black hole whose event horizon is of genus $g \geq 2$. The electromagnetic field strength is

$$F = -\frac{q}{R^2} dt \wedge dR \tag{15.29}$$

in the electric case, and

$$F = q \sinh\theta d\theta \wedge d\phi \tag{15.30}$$

in the magnetic case. Toroidal black holes (genus $g = 1$) have the metric

$$ds^2 = -\left(R^2/l^2 - 2m/R + q^2/R^2\right) dt^2 + \frac{dR^2}{R^2/l^2 - 2m/R + q^2/R^2} + R^2\left(d\theta^2 + d\phi^2\right) \tag{15.31}$$

with

$$F = -\frac{q}{R^2} dt \wedge dR \qquad F = q d\theta \wedge d\phi \tag{15.32}$$

in the electric and magnetic cases respectively, where θ and ϕ are periodically identified. Note that the entire spacetime has topology $R^2 \times H_g^2$ for a genus $g \geq 1$ black hole.

Using the quasilocal formalism developed for anti de Sitter spacetimes [17] it is straightforward to show that

$$M = m(|g-1| + \delta_{g,1}) \tag{15.33}$$

is the conserved mass parameter associated with the Killing vector $\partial/\partial t$. for genus $g \geq 1$ [18, 19]. Similarly,

$$Q = q(|g-1| + \delta_{g,1}) \tag{15.34}$$

is the conserved charge Q associated with a genus g black hole.

The genus g metric function

$$V(r) \equiv R^2/l^2 - (1 - \delta_{g,1} - 2\delta_{g,0}) - 2m/R + q^2/R^2 \tag{15.35}$$

has at most two roots for positive R, corresponding to an inner and outer horizon, as with the usual $g = 0$ Reissner-Nordstrom anti de Sitter metric. For $g = 1$, provided

$$27\, l^2\, m^4 \geq 16\, q^6 \tag{15.36}$$

there are two horizons, with the extremal case saturating the inequality. For $g \geq 2$ event horizons exist provided

$$m^2 \leq \frac{l^2}{27} \frac{16 - 24\, e^2 b - 16 b\sqrt{1 - e^2 b}\, e^2 + 6 b^2\, e^4 + 16\sqrt{1 - e^2 b}}{e^6} \tag{15.37}$$

where $e = \frac{2\sqrt{2}q}{3m}$. There is no (obvious) upper limit on e, and event horizons can exist for arbitrarily large values of q relative to m.

The causal structure of these spacetimes is similar to that of Reissner-Nordstrom anti de Sitter spacetime with spherical topology. The three causal diagrams in the neutral, sub-extremal and extremal cases are shown in Fig. 15.3. A curious feature of these higher-genus black holes is that the mass parameter need not be positive in order for an event horizon to exist [20], even if the black hole is uncharged.

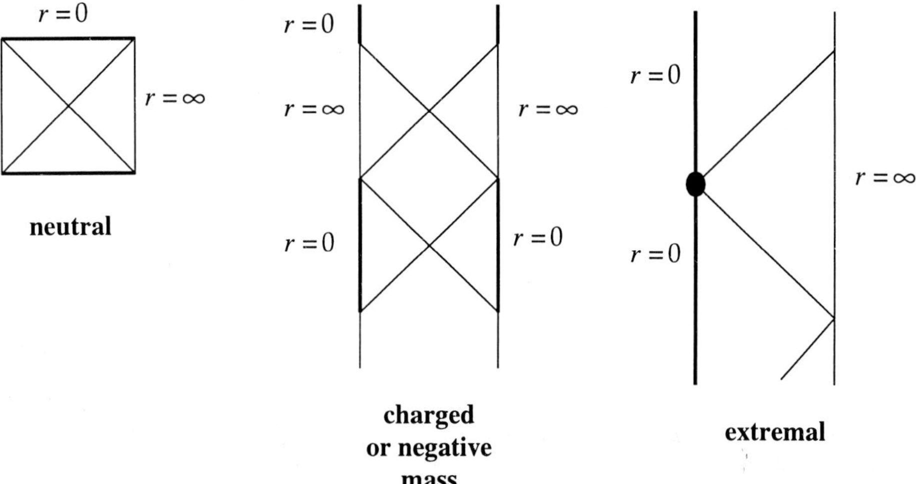

Figure 15.3: Causal diagrams for topological black holes in the neutral, subextremal and extremal cases. The subextremal case can be that of a charged black hole or of a negative mass black hole.

An extensive study of the thermodynamic properties of these black holes was recently carried out by Brill *et. al.* [19]. The entropy/area relation

$$S = \frac{1}{4}A \tag{15.38}$$

was found to hold, although this result has been disputed by Vanzo [21]. Their heat capacities $C_X = T\left(\frac{\partial S}{\partial T}\right)_X$ were also computed, and found to be of indefinite sign if $g \geq 2$.

It has also been demonstrated that topological black holes of arbitrary genus can form from the gravitational collapse of pressureless dust [22]. Although the procedure is somewhat analogous to that in the spherical case, there are a few interesting discrepancies. For non-trivial topologies, collapse from rest will not take place unless

$$\Lambda > 8\pi G\rho_0, \tag{15.39}$$

where ρ_0 is the initial density of the cloud. A more general but qualitatively similar condition holds if the dust cloud is given some initial velocity. The exterior spacetime formed is that given by (15.28) with $q = 0$. Another interesting feature is that collapse can take place even for dust which violates the weak energy condition, *i.e.* $\rho_0 < 0$. Provided the initial negative energy density is not too large in magnitude, a cloud of negative energy dust can collapse to a topological black hole of negative mass [20].

Finally, topological black holes may be pair produced in the presence of a domain wall of suitable topology [14, 18]. The mechanism is analogous to that discussed for other cosmological black holes [23, 24].

15.4 Inside Topological Black Holes?

Both charged and negative mass topological black holes have inner as well as outer event horizons. As can be seen from figure 15.3, the maximal extensions of such black hole spacetimes can be imagined as a collection of different asymptotically anti de Sitter universes connected by different charged (or negative mass) black holes. The presence of the inner horizon causes any radiation (either scalar, electromagnetic or gravitational in nature) entering this kind of black hole to be indefinitely blue-shifted at the inner (or Cauchy) horizon.

Does the phenomenon of mass inflation [25] take place for topological black holes? There are in general three necessary ingredients for mass inflation to take place: the existence of a Cauchy horizon, the presence of radiation which enters the black hole at some rate which decays at late times, and the possibility of cross-flow of this radiation with that emitted from a collapsing object that is forming the black hole. The first condition is clearly satisfied by either charged or negative mass black holes. Since both kinds of black holes can be formed from gravitational collapse, it is clear that the third condition can be satisfied provided the second one is.

An understanding of the nature of radiative falloff outside topological black holes is therefore crucial insofar as investigating their interior is concerned. For asymptotically flat spacetimes, the radiative falloff at late times obeys a power-law [26]. However this situation changes for other kinds of spacetimes. For example Mellor and Moss have shown that the radiation from perturbations in a de Sitter background exponentially decreases [27]. Strictly speaking this result has nothing to do with late time falloff since the global geometry extends beyond the cosmological horizon. However it does indicate that the radiative falloff behavior is sensitive to the asymptotic structure of the spacetime. Indeed, for conditions more general that simple asymptotic flatness, the tail can be something other than an inverse power-law [28].

A recent investigation [29] of radiative falloff in Schwarzschild anti de Sitter spacetime indicated that the late time falloff behavior is considerably more complicated than that in either asymptotically flat or asymptotically de Sitter spacetimes. This is a consequence of spatial infinity being timelike – although the proper distance to any point at large R in such a spacetime diverges as $R \to \infty$, a light ray can travel to arbitrarily large R in a finite amount of time.

A useful means for probing the falloff behavior of late times is to study the conformally coupled scalar wave equation. The scalar wave equation has qualitatively the same behavior and its more complicated electromagnetic and gravitational tensorial counterparts, and the conformal coupling ensures that the effective potential in which the wave moves most closely resembles the previously studied asymptotically flat case [29].

The (conformally coupled) scalar wave equation in $(3+1)$ dimensions is

$$\nabla^2\Psi = \xi R \Psi , \tag{15.40}$$

where ξ is an arbitrary constant. If $\xi = \frac{1}{6}$ this equation is conformally invariant. The form of the metric for topological black holes is

$$ds^2 = -V(r)\, dt^2 + \frac{dr^2}{V(r)} + r^2\, d\Omega_g{}^2 , \tag{15.41}$$

where $V(r)$ is the lapse function given by (15.35) and $d\Omega_g{}^2$ is the metric of a genus g Riemann surface. Assuming the separability condition

$$\Psi = \frac{1}{r}\,\psi(t,r)\,\mathcal{Y}(\Omega_g)_l \tag{15.42}$$

it is straightforward to show that equation (15.40) gives

$$-\partial_{tt}\psi(t,r) + V(r)\,\partial_r\left[N(r)\,\partial_r\psi(t,r)\right] - V(r)\,V_e(r)\,\psi(t,r) = 0 \tag{15.43}$$

where

$$V_e(r) \equiv \xi R + \frac{1}{r}\frac{d}{dr}V(r) + \frac{l(l+1)}{r^2} \tag{15.44}$$

defines the function $V_e(r)$. The functions $\mathcal{Y}_l$ are the genus g analogues of the spherical harmonics [15], and for $g \geq 2$, say, satisfy the equation

$$\hat{L}^2[\mathcal{Y}_l] \equiv \left[\frac{1}{\sinh\theta}\frac{\partial}{\partial\theta}\left(\sinh\theta\frac{\partial}{\partial\theta}\right) + \frac{1}{\sinh^2\theta}\frac{\partial^2}{\partial\phi^2}\right]\mathcal{Y}_l = -l\,(l+1)\,\mathcal{Y}_l \tag{15.45}$$

and are referred to as conical functions.

For simplicity the scalar wave will be assumed to be independent of (θ,ϕ). One can then rewrite the wave equation (15.43) as

$$\partial_{tt}\psi(t,r(x)) - \partial_{xx}\psi(t,r(x)) + \mathcal{V}(r(x))\,\psi(t,r(x)) = 0 \tag{15.46}$$

where $\mathcal{V}(x) \equiv V(r(x))V_e(R(x))$ with $l=0$ and

$$x \equiv \int \frac{dr}{N(r)} \tag{15.47}$$

is the so-called tortoise coordinate.

The function $\mathcal{V}(r)$ plays the role of a potential barrier which is induced from the background spacetime geometry. Equation (15.46) has the familiar form of a potential scattering problem, although $\mathcal{V}$ has a rather complicated dependence on the tortoise co-ordinate x. It can be integrated numerically in a straightforward fashion by using finite difference methods. The D'Alembert operator $\partial_{tt} - \partial_{xx}$ is first discretized as

$$\frac{\psi(t-\Delta t, x) - 2\,\psi(t,x) + \psi(t+\Delta t, x)}{\Delta t^2} \tag{15.48}$$
$$-\frac{\psi(t, x-\Delta x) - 2\,\psi(t,x) + \psi(t, x+\Delta x)}{\Delta x^2} \quad + \quad O(\Delta t^2) + O(\Delta x^2)$$

using Taylor's theorem. In order to formulate a well-posed Cauchy problem initial conditions must be chosen. For simplicity these can be taken to be

$$\psi(t=0,x) \;=\; 0 \qquad \text{and} \qquad \partial_t\psi(t=0,x) \;=\; u(x)\;. \tag{15.49}$$

Because the field ψ is initially zero, its subsequent evolution is solely the result of the initial impulse of the field $\partial_t\psi$. Discretizing the second condition in (15.49) yields

$$\frac{\psi(\Delta t, x) - \psi(\text{-}\,\Delta t, x)}{2\,\Delta t} \;=\; u(x) + O(\Delta t^2)\;, \tag{15.50}$$

where a Gaussian distribution with finite support for $u(x)$ is employed. Defining

$$\psi(m\,\Delta t, n\,\Delta x) \;\equiv\; \psi_{m,n}\;, \tag{15.51}$$
$$V(n\,\Delta x) \;\equiv\; V_n\;, \tag{15.52}$$
$$u(n\,\Delta x) \;\equiv\; u_n\;, \tag{15.53}$$

where the mesh size has to satisfy the condition $\Delta x > \Delta t$ so that the numerical rate of propagation of data is greater than its analytical counterpart.

If the black hole geometry is asymptotically flat, the tortoise coordinate x goes from negative infinity to positive infinity. When the background is asymptotically anti de Sitter the initial data no longer enjoy this privilege because the tortoise coordinate goes from minus infinity to zero only. In other words, rightward propagating data cannot travel in this direction forever. As with the semi-infinite vibrating string problem, boundary conditions at spatial infinity (i.e. $x = 0$) are needed in the asymptotically anti de Sitter background in order to formulate the problem appropriately. Here the boundary conditions employed will be either

$$\psi(t,x=0) \;=\; 0 \qquad \text{and} \qquad \partial_x\psi(t,x=0) \;=\; 1 \tag{15.54}$$

which are referred to as the Dirichlet boundary conditions, or

$$\psi(t,x=0) \;=\; 1 \qquad \text{and} \qquad \partial_x\psi(t,x=0) \;=\; 0\;. \tag{15.55}$$

which are the Neumann boundary conditions.

Both of these boundary conditions will be used in investigated the radiative falloff of conformally coupled waves in the topological black hole spacetimes (15.41). For simplicity, only neutral topological black hole spacetimes will be considered.

15.5 Radiative Falloff Outside Neutral Topological Black Holes

Before presenting the results of the numerical analysis of equation 15.46 it will be worthwhile recapitulating what takes place in Schwarzschild and Schwarzschild anti de Sitter spacetimes [29]. The wave equation (15.46) in all cases is solved numerically using the scheme discussed in the previous section, and $\xi = \frac{1}{6}$ throughout.

The general form of the potential $\mathcal{V}(x)$ for a background Schwarzschild spacetime in both linear and logarithmic coordinates (with the $l = 1$ spherical harmonic) is shown in Fig. 15.4. The bottom diagram in 15.4 is a logarithmic plot of the magnitude of the scalar wave at the point $RO = 20M$ as a function of time, where the compact initial Gaussian pulse is at the point $Ro = 10M$ (or $x = 12.76\,M$). The field vanishes until the pulse has propagated outwards to the point RO. It reaches its maximum value, after which it undergoes a "ringing" effect due to the presence of quasinormal modes [28]. This ringing dies out after $t \approx 200$, after which the field decays according to a smooth power law which from linear regression, is found to have a slope of - 5.026 in agreement with the analytic prediction of an inverse power-law falloff with exponent $2\,l + 3$ [26].

The situation for Schwarzschild anti de Sitter (SAdS) spacetime is somewhat different, and is shown in figure 15.5. As with the Schwarzschild black hole (figure 15.4), the potential function $V(x)$ attains a maximum not far away from the event horizon Rb, given by the largest positive solution to $V(R_b) = 0$. However unlike the Schwarzschild case, the tortoise coordinate x for the SAdS background is bounded above. The top part of figure 15.5 illustrates the shape of the potential function $\mathcal{V}(x)$ for a variety of values of $|\Lambda| = 3/\ell^2$ and the spherical harmonic parameter l. Eventually all the outgoing conformal waves that leave the black hole region will return towards it due to the boundary condition at $x = 0$. The returning wave will then reflect off of the potential barrier back toward spatial infinity for both the Dirichlet and Neumann boundary conditions. In this and all subsequent diagrams the quantities "delx" and "delt" on the graphs refer to the step sizes Δx and Δt of the variables x and t respectively, where $\Delta x > \Delta t$ holds as noted above.

For small $|\Lambda|$ and small times, the falloff behavior resembles the Schwarzschild case. Initially there is a ringing effect (due to the quasi-normal modes) followed by inverse power-decay behavior. However this inverse power-decay does not last very long because of the return of the outgoing wave from spatial infinity. As l gets larger the ringing increases in frequency as shown in figure 15.6. As $|\Lambda|$ increases, the transient

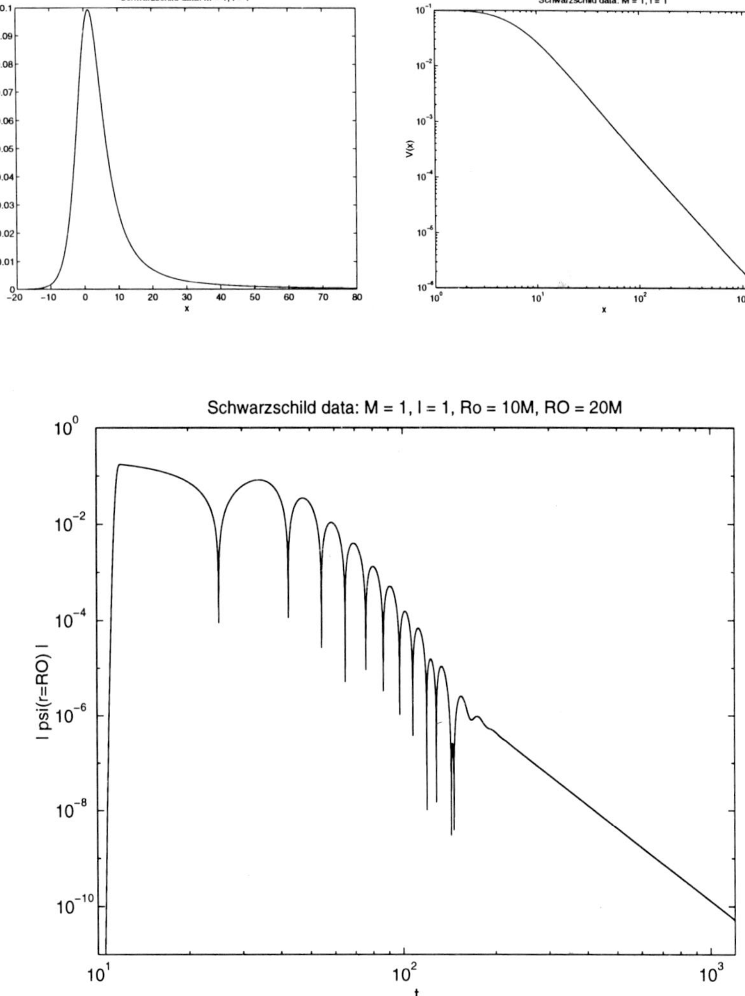

Figure 15.4: The potential $\mathcal{V}(x)$ in a Schwarzschild background and the resultant decay of a scalar wave. Prior to $t \approx 200$ the decay is accompanied by 'ringing' of the quasi-normal modes, after which the falloff rate is that of an inverse power-law.

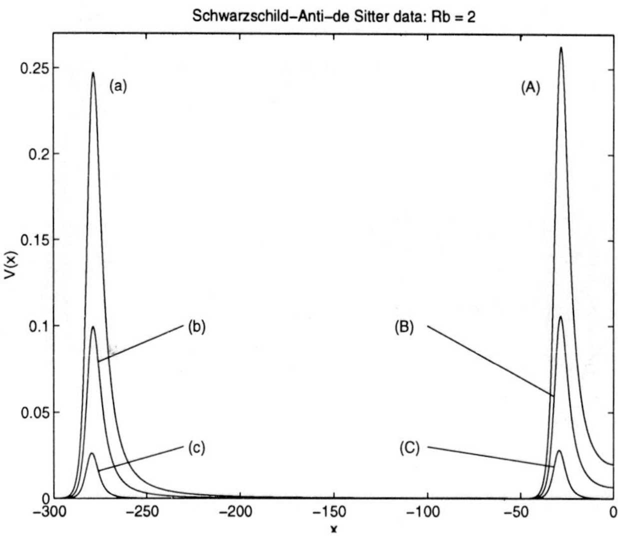

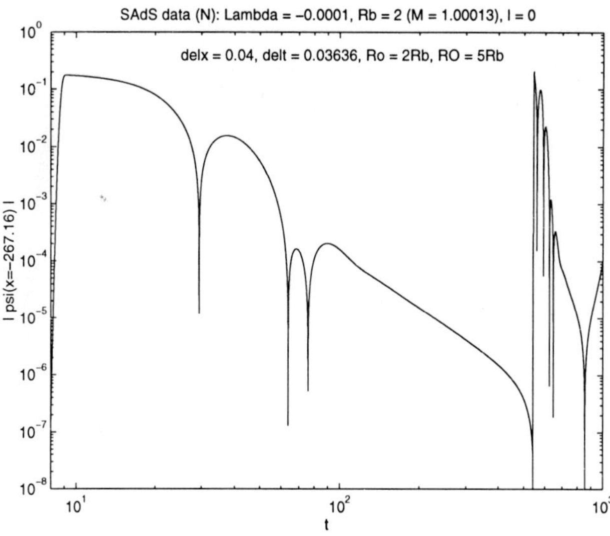

Figure 15.5: Potential functions $V(x)$ for the SAdS background. The six potentials are generated with the parameters (a) $\Lambda = -10^{-4}, l = 2$, (b) $\Lambda = -10^{-4}, l = 1$, (c) $\Lambda = -10^{-4}, l = 0$, (A) $\Lambda = -10^{-2}, l = 2$, (B) $\Lambda = -10^{-2}, l = 1$, (C) $\Lambda = -10^{-2}, l = 0$. The bottom diagram shows the $l = 0$ scalar wave falloff pattern using Neumann condition.

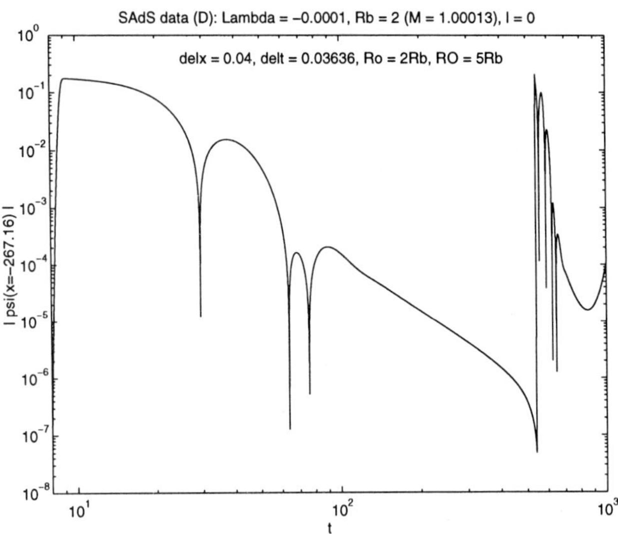

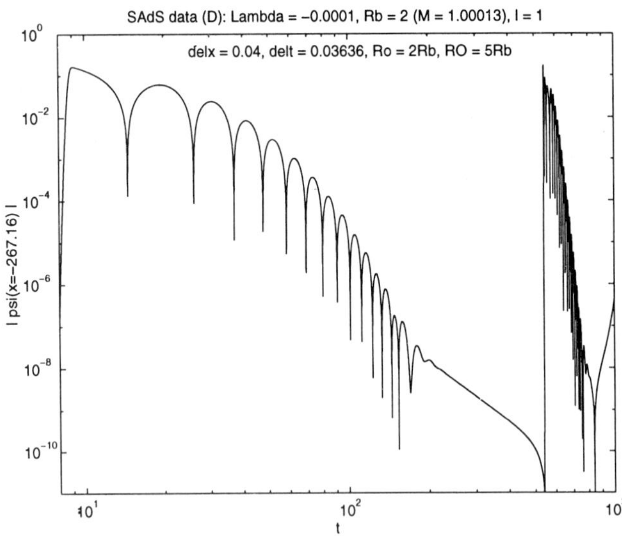

Figure 15.6: The $l = 0$ (top) and $l = 1$ (bottom) scalar wave falloff patterns using the Dirichlet condition.

power-law effect is eliminated entirely and new behavior emerges. For small l there can be a pure ringing effect within an exponentially decaying envelope, as shown in the top part of figure 15.7. As l increases, the ringing itself undergoes a complicated oscillation effect which mildly decays, as shown in the bottom part of figure 15.7. These results are all largely insensitive to the use of either Dirichlet or Neumann boundary conditions [29].

Turning next to the (neutral) topological black hole (TBH) case, figure 15.8 compares the potential $\mathcal{V}(x)$ for $l = 0$ for the SAdS (top) and TBH (bottom) cases. The event horizon is at $x = -\infty$, and spatial infinity is at $x = 0$. Each potential has a maximum at some finite x and decays exponentially toward the horizon. However the TBH potential does not have a point of inflection on the rightward side of the maximum, unlike the SAdS case. Note that even if M and Λ for a given topological black hole are equal to that of a given SAdS black hole, the location of the event horizons Rb will differ because $V[\text{TBH}] = V[\text{SAdS}]-2$. One is then faced with the problem of how to meaningfully compare the falloff behavior of a given TBH with that of a "similar" SAdS black hole for differing values of M and Λ. This can be overcome by choosing to compare the SAdS and TBH cases for equal values of the parameter $\gamma \equiv 3\sqrt{3}M/\ell = 3M\sqrt{|\Lambda|}$, which is a dimensionless measure of the black hole mass. This parameter governs the location of the event horizon $Rb = r_h\ell$, where

$$r_h^3 - (1 - \delta_{g,1} - 2\delta_{g,0})r_h = \frac{2}{3\sqrt{3}}\gamma \tag{15.56}$$

defines r_h, for a genus g black hole. For the SAdS case ($g = 0$), $0 < \gamma < \infty$, whereas for the TBH case ($g \geq 2$) $-1 < \gamma < \infty$ because subextremal negative mass black holes can also be included. Here only the range $0 < \gamma < \infty$ will be considered. For a given value of γ, I shall compare the SAdS, TBH ($g \geq 2$) and toroidal ($g = 1$) cases over a range of values of Λ.

Consider first the cases where the genus $g \geq 2$. These cases all have the same qualitative behavior, shown in figures 15.9 – 15.11 For large γ, the behavior of the TBH and SAdS cases is qualitatively similar over a wide range of values of $|\Lambda|$ with some quantitative differences as shown in figure 15.9. Each exhibits a ringing effect which falls off exponentially with time. The ringing frequency is slightly higher and the falloff rate slightly weaker in the SAdS case that in the TBH case. For each, as $|\Lambda|$ increases the ringing frequency and the falloff rate both increase.

These differences become more pronounced as γ decreases. From figure 15.10, in which $\gamma = 1$, it is clear that the approximate exponential falloff rate in the TBH case increases substantially relative to the SAdS case. In addition the ringing frequency in the SAdS case rises relative to that of the TBH case, although not as dramatically as it might first appear due to the difference in scale on the t-axis of the plots. These effects become slightly more pronounced as $|\Lambda|$ increases.

For very small γ substantial qualitative differences between the two cases emerge. The sequence of ringing, power-law falloff and resurgence of the wave noted above

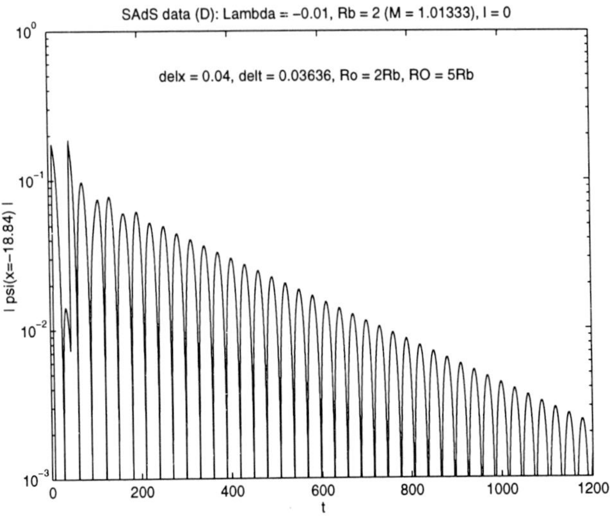

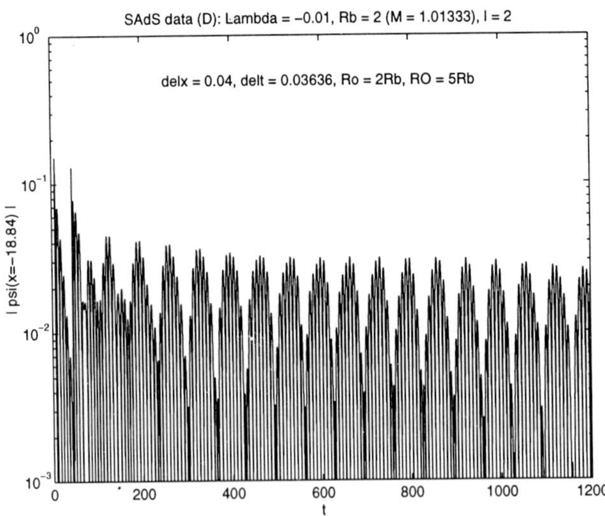

Figure 15.7: The $l = 0$ (top) and $l = 2$ (bottom) scalar wave falloff patterns using for large $|\Lambda|$.

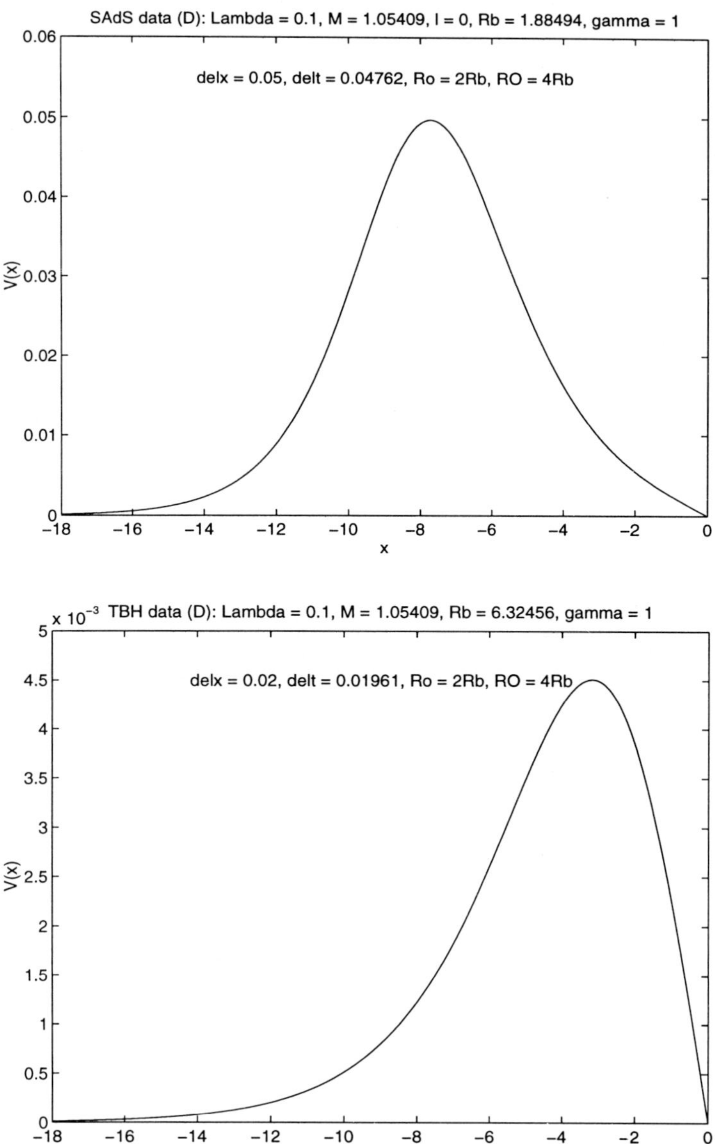

Figure 15.8: A comparison of the potential $\mathcal{V}(x)$ for the SAdS (top) and TBH (bottom) cases.

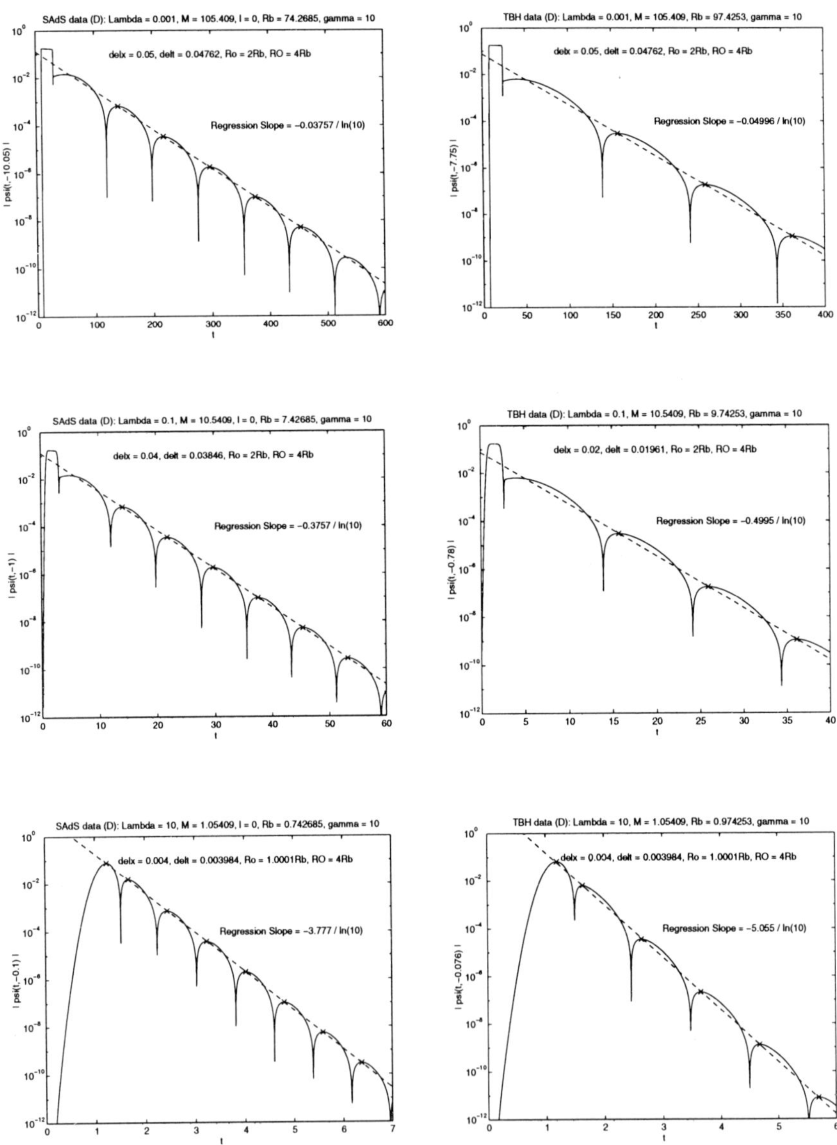

Figure 15.9: Falloff rates for the SAdS (left) and TBH (right) cases for $\gamma = 10$ for a range of values of $|\Lambda|$.

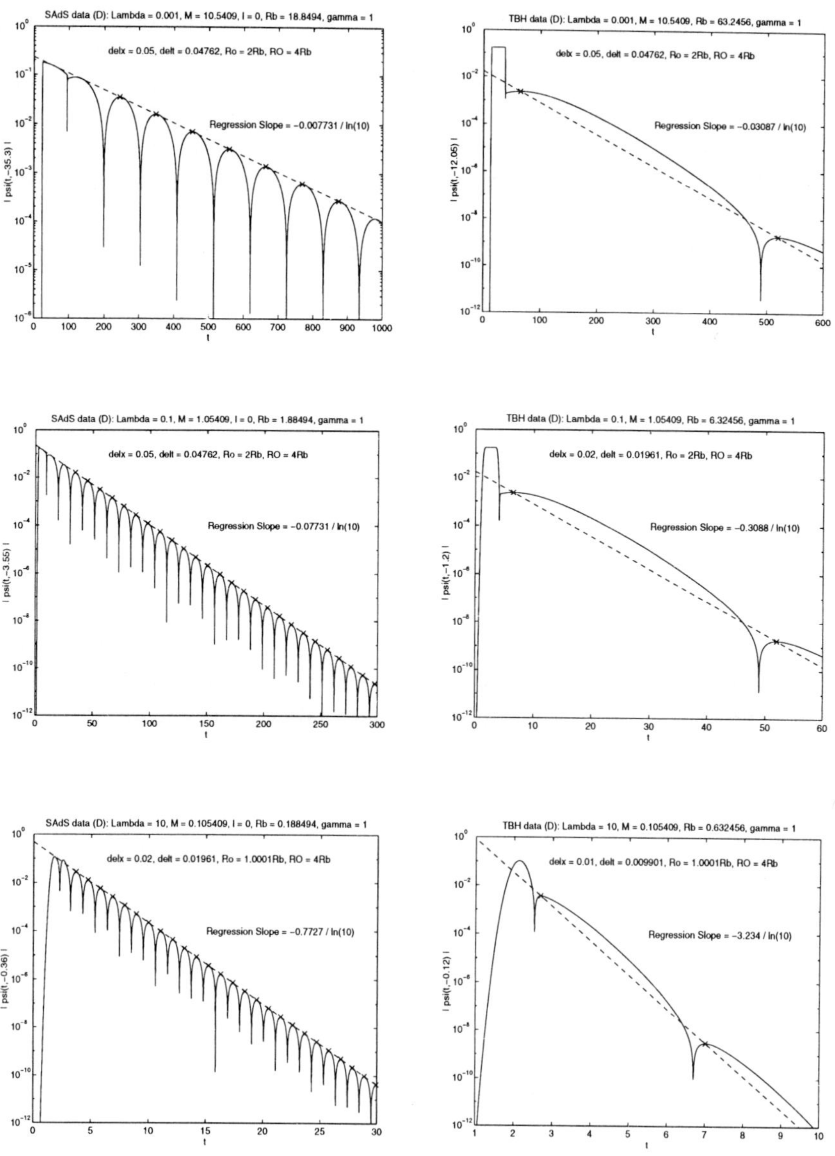

Figure 15.10: Falloff rates for the SAdS (left) and TBH (right) cases for $\gamma = 1$ for a range of values of $|\Lambda|$.

from figure 15.6 for the SAdS case returns. As $|\Lambda|$ increases the intermediate power-law falloff is obliterated at late times, replaced by a slowly decaying set of oscillatory ringing behavior. However in the TBH case the ringing virtually disappears, being replaced by a smooth exponential falloff. Due to limitations in computer memory and running time, it is not possible to tell if the ringing effect has actually vanished or if the frequency is just extremely low, although the upper right graph in figure 15.11 would suggest that it has actually vanished. The falloff rate grows with increasing $|\Lambda|$.

A comparison of the SAdS and toroidal (genus $g = 1$) cases is given in figures 15.12 – 15.14. For large γ the toroidal case is similar to that of the genus $g \geq 2$ cases, with a slightly less rapid falloff and more frequent ringing for the toroidal case. The difference is most pronounced for small γ, with the toroidal case still exhibiting some ringing for $\gamma = 0.1$. For clarity figure 15.15 shows a comparison at $\gamma = 1$ between the higher genus (TBH) and toroidal cases.

For all of the above results, the radiative falloff behavior for topological black holes is relatively insensitive to the use of Dirichlet or Neumann boundary conditions.

15.6 Summary

The radiative falloff behavior described in the previous section indicates that (charged or negative mass) topological black holes can indeed undergo mass inflation, provided the exponential falloff rate is sufficiently small. In this case the blueshift effect at the Cauchy horizon will overwhelm the exponential decay of the incoming radiation, triggering mass inflation. However it is quite conceivable that that the falloff rate could be so strong as to cut off the mass inflation process for some range of the parameter set $(\gamma, |\Lambda|)$.

Whether or not this can take place is presently under investigation. However a similar phenomenon has been observed in $(2+1)$ dimensions for the BTZ black hole. As noted above, for $|\Lambda|J^2/M^2 > .64$, the exponential falloff rate outside a BTZ black hole is so large that mass inflation is cut off. A detailed study of the nature of the transition at this point is presently being carried out.

Topological black holes present us with an interesting new set of possibilities to explore in our quest to understand the physics of black holes and the role they play in quantum gravity. Although astrophysical applications of topological black holes are not immediately apparent, they will necessarily play some role in any theory of quantum gravity which includes topology changing processes.

Acknowledgements

This work was supported by the Natural Sciences and Engineering Research Council of Canada. I would like to thank J.S.F. Chan, J. Creighton, N. Kaloper and S. Solo-

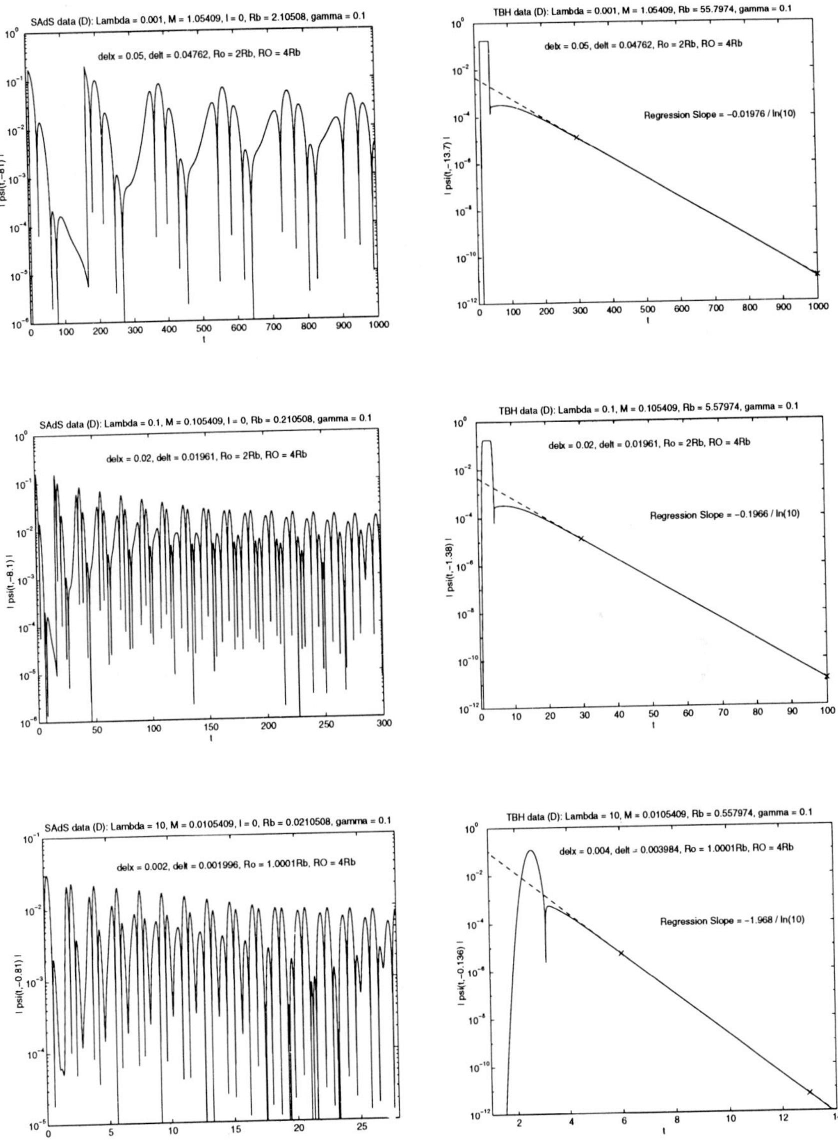

Figure 15.11: Falloff rates for the SAdS (left) and TBH (right) cases for $\gamma = 0.1$ for a range of values of $|\Lambda|$.

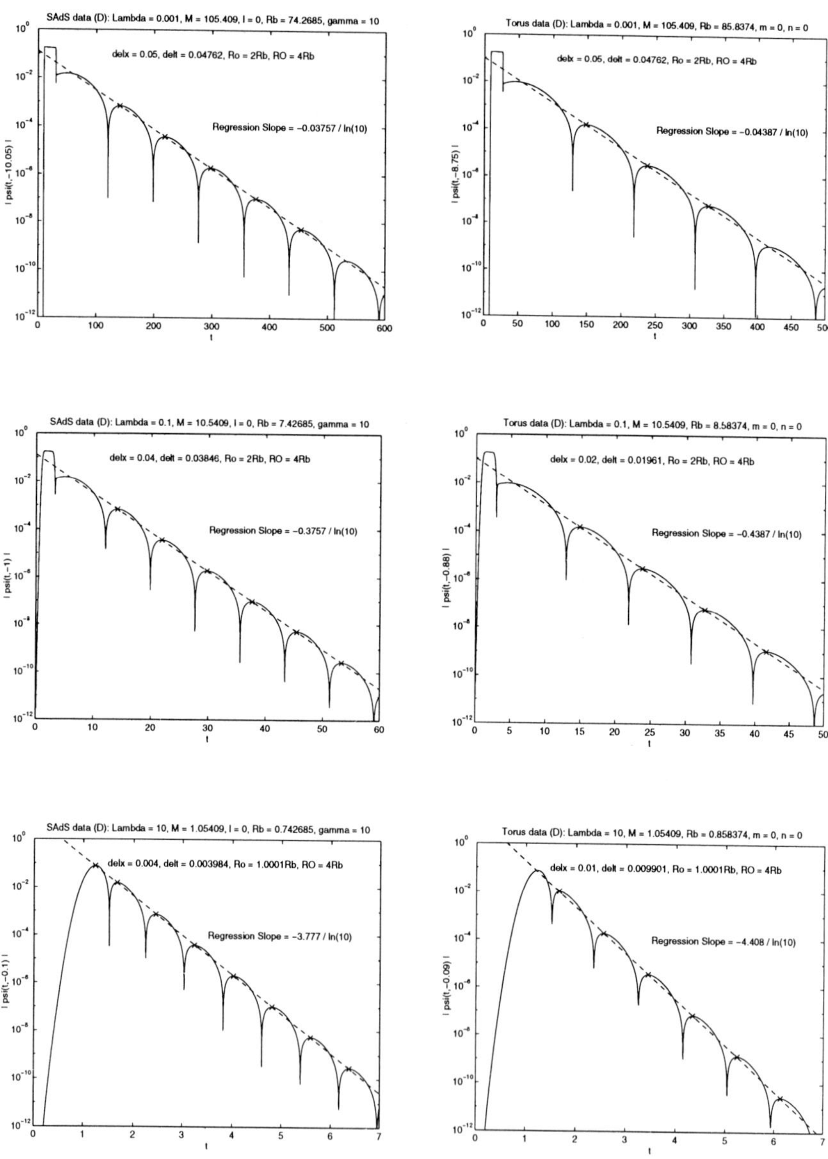

Figure 15.12: Falloff rates for the SAdS (left) and Toroidal (right) cases for $\gamma = 10$ for a range of values of $|\Lambda|$.

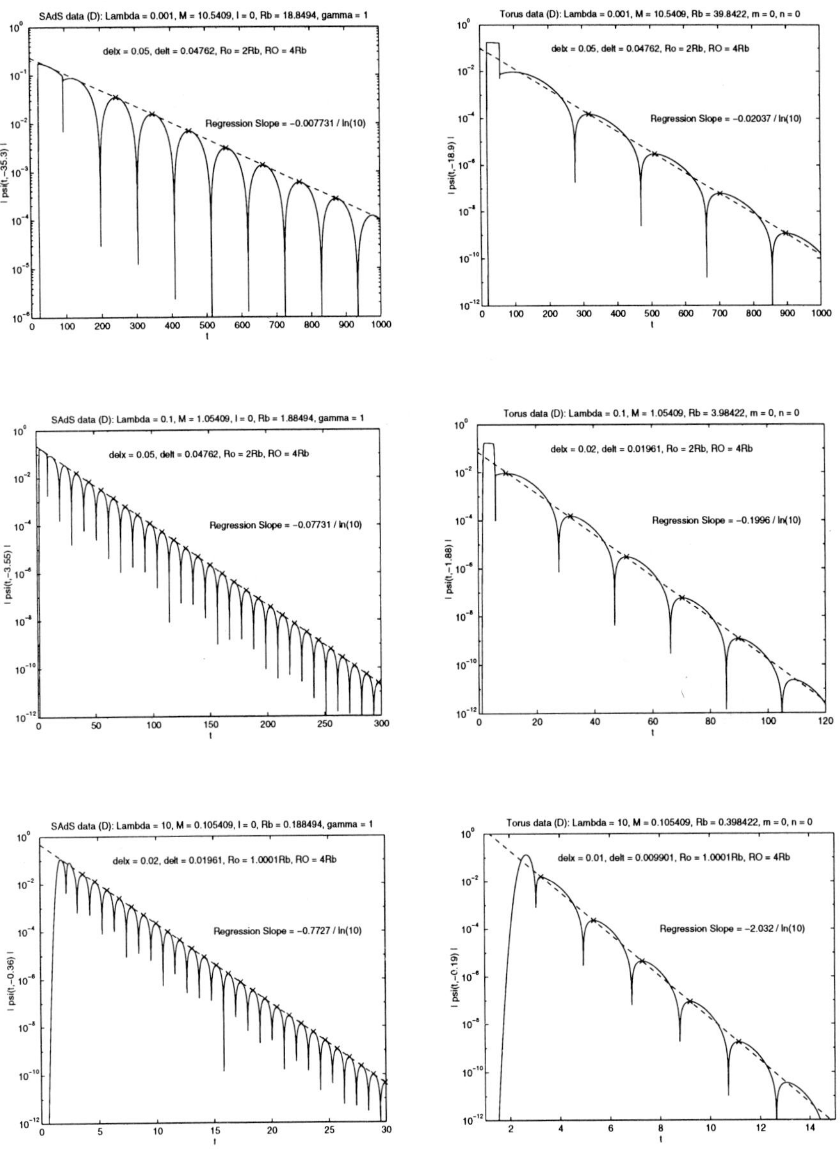

Figure 15.13: Falloff rates for the SAdS (left) and Toroidal (right) cases for $\gamma = 1$ for a range of values of $|\Lambda|$.

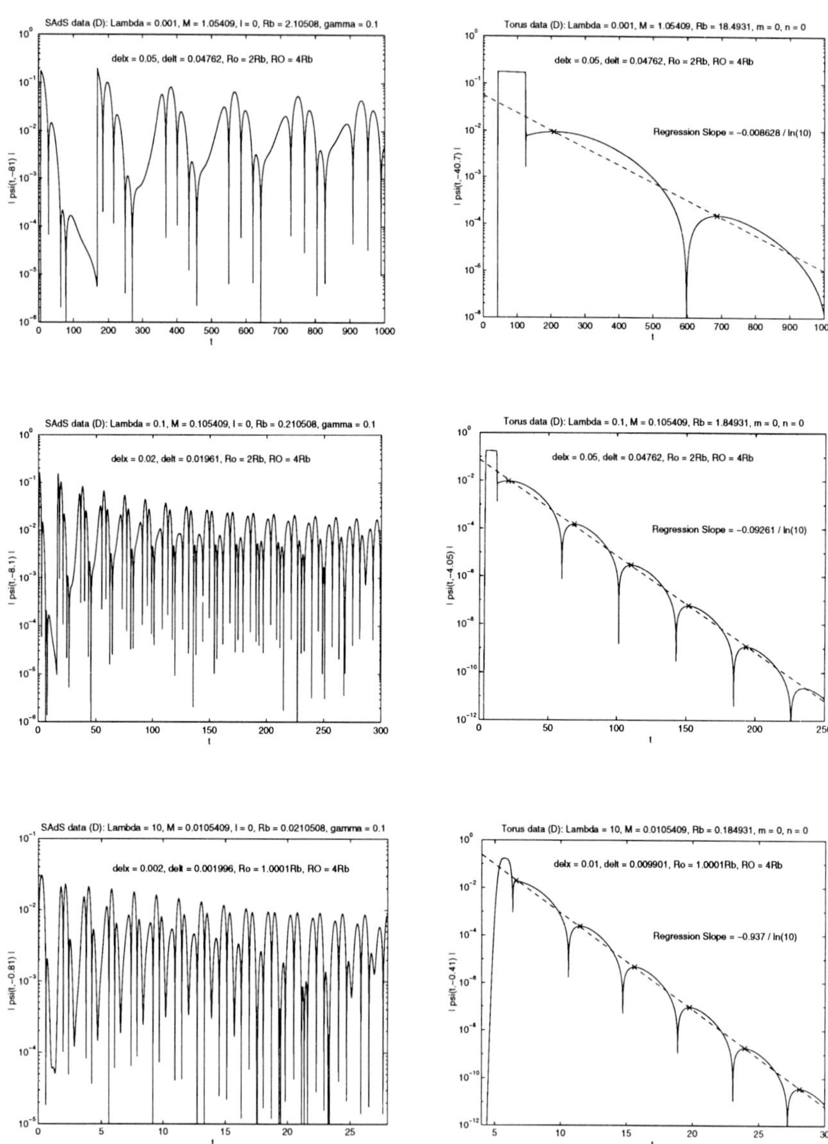

Figure 15.14: Falloff rates for the SAdS (left) and Toroidal (right) cases for $\gamma = 0.1$ for a range of values of $|\Lambda|$.

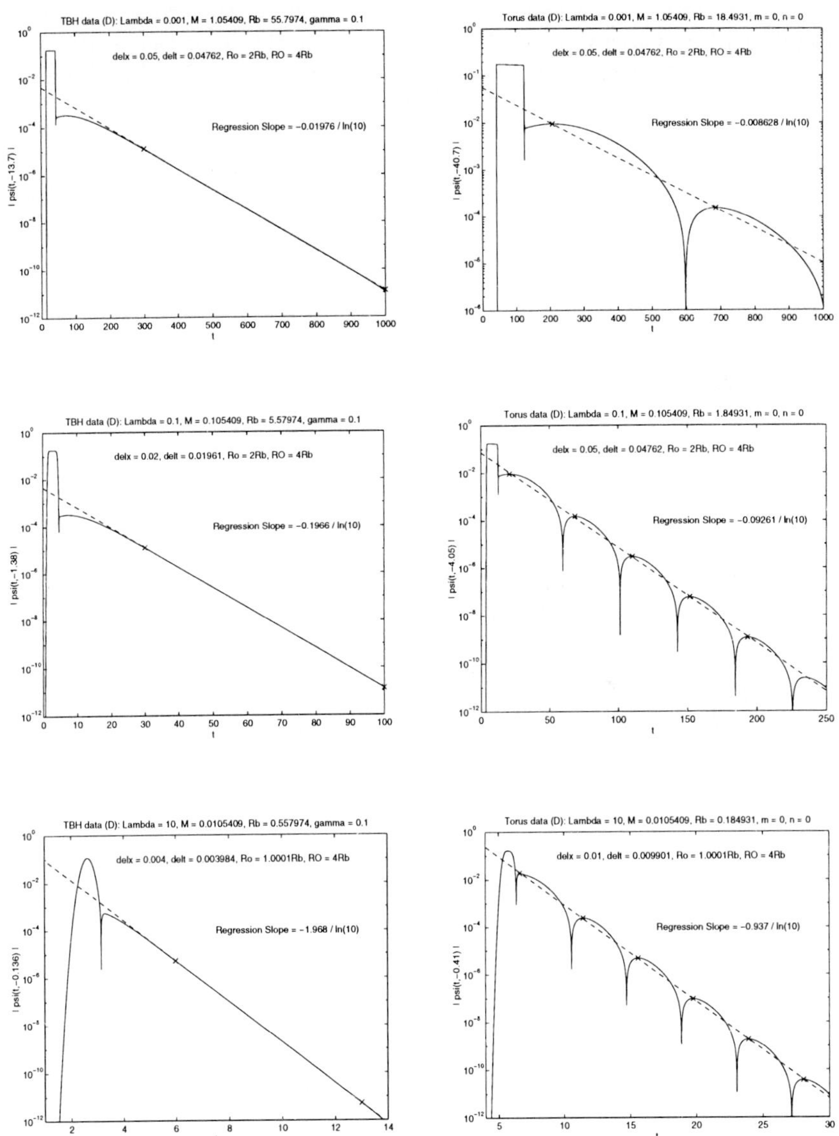

Figure 15.15: Falloff rates for the higher genus TBH (left) and Toroidal (right) cases for $\gamma = 0.1$ for a range of values of $|\Lambda|$.

dukhin for interesting discussions on various aspects of this work. I am particularly grateful to J.S.F. Chan who developed the code for obtaining the results described in section V. I would also like to thank L. Burko and A. Ori for their kind hospitality at the Technion Centre where these results were presented.

Bibliography

[1] M. Banados, C. Teitelboim and J. Zanelli, Phys. Rev. Lett. **69**, 1849 (1992).

[2] M. Banados, M. Henneaux, C. Teitelboim and J. Zanelli, Phys. Rev. D **48**, 1506 (1993).

[3] R.B. Mann and S.F. Ross, Phys. Rev. D **47**, 3319 (1993).

[4] For a review see S. Carlip, Class. Quant. Grav. **12**, 2853 (1995); for a more general review on the subject of lower dimensional black holes see R.B. Mann, *Proceedings of the Winnipeg Conference on Heat Kernels and Quantum Gravity*, ed. S. Christensen (Texas A& M Press, 1996) p. 303.

[5] E. Poisson and W. Israel, Phys. Lett. B **233**, 74 (1989); E. Poisson and W. Israel, Phys. Rev. D **41**, 1796 (1990).

[6] S.F.J. Chan, K.C.K. Chan and R.B. Mann, Phys. Rev. D **54**, 1535 (1996).

[7] B. Wang, R. Su and P.K.N. Yu Phys. Rev. D **54**, 7298 (1996).

[8] T. M. Helliwell and D. A. Konkowski, Phys. Rev. D **51**, 5517 (1995); Phys. Rev. D **54**, 7898 (1997).

[9] R.B. Mann and J.S.F. Chan, Phys. Rev. D **55**, 7546 (1997).

[10] N. Kaloper, Phys. Rev. D **48**, 4658 (1993).

[11] J.S.F. Chan and R.B. Mann, Phys. Rev. D **51**, 5428 (1995).

[12] S. Aminneborg, I Bengtsson, S. Holst and P. Peldan, Class. Quantum Grav. **13**, 2707 (1996); D. Brill, Helv. Phys. Acta **69**, 249 (1996).

[13] M. Bañados, Report No. gr-qc/9703040 (1997).

[14] R.B. Mann, Class. Quantum Grav. **14**, L109 (1997).

[15] N. L. Balasz and A. Voros, Phys. Rep. **143**, 109 (1986).

[16] S. Holst and P. Peldan, Report No. gr-qc/9705007 (1997).

[17] J.D. Brown, J. Creighton and R.B. Mann, Phys. Rev. D **50** 6394 (1994).

[18] R.B. Mann, WATPHYS-TH97/05, gr-qc/9705223.

[19] D. Brill, J. Louko and P. Peldan, Phys. Rev. D **56**, 3600 (1997).

[20] R.B. Mann, WATPHYS-TH97/03, gr-qc/9705007, to be published in Class. Quantum Grav.

[21] L. Vanzo, UTF-400, REPORT NO. gr-qc/9705004.

[22] W.L. Smith and R.B. Mann, Phys. Rev. D **56**, 4942 (1997).

[23] R.B. Mann and S.F. Ross, Phys. Rev. D **52**, 2254 (1995).

[24] R.R. Caldwell, A. Chamblin and G.W. Gibbons, Phys. Rev. D **53**, 7103 (1996).

[25] E. Poisson and W. Israel, Phys. Lett. **B 233**, 74 (1989); Phys. Rev. D **41**, 1796 (1990).

[26] R. H. Price, Phys. Rev. D **5**, 2419 (1972).

[27] F. Mellor and I. Moss, Phys. Rev. D **41**, 403 (1990).

[28] E. S. C. Ching, P. T. Leung, W. M. Suen and K. Young, Phys. Rev. D **52**, 2118 (1995).

[29] R.B. Mann and J.S.F. Chan, Phys. Rev. D **55**, 7546 (1997).

THE CAUCHY HORIZON IN HIGHER–DERIVATIVE GRAVITY THEORIES

A. Bonanno

Istituto di Astronomia, Universitá di Catania, Viale Andrea Doria 6, 95125 Catania, Italy
and
I.N.F.N. Sezione di Catania, Corso Italia 57, 95129 Catania, Italy

Abstract

A class of exact solutions of the field equations with Higher derivative terms is presented when the matter field is a pressureless null fluid plus a Maxwellian static electric component. It is found that the stable solutions are black holes in anti de Sitter background. The issue of the stability of the Cauchy horizon is discussed.

16.1 Introduction

Why should one wish to study the internal structure of a black hole? Aside from sheer curiosity there is the intellectual challenge of testing our present physical theories up to their extreme boundary of validity.

In fact the spacetime outside a black hole is indeed a rather uninteresting subject. Thanks to the no-hair theorem we know the late time structure of the spacetime after the star has radiated away all the asphericities [1]. Instead, the ultimate fate of the star that undergoes a gravitational collapse, is still an open issue.

The plausible, still unproven strong cosmic censorship conjecture states [2] that the singularity at $r = 0$ is spacelike. The spacetime near the singularity can probably be described by the general mixmaster type solution [3].

It is however puzzling that even an infinitesimally small amount of angular momentum generates the Kerr-Newman singularity which is instead timelike. In this case the situation is perhaps more dramatic due to the presence of a null hypersurface, the Cauchy horizon (CH), boundary of predictability of the evolution of the fields. Like in the Reissner-Nordström solution, the existence of this null hypersurface makes the existence of a timelike singularity in a region beyond it physically irrelevant.

As it was pointed out by Penrose, [2] the Cauchy horizon is also a surface of infinite blueshift. A freefalling observer crossing the Cauchy horizon will measure infalling radiation to have infinite energy density. When only ingoing radiation is present a weak nonscalar singularity forms, which is classified as a whimper singularity [4].

Whimper singularities are unstable to perturbations that can transform them into stronger, scalar, singularities.

This scenario can be examined more closely by assuming spherical symmetry. When an outflow crossing the Cauchy horizon is included in the analysis, the (Bondi) mass function is found to diverge exponentially[5] at late advanced times v

$$m \sim e^{\kappa v} \tag{16.1}$$

where κ is the surface gravity of the inner horizon located at $v = +\infty$. The only non-vanishing component of the Weyl tensor, the invariant Ψ_2, is proportional to the mass function and a scalar curvature singularity forms along the Cauchy horizon [6, 7]. This phenomenon, named "mass inflation", has become a topic of increasing interest and lot of effort is devoted to study this subject [8].

The Reissner-Nordström black hole shares the same causal structure as the Kerr-Newman black hole, so it is not surprising that the basic picture derived in spherical symmetry should be qualitatively the same for the more general black hole. Some investigations suggest that the general scenario derived for spherical symmetry does not change dramatically in a non-spherical black hole [9, 10, 11]. In fact one key result of the mass inflation picture is that the rate of the exponential divergence of the mass function is entirely determined by the surface gravity which is constant along the generators of the Cauchy horizon.

The relevant question is then to understand how much the classical description of the spacetime near the singularity changes when quantum effects are taken into account. In particular, would quantum effects enforce the strong censorship conjecture? A first analysis as been already performed in [12] where the semiclassical approach has been used to show that the classical picture remains valid up to few Planck lengths from the CH. The mass function at that "time" has already grown up to the mass of the observable Universe!

In fact, the classical description of the black hole interior is simplified by causality: behind the event horizon the coordinate r is timelike, so a descent into a black hole is a progression in time. The evolution down to any particular radius is only influenced by the initial data at larger radii. For this reason, a mean field, semiclassical description of quantum effects cannot drastically change the classical picture. This scenario can instead change in two-dimensional models, as shown by [13].

However, there is no reason why one should assume that a perturbative calculation can be applied when the curvature has reached planckian level. Only in a framework of a quantum gravity calculation one could hope to address this issue.

In recent years it has become clear that a very convenient way to quantize gravity is to describe it as a quantum effective field theory [14, 15, 16]. As a result one is lead to consider a more general action that is a functional of any geometrical invariant $\Re$ which can be constructed from the principle of general covariance, $\Re =$

$\{R, R_{\alpha\beta}R^{\alpha\beta}, C_{\alpha\beta\gamma\delta}C^{\alpha\beta\gamma\delta}, ...\}$, and of the matter field ϕ and its derivatives

$$S[\phi, g] = \int d^4x\sqrt{-g}\mathcal{L}(\Re,\ \phi,\ \nabla^2\Re, \nabla^2\phi,\ \nabla^{2j}\Re,\ \nabla^{2j}\phi, ...,) \tag{16.2}$$

This new theory is analogous to the Euler-Heisenberg lagrangian, low-energy effective theory of the QED. The important point is that both QED and the Euler-Heisenberg theory represent *the same theory* at different scale lengths. One would not include those new operators in the QED lagrangian since they are suppressed as inverse power of the ultraviolet cutoff in the infrared region showing their "irrelevant" character in this case. They can become instead significant if we move the original cut-off of the theory in the infrared region, where the theory must describe phenomena at much lower energy scales.

This situation is similar in gravity. Due to the smallness of the ratio between the Newton constant and the Fermi constant, modifications of Einstein's theory as in eq. (16.2), are experimentally indistinguishable from the standard theory at ordinary energy scales [17]. However at higher energy scales it can alter the physics content of the theory in a significant way. They might eventually destroy the unitarity of the S-matrix, signaling that a new physics sets in at some energy scale. The theory is perfectly well defined below that threshold, and it makes precise predictions. This happens also in the so-called renormalizable theories. The Standard Model, for instance, has to be regarded as an effective field theory which breaks down at the TeV scale, where the Higgs would enter in a strongly self-coupled phase.

Higher-derivative gravity theories arise also through the coupling of a quantized field with the classical background geometry [18] through the renormalization process. They are general functions of $\Re$ and are needed to cancel the ultraviolet divergences for any given order in perturbation theory.

However, the standard perturbative approach based on the scaling property of the Green's functions under a rescaling of the metric [19] handles only a finite number of operators, those which are important for the ultraviolet fixed point. In this way one cannot follow the evolution of the irrelevant, i.e. non-renormalizable operators. A coupling is irrelevant, marginal or relevant if it, respectively, it gets smaller, it does not run, it grows when the cut-off is lowered from the ultraviolet towards the infrared [20]. The irrelevant operators, even if they are not present in the bare lagrangian, mix their evolution with the renormalized trajectory of the relevant couplings. Although near the gaussian ultraviolet fixed point their running is suppressed, they can behave in a quite different manner in other scaling region and they can eventually drive the system in a different continuum limit.

It is therefore important to trace the evolution of all the coupling constants generated by the renormalization procedure in order to study the phase structure of the theory. In this way one recovers the predictability power of the theory by retaining only few relevant operators at a given fixed point. If for example a new scale other than the cut-off is present in the problem, the scaling laws change near that scale and the standard ultraviolet relevant interactions might not be sufficient to describe the physics at a crossover between ultraviolet and infrared.

The renormalization group approach used in Statistical Mechanics is the best tool to study problems where many scales are coupled together [21]. A powerful way of obtaining a differential form of the RG transformation has been formulated by Wegner and Houghton [22]. Starting from a bare action S_k at the cut-off k one first calculates $S_{k-\Delta k}$ in

$$e^{-S_{k-\Delta k}[\phi]} = \int \mathcal{D}[\psi] e^{-S_k[\phi+\psi]} \tag{16.3}$$

by using the loop expansion, where ψ and ϕ respectively have non-zero Fourier components only in the momentum shells $k - \Delta k < p \leq k$ and $p \leq k - \Delta k$. The differential RG transformation is obtained by taking the limit of an infinitesimal shell $\Delta k/k \to \delta k/k$. The higher loop contributions in Eq. (16.3) are suppressed as powers of $\Delta k/k$ in the limit $\Delta k \to 0$ for finite k and an exact, non-perturbative, one-loop RG equation is obtained

$$k\frac{dS_k[\phi]}{dk} = -\frac{1}{2}\langle \ln \frac{\delta^2 S[\phi]}{\delta\phi^2}\rangle + \langle \frac{\delta S[\phi]}{\delta\phi}\left(\frac{\delta^2 S[\phi]}{\delta\phi^2}\right)^{-1}\frac{\delta S[\phi]}{\delta\phi}\rangle \tag{16.4}$$

where the brackets indicates the sum over the Fourier components within the shell. This functional equation rules the evolution of all the interaction terms that are generated in the renormalization procedure. It has been applied in [23] to discuss the evolution of higher-derivative (HD) operators generated by the inflaton field. In particular it has been shown that even if they are not present at the cut-off - which has been chosen below the Planck scale! - they show up above the mass threshold and influence the renormalized flow.

A consistent quantum gravity calculation in the framework of the effective theories, must be performed by including the HD operators from the beginning [16]. Within a perturbative (weak field) calculation one can use the Einstein-Hilbert truncation [24], but in a strongly non perturbative situation one should consider more general lagrangians and, with the help of the renormalization group equations, look for the consistent field configurations that effectively dominate the path integral at that scale of energy.

The first step towards this goal is to understand the tree-level structure of the spacetime in the interior of a black hole, when higher-derivative operators are included from the beginning.

Those new operators can produce new istantons configuration in the path integral formulation of quantum gravity, with stronger weight than the solutions of the standard Einstein theory. If one wants to perform a saddle evaluation of the path integral the first step is to determine the stationary points of the (euclidean) action. Expansion around those saddle points are performed by writing

$$\gamma_{\mu\nu} = g_{\mu\nu} + h_{\mu\nu} \tag{16.5}$$

where, as usual, treating $g_{\mu\nu}$ as a background field and $h_{\mu\nu}$ as a quantum field one generates the usual perturbation expansion that can be expressed in terms of Feynman diagrams. The strong assumption that is usually believed, is that vacuum state for the renormalized system is still given by the $g_{\mu\nu}$ field configuration. Is this assumption still valid in the renormalized system when new higher dimensional operators are generated by the renormalization process? If not, we have chosen a wrong, unstable, saddle point. Strominger has shown [25] that for quadratic gravity, the Minkowsky solution is the only one having zero ADM energy. However that is not necessarily true for a more general non-linear gravity theory. As it has recently been pointed out by [26] where a a completely new vacuum structure has appeared already for $R+R^3$ lagrangians.

In this work we shall reduce the original HD theory to the standard second order form by means of the method outlined in [26, 27]. We shall obtain an equivalent (locally isomorphic) second order theory with an additional, non-geometric degree of freedom represented by a non-ghostlike scalar field. We call "vacuum" a solution which is stable with respect to the elementary excitation of this new field. We show that for a general non-linear Lagrangian the theory has non-trivial vacuum solutions representing BH embedded in Anti de Sitter spacetime.

We shall also comment on the stability of the Cauchy horizon in those solutions.

16.2 Basic equations

Let us consider the following action

$$S = \int d^4x\sqrt{-g}(l_g(R) + l_m(\mathcal{A}, g)) \tag{16.6}$$

where $l_g(R)$ is a non-linear function of the scalar curvature R which we shall suppose to be an analytic function of R

$$l_g(R) = \sum_n^{\infty} \frac{\epsilon_n}{n!} R^n \tag{16.7}$$

ϵ_n are the coupling constants of the generic power of R and $l_m(\mathcal{A})$ is the lagrangian density of the matter field $\mathcal{A}$. We use the signature (-+++) and we set $c = \hbar = 8\pi G$. With these units $\epsilon_0 = -\Lambda$ corresponds to the vacuum energy, being Λ the cosmological constant and $\epsilon_1 = 1/2$. The gravitational field equations for such a system are the following fourth order differential equations

$$\begin{aligned} \frac{\delta}{\delta g_{\mu\nu}}[\sqrt{-g}l_g(R)] &= l'_g(R)R_{\mu\nu} - \frac{1}{2}l_g(R)g_{\mu\nu} - \nabla_\mu\nabla_\nu l'_g(R) + g_{\mu\nu}\Box l'_g(R) \\ &= T_{\mu\nu}(\mathcal{A}, g) \end{aligned} \tag{16.8}$$

where $T_{\mu\nu}$ is the stress-energy tensor of the matter field which obeys $\nabla_\mu T^{\mu\nu} = 0$.

In order to analyze the above theory it is convenient [26] to introduce an auxiliary field ψ so that the action now reads

$$S = \int d^4x\sqrt{-g}\Big(l'_g(R)(R-\psi) - l_g(\psi) + l_m(A,g)\Big) \tag{16.9}$$

with the equation of motion

$$l''_g(\psi)(R-\psi) = 0. \tag{16.10}$$

The new action, written with the help of the auxiliary field, is (at the classical level) equivalent to the original theory provided that we are not at the critical points $l''_g(R) = 0$, but it is not of the second order form. In order to reduce the action (16.10) in a canonical second order form, one introduces a pair of new variables $(\bar{g}_{\mu\nu}, \omega)$ related to $g_{\mu\nu}$ by

$$e^{\omega} = l'_g(R); \qquad \bar{g}_{\mu\nu} = e^{\omega} g_{\mu\nu} \tag{16.11}$$

where $l'_g > 0$ is assumed and $\Psi(\omega)$ becomes a solution (not necessarily unique), of

$$l'_g(\Psi(\omega)) - e^{\omega} = 0. \tag{16.12}$$

This transformation has already been used in the literature, and it has been generalized to the case of Lagrangians depending on $\nabla^{2k}R$ (see [27] for a general review on the subject). Under the above conformal transformation the Ricci scalar density transforms like

$$\sqrt{-g}R = \sqrt{-\bar{g}}e^{-\omega}\Big(\bar{R} - \frac{3}{2}(\bar{\nabla}\omega)^2 - 3\bar{\nabla}^2\omega\Big). \tag{16.13}$$

By dropping a total derivative term, the action has been reduced to the canonical form

$$S = \int d^4x\sqrt{-\bar{g}}\Big(\bar{R} - \frac{3}{2}\bar{g}^{\mu\nu}\partial_\mu\omega\partial_\nu\omega - 2V(\omega) + e^{-2\omega}l_m(\mathcal{A}, e^{-\omega}\bar{g})\Big) \tag{16.14}$$

where the potential term

$$V(\omega) = \frac{1}{2}e^{-2\omega}\Big(e^{\omega}\psi(\omega) - l_g(\psi(\omega))\Big). \tag{16.15}$$

depends on the specific original higher-derivative theory. The field equations thus read

$$\bar{G}_{\mu\nu} = t_{\mu\nu} + \bar{T}_{\mu\nu} \tag{16.16}$$

where $\bar{G}_{\mu\nu}$ is the Einstein tensor of the reduced theory

$$\bar{G}_{\mu\nu} = \bar{R}_{\mu\nu} - \frac{1}{2}\bar{g}_{\mu\nu}\bar{R} \tag{16.17}$$

$t_{\mu\nu}$ is the stress energy-tensor of the scalar field

$$t_{\mu\nu} = \frac{3}{2}\Big[\partial_\mu\omega\partial_\nu\omega - \frac{1}{2}\bar{g}_{\mu\nu}\bar{g}^{\alpha\beta}\partial_\alpha\omega\partial_\beta\omega\Big] - \bar{g}_{\mu\nu}V(\omega) \tag{16.18}$$

and $\bar{T}_{\mu\nu}$ an effective stress-energy tensor generated by the interaction between the original matter field and the non-geometric gravitational new degrees of freedom represented by the scalar ω,

$$\bar{T}_{\mu\nu} = e^{-\omega} T_{\mu\nu}(\mathcal{A}, e^{-\omega}\bar{g}). \tag{16.19}$$

One should also note that in general the two terms on the left hand side of the equation (16.16) are not separately conserved, but it holds instead the conservation law $\nabla_\mu \bar{G}^{\mu\nu} = 0$. We stress that the spin-0 field introduced with the help of eq.(16.11) has physical meaning since it corresponds to an additional degree of freedom already present in the starting non-linear lagrangian. Therefore the advantage of having used this approach is that this new degree of freedom has been now made explicit in the reduced lagrangian.

Since we are interested in studying the structure of the CH in the presence of HD terms, we consider a system where the gravitational field is coupled with a pressureless null fluid plus an electromagnetic contribution coming from a static electric field generated by a charge of strength e. The stress-tensor for the matter field is therefore given by

$$T_{\mu\nu} = \rho l_\mu l_\nu + E_{\mu\nu} \tag{16.20}$$

where $E^\nu_\mu = e^2/8\pi r^4 \mathrm{diag}(1, 1, -1, -1)$ is the Maxwellian component of the electric field, ρ is the energy density of the radiation and l_μ is a null vector tangent to the radial null geodesic. In the following we shall specify our calculations in the case of spherical symmetry. One can introduce a null tetrad $\{l_\mu, n_\mu, m_\mu, \bar{m}_\mu\}$ with $l_\mu n^\mu = -1 = -m_\mu \bar{m}^\mu$. The metric tensor then reads

$$g_{\mu\nu} = 2\Big[(\bar{m}_{(\mu} m_{\nu)} - l_{(\mu} l_{\nu)}\Big]. \tag{16.21}$$

and

$$E_{\mu\nu} = \frac{e^2}{4\pi r^4}[l_{(\mu} n_{\nu)} + \bar{m}_{(\mu} m_{\nu)}] \tag{16.22}$$

In our case the matter Lagrangian is conformally invariant, and the equation of motion for the matter field completely decouple from those for the scalar field. We analyze the case of $\omega =$ const. and the solutions of the equation of motion correspond to constant field configuration that are extrema for the potential

$$\frac{dV}{d\omega} = 0 \tag{16.23}$$

The vacuum case -no matter field- has been analyzed in [26]. In our case we see that the conformal transformation in (16.11) generates a conformal rescaling of the null tetrad in (16.21) with $l_\mu \rightarrow \tilde{l}_\mu = \exp(\omega/2) l_\mu$. It is then easily seen that the stress-energy tensor in the reduced theory in eq.(16.19) has the same functional form of that in the original theory, with the "renormalized" energy density and charge

$$\rho \rightarrow \bar{\rho} = \exp(-2\omega)\rho \qquad e^2 \rightarrow \bar{e}^2 = \exp(-2\omega) e^2. \tag{16.24}$$

The solutions of the new field equations (16.16) can therefore be classified by looking at the minima of the potential term (16.15). Even if in the original theory a cosmological term is not present, the solutions of the reduced theory are given by a Vaidya like spacetime in a de Sitter (dS) or Anti-de Sitter (AdS) background, depending on the value of the potential at minimum. In particular, one can use the Eddington-Finkelstein coordinate so that an explicit for of the metric element in the reduced theory reads

$$d\bar{s}^2 = dv(2dr - fdv) + r^2 d\Omega^2. \tag{16.25}$$

where

$$f = 1 - \frac{2m(v)}{r} + \frac{\bar{e}^2}{r^2} - \frac{\Lambda}{3} r^2 \tag{16.26}$$

is the so called mass function, $\Lambda = V(\omega_c)$ and ω_c is the location of any extrema of the potential. In a local chart adapted to the inner horizon the Cauchy horizon is located at $v = +\infty$, and $f = 0$. If we compare this metric with the solution we would have gotten in the Einstein gravity, we note that, according to eq.(16.24) the charge and the energy densities are screened or antiscreened depending on the sign of ω_c. It should be also noted that, while in standard gravity one has always AdS solution of the kind (16.25) in our case the above metric is a solution of the field equations only for special values of the cosmological constant.

The global spacetime structure of those solution is well known. An updated review can be found in the contribution by Chris Chambers in these proceedings [28], while recents results on topological (AdS) black holes can be found in the lecture of R. Mann [29]. See also there for the figures displaying the conformal structure of the two spacetimes.

Here we shall simply stress that the black hole spacetimes in both dS and AdS solutions may have Cauchy horizons depending on the roots of the equation

$$f(r) = 0 \tag{16.27}$$

If the dS case is considered eq.(16.27) may have three positive roots and the spacetime has therefore a cosmological horizon besides an inner and an event horizon. Thanks to the work of Brady and Poisson [30] we know the Cauchy horizon is stable if the surface gravity of the cosmological horizon is greater than the surface gravity of the inner horizon

$$\kappa_{CS} > \kappa_{CH} \tag{16.28}$$

The surface gravity of the horizon in the original theory is obtained by a simple scale transformation (from the definition of surface gravity), and one can conclude that the stability criteria (16.28) holds in the original theory as well.

However, a black hole in dS background does not represent a vacuum solution since the field sets on a maximum of the potential. The stable vacuum solutions for the excitations of the ω field are instead given by black holes in AdS background and for those solutions the Cauchy horizon is a subtle question. In fact since the spatial infinity is now timelike, the radiative falloff at late times is not well understood [31]. For a power law decay of the kind

$$\delta(v) = 1/(\alpha v)^p \tag{16.29}$$

with $p \geq 3$ and α has to be introduced on dimensional grounds, the Cauchy horizon is unstable. In fact in general the energy density ρ_{obs} measured by an observer that crosses the Cauchy horizon would diverge as

$$\rho_{obs} \sim \dot{\delta} \exp(2\kappa_{CH} v) \tag{16.30}$$

thus signaling the instability of the Cauchy horizon. The CH in the original theory would also be unstable since the conformal transformation in eq. (16.11) is regular. However for an exponential decay rate of the kind

$$\delta \sim \exp(-2\alpha v) \tag{16.31}$$

one would not see any divergence for

$$\alpha > \kappa_{CH} \tag{16.32}$$

In a more realistic situation, one should use the eq. (16.29) to mimic the radiation falling into the event horizon. In fact outside the event horizon there is no reason to suppose that the background being AdS. Our effective "cosmological" constant is in fact dynamically generated during the gravitational collapse for the presence of the quantum fluctuations when the curvature has reached planckian level. The initial data coming from larger radii is still given by the classical evolution and therefore the Cauchy horizon would still be unstable even in the presence of a negative (or positive) Λ term in the metric.

16.3 R^2, $R^3 + R^2$ gravity

Let us analize some cases in detail. If we consider R^2 gravity, the density lagrangian of the gravitational field reads

$$l_g(R) = \epsilon_0 + \epsilon_1 R + \frac{1}{2}\epsilon_2 R^2. \tag{16.33}$$

then, eq. (16.11) can be inverted to yeld $\psi = \ln(1/2 + \epsilon_2 \psi)$ provided $\psi > -1/2\epsilon_2$. In particular the potential term becomes

$$V(\omega) = \frac{1}{2\epsilon_2}(1 - \epsilon_1 e^{-\omega})^2 - e^{-2\omega}\epsilon_0 \tag{16.34}$$

We see that for a positive cosmological constant in the original theory the potential term is not bounded from below and there are no stable solutions of the quadratic gravity theory. In the case of a negative cosmological constant, the potential term has only one minimum for $\omega_c = -\ln 2$ and we find the BH solutions given by eq.(16.26). We note that the "renormalized" charge and energy density have now an increased strength of a factor 4 according to (16.24).

Another interesting case is that of R^3 gravity where we suppose that the original lagrangian has the following functional form

$$l_g(R) = \epsilon_1 R + \frac{1}{2}\epsilon_2 R^2 + \frac{1}{3!}\epsilon_3 R^3 \tag{16.35}$$

in this case we can invert the relation (16.11) in order to obtain the two solutions

$$\psi\pm(\omega) = -\frac{1}{\epsilon_3}\Big(-\epsilon_2 \pm \sqrt{\epsilon_2^2 + 2(e^\omega - \epsilon_1)\epsilon_3}\Big) \tag{16.36}$$

in the two branches $\psi > -\epsilon_2/\epsilon_3$ and $\psi < -\epsilon_2/\epsilon_3$ and for the reality condition we also consider $\omega > \omega_* = \ln(1/2 - \epsilon_2^2/\epsilon_3)$. The potential term is therefore given by

$$V(\psi_\pm(\omega)) = \frac{\epsilon_2}{2}\psi_\pm^2(\omega)e^{-2\omega}\Big(1 + \frac{2\epsilon_3}{3\epsilon_2}\psi(\omega)\Big) \tag{16.37}$$

In the first case the potential $V(\psi_+(\omega)$ has one minimum for $\omega_c = -\ln 2$ with $V(\omega_c) = 0$ and a local maximum at another $\omega_c < 0$ with $V(\omega_c) > 0$. Therefore in the ψ_+ branch we can conclude that the solution of the field equations is given by a Vaidya like metric in a dS background. In the ψ_- branch we have only one local minimum that corresponds to a negative value of the potential, and therefore the metric is still a Vaidya like, but in AdS background.

Other interesting cases are given by Lagrangian like $l_g(R) = R + bR\ln(a+R)$, with a and b constants, which would arise from a one-loop contribution in an interacting stress-energy tensor. In any case one can use the above outlined procedure, and derive the explicit expression of the dS or AdS black hole solution.

The use of the conformal transformation is not the only method to obtain the solution of the original theory. There is also a direct procedure which is of course equivalent. Let us consider the field equations given in (16.8) and set

$$R_{\mu\nu} = C(R)\Big(\rho l_\mu l_\nu + E_{\mu\nu}\Big) + \frac{1}{4}Rg_{\mu\nu} \tag{16.38}$$

with constant Ricci scalar R. The following statement is true:

The solutions of the field equations in (16.8) with the stress energy-tensor in (16.20) and the Ricci tensor of the form (16.38) have

$$Rl_g'(R) - 2l_g(R) = 0 \tag{16.39}$$

$$l_g'(R)C(R) = 1. \tag{16.40}$$

The above conditions have to be considered as implicit relations that constrain the possible values of the constant R. They can be deduced by direct substitution of (16.38) in (16.8) and by projection along $l^\mu n^\nu$ and $n^\mu n^\nu$ respectively. One also notes that the first of the above relations in (16.38) is equivalent, aside from a constant factor, to the extrema condition for the potential, and the second relation can be read off as $C(R) = e^{-\omega}$ which is the first of the relations in (16.11). This shows the equivalence of the two approaches in obtaining solutions of the HD theory. However the conformal transformation method is simpler and elegant, while the direct method has the advantage that can also be used when $l_g''(R) = 0$. The set of consistency equations in (16.38) can also be deduced by considering the following stress energy tensor

$$T_{\mu\nu} = C(R)^{-1}\rho l_\mu l_\nu + E_{\mu\nu} \tag{16.41}$$

and with the Ricci tensor given by

$$R_{\mu\nu} = \rho l_\mu l_\nu + C(R)E_{\mu\nu} + \frac{1}{4}Rg_{\mu\nu}. \tag{16.42}$$

The above relations can be used to investigate the full global content of the theory. Once eq.(16.38) has been solved, one introduce locally the conformal frame where the new degrees of freedom is made explicit. Then one can address the question of local stability and mass of the field.

16.4 Conclusions

As we have stressed in the introduction, the non-linear modifications of the standard Einstein theory are needed for taking into account of the quantum effects when the Weyl curvature has reached Planckian levels. We have seen that stable black hole solutions appears in these cases. What is the possible role of these new solutions in the standard mass inflation scenario? It is therefore interesting to understand the structure of the CH singularity in a more consistent calculation where these new terms are dynamically generated during the collapse. We have argued that the role of those new terms would not change the (unstable) character of the Cauchy horizon. We therefore think that the mass inflation phenomenon also takes place in such a background. However since the appearance of an effective Λ term changes both the location of the horizon and the surface gravity of the inner horizon, a detailed calculation must be performed in order to address this issue in detail.

Acknowledgements

The author thans the organizer of "The Internal Structure of Black Holes and Spacetimes Singularity" workshop, at Haifa for the very pleasant meeting and for the friendly hospitality at Technion. My warmest thanks go to Amos Ori, Lior Burko and Liz Youdim for having organized this interesting meeting. I have personally benefited from discussions with Lior Burko, Chris Chambers, Valery Frolov, Robert Mann and Amos Ori whom I gratefully acknowledge. I would also thank Daniela Recupero for careful reading the manuscript.

Bibliography

[1] R. H. Price, Phys. Rev D **5**, 2419 (1972); **5**, 2439 (1972).

[2] R. Penrose, in *Battelle Rencontres,* Eds. C. M. De Witt and J. A. Wheeler, (W. A. Benjamin, New York, 1968), p. 222.

[3] V. A. Belinskii, E. M. Lifshitz and I. M. Khalatnikov, Adv. Phys. **19**, 525 (1970); C. W. Misner, Phys. Rev. Lett. **22**, 1071 (1969).

[4] G. F. R. Ellis and A. R. King, Commun. Math. Phys. **38**, 119 (1974). A. R. King, Phys. Rev. D **11**, 763 (1975).

[5] E. Poisson and W. Israel, Phys. Rev. D **41**, 1796, (1990).

[6] A. Ori, Phys. Rev. Lett. **67**, 781 (1991).

[7] A. Bonanno, S. Droz, W. Israel and S. M. Morsink, Phys. Rev. D **50**, 7372, (1994).

[8] These proceedings.

[9] A. Ori, Phys. Rev. Lett. **68**, 2117 (1992).

[10] P. R. Brady and C. M. Chambers, Phys. Rev. D **51**, 4177, (1995).

[11] A. Bonanno, Phys. Rev. D **53**, 7373 (1996).

[12] W. G. Anderson, P. R. Brady, W. Israel, and S. M. Morsink, Phys. Rev. Lett **70**, 1041, (1993).

[13] D. Marković and E. Poisson, Phys. Rev. Lett **74**, 1280, (1995).

[14] S. L. Adler, Rev. Mod. Phys. **54**, 729, (1982).

[15] J. F. Donoghue, Report No. gr-qc/9512024 (unpublished).

[16] S. Weinberg, Talk given at the Conference on Historical Examination and Philosophical Reflections on the Foundations of Quantum Field Theory, Boston, MA, 1-3, Texas U. Report UTTG-05-97, Mar 1997, 170, and Report No. hep-th/9702027 (unpublished).

[17] K. Stelle, Phys. Rev. D **16**, 353, (1978).

[18] N. D. Birrel and P. C. W. Davies, *Quantum Fields in Curved Space* (Cambridge University Press, Cambridge, 1982).

[19] B. Nelson and P. Panangaden Phys. Rev. D **25**, 1019, 1982.

[20] The correct definition of irrelevant and marginal operators should be given after linerization of the renormalization group equation around a fixed point. We are not interested here in these technical issues.

[21] K. Wilson and J. Kogut, Phys. Rep. **12 C**, 75, 1974.

[22] F. J. Wegner and A. Houghton, Phys. Rev. A **8**, 401 (1972).

[23] A. Bonanno, D. Zappalá, Phys. Rev. D **55**, 6135, (1997).

[24] M. Reuter, preprint DESY 96-080 and Report No. hep-th/9605030 (unpublished).

[25] A. Strominger, Phys. Rev. D **30**, 2257 (1984).

[26] A. Hindawi, B. A. Ovrut and D. Waldram, Phys. Rev. D **53**, 5597 (1996).

[27] G. Magnano and L. M. Sokolowski, Phys. Rev. D **50**, 5039 (1994).

[28] C. M. Chambers, these proceedings.

[29] R. B. Mann, these proceedings.

[30] P. R. Brady and E. Poisson, Class. Quantum Grav. **9**, 121, (1992).

[31] R. B. Mann and J. S. F. Chan, Phys. Rev. D **55**, 7546 (1997).

THE FINAL STATE OF COLLAPSE IN EINSTEIN THEORY OF GRAVITATION

S. Jhingan and P. S. Joshi

Theoretical Astrophysics Group, Tata Institute of Fundamental Research, Homi Bhabha Road, Colaba, Bombay 400005, India.

Abstract

We study here the structure of singularity forming in gravitational collapse of spherically symmetric inhomogeneous dust. Such a collapse is described by the Tolman-Bondi-Lemaître metric, which is a two-parameter family of solutions to Einstein equations, characterized by two free functions of the radial coordinate, namely the 'mass function' $F(r)$ and the 'energy function' $f(r)$. The main new result here relates, in a general way, the formation of black holes and naked shell-focusing singularities resulting as the final fate of such a collapse to the generic form of regular initial data. Such a data is characterized in terms of the density and velocity profiles of the matter, specified on an initial time slice from which the collapse commences. Several issues regarding the strength and stability of these singularities, when they are naked, are examined with the help of the analysis developed here. In particular, it is seen that strong curvature naked singularities can develop from a generic form of initial data in terms of the initial density profiles for the collapsing configuration. We also establish here that similar results hold for black hole formation as well. We also discuss here the physical constraints on the initial data for avoiding shell-crossing singularities; and also the shell-focusing naked singularities, so that the collapse will necessarily end as a black hole, preserving the cosmic censorship. These results generalize several earlier works on inhomogeneous dust collapse as special cases, and provide a clearer insight into the phenomena of black hole and naked singularity formation in gravitational collapse.

17.1 Introduction

The classic paper by Oppenheimer and Snyder [1] analyzes the gravitational collapse of a spherically symmetric massive object to conclude that the collapse of a homogeneous dust cloud would result into a black hole in the spacetime. Such a black hole is characterized by the presence of an event horizon, and the spacetime singularity at the center, which is covered by the horizon. This scenario provides the basic motivation for the black hole physics, and the cosmic censorship conjecture [2], which states that even when the assumptions contained in the above case are relaxed, in the form of either perturbations in the symmetry or form of matter etc., the outcome would still be a black hole in generic situations. As no proof, or a rigorous mathematical formulation of the censorship hypothesis has been available so far, the course of action that has emerged in past decade or so has been to carry out a detailed investigation of different gravitational collapse scenarios in general relativity to understand the final fate of collapse (see e.g. [3] and references therein).

Within such a context, an important situation that immediately offers itself for analysis is the introduction of inhomogeneities in the matter distribution. It is clear that the assumption of homogeneity is only an idealization, and realistic density profiles, for massive objects such as stars will have inhomogeneous density distribution, peaked typically at the center of the object. One would like to examine the final fate of gravitational collapse of such an inhomogeneous dust cloud, and examine in what ways there are departures in conclusions as opposed to the homogeneous situation. The Einstein equations can be fully solved for the stress-energy tensor of the form of inhomogeneous dust, and such a collapse is described by the Tolman-Bondi-Lemaître (TBL) metric. This is a two-parameter family of solutions, characterized by two free functions of the radial coordinate which are the 'mass function' $F(r)$ and the 'energy function' $f(r)$ of the cloud. While the first quantity describes here the initial density distribution for the collapsing cloud, the second function characterizes the initial velocity profile of the collapsing shells.

Our main purpose here is to relate, in a general way, the formation of black holes and naked shell-focusing singularities resulting as the final fate of such a collapse to the generic form of regular initial data. Such a data is characterized in terms of the density and velocity profiles of the matter, as given by the two functions above, specified on an initial time slice from which the collapse commences. An important issue, when naked singularities do form in gravitational collapse, is that of their curvature strength. We discuss here several issues regarding the strength and stability of such singularities, with the help of the analysis developed here. In particular, it is seen that strong curvature naked singularities can develop from a generic form of initial data in terms of the initial density profiles for the collapsing configuration. We also establish here that similar results hold for black hole formation as well. Thus, we show that both the black hole as well as naked singularity formation are related to the nature of the regular initial data defined at the onset of gravitational collapse. We also discuss here the physical constraints on initial data for avoiding shell-crossing singularities; and also the shell-focusing naked singularities, so that the collapse will necessarily end as a black hole in conformity with the cosmic censorship conjecture. These results generalize several earlier works on inhomogeneous dust collapse as special cases, and provide a clearer insight into the phenomena of the black hole and naked singularity formation in gravitational collapse. For example, while some of the earlier works (see e.g. [4, 5, 6, 7]) discuss the formation of naked singularities in dust collapse, either under various special assumptions or under general conditions, they do not address the important physical problem of relating this naked singularity formation to the nature of regular initial data at the onset of the gravitational collapse. It is crucial, from the physical point of view, to characterize and identify the properties of initial data which distinguish between the formation of black holes and the occurrence of naked singularity. Such an analysis could also pave the way for any possible formulation and proof for the cosmic censorship conjecture. A beginning in such a direction was made in [8, 9]. We present here a more general and complete analysis, obtaining several new results in connection with the naked singularity and black hole formation in gravitational collapse, in the specific context of the genericity of the initial data.

The main issue of concern here to us is, given a generic density distribution at the onset of collapse what *all* is possible as the final outcome. It is seen that given such a generic density profile, one can always have either a black hole or a strong curvature naked singularity, depending on the nature of the initial data. Certain points related to genericity are also discussed.

Section 17.2 below provides a brief review of the TBL models, giving the main set of equations used. Section 17.3 specifies the generic initial data to be used for collapse, and discusses the formation of singularities in collapse situations. Shell-crossing singularities are discussed in Section 17.4. The formation of black holes and naked singularities is discussed in Section 17.5, and the last section gives certain concluding remarks on our results.

17.2 Tolman-Bondi-Lemaître Dust Models

The TBL spacetimes are spherically symmetric manifolds ($\mathcal{M}$,g), with metric of the form,

$$ds^2 = -dt^2 + e^{2\omega}dr^2 + R^2 d\Omega^2, \tag{17.1}$$

and energy-momentum tensor of the form of a perfect fluid with equation of state p=0, given by

$$T^{ij} = \epsilon\, \delta^i_t \delta^j_t\,. \tag{17.2}$$

Here ϵ, ω and R are functions of r and t, and $d\Omega^2$ is the metric on the 2-sphere. The Einstein equations become

$$\dot{R}^2 = \frac{F(r)}{R} + f(r), \tag{17.3}$$

$$e^{2\omega}\left(1 + f(r)\right) = R'^2, \tag{17.4}$$

$$\epsilon(r,t) = \frac{F'(r)}{R^2 R'}, \tag{17.5}$$

where the dot and prime signify partial derivatives with respect to t and r respectively. It is seen that the energy density blows up either at the "shell-focusing singularity" $R = 0$, or when $R' = 0$ which corresponds to a "shell-crossing singularity" in the spacetime. The functions $F(r)$ and $f(r)$ are free functions of integration which completely specify the initial data in the model. Here $F(r)$ can be interpreted as the weighted mass given by

$$F(r) = \int_{\mathcal{B}_{r,t}} (1+f)\epsilon(r,t)dv_t = 4\pi \int_0^r \rho r^2 dr \tag{17.6}$$

where $\rho = \epsilon(0, r)$, and $\mathcal{B}_{r,t}$ is a ball of coordinate radius r centered on r=0 in the hypersurface $\Sigma_t(t = \text{const.})$. The time slicing ($t$=const.) has been so chosen such that r=const. labels a matter shell with t measuring the proper time elapsed along its geodesic path, that is we write the metric in comoving coordinates. Equation (17.3) can be thought of as the general relativistic generalization of the Newtonian energy equation [10], which leads to the interpretation of $F(r)$ as the effective mass contained in the sphere of radius r; and $f(r)$ as effective total energy per unit mass of a fluid element labeled by the comoving radial coordinate r. The model is said to be bound, unbound, or marginally bound if $f(r)$ is less than zero, greater than zero, or equal to zero respectively.

The integrated form of equation (17.3) is given by

$$t - t_s(r) = -\frac{R^{3/2}G(-fR/F)}{\sqrt{F}} \tag{17.7}$$

where $G(y)$ is a real, positive and smooth function which is bounded, monotonically increasing and strictly convex, and is given by

$$G(y) = \begin{cases} \frac{\arcsin\sqrt{y}}{y^{3/2}} - \frac{\sqrt{1-y}}{y}, & 1 \geq y > 0, \\ \frac{2}{3}, & y = 0, \\ \frac{-\text{arcsinh}\sqrt{-y}}{(-y)^{3/2}} - \frac{\sqrt{1-y}}{y}, & 0 > y \geq -\infty. \end{cases} \tag{17.8}$$

and $t_s(r)$ is a constant of integration which can be fixed by the choice of scaling on the initial surface ($t = 0$). Using this scaling freedom, if we choose so that $R(0, r) = r$ then,

$$t_s(r) = \frac{r^{3/2}G(-fr/F)}{\sqrt{F}} \tag{17.9}$$

The time $t = t_s(r)$ corresponds to the value $R = 0$ where the area of the matter shell at a constant value of the coordinate r vanishes, which corresponds to the physical spacetime singularity. Thus the range of coordinates is given by

$$0 \leq r < r_c, -\infty \leq t < t_s(r), \tag{17.10}$$

where $r = r_c$ denotes the boundary of the dust cloud where the solution is matched to the exterior Schwarzschild solution.

The argument of the function $G(-fR/F)$ changes it's sign depending on the sign of $f(r)$, since both R and $F(r)$ take only non-negative values. If $f(r) \geq 0$ for all $r \geq 0$, then from equation (17.3) we get $\dot{R} \neq 0$ for all t. From the integrated form of equation (17.3) we can show that if for some $r > 0$ we have $f(r) > 0$ then this particular shell is strictly unbounded, and from a physical point of view, the collapse can be interpreted as not being gravitational, but due to specific initial conditions (infinitely far in the past) [6]. Therefore, we shall discuss here only the more physical case with $f(r) < 0$ (however, it will be possible to generalize the conclusions here to other cases as well by similar methods as used here).

It is seen that if the condition

$$R' > 0 \tag{17.11}$$

is satisfied then the dust collapse always avoids shell-crossing singularities. The restrictions implied by this condition on the initial data will be discussed in sections 17.4 and 17.5. The weak energy condition, i.e. $T_{ij} = \epsilon(r,t)V^iV^j \geq 0$, for all non-spacelike vectors, V^i is assumed everywhere in spacetime. This imposes a condition on the initial data that the energy density $\epsilon(r,t)$ is non-negative everywhere.

17.3 Generic Initial Data and Formation of Singularities

We call a spacetime to be singular [11, 12] if it contains an incomplete nonspacelike geodesic,

$$\gamma : [0, x) \to \mathcal{M}$$

(where $\mathcal{M}$ is the spacetime manifold) such that there is no extension

$$\theta : \mathcal{M} \to \mathcal{M}'$$

for which $\theta \circ \gamma$ is extendible. The spacetime should be inextendible, that is, we should not be able to extend the above incomplete curve by embedding the spacetime into a larger manifold. This essentially means that we do not admit singularities created by removing pieces from the spacetime manifold. Important theorems showing the existence of singularities in a spacetime under "reasonable" physical conditions, which might form either in gravitational collapse or cosmology, were proved by Penrose, Hawking, and Geroch [13, 14]. While these theorems prove the existence of singularities under a fairly broad spacetime framework, they provide no information on the nature and structure of the singularities they predict, either in collapse scenarios or cosmology. In particular, they do not imply whether the singularities resulting as the final fate of gravitational collapse will be necessarily covered, or otherwise, by the event horizons of gravity. Our purpose here is to analyze the nature and structure of curvature singularities occurring in the TBL collapse models. A singularity will be called naked if there are future directed nonspacelike curves in the spacetime which terminate at the singularity in the past, otherwise it will be hidden behind a black hole. Considering equation (17.7), the apparent horizon within the collapsing dust cloud lies at $R = F(r)$, i.e.

$$t_{ah}(r) = t_s(r) - FG(-f) \tag{17.12}$$

It can be easily seen from the above equation that all the points on the singularity curve $t_s(r)$, other than the central point ($r = 0$) are covered by the apparent horizon. This is because, since both the functions $F(r)$ and $G(r)$ are strictly positive for $r > 0$, with $F(r) = 0$ at $r = 0$, therefore for all $r > 0$ $t_s(r) > t_{ah}(r)$ and $t_s(0) = t_{ah}(0)$. Thus, all the other points on the singularity curve, except the point $r = 0$, are covered by the apparent horizon. It is the central singularity point $r = 0$ whose nature, in terms of being naked or covered, depends on the data specified at the initial surface.

In the TBL models the initial data, to be specified on the initial hypersurface Σ_i, consists of two free functions $F(r)$ and $f(r)$ respectively and such a specification uniquely determines the solution to field equations inside $D^+(\Sigma_i)$, which is the future Cauchy development of the hypersurface Σ_i. To study the formation of singularity for this initial data, we consider the initial functions in a "generic" (most general expandable) form

$$F(r) = \sum_{n=0}^{\infty} F_n r^{n+3}, \quad f(r) = \sum_{n=2}^{\infty} f_n r^n. \tag{17.13}$$

The choice of n here is made to avoid any singular behavior of functions, and to ensure the finiteness of density, on the initial surface. The functions chosen here are assumed to be C^2 only, and not necessarily analytic (C^ω); since the only requirement of the theory is to have C^2 functions rather than restricting to the stronger differentiability condition of analyticity. This way, we are able to deal with a broader class of spacetimes. Additional differentiability requirements higher than C^2 will make the situation functionally less generic. We now recall briefly some definitions [7] which are necessary for further analysis. Especially, the partial derivative for the area function R is written as,

$$\begin{aligned} R' = & \; r^{\alpha-1}[(\eta - \beta)X + [\Theta - (\eta - \tfrac{3}{2}\beta)X^{3/2}G(-PX)] \times [P + \tfrac{1}{X}]^{1/2}] \\ \equiv & \; r^{\alpha-1}H(X,r) \end{aligned} \tag{17.14}$$

where we have used the following

$$u = r^\alpha, \ X = (R/u), \ \eta(r) = r\frac{F'}{F}, \ \beta(r) = r\frac{f'}{f},$$

$$p(r) = r\frac{f}{F}, \quad P(r) = pr^{\alpha-1},$$

$$\Lambda = \frac{F}{r^{\alpha}}, \quad \Theta \equiv \frac{t_s'\sqrt{\Lambda}}{r^{\alpha-1}} = \frac{1+\beta-\eta}{(1+p)^{1/2}r^{3(\alpha-1)/2}} + \frac{(\eta - \frac{3}{2}\beta)G(-p)}{r^{3(\alpha-1)/2}} \tag{17.15}$$

The function $\beta(r)$ is defined to be zero when $f(r)$ is zero (the marginally bound case). The parameter α (which satisfies $\alpha \geq 1$) is introduced here for examining the structure of the central singularity at $r = 0$. In terms of these new variables, we can write for energy density as

$$\epsilon = \frac{\eta\Lambda}{R^2 H} \tag{17.16}$$

The weak energy condition implies $H(X, r) \geq 0$, and $\eta\Lambda \geq 0$. The actual value of α is uniquely determined by the initial data prescribed, and is the key factor in deciding whether or not we have a strong curvature (covered or naked) singularity. The expression for energy density on the initial surface (which is the scaling surface Σ_i, on which $R = r$) is $\epsilon = F'/r^2$, and the weak energy condition implies that $F' \geq 0$. The requirement that initial surface should not contain any trapped surfaces $(F(r)/R(r,t) > 1)$ also gives a restriction on the choice of initial data.

The equation for radial null geodesics in the spacetime (17.1) is given by

$$\frac{dt}{dr} = \pm\frac{R'}{(1+f(r))^{1/2}} \tag{17.17}$$

This can also be written in the form,

$$\frac{dR}{du} = (1 - \frac{\sqrt{f + \Lambda/X}}{\sqrt{1+f}})\frac{H(X,r)}{\alpha} \equiv U(X,r) \tag{17.18}$$

If the terms in the numerator and denominator of above equation are expandable and contain linear terms then we can apply standard techniques of analyzing linear differential equations to classify the singularity points. But even for the simplified form of geodesics, namely the radial null trajectories considered here, this is not the case. Hence we study the detailed behavior of characteristic curves in the vicinity of the singularity. Defining quantity

$$X_0 = \lim_{\substack{R\to 0\\ r\to 0}} \frac{R}{r^{\alpha}} = \lim_{\substack{R\to 0\\ u\to 0}} \frac{R}{u} = \lim_{\substack{R\to 0\\ u\to 0}} \frac{dR}{du} \tag{17.19}$$

which gives the tangent to characteristics which terminate in the past at the singularity, it can be shown [7] that if the equation

$$V(X) = U(X,0) - X = (1 - \frac{\sqrt{f_0 + \Lambda_0/X}}{\sqrt{1+f_0}})\frac{H(X,0)}{\alpha} - X = 0 \tag{17.20}$$

admits a real positive root, then the central singularity at $r = 0, R = 0$ is naked (at least locally). The global visibility of such a singularity will depend on the overall behavior of the concerned functions within the dust cloud in the range $0 < r < r_c$. In the case otherwise, a black hole will be formed as the end product of collapse. The parameter α is uniquely fixed by demanding that $\Theta(X,r)$ goes to a nonzero finite value in the limit $r \to 0$. Consider the expression for Θ

$$\Theta \equiv \frac{t'_s\sqrt{\Lambda}}{r^{\alpha-1}} = \frac{1+\beta-\eta}{(1+p)^{1/2}r^{3(\alpha-1)/2}} + \frac{(\eta-\frac{3}{2}\beta)G(-p)}{r^{3(\alpha-1)/2}} = \frac{\Psi(r)}{r^{3(\alpha-1)/2}} \tag{17.21}$$

where

$$\Psi(r) = \Psi_0 + \Psi_1 r + \Psi_2 r^2 + \Psi_3 r^3 + \Psi_4 r^4 + \cdots \tag{17.22}$$

The value of α is decided by the first nonvanishing term in the expansion of $\Psi(r)$ which can take zero or non-zero finite values. For $\Psi(r)$ finite we always have the choice of α to be unity. When $\Psi(r)$ vanishes, we have to suitably choose the value of α in order for the equation (17.18) to be meaningful. There are also possibilities when $\Psi(r)$ vanishes identically (Oppenheimer-Snyder model) and we know for that case there are no future directed outgoing geodesics in the space time meeting singularity in past. Each term in the expansion of $\Psi(r)$ is completely specified in terms of the initial data; so given such data at the initial surface in terms of the density and velocity distributions, as specified by the functions F and f above, we can tell using the criterion above whether the final fate of collapse is going to be a black hole or a naked singularity.

The nonspacelike geodesics of the spacetime (17.1) are given by,

$$K^t = \frac{dt}{dk} = \frac{\mathcal{P}}{\mathcal{R}} \tag{17.23}$$

$$K^r = \frac{dr}{dk} = \frac{\sqrt{1+f}\sqrt{\mathcal{P}^2 - l^2 + BR^2}}{RR'} \tag{17.24}$$

$$(K^\theta)^2 + sin^2\theta(K^\phi)^2 = l^2/R^4 \tag{17.25}$$

where the $K^i = dx^i/dk$ denote the tangent to the geodesics. Here k is affine parameter along geodesics, l is the impact parameter ($l = 0 \Rightarrow$ radial trajectories), and the values $B = 0, -1$ correspond to null and timelike curves respectively. The function $\mathcal{P}$ satisfies the differential equation

$$\frac{d\mathcal{P}}{dk} + (\mathcal{P}^2 - l^2 + BR^2)[\frac{\dot{R}'}{RR'} - \frac{\dot{R}}{R^2}] - (\mathcal{P}^2 - l^2 + BR^2)^{1/2}\mathcal{P}\frac{\sqrt{1+f}}{R^2} + B\dot{R} = 0 \tag{17.26}$$

For the clarity of discussion, we consider only radial null geodesics here, in which case the above equation reduces to

$$\frac{1}{\mathcal{P}^2}\frac{d\mathcal{P}}{dk} + [\frac{\dot{R}'}{RR'} - \frac{\dot{R}}{R^2}] - \frac{\sqrt{1+f}}{R^2} = 0 \tag{17.27}$$

We shall call the singularities to be *strong* (see e.g. [12, 15]) if along nonspacelike trajectories the following is satisfied,

$$\lim_{k\to 0} k^2\psi(r) = \lim_{k\to 0} k^2 R_{ij}V^iV^j \neq 0 \tag{17.28}$$

Such a rate of divergence indicates a very powerful curvature growth in the limit of approach to the singularity (which is the same as in the case of the big bang singularity of cosmology), and ensures that all the volume forms, as defined by the Jacobi fields along the nonspacelike geodesics, are crushed to zero size in the limit of approach to the singularity. Physically, this can be interpreted as indicating that all the objects falling into the singularity are crushed to zero size. The significance of this would be that there could not possibly be any extension of the spacetime through such a singularity, as opposed to a gravitationally weak singularity such as a shell-cross through which the spacetime could possibly be extended and continued further.

In case of TBL models, we have

$$\lim_{k\to 0} k^2\psi(r) = \lim_{k\to 0}\frac{F'(K^t)^2}{R^2R'} \tag{17.29}$$

For radial null geodesics, we can write using the expressions above for K^t, R', and using the L'Hospital rule,

$$\lim_{k\to 0} k^2\psi(r) = \lim_{k\to 0}\frac{4\eta_0\Lambda_0 H_0}{{X_0}^2[\sqrt{1+f_0}(3\alpha-\eta_0) - N_0]^2} \tag{17.30}$$

Where the quantity N_0 is the limiting value of $-r\dot{R}'$, and Λ_0 is given by

$$\Lambda_0 = \begin{cases} 0, & \alpha < 3, \\ F_0, & \alpha = 3, \\ \infty, & \alpha > 3. \end{cases} \tag{17.31}$$

Since Λ_0 occurs in the numerator of expression for $k^2\psi(r)$, we have strong curvature singularities only for $\alpha \geq 3$. All the same, it follows from equation (17.18) that whenever α is greater than 3 we always have black hole, since we cannot have any outgoing geodesics meeting the singularity in the past in that case.

17.4 Shell-Crossing Singularities

Some counterexamples to cosmic censorship were proposed [16] using the so called shell-crossing singularities by Yodzis et. al., who showed the existence of such singularities and also that they are naked. However, these singularities are gravitationally weak (both in Tipler [15] as well as the Królak [17] sense); and there have been proposals for extending the spacetime through such singularities, in particular, by Papapetrou and Hamui [18] and also others. These singularities are generally not considered as being any serious counterexample to cosmic censorship conjecture, as the spacetime may be extended in a distributional sense through such a mild singularity. However, it is of importance to see what are the restrictions imposed on the initial data and the spacetime by the avoidance of such singularities at any point other then at the center, before the central shell-focusing singularity occurs; and to examine how 'physical' or 'unphysical' such restrictions are. That is, we would like to know what constraints the avoidance these singularities places on the functional form of initial data. We will impose the criterion for avoidance of shell-cross singularities on the initial data for making it physically more reasonable.

Shell-crossings are singularities where we observe crossing of dust shells, and the density diverges at these points. Consider the expression for energy density (17.5) in the TBL models. We have a shell-crossing singularity at $R' = 0$ and at this point there is a divergence in density and also certain curvatures. But for the spherically symmetric dust collapse it is possible to find a coordinate system so that we have a regular C^1 extension of the metric through these singularities, but which need not be C^2 [6, 19]. Calculating the expression for R' using equation (17.7),

$$\begin{aligned}[\frac{3q}{2}(\frac{R}{r})^{1/2}G(sR/r) + qs(\frac{R}{r})^{3/2}G^1(sR/r)]R' &= rqs'[G^1(s) - (\frac{R}{r})^{5/2}G^1(sR/r)] \\ +q(\frac{R}{r})^{3/2}[\frac{3G(sR/r)}{2} + p(\frac{R}{r})G^1(sR/r)] &- \frac{rq'}{2}[G(s) - (\frac{R}{r})^{3/2}G(sR/r)] \end{aligned} \quad (17.32)$$

where we have used notation,

$$G^1(x) = dG(x)/dx, \quad s(r) = -\frac{rf(r)}{F(r)}, \quad q(r) = \frac{F(r)}{r^3} \quad (17.33)$$

Clearly, if $G(x)$ is strictly positive and convex, and $R/r \leq 1$ for $t > 0$, we cannot have shell crossings [6] if

$$s' \geq 0 \ , \ q' \leq 0 \quad (17.34)$$

The restrictions imposed by the above equations are physically reasonable in the following sense. The condition $q' \leq 0$ implies that the density should be constant or decreasing away from the center, which gives a restriction on the nonvanishing terms in the expansion for density. Now analyzing the condition $s' \geq 0$, the kinetic energy term T and the potential energy term $V(0,r)$ on the initial hypersurface are given by

$$T \equiv \dot{R}^2, \quad V(0,r) \equiv -\frac{F(r)}{r} \tag{17.35}$$

Then from equation (3) we can write,

$$T = -V(r) + f(r) \tag{17.36}$$

Using the above two expressions we can write,

$$s(r) = \frac{f(r)}{V(0,r)} = \frac{1}{1 - T/f(r)} \tag{17.37}$$

The function $s(r)$ takes values between 0 and 1. Since $s'(r) \geq 0$ (it is zero for homogeneous and marginally bound case), it is clear from the above equation that as we move away from the center to the boundary of the star, contribution of the kinetic energy term to total energy decreases. Physically this implies that we do not give additional velocity to the outer layers of the matter in the cloud, in comparison to the inner layers, to avoid shell crossings.

17.5 Naked Singularities and Black Holes

The singularity theorems prove only the existence, and do not give any information on the nature and behavior of the singularities they predict. This leaves an important gap between the theoretical existence and physical presence of the singularities of cosmology and gravitational collapse. As mentioned earlier, one of the main differences lie in the issue related to genericity. As we discussed, in dust models the most general solution to Einstein equations contains two free functions, which are functions of the comoving radial coordinate r. If the final outcome of gravitational collapse is specified in terms of both the functions (in a general form), the solution can be called as functionally generic (see e.g. [20] also), at least within the context of the dust models we are discussing.

We have discussed in Section 17.3 the conditions on the initial data for the formation of a naked singularity or a black hole. As was pointed out there, the behavior of the first point $r = 0$ of the singularity curve depends on the initial data, and all other points on the curve are covered by the apparent horizon independently of the initial conditions.

Consider now a generic density profile, that is,

$$\rho(r) = \sum_{n=0}^{\infty} \rho_n r^n \tag{17.38}$$

There are physical reasons to avoid the ρ_1 term from the above expansion, namely that if this term is nonzero, there will be a cusp in the density at the center of the cloud. However, for the sake of generality, we shall not assume this term to be necessarily zero. Corresponding expression for $F(r)$, and the expression for $f(r)$ are then of the following form,

$$F(r) = \sum_{n=0}^{\infty} F_n r^{n+3}, \quad f(r) = \sum_{n=2}^{\infty} f_n r^n. \tag{17.39}$$

To analyze the central singularity $r = 0$, we now evaluate the previously defined quantities $\eta(r)$, $\beta(r)$ and $p(r)$ for these density and velocity profiles, which are given by

$$\eta(r) = 3 + \frac{rF_1}{F_0} + r^2[\frac{2F_2}{F_0} - \frac{F_1{}^2}{F_0{}^2}] + r^3[\frac{F_1{}^3}{F_0{}^3} - \frac{3F_1F_2}{F_0{}^2} + \frac{3F_3}{F_0}] + O[r]^4 \tag{17.40}$$

$$\beta(r) = 2 + \frac{rf_3}{f_2} + r^2[\frac{2f_4}{f_2} - \frac{f_3{}^2}{f_2{}^2}] + r^3[\frac{3f_5}{f_2} - \frac{3f_3f_4}{f_2{}^2} + \frac{f_3{}^3}{f_2{}^3}] + O[r]^4 \tag{17.41}$$

and

$$\begin{aligned} p(r) &= \frac{f_2}{F_0} + r\frac{f_2}{F_0}[\frac{f_3}{f_2} - \frac{F_1}{F_0}] + r^2\frac{f_2}{F_0}[\frac{f_4}{f_2} - \frac{f_3F_1}{f_2F_0} + \frac{F_1{}^2}{F_0{}^2} - \frac{F_2}{F_0}] \\ &+ r^3\frac{f_2}{F_0}[\frac{f_5}{f_2} - \frac{f_4F_1}{f_2F_0} + \frac{f_3F_1{}^2}{f_2F_0{}^2} - \frac{f_3F_2}{f_2F_0} - \frac{F_1{}^3}{F_0{}^3} + \frac{2F_1F_2}{F_0{}^2} - \frac{F_3}{F_0}] + O[r]^4 \end{aligned} \tag{17.42}$$

respectively. Now consider the expression for $\Theta(r)$

$$\Theta(r) = \frac{1}{r^{3(\alpha-1)/2}}[\Psi_0 + \Psi_1 r + \Psi_2 r^2 + \Psi_3 r^3] + \frac{O[r]^4}{r^{3(\alpha-1)/2}} \tag{17.43}$$

where the expressions for various coefficients of Ψ are the following,

$$\Psi_0 = 0 \tag{17.44}$$

$$\Psi_1 = [\frac{1}{(1+f_2/F_0)^{1/2}}(\frac{f_3}{f_2} - \frac{F_1}{F_0}) + (\frac{F_1}{F_0} - \frac{3f_3}{f_2})(\frac{sin^{-1}\sqrt{-f_2/F_0}}{(-f_2/F_0)^{3/2}} + \frac{\sqrt{1+f_2/F_0}}{(f_2/F_0)})] \tag{17.45}$$

$$\Psi_2 = [\frac{1}{(1+f_2/F_0)^{1/2}}(-\frac{f_3{}^2}{f_2F_0} + \frac{f_3F_1}{F_0{}^2} + \frac{2f_4}{f_2} - \frac{f_3{}^2}{f_2{}^2} - \frac{2F_2}{F_0}) + (\frac{2F_2}{F_0} - \frac{F_1{}^2}{F_0{}^2}$$

$$-\frac{3f_4}{f_2} + \frac{3f_3{}^2}{2f_2{}^2}) \times (\frac{sin^{-1}\sqrt{-f_2/F_0}}{(-f_2/F_0)^{3/2}} + \frac{\sqrt{1+f_2/F_0}}{(f_2/F_0)})(\frac{3}{2}\frac{sin^{-1}\sqrt{-f_2/F_0}}{(-f_2/F_0)^{1/2}}$$

$$-\frac{3+f_2/F_0}{2(1+f_2/F_0)^{1/2}})(\frac{f_3F_0 - f_2F_1}{f_2{}^2})(\frac{F_1}{F_0} - \frac{3f_3}{f_2})] \tag{17.46}$$

$$\Psi_3 = [\frac{1}{(1+f_2/F_0)^{1/2}}(-\frac{5f_3f_4}{f_2F_0} + \frac{f_3{}^3}{f_2{}^2F_0} + \frac{2f_3F_2}{F_0{}^2} - \frac{3f_3F_1{}^2}{2F_0{}^3} + \frac{F_1f_3{}^2}{2f_2F_0{}^2}$$

$$+\frac{3f_3{}^3}{8f_2F_0{}^2} + \frac{F_1F_4}{2F_0{}^2} - \frac{F_1f_3{}^2}{8F_0{}^3} + \frac{3f_5}{f_2} - \frac{3F_3}{F_0} - \frac{3f_3f_4}{f_2{}^2} + \frac{3F_1F_2}{F_0{}^2} + \frac{f_3{}^3}{f_2{}^3}$$

$$-\frac{F_1{}^3}{F_0{}^3}) + [\frac{5}{4}(\frac{3sin^{-1}\sqrt{-f_2/F_0}}{(-f_2/F_0)^{3/2}} - \frac{(f_2/F_0)^2 + 4f_2/F_0 + 3}{f_2/F_0(1+f_2/F_0)^{3/2}})(\frac{f_3}{f_2} - \frac{F_1}{F_0})^2$$

$$+(\frac{3sin^{-1}\sqrt{-f_2/F_0}}{(-f_2/F_0)^{1/2}} - \frac{(3+f_2/F_0)}{(1+f_2/F_0)^{1/2}})(\frac{-F_2}{f_2} + \frac{F_1{}^2}{F_0f_2} - \frac{F_1f_3}{f_2{}^2} + \frac{F_0f_4}{f_2{}^2})]$$

$$\times[\frac{F_1}{F_0} - \frac{3f_3}{2f_2}] + [(\frac{3}{2}\frac{sin^{-1}\sqrt{-f_2/F_0}}{(-f_2/F_0)^{1/2}} - \frac{3+f_2/F_0}{2(1+f_2/F_0)^{1/2}})(\frac{f_3F_0 - f_2F_1}{f_2{}^2})\times$$

$$(\frac{2F_2}{F_0} - \frac{F_1{}^2}{F_0{}^2} - \frac{3f_4}{f_2} + \frac{3f_3{}^2}{2f_2{}^2})] + (\frac{sin^{-1}\sqrt{-f_2/F_0}}{(-f_2/F_0)^{3/2}} + \frac{\sqrt{1+f_2/F_0}}{(f_2/F_0)})\times$$

$$(\frac{F_1{}^3}{F_0{}^3} - \frac{3F_1F_2}{F_0{}^2} + \frac{3F_3}{F_0} - \frac{9f_5}{2f_2} + \frac{9f_3f_4}{2f_2{}^2} - \frac{3f_3{}^3}{2f_2{}^3})] \tag{17.47}$$

These coefficients completely specify the form of Θ, and α is uniquely determined by the condition that, in the limit $r \to 0$, Θ takes a nonzero finite value. The existence of naked singularities (black holes) is determined by the existence (absence) of real positive roots to the equation $V(X) = 0$. We can write equation (17.20) as (using the fact that $f_0 = 0$)

$$V(X) = (1 - \sqrt{\frac{\Lambda_0}{X}})\frac{H(X,0)}{\alpha} - X = 0 \tag{17.48}$$

where the quantity Λ_0 is specified by equation (17.31), and $H(X,0)$ can be obtained using equation (17.14) as

$$H(X,0) = X + \frac{\Theta_0}{X^{1/2}} \tag{17.49}$$

The equation $V(X) = 0$, for which we need to investigate the existence of real positive roots, then can be reduced to a quartic of the form

$$(\alpha - 1)x^4 + \sqrt{\Lambda_0}x^3 - \Theta_0 x + \sqrt{\Lambda_0}\Theta_0 = 0 \tag{17.50}$$

where $x = \sqrt{X}$. The roots, positive or otherwise, of this equation can be completely specified in terms of the functions Θ_0 and Λ_0 at the initial hypersurface. To analyze the existence of real positive roots in the above equation, consider a general quartic of the form [21]

$$ax^4 + 4bx^3 + 6cx^2 + 4dx + e = 0 \tag{17.51}$$

with the following definitions, $Q = ac - b^2$, $I = ae - 4bd + 3c^2$, $J = ace + 4bcd - ad^2 - eb^2 - c^3$ and $\triangle = I^3 - 27J^2$. The conditions for the existence of real roots are then, $\triangle < 0$ implies the existence of two real and two imaginary roots and $\triangle > 0$ implies that all roots are imaginary unless $Q < 0$ and $(a^2 I - 12Q^2) < 0$ (in which case all roots are real). For the quartic under consideration, we have

$$\triangle = (\alpha - \frac{3}{4})^3 {\Lambda_0}^2 {\Theta_0}^3 - \frac{27}{(16)^2}[{\Lambda_0}^{3/2} + (\alpha - 1){\Theta_0}^2]^2 \tag{17.52}$$

The above analysis only checks for roots to be real or imaginary; positivity or otherwise of the roots has to be checked separately for different cases. As explained in Section 17.3, for $\alpha > 3$ we only have black holes since we cannot have outgoing geodesics meeting the singularity in the past. Thus, for $\alpha \leq 3$ we can have real positive roots satisfying equation (17.50). Consider first the case $\alpha < 3$. From equation (17.31) we then have $\Lambda_0 = 0$, and the expression for $\triangle$ simplifies to

$$\triangle = -\frac{27}{(16)^2}(\alpha - 1)^2 {\Theta_0}^4 \tag{17.53}$$

Since both α and Θ_0 take only real values, it follows that $\triangle < 0$ for all values of Θ_0, which ensures the existence of real roots. The roots equation $V(X) = 0$ gets simplified for the case $\alpha < 3$ as below,

$$x^3 = \frac{\Theta_0}{(\alpha - 1)}, \quad i.e. \; X = [\frac{\Theta_0}{(\alpha - 1)}]^{2/3} \tag{17.54}$$

Clearly, for the range $1 < \alpha < 3$, we always have real positive root(s) to the above equation, ensuring the nakedness of the singularity. However, the naked singularities in this case are not strong in the sense described earlier, since $\Lambda_0 = 0$.

The criterion for the existence of strong curvature singularity is that $\alpha \geq 3$, and $\alpha = 3$ is the most interesting case since $\alpha > 3$ give rise to black holes. To analyze this case, using (17.50), the roots equation in this case is given as

$$2x^4 + \sqrt{F_0}x^3 - \Theta_0 x + \sqrt{F_0}\Theta_0 = 0 \tag{17.55}$$

comparing various coefficients with the expression for a general quartic (51), we have,

$$a = 2, \ b = \frac{\sqrt{F_0}}{4}, \ c = 0, \ d = -\frac{\Theta_0}{4}, \ e = \sqrt{F_0}\Theta_0 \tag{17.56}$$

and we have,

$$\triangle = -\frac{27}{(16)^2}{\Theta_0}^2(4{\Theta_0}^2 - 104{F_0}^{3/2}\Theta_0 + {F_0}^3) \tag{17.57}$$

If the expression in parenthesis on the right side is greater than zero, we necessarily have two real roots to the above equation. To analyze the expression in parenthesis, consider

$$g(\Theta_0, F_0) = 4{\Theta_0}^2 - 104{F_0}^{3/2}\Theta_0 + {F_0}^3 \tag{17.58}$$

This equation admits roots $\Theta_1(= \frac{{F_0}^{3/2}}{2}[26 - 5\sqrt{27}])$ and $\Theta_2(= \frac{{F_0}^{3/2}}{2}[26 + 5\sqrt{27}])$, and both are positive. We can easily see that the function takes negative values for the range $\Theta_1 < \Theta_0 < \Theta_2$, and is positive outside this range. Hence the range $\Theta_0 < \Theta_1$ and $\Theta_0 > \Theta_2$ admits atleast two real roots. If $\triangle > 0$, we can have all the roots real if $Q < 0$ and the quantity $a^2 I - 12Q^2 < 0$. The first condition is always satisfied, since $Q = -F_0/16$ (where $F_0 > 0$), and the second condition is satisfied if $3\sqrt{F_0}(3\Theta_0 - \frac{{F_0}^{3/2}}{64}) < 0$, i.e. $\Theta_0 < \frac{{F_0}^{3/2}}{192}(= \Theta_3, say)$. Since $\Theta_3 < \Theta_1 < \Theta_2$ we cannot have real roots for the range $\Theta_1 < \Theta_0 < \Theta_2$. To analyze the positivity of given real roots consider equation (17.55) which can be written as

$$\Theta_0 = x^3(\frac{2x + \sqrt{F_0}}{x - \sqrt{F_0}}) \tag{17.59}$$

The Fig. 17.1 explicitly shows the positivity of roots for equation (17.55).

It is easy to analyze from equation (17.15) that the regions, in the above figure, where Θ_0 admits negative values necessarily give rise to shell crossings near singularity and the region admitting positive values of Θ_0 is free from shell crossings for the given initial data. For a further discussion on shell-crossing singularities, and considerations involving more general class of functions than considered here, we refer to [22].

To see more explicitly as to how the existence of real positive roots is related to the data specified at the initial surface, (depending on the choice of $F(r)$ and $f(r)$), we now analyze some important cases. The parameter α will decide the strength of the final singularity and we will be mainly concerned here for the case $\alpha = 3$, i.e. strong

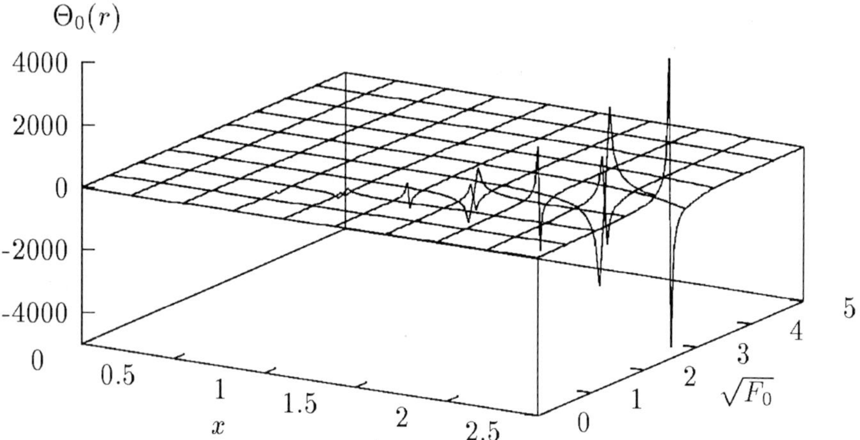

Figure 17.1: The graph showing existence of real positive roots of equation (17.55). There are no roots of the form $x = \sqrt{F_0}$ since this implies $F_0 = 0$ which is not true. At all other points on the hypersurface x admits positive values with non-zero Θ_0 $(x \neq 0)$.

curvature singularities (naked or covered). The value of α can be uniquely fixed by the limiting value of $\Theta(r)$, which is specified in terms of various coefficients in the expansion of $\Psi(r)$. In the expansion of $\Psi(r)$, if the first two terms Ψ_1 and Ψ_2 vanish ($\Psi_0 \equiv 0$), and Ψ_3 or a higher term is the first nonvanishing term, then necessarily it is a strong curvature singularity.

The two important specific questions which need to be addressed in this connection are, (i) Given an arbitrary density distribution on the initial surface, is it possible to keep the first two terms in the expansion of $\Psi(r)$ zero, and later terms non-zero, by means of a suitable choice of a velocity profile for the cloud, so that the resulting singularity is strong curvature type, and (ii) How 'generic' is the occurrence of the strong curvature naked singularities.

To investigate the answer to the first question, we start by considering a specific example. Consider a density profile of the form

$$\rho(r) = \rho_0 + \sum_{n=2}^{n=\infty} \rho_n r^n, \quad \rho_2 \neq 0 \tag{17.60}$$

where ρ_0 is the central density of the cloud. We will assume the density to be decreasing outwards from the center, which is a reasonable condition to take for realistic density profiles, and which is also a necessary condition to avoid any possible shell-crossing singularities. The choice of such a density distribution has some physical significance. Firstly, as commented earlier, the linear term in the density profile is chosen to be zero, because a non-vanishing linear term would mean a cusp in the density at the center with a non-zero pressure gradient at the center. This may not be desirable on physical grounds. Next, for physically realistic density configurations, the ρ_2 term is generally believed to be non-zero, with $\rho_2 < 0$, for reasons related to the stability of the stars (see e.g. [23, 24]). The subsequent terms in the expansion may be vanishing or otherwise, which will not affect our considerations here, because as we shall see, it is the first non-vanishing term in the expansion which determines the nature of the singularity. Another useful point to note is that for a marginally bound collapse ($f = 0$), such a density profile necessarily results into a weak naked singularity [8]. Hence, it is of importance to know whether the collapse of the same density profile can result into a strong curvature naked singularity when the collapse is not marginally bound.

For the density profile given above, the corresponding expression for $F(r)$ is given by

$$F(r) = F_0 r^3 + \sum_{n=2}^{n=\infty} F_n r^{n+3} \tag{17.61}$$

As for the other free function $f(r)$, the appropriate differentiability for the metric requires it to be at least a C^2 function. Therefore, a general choice for $f(r)$ is,

$$f(r) = \sum_{n=2}^{n=\infty} f_n r^n \tag{17.62}$$

We now fix $f(r)$, which specifies the velocity profile for the cloud, by imposing the following two requirements: (i) the second term in the expansion of $f(r)$, i.e. f_3 is zero and, (ii) the coefficients $\tilde{F}_2$ and $\tilde{f}_4$ as given by

$$\tilde{F}_2 = \frac{F_2}{F_0}, \quad \tilde{f}_4 = \frac{f_4}{f_2} \tag{17.63}$$

be related as

$$\tilde{F}_2 = \tilde{f}_4 \frac{[\frac{1}{(1+f_2/F_0)^{1/2}} - \frac{3}{2}G(-f_2/F_0)]}{[\frac{1}{(1+f_2/F_0)^{1/2}} - G(-f_2/F_0)]} \tag{17.64}$$

Under this situation, we evaluate the expressions for $\eta(r)$ and $\beta(r)$, which turn out to be

$$\eta(r) = 3 + r^2[2\tilde{F}_2] + r^3[\frac{3F_3}{F_0}] + O[r^4] \tag{17.65}$$

$$\beta(r) = 2 + r^2[2\tilde{f}_4] + r^3[\frac{3f_5}{f_2}] + O[r^4] \tag{17.66}$$

respectively. Then, considering the general expression for $\Theta(r)$,

$$\Theta(r) = \frac{\Psi(r)}{r^{3(\alpha-1)/2}} \tag{17.67}$$

the various coefficients are evaluated to be as follows

$$\Psi_0 = 0 \tag{17.68}$$

$$\Psi_1 = 0 \tag{17.69}$$

$$\Psi_2 = 0 \tag{17.70}$$

$$\Psi_3 = 3(\frac{f_5}{f_2})[\frac{1}{(1+f_2/F_0)^{1/2}} - \frac{3}{2}G(-f_2/F_0)] \tag{17.71}$$

Since Ψ_3 is the first nonvanishing term in the expansion of $\Psi(r)$, we have $\alpha = 3$, which makes it to be a strong curvature singularity. Also, whenever $\alpha = 3$ we know that Λ_0 is a nonzero quantity, so from the roots equation $V(X) = 0$ we have,

$$(1-\sqrt{\frac{\Lambda_0}{X}})\frac{H(X,0)}{3} - X = 0 \tag{17.72}$$

where $H(X,0)$ is given by

$$H(X,0) = X + \frac{\Theta_0}{X^{1/2}} \tag{17.73}$$

Writing again $x = \sqrt{X}$, this becomes a quartic equation,

$$2x^4 + \sqrt{F_0}x^3 - \Theta_0 x + \Theta_0\sqrt{F_0} = 0 \tag{17.74}$$

Depending on the choice of the values of the initial variables, that is f_5, f_2 and F_0, we can have the above quartic with or without real positive roots (Fig. 17.1). Hence we can have both the possibilities as we would desire, namely the black holes as well as naked singularities, depending on the choice of initial data. As we have already pointed out earlier, in either case this is a strong curvature singularity.

We have illustrated above how strong curvature singularities arise as a result of collapse when the coefficient ρ_2 is the first nonvanishing term in the expansion of the density profile. We chose in the above the ρ_1 term in the expansion of density to be identically zero, since it would represent a cusp in the density at the center of the star and may be objectionable on physical grounds. However, it is not conclusively established that such density cusps at the center are ruled out by astrophysical considerations (see e.g. [25]) Hence, if we assume that the ρ_1 term is non-vanishing and arises as a perturbation (which can possibly happen in the turbulent interiors of stars), in that case as well we can show that strong curvature singularities can arise as the final fate of collapse. For that purpose, using similar techniques as above, one again chooses the velocity profile suitably, imposing required constraints on $f(r)$. Next, when ρ_3 is the first non-vanishing term in the density expansion, then it is already known that we have a strong curvature singularity for the marginally bound collapse. This case is discussed earlier in detail [9], showing the formation of naked singularities and black holes. In a subsequent paper, the causal structure of space-time has been analyzed, using the dynamics of trapped surfaces, for this particular example [26] (see Fig. 17.2)

Interesting situation arises when all the first three terms ρ_1, ρ_2 and ρ_3 in the expansion of density profile are zero, and the first non-vanishing term in the expansion of density is ρ_n $(n > 3)$. We know that in the marginally bound case this situation always corresponds to a black hole [8]. Now, considering the general non-marginally bound case, the density profile is of the form

$$\rho(r) = \rho_0 + \sum_{n=4}^{n=\infty} \rho_n r^n \tag{17.75}$$

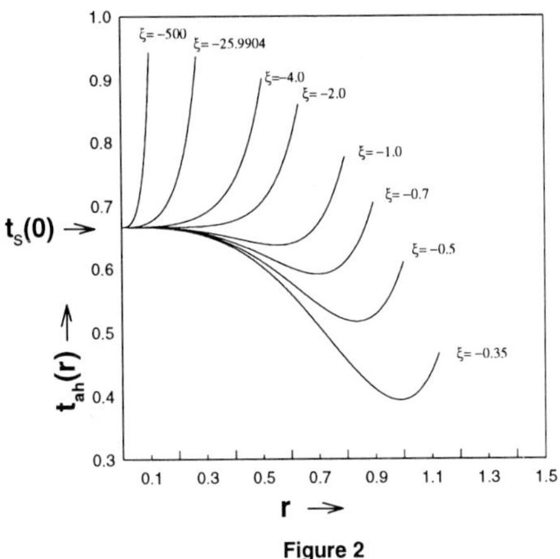

Figure 17.2: A plot of the apparent horizon curves (17.12) for marginally bound case and , $\rho(r) = \rho_0 \left(1 - \frac{r^3}{r_c^3}\right)$, obtained by setting $\rho_0 = 1$ and varying r_c (boundary of the cloud). For $\xi < -2\, \xi (= F_3/{F_0}^{5/2})$ the center is the first point to get trapped, whereas for $\xi > -2$ some surface in the interior of the star is the first one to get trapped, and the trapped region moves both inwards and outwards. But for all other points on the singularity curve $(r > 0)$, $t_s(r) > t_{ah}(r)$, independent of the initial data.

The corresponding expression for $F(r)$ can be written as

$$F(r) = F_0 r^3 + \sum_{n=4}^{\infty} F_n r^{n+3} \tag{17.76}$$

Consider then the following choice of the energy function, or equivalently the particle velocity profile, as given by

$$f(r) = f_2 r^2 + \sum_{n=5}^{n=\infty} f_n r^n \tag{17.77}$$

We can see that for this choice of $f(r)$ (i.e. $f_3 = 0,\ f_4 = 0$) we will have Ψ_0, Ψ_1 and Ψ_2 identically zero, and Ψ_3 is nonzero with

$$\alpha = 3, \quad \Psi_3 = \frac{3f_5}{f_2}\left[\frac{1}{(1+f_2/F_0)^{1/2}} - \frac{3}{2}G(-f_2/F_0)\right] \tag{17.78}$$

The expression for quartic remains the same equation (17.55), except that the functional form of Θ_0 changes now. Hence, we can have again both the possibilities, namely the black holes and strong curvature naked singularities depending on the choice of initial free functions.

After starting the gravitational collapse from a regular initial data, it is desirable to ensure that no shell-crossing singularity forms in the cloud before formation of central shell-focusing singularity. The necessary and sufficient condition to avoid shell-crossings in the cloud is given by equation (17.11), which can again be written as (equation (17.34)),

$$q'(r) \leq 0\ ,\ s'(r) \geq 0$$

The condition $q'(r) \leq 0$ implies that the density should decrease or remain constant as we move away from the center of the cloud. This condition is satisfied as long as the sum of all non-vanishing derivatives is negative (or zero), keeping the overall density positive (to satisfy the weak energy condition); or if all the coefficients in the density profile are negative. We note that the condition for existence of strong curvature singularity derived here depends on the first non-vanishing term of density profile, which is taken negative in our case, and the remaining terms would not contribute to the final outcome. Hence, rest of the terms in the density profile can always be chosen to satisfy this condition by the initial data $F(r)$ chosen in the analysis. Now consider the second constraint i.e., $s'(r) \geq 0$. As discussed earlier, as we go out from the center of the cloud, this condition implies that the contribution of kinetic energy to the total energy decreases. This reduces to the following condition on the initial data,

$$F(r) \leq C[-rf(r)] \tag{17.79}$$

Since $F(r)$ and $-rf(r)$ admit only positive values, C here is a non-zero positive constant. A lower bound on the values of C easily follows using the boundary condition at the surface of the star, i.e.

$$C \geq \frac{M}{-r_c f_c} \tag{17.80}$$

where M is the total mass of the cloud and the subscript c denotes values measured at the boundary of the cloud. In equation (17.80), M/r_c and f_c denote respectively the potential energy and the total energy of the shell at the boundary of the cloud respectively. In general, from equations (17.3) and (17.79) we can see that we must have $C \geq 1$ always, equality being satisfied for shells which are rest. Also, from equation (17.80), the equality implies that boundary of the star is at rest.

Let us consider a specific example to understand the possibility of existence of regular initial data $(F(r), f(r))$ for which there are no shell-crossing singularities in the cloud till the first shell-focusing singularity forms at the center of the cloud, i.e.

$$t_s(0) < t_{sc}(r)|_{r>0}$$

At the center of the cloud we have $r = 0, t_s(0) = t_{sc}(0)$. Consider a density profile of the form (17.60). Corresponding expression for mass function is given by (17.61). For energy function $f(r)$ of the form (17.62) subject to suitable constraints ($f_3 = 0$) and (17.64), we have shown earlier that it gives rise to strong curvature naked singularity. The condition (17.79) for avoiding shell-crossing singularities reduces to

$$F_0 + F_2 r^2 + \cdots \leq C[-f_2 - f_4 r^2 - f_5 r^3 + \cdots]$$

where $C \geq 1$. The above condition will hold in general, everywhere in a cloud of arbitrary radius, at least for that class of functions, $f(r)$, where each coefficient in the expansion of $f(r)$ satisfies the above inequality with corresponding coefficients in the expansion of $F(r)$. This can be easily done in this case, since f_5 can be given any finite value (keeping $-1 < f(r) < 0$) which makes $\Psi_3 \neq 0$ in (17.71), and since both f_5 and f_2 are less then zero therefore $\Psi_3 > 0$, which is consistent with $\Theta_0 > 0$. Since $C \geq 1$, the argument of $G(y)$ $(y = -f_2/F_0)$ in (17.64) also satisfies the condition $0 < y \leq 1$.

Now consider another example where the density profile is given by (17.75). Here we make a choice of $f(r)$ such that the coefficients f_3 and f_4 are identically zero. Clearly, the condition (17.79) is trivially satisfied as the corresponding coefficients in the expansion of $F(r)$ are also zero. So we are free to make a suitable choice of $|f_5|$, and $|f_2|$ (non-zero) for any given $F(r)$ such that the required inequality (17.80) is satisfied. Hence the condition for avoiding shell-crossing singularities can always be satisfied in general without affecting the general conclusions regarding the generic existence of strong curvature singularities.

This completes our answer to the first question, which is in the affirmative. That is, for any given density distribution at the initial epoch of time, we can always keep the coefficient Ψ_3 as the first non-vanishing term in the expansion of $\Psi(r)$, by means of a suitable initial choice of the velocity profile $f(r)$ for the cloud. In other words, given any generic density profile, we can always choose the rest of the initial data, that is the particle velocity profile, in such a manner that the final fate of the collapse results in a strong curvature singularity which is either hidden inside a black hole, or naked, communicating with faraway observers in the spacetime.

Coming to question (ii), the following consideration may provide some insight into the issues regarding genericity. Consider the functional form of the initial data, as we have discussed above. The functions $F(r)$ and $f(r)$ both are specified in terms of infinitely many free functions in the form of the coefficients of the expansion terms such as $F_0, F_1, F_2, ...$, and $f_2, f_3, ...$ etc. Hence, we have here a $2 \otimes \infty$ dimensional initial data space. Now keeping $F(r)$ completely free and fixing finite number of coefficients in $f(r)$, as we have done above, to demonstrate the existence of strong naked singularities still leaves us with a $2 \otimes \infty$ dimensional parameter space, and hence the final result in terms of either a naked singularity or black hole is stable under a range of perturbations in the newly constrained parameter space. We would like to contrast this situation with the possible scenario when we could possibly have only a finite dimensional initial data space which generates a strong naked singularity, as a subset of an infinite dimensional initial data space. Then, since the strong curvature singularity occurs only for a finite number of coefficients in the space of finitely many coefficients characterizing the complete space of initial data, we could possibly say that any generic point in the complete space will not lie on such a hypersurface of strong singularity, and hence the data generating strong singularities must be a set of "measure zero" in some sense. Such an argument gives an idea on the genericity of the occurrence of strong curvature singularities in gravitational collapse which could be either naked or covered.

17.6 Discussion and Conclusion

We have used here collisionless fluid, that is dust with $p = 0$ equation of state, as the model to analyze the issue of the final fate of gravitational collapse of a massive cloud, if the collapse started from a "physically reasonable" initial data in terms of the initial density and velocity profiles of the cloud. We have imposed only rather general and physically reasonable conditions such as the matter satisfies the weak energy condition, and that the collapsing dust avoids shell-crossing singularities or caustics (this is essentially because our main focus of interest is the nature of the shell-focusing singularity occurring at the center). For the sake of clarity, and to generate maximum possible physical insight, we have confined here the attention to the class of Taylor expandable density and velocity profiles only and considered mainly the occurrence of strong curvature singularities which are naked or hidden inside black holes.

An important limitation of the analysis presented here could be thought of as the choice of the dust equation of state. The question as to whether the introduction of pressure will significantly modify the conclusions given here needs to be looked into carefully. It may be pointed out, however, that one of the motivations for not considering the general equation of state presently is the possibility indicated by recent work [27] that the specific properties of matter fields may not turn out to be the key factor in deciding the final fate of gravitational collapse. The indications by the work such as above are that the nature of the central singularity should essentially depend on the choice of initial data, and also there is some kind of a pattern in the behavior of the central point ($r = 0$), and the other points on the singularity curve, regardless of the exact form of the matter used [4, 7, 27, 28]. Again, while deciding on the issue of how significant (or insignificant) the assumption of dust is, the arguments such as those by Hagedorn [29] need to be considered that a very soft equation of state is a good approximation under extreme physical conditions of the advanced state of collapse. It is possible that even if we include pressures, the collisions also will give a contribution to the energy-momentum tensor, and will accelerate the formation of singularities [30, 31]. Hence the collisionless assumption could represent fairly general features of collapse and deserves serious consideration from the point of view of possible physical applications. From such a perspective, if we assume the formation of singularity in collapse, as implied by the singularity theorems as well as the physical considerations on the final fate of collapsing massive stars which have exhausted their nuclear fuel, then we have tried to argue here that "actual and real" strong curvature singularities (naked or covered) do appear in a rather "generic" way as the end product of gravitational collapse.

Acknowledgements

We thank I. H. Dwivedi and T. P. Singh for comments and useful discussions. We thank the organizers of the Haifa meeting, and in particular Amos Ori, who provided an excellent atmosphere to discuss these ideas.

Bibliography

[1] J. R. Oppenheimer and H. Snyder, Phys. Rev. **56**, 455 (1939).

[2] R. Penrose, Riv. del. Nuovo Cim. **1**, 252 (1969).

[3] P. S. Joshi, *Global aspects in gravitation and cosmology*, (Clarendon press, Oxford, 1994), and references therein.

[4] D. Christodoulou, Comm. Math. Phys. **93**, 171 (1984).

[5] D. M. Eardley and L. Smarr, Phys. Rev. D **19**, 2239 (1979).

[6] R. P. A. C. Newman, Class. Quantum Grav. **3**, 527 (1986).

[7] P. S. Joshi and I. H. Dwivedi, Phys. Rev. D **47**, 5357 (1993).

[8] P. S. Joshi and T. P. Singh, Phys. Rev. D **51**, 6778 (1995).

[9] T. P. Singh and P. S. Joshi, Class. and Quantum Grav. **3**, 559 (1996).

[10] H. Bondi, Mon. Not. R. Astron. Soc. **107**, 410 (1947).

[11] R. Geroch, Ann. Phys. **48**, 526 (1968).

[12] C. J. S. Clarke, *The analysis of space-time singularities*, (Cambridge University Press, Cambridge, 1993), and references therein.

[13] R. Penrose, Phy. Rev. Lett. **14**, 57 (1965); S. W. Hawking and R. Penrose, Proc. R. Soc. London **A 300**, 187 (1970).

[14] S. W. Hawking and G. F. R. Ellis, *The large scale structure of space-time*, (Cambridge university press, Cambridge, 1973).

[15] F. J. Tipler, Phys. Lett. **A 64**, 8 (1977).

[16] P. Yodzis, H. J. Seifert and H. Muller zum Hagen, Comm. Math. Phys. **34**, 134 (1973); P. Szekeres, Phys. Rev. D **12**, 2941 (1975).

[17] C. J. S. Clarke and A. Królak, J. Geom. Phys. **2**, 127 (1985).

[18] A. Papapetrou and A. Hamoui, Ann. Inst. H. Poincare **A IV**, 343 (1967).

[19] P. Szekeres and Lun Anthony, *What is a shell crossing singularity?*, Journal of the Australian Mathematical Society **B** (1996).

[20] A. Ori and É. É Flanagan, Phys. Rev. D **53** R1754 (1996).

[21] W. S. Burnside and A. W. Panton, *The theory of equations*, Vol **1** (1960).

[22] I. H. Dwivedi and P. S. Joshi, Class. Quantum Grav. **14**, 1223 (1997).

[23] D. O. Gough, in *Astrophysical Fluid Dynamics*, edited by J. P. Zahn and J. Zinn-Justin (Elsevier Science, New York, 1993), p. 339.

[24] H. M. Antia, Phys. Rev. D **53**, 3472 (1996).

[25] L. Spitzer Jr., *Dynamical evolution of globular clusters*, (Princeton university press, Princeton, 1987).

[26] S. Jhingan, P. S. Joshi and T. P. Singh, Class. Quantum Grav. **13**, 3057 (1996).

[27] P. S. Joshi and I. H. Dwivedi, Comm. Math. Phys. **166**, 117 (1994).

[28] P. S. Joshi and I. H. Dwivedi, Comm. Math. Phys. **146**, 133 (1992); Gen. Rel. Grav. **24**, 129; A. Ori and T. Piran, Phys. Rev. Lett. **59**, 2137. (1987); Phys. Rev. D **42**, 1068 (1992).

[29] R. Hagedorn, Nuovo Cimento **56 A**, 1027 (1968).

[30] C. Misner, K. S. Thorne and J. A. Wheeler, *Gravitation*, (San Francisco, Freeman, 1973); R. Penrose R., Nature **236**, 377 (1972).

[31] S. L. Shapiro and S. L. Teukolsky, American Scientist, **79**, 330 (1991); Phys. Rev. Lett. **66**, 994 (1991); in *Proceedings of the 13th International Conference on General Relativity and Gravitation*, Ed. R. J. Gleiser, C. N. Kozameh and O. M. Moreschi (Institute of Physics Publishing, Bristol, 1993), and references therein.

LATE TIME DYNAMICS OF SCALAR PERTURBATIONS OUTSIDE ROTATING BLACK HOLES

Leor Barack

Department of Physics,
Technion—Israel Institute of Technology, 32000 Haifa, Israel

Abstract

We present a new analytic approach for the investigation of linear massless scalar test-fields, propagating on the exterior background of black holes. This method, perturbative in its mathematical nature, is applicable to Kerr black holes, for which complications due to the lack of spherical symmetry prevented previous study. Our analysis also throws some new light on the mechanism of radiative power-law tails formation in general. When applied to the spherical Schwarzschild black hole, this scheme produces the well known results; in particular the late-time power-law tails are reconstructed, both at null infinity and along $r = const$ lines. Applying the scheme to the Kerr exterior, the non-spherical dynamics is treated in terms of interaction between spherical modes. For a generic initial pulse of certain multipole order l and certain magnetic number m, we find that any l'-mode of even $|l'-l|$, satisfying $l' \geq m$, will be observed along null infinity. Each of these modes will be characterized by a Schwarzschild-like tail, though with a different amplitude. As a consequence, the dominant mode along null infinity at late times is the $|m|th$ [or the $(|m|+1)th$ if $l-m$ is odd]. This observation agrees with the results of a recent numerical study by Krivan, Laguna, and Papadopoulos. Also discussed is the possibility of extending the analysis to gravitational perturbations in Kerr.

18.1 introduction

18.1.1 Review of the problem

It is well established that the gravitational field outside a black hole created by a generic gravitational collapse relaxes to the stationary Kerr-Newman field. This result, valid under a few plausible assumptions, has been formulated long ago in a series of theorems, collectively referred to as the "*no-hair theorem*" (see, for example, [1]). The theorem implies, in particular, that all the initial characteristics of the collapsing object are radiated away during the collapse, except for its mass, charge, and angular momentum. These three quantities, characterizing the Kerr-Newman field, are conserved by conservation laws; any other quantity will vanish by the (late) time the gravitation field approaches its stationary state.

The underlying mechanism for this relaxation process was first demonstrated by Price [2] for a nearly spherical collapse. Price analyzed the dynamics of massless integer-spin test fields, evolving on a fixed Schwarzschild background, and showed that when viewed from a fixed spatial point outside the black hole, the waves die off at late time with an inverse power-law tail (in the Schwarzschild time t),whose power index depends only on the multipole number l of the mode in consideration. Later it was shown by analytic and numerical methods ([3]), that at late time the nearly-spherical collapse exhibit tails also at $\mathcal{J}^+$ (future null infinity) and along the event horizon. The presence of tails was explained as a consequence of back-scattering of the outgoing radiation off spacetime curvature at very large distances.

By virtue of previous studies (to be described below) we now have the following schematic picture concerning the dynamics of waves outside a nearly-spherical collapsing object: For clarity and simplicity consider a perturbation in the form of a compact pulse of radiation somewhere outside the collapsing object, emitted at some time during the collapse. This pulse may represent radiation emerging from the surface of the collapsing object, as well as any other form of perturbation on the background geometry. We shall refer to it as the "initial pulse". We may then indicate three successive stages of the wave evolution. First, the exact shape of the waves depends on the detailed form of the initial pulse. This relatively short stage is followed by the "quasinormal ringing" stage, during which the waves undergo exponentially-decaying oscillations with (complex) frequencies completely determined by the mass and electric charge of the central object. Finally, as the quasinormal ringing decays exponentially in time, it leaves behind an inverse power-law decaying tail of radiation. During the last two stages, the details of the initial pulse affect the shape of the waves only through a global amplitude factor.

Historically, the study of wave dynamics outside black holes was motivated by the will to construct a detailed description of the relaxation process to a stationary "no-hair" state. In this context, the study provided insight to issues as nonspherical stellar dynamics and black holes formation. However, further motivations exist. For example, quasinormal ringing is expected to characterize the radiative outflux from a realistic astrophysical system right after a black hole is formed. Observing such a pattern of gravitational radiation (e.g. with the LIGO experiment, see [4]) may directly indicate the existence of black holes [5]. Analysis of wave dynamics outside black holes also has crucial relevance to the study of their internal structure, as these waves provide the input for the internal wave problem. For example, it has been shown [6] that the form of the slowly decaying wave tail along the event horizon affects the strength of the mass-inflation singularity at the Cauchy horizon inside charged and rotating black holes. Finally, the study of waves evolving on curved spacetimes is interesting from a theoretical point of view: such waves, massless just as well as massive, do not propagate along light cones solely; rather they also spread inside them. Regardless of the presence of an event horizon, it is this feature of the evolution which is responsible for the phenomenon of radiative tails.

The existence of late time tails was demonstrated for linear perturbations of both Schwarzschild [2],[3] and Reissner–Nordström [3] exteriors. Remarkably, numerical analyses of the strong non-linear dynamics of self-gravitating fields yield the same late time decay rates as for the linear fields (see [7],[8]). This, of course, encourages the application of the perturbative approach, even though the underlying theory is intrinsically non-linear.

When studying the dynamics of realistic fields on curved spacetimes, a scalar field is often employed as a simplifying model (a "toy-model"). Then, usually, the Klein–Gordon equation for the scalar waves is used. The physical problem is then mathematically formulated as a problem of wave scattering off an effective potential. It was demonstrated in several previous studies (e.g. [2],[3]) that the evolution of physical fields (electromagnetic and gravitational) shares a lot of common features with its scalar analogue; In particular, the three stages of evolution mentioned earlier are qualitatively the same on both cases. We are thus encouraged to consider the relatively simple scalar problem when developing the calculation scheme.

Ching *et al.* [9] explored the evolution of scalar waves on a wide class of asymptotically flat spherically symmetric spacetimes, represented by effective potentials of the form $(\ln r)^\beta/r^\alpha$ (where $\beta = 0, 1$ and $\alpha > 2$ are parameters). This class includes the Schwarzschild and Reissner–Nordström geometries. They found that, generically, the scalar wave behavior at late time is not characterized by a strict power-law tail, but rather it has the form $(\ln t)^\beta \times$(inverse-power in t). In this sense, the Schwarzschild and Reissner–Nordström geometries belong to a special subgroup of spacetimes. For the monopole moment of the scalar radiation, it was shown by Gómez *et al.* [10] that the form of the tails (whether "logarithmic" or not) depends on whether or not the Newman–Penrose constant for the field vanishes.

The analysis by Ching *et al.* follows a technical scheme, first introduces by Leaver [11], in which the waves are first Fourier-decomposed, then evaluated in the complex frequency plane. In this technique, the late time tails are explained in terms of a branch cut in the Green's function in the frequency plane. The fact that the branch cut is due to the form of spacetime structure at asymptotically large radius (see [9] for details) implies, again, that the tails originate from scattering off the curvature at large distances. This observation, in turn, suggests that the development of tails is independent of the existence of an event horizon, and thus that the tail phenomenon is of a more universal nature: it may also be a feature of realistic stellar dynamics. This, indeed, was suggested by Gundlach *et al.* [3] and Ching *et al.* [9], and will be further discussed below.

Recently [12], Brady *et al.* studied scalar waves dynamics in the non-asymptotically-flat exteriors of Schwarzschild–de Sitter and Reissner– Nordström–de Sitter black holes. Contrary to the asymptotically-flat geometries, no power-law tails are observed in this cases. Instead, the waves decay exponentially at late time.

Now, realistic black holes, formed by generic gravitational collapses, are expected to spin, as do astrophysical stars. On the contrary, models of black holes with electric charge and/or cosmological constant are mainly hypothetical. Therefore it is natural to ask how the tail phenomenon is affected by the presence of angular momentum in the background geometry. We may suspect that the late time behavior of waves, evolving outside a rotating Kerr–Newman black hole, will be qualitatively of the same nature as in the spherically-symmetric spacetimes. This is because the Kerr–Newman geometry asymptotically approaches that of Schwarzschild at very large radii, and presumably it is this far region of spacetime whose structure determines the form of the late time radiation.

Yet, in spite of the obvious interest, no analytic scheme has been constructed to describe wave dynamics outside rotating black holes. The basic obstacle is, of course, the fact that the axially-symmetric Kerr–Newman black hole possesses three non-trivial dimensions, instead of only two in the spherical cases. This makes both analytic and numerical investigation significantly more complicated.

A separable wave equation, governing integer spin perturbations (i.e. scalar, electromagnetic, or gravitational) of Kerr black holes was introduced by Teukolsky [13]. A full separation of Teukolsky's equation is possible only by means of the Fourier transformation, that is by first writing $\psi = \int e^{-i\omega t}\varphi_\omega d\omega$, where ψ is any component of the (time-dependent) wave function, t is the Boyer-Lindquist time coordinate, and φ_ω represents the Fourier components of the wave, which are time-independent but depend on their frequencies ω. Accordingly, we can mark three possible approaches to the problem of wave dynamics outside rotating black holes. (1) One may attempt to carry out a full 2+1 numerical investigation of the corresponding Teukolsky equation (the azimuthal coordinate is trivially separated). There has been an initial progress in this direction [14], though much work is still needed. Or, (2) the wave equation may be fully separated, using Teukolsky's equation, and then the frequency-dependent solutions of the resultant radial equation be evaluated (either analytically or numerically). Krivan *et al.* [14] give arguments why this approach would be impractical for the propose of late time numerical integration. Finally, (3) one may give up the separation of the t coordinate, then confront a three-dimensional initial-value problem for the evolution of the scalar wave (the azimuthal coordinate is again trivially separated). Our analysis adopts the third approach, as we describe below.

18.1.2 Outline of our analytic approach

We present a new analytic method for evaluating the late time solutions of the initial value problem for linear massless scalar fields, propagating outside both rotating and non-rotating black holes. As a first step, the analytic scheme has been developed and tested for the Schwarzschild background, where comparison to previous results by other approaches is possible. In the second and most interesting part of the analysis, the same scheme is applied to the rotating Kerr black hole. In this paper we demonstrate in some details the application of the scheme to Schwarzschild. We then show briefly how the same technique may be used to analyze waves in Kerr, and present the main results of this calculation. Further details of the application of our scheme to Kerr will be given is [15]

The technical scheme can be briefly outlined as follows: The wave function is first written in terms of its spherical harmonics expansion. When applied to spherically-symmetric spacetimes (e.g. Schwarzschild) the Klein–Gordon equation thereby reduces to a set of independent wave equations (in $1 + 1$ dimensions) for the various modes. Next, each mode is decomposed into an infinite sum of functions, defined such that the first one describes the evolution of the scalar field over a flat space, while the following functions represent (in a recursive way) the effect of scattering due to spacetime curvature. We can then write a closed formal expression for each of these functions, using the Green's function method. Finally, applying combined numerical experiments and analytic arguments, we find indications that (i) the sum of the above set of functions converges at null infinity, and (ii) the second function of the series (the first to describe spacetime curvature) well represents the asymptotic overall form of the scalar field at null infinity. Since this second term is rather simple to calculate, we are thus able to obtain an analytic expression for the late-time behavior of the scalar field at null infinity.

The same analytic scheme is also applicable to the case of rotating black holes. Here, angular separation of variables is problematic (see below), what complicates the analysis significantly. However, our method allows us to handle this difficulty, as we shall briefly describe later.

The rest of this paper is arranged as follows: In Section 18.2 we demonstrate our analytic approach for the case of the Schwarzschild black hole. We first formulate the corresponding initial value problem, then present in some details the technical scheme we used to solve it, and the main results obtained. Numerical results are presented for comparison. In Section 18.3 we describe in brief the application of the scheme to Kerr black holes, and present the main results obtained so far. We also discuss the possibility of extending the analysis to gravitational perturbations in Kerr.

18.2 The Analytic Approach: Schwarzschild Black Holes

18.2.1 The initial-value problem

In this section we introduce our analytic scheme and summarize the results of its application to Schwarzschild black holes. The full details of the analysis are presented in a paper in preparation [15].

We consider the evolution of initial data, representing a generic pulse of massless scalar radiation, on a fixed Schwarzschild background. The scalar field is assumed to satisfy the (minimally-coupled) Klein–Gordon equation [16]

$$\Box\Phi \equiv \Phi_{;\mu}^{\;;\mu} = 0, \tag{18.1}$$

where Φ represents the scalar wave, and covariant differentiation is denoted by a semicolon.

Decomposing the field into spherical harmonics,

$$\Phi(t,r,\theta,\varphi) = \sum_{l=0}^{\infty} \sum_{m=-l}^{l} \phi^l(t,r) Y_{lm}(\theta,\varphi), \tag{18.2}$$

we obtain an independent field equation for each of the components $\phi^l(t,r)$:

$$\left(1-\frac{2M}{r}\right)^{-1} \phi^l_{,tt} - \left(1-\frac{2M}{r}\right)\phi^l_{,rr} - \frac{2}{r}\left(1-\frac{M}{r}\right)\phi^l_{,r} + \frac{l(l+1)}{r^2}\phi^l = 0. \tag{18.3}$$

Here t, r, θ and φ are the standard Schwarzschild coordinates, M is the mass of the black hole, and l is the multipolar number of the mode under consideration.

To simplify eq. (18.3) we define a new wave function $\Psi^l(t,r) \equiv r\,\phi^l(t,r)$, and introduce the double-null characteristic coordinates (the Eddington–Finkelstein coordinates) $v \equiv t + r_*$ and $u \equiv t - r_*$, where $r_*(r) \equiv r + 2M \ln\left(\frac{r-2M}{2M}\right)$. The "tortoise" coordinate r_* ranges from $-\infty$ (the event horizon) to $+\infty$ (space-like infinity). The wave equation now reads

$$\Psi^l_{,uv} + V^l(r)\Psi^l = 0, \tag{18.4}$$

with

$$V^l(r) = \frac{1}{4}\left(1-\frac{2M}{r}\right)\left[\frac{l(l+1)}{r^2} + \frac{2M}{r^3}\right]. \tag{18.5}$$

V^l serves as an effective potential in the wave equation (18.4). It represents the effect of spacetime curvature (terms proportional to M), as well as the centrifugal effects on the non-spherical ($l > 0$) modes of the wave [terms proportional to $l(l+1)$]. The effective potential is sketched in figure 18.1 as a function of r_* for $l = 0, 1, 2$.

The effective potential possesses three essential features which play a crucial rule in our analysis:

- It is *localized* (in a sense that becomes clear from figure 18.1 above) in a small region near $r_* = 0$ (around $r \sim 3M$).
- Its amplitude decays exponentially in r_*/M at small r_* values (making the potential effectively zero near the event horizon).
- At large radii it decays slowly, as r_*^{-2}.

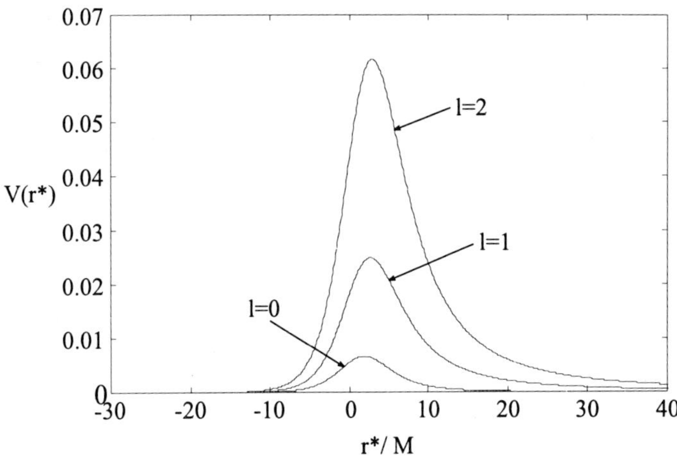

Figure 18.1: The effective potential $V^l(r_*)$ [for $l = 0, 1, 2$].

Evidently, the late time behavior is affected mostly by the shape of the potential at large radii. In fact, it is suggested [3] [7] [9] that to the leading order in $1/t$ the dynamics of the waves is solely determined by the structure of the potential at the asymptotically far region of spacetime. This is in contrast to the early evolution (e.g. the quasinormal ringing stage), which is strongly affected by the fine details of the potential shape at small radii [17].

To set-up the initial value problem, we specify initial data on two characteristic surfaces outside the event horizon, as sketched in the Penrose diagram of figure 18.2. We choose initial data in the form of some compact outgoing pulse specified on the ingoing null surface $v = 0$:

$$\begin{cases} \Psi(u = u_0) = 0 \\ \Psi(v = 0) = \Gamma(u) \end{cases}, \tag{18.6}$$

where $\Gamma(u)$ is some function with a compact support (the "pulse") starting at $u = u_0$. u_0 is a free parameter in our analysis, its value being of some importance later.

The evolution equation (18.4), together with the initial conditions (18.6), establish a well defined initial value problem for the scalar field anywhere in the domain S outside the event horizon.

18.2.2 The iterative expansion

Our goal is to find the solution of the above initial-value problem at late-time. This is done in two steps. At the first and most crucial one, we deduce the late-time behavior of the wave along $\mathcal{J}^+$ (that is, for $v \to \infty$, $u =$const$\gg M$). This requires a full 1+1 dimensions analysis, whose details will be briefly described in this section. Then, a

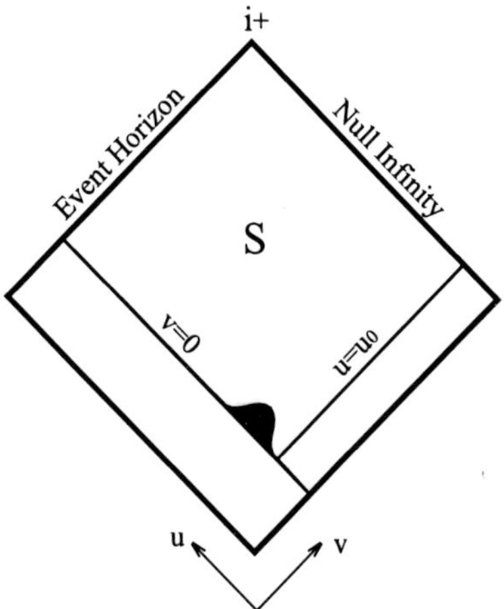

Figure 18.2: The set-up of initial data. Shown is the Penrose diagram representing the external Schwarzschild geometry.

simpler local, one-dimensional, analysis near i^+ (future timelike infinity) can be used to determine the late-time behavior of the wave along any line of constant r outside the event horizon. This second step, described in short in section 18.2.4, is based on necessary information deduced in the first step.

To analyze the wave equation (18.4) we first look at the $1/r_*$ expansion of the effective potential:

$$V^l(r) = V_0^l + \delta V^l, \tag{18.7}$$

where

$$V_0^l = \frac{l(l+1)}{4r_*^2}, \tag{18.8}$$

and

$$\delta V^l = M\frac{\{1 + l(l+1)[\ln(r_*/2M) - 1]\}}{2r_*^3} + O\left(\frac{ln^2(r_*/2M)}{r_*^4}\right). \tag{18.9}$$

The first term of the expansion represents the Minkowski potential V_0^l, which is purely centrifugal. This simply manifests the fact that The Schwarzschild spacetime asymptotically approaches Minkowski at large radius. The rest of the terms (the sum of which is denoted by δV^l) are proportional to the mass M, thus correspond to corrections induced by spacetime curvature.

We now decompose the wave function to an infinite series

$$\Psi = \Psi_0 + \Psi_1 + \Psi_2 + \cdots, \tag{18.10}$$

such that the components Ψ_n are defined by the following recursion formula:

$$\Psi_{n,uv} + V_0 \Psi_n = \begin{cases} 0 & ,n=0 \\ -(\delta V)\Psi_{n-1} & ,n>0 \end{cases} \qquad \text{for } r_* > r_0 \ , \tag{18.11a}$$

and

$$\Psi_{n,uv} = \begin{cases} 0 & ,n=0 \\ -V\Psi_{n-1} & ,n>0 \end{cases} \qquad \text{for } r_* < r_0 . \tag{18.11b}$$

Here $r_0 > 0$ is a free parameter of order M, that divides spacetime into two regions. On both regions, V_0 is finite (and therefore δV is finite either). The initial data for the Ψ_n functions are chosen as follows:

$$\Psi_n(u = u_0) = 0 \ (\forall n \geq 0) \ ; \qquad \Psi_n(v = 0) = \begin{cases} \Gamma(u) & ,n=0 \\ 0 & ,n>0 \end{cases} . \tag{18.12}$$

By a formal summation of eqs. (18.11, 18.12) over n, we recover eqs. (18.4, 18.6) for Ψ. This suggests that, provided that the series (18.10) converges, it will yield the correct function Ψ.

Equations (18.11, 18.12) constitute an infinite hierarchy of initial value problems for the functions Ψ_n. Each of the evolution equations consists of a simple homogeneous part, which is the free evolution equation in Minkowski's spacetimes, with or without the centrifugal potential (for $r > r_0$ or $r < r_0$ respectively). In addition, each of the functions Ψ_n (excluding Ψ_0) has a source term proportional to the previous function in the series. Therefore, in principle, if the solution for Ψ_0 is found, and the appropriate Green's function is constructed, then we should be able to solve for the functions Ψ_n one by one, using the Green's function method.

Numerical simulations strongly suggest that at $\mathcal{J}^+$ the series (18.10) converges to the correct function Ψ. Moreover, in the case $u_0 \gg M$, which will concern us here, we find (both numerically and analytically) that the late-time behavior of Ψ in Schwarzschild at $\mathcal{J}^+$ is well approximated by Ψ_1 (the correction coming from $n \geq 2$ is smaller by a factor proportional to M/u_0). Thus, in our scheme, constructing an analytic expression for Ψ_1 will suffice to determine the essential features of the late-time dynamics at $\mathcal{J}^+$.

The analysis shows that at $\mathcal{J}^+$, Ψ_1 depends only weakly on the value of the parameter r_0. Specifically, it is independent at all in r_0 to the leading order in M/u and in M/u_0. Now, though the analysis with any non-negative value of r_0 is technically possible, a model in which we take the limit $r_0 \to 0$ is considerably simpler. In this model, the potential diverges at the sphere $r_* = 0$. Nevertheless, there is still a regular solution corresponding to this model (see below).

The analysis shows that the model in which $r_0 \to 0$ cannot describe properly the wave behavior at small r values. This is especially clear at the region $r_* \leq 0$ of spacetime, where in this model all functions Ψ_n vanish[18], and thus so is Ψ. Surely, this cannot represent the physical situation, since radiation coming far away from the black hole is known to be present in the vicinity and along the event horizon. However, at this stage of the analysis, we shall be concerned only in the behavior at $\mathcal{J}^+$. Here, the wave is not affected by the behavior at small radii (to the leading order in $1/u$ and in $1/u_0$); in particular, it is indifferent to the value of r_0.

Therefore, our strategy would be to study the simple model with $r_0 \to 0$ in order to yield the correct late-time results only along $\mathcal{J}^+$. Then, based on these results, apply a different method, to be described in section 18.2.4, in order to complete the picture of late-time behavior.

In the rest of this paper we will concentrate on the model with $r_0 = 0$. However, we shall first present here the major preliminary results of the analysis for a model with a positive value of r_0. This general analysis was carried out only to a limited extent, with the aim of clarifying several important features of our model, mainly (*i*) the issue of convergence at small radii, and (*ii*) the effect of radiation scattered at small radii on the behavior at $\mathcal{J}^+$ (more details of this part of the analysis are given in [15]). It was demonstrated, using combined analytic and numerical methods, that:

- As stated above, the form of the solution at $\mathcal{J}^+$ is independent of r_0 to the leading order in $1/u$ and in $1/u_0$. In particular, our expansion seems to converge at $\mathcal{J}^+$ regardless of the value of r_0, provided only that u_0/M is large enough.

- There is a range of values of the parameter r_0, for which our expansion seems to converge anywhere, including at small radii and down to the horizon (assuming again that u_0/M is large). The rate of convergence can be controlled by tuning the value of r_0. It was found, for example, that the choice $r_0 \sim M$ (corresponding to $r(r_0) \sim 2.5M$) yields a good convergence rate. (So far, this result has been indicated to be valid only for the leading order in $1/v$ at fixed r. More study of this subject is being carried out currently.)

18.2.3 Solution of the initial-value problem

In what follows we shall focus attention on the model with $r_0 = 0$. As mentioned earlier, in this model the demand for regularity at $r_* = 0$ automatically implies that all functions Ψ_n vanish there, and hence also at $r_* < 0$. Thus the sphere $r_* = 0$ effectively serves in this model as a reflective mirror, or an impenetrable potential wall. Accordingly, we may restrict attention only to the part of spacetime shown in figure 18.3, that is the region $r_* \geq 0$. Here we would like to solve the set of equations (18.11a) for the functions Ψ_n, with the initial conditions (18.12), supplemented by

$$\Psi_n(r_* = 0) = 0. \tag{18.12a}$$

More specifically, we wish to explore the late-time solutions along $\mathcal{J}^+$. (We recall that we should not expect the solutions of this model to represent the correct solutions at small r values. Even at $\mathcal{J}^+$, the model yields the correct result only to the leading order in $1/u$ and in $1/u_0$).

By definition, Ψ_0 obeys the wave equation

$$\Psi^l_{0,uv} + V^l_0(r)\Psi^l_0 = 0, \tag{18.13}$$

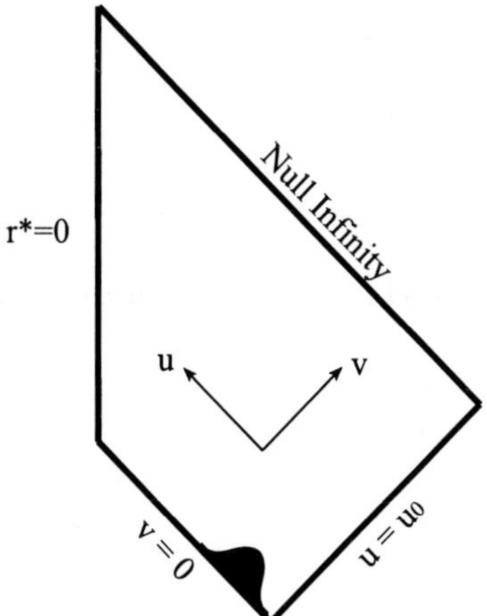

Figure 18.3: The set up of initial data (model with $r_0 = 0$). Shown is the relevant part of the Penrose diagram.

where V_0^l is the purely centrifugal potential defined in eq. (18.8). Therefore, Ψ_0 is nothing but the solution of the corresponding problem in Minkowski, where $r_* = r = 0$ is the origin of coordinates. In this respect, the Ψ_n's decomposition is actually an expansion of Schwarzschild about Minkowski.

The general solution of eq. (18.13) is given by

$$\Psi_0^l = \sum_{n=0}^{l} A_n^l \frac{g^{(l-n)}(u)}{(v-u)^n} + \sum_{n=0}^{l} B_n^l \frac{h^{(l-n)}(v)}{(v-u)^n}, \tag{18.14}$$

where $g(u)$ and $h(v)$ are arbitrary functions, A_n^l and B_n^l are (known) coefficients, and the parenthetical superindices on $g(u)$ and $h(v)$ indicate the number of times these functions are differentiated.

The functions $g(u)$ and $h(v)$ are uniquely determined by the initial conditions (18.12). For our choice of initial date (a compact outgoing pulse) the second sum in (18.14) vanishes, and the function $g(u)$ always satisfies

$$\begin{cases} g(u \geq 0) \equiv 0 \\ g(u < 0) \propto u^{l+1} \quad \text{near } u = 0. \end{cases} \tag{18.15}$$

(Exceptional is the case $l = 0$, for which $g(u)$, and thus Ψ_0 itself, vanish identically right after the initial pulse ceases.)

We conclude that for any mode, Ψ_0 is sharply "cut off" at $u = 0$ [19]. This is shown in the diagram of figure 18.4.

Since Ψ_0 vanishes identically at $u > 0$, it is obvious that Ψ_0 does not contribute to the late time radiation. Rather it serves as a source to higher terms in our iterative

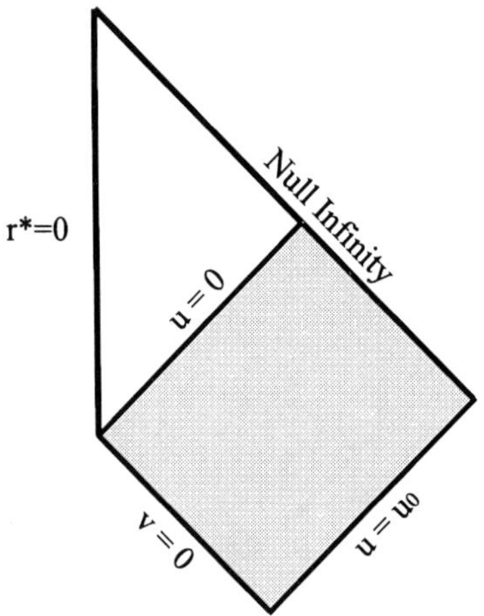

Figure 18.4: Compactness of Ψ_0. The shaded area is the region where Ψ_0 survives.

expansion. For Ψ_1 we have (by definition)

$$\Psi_{1,uv} + V_0 \Psi_1 = -(\delta V)\Psi_0, \tag{18.16}$$

with the initial conditions $\Psi_1(u = u_0) = \Psi_1(v = 0) = 0$ (together with (18.12a)).

The solution of eq. (18.16) at $u > 0$ may be formally written as follows:

$$\Psi_1^l(u > 0, v) = -\int_{u_0}^{u} du' \int_{u}^{v} dv' G_M^l(u, v, u', v')\, \delta V^l(u', v')\, \Psi_0^l(u', v'). \tag{18.17}$$

Here, G_M^l is the Green's function in Minkowski's spacetime, i.e. the solution of $G^l_{,uv} + V_0^l G^l = \delta(u - u')\delta(v - v')$ for $u' > u_0$ and $v' > 0$, with the initial conditions $G(u = u_0) = G(v = 0) = 0$, and the boundary condition $G(r_* = 0) = 0$.

The detailed form of the Green's function is given in [15]. An important result is that, for a given evaluation point (u, v), G_M vanishes for source points with $v' < u$. For that reason, the integration over v' in (18.17) begins at u, and the double-integration region is represented by the rectangle shown in figure 18.5. We note that this is a special feature of the model with $r_0 \to 0$, and it contributes significantly to its simplicity.

We observe that when evaluated at late time, Ψ_1 is influenced only by sources located at large radii. As $t \to \infty$, it is only the asymptotically far region of spacetime that affects Ψ_1. This is a manifestation of the above assumption that the asymptotically late time behavior of the waves is independent of the structure of spacetime at small radii.

The details of the calculation of the double-integral in eq. (18.17) are presented in

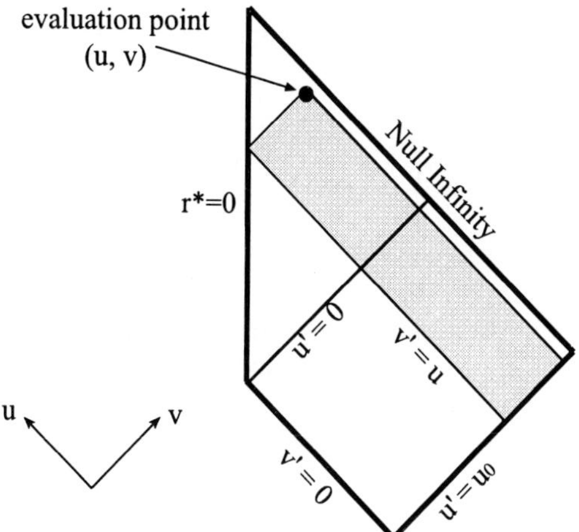

Figure 18.5: Rectangle of integration. The field at (u, v) is affected only by sources located within the shaded area.

[15]. The final result is

$$\Psi_1^l(u>0,\mathcal{J}^+) = a_l M \int_{u_0}^{0} \frac{g(u')}{(u-u')^{l+2}} du' + O\left(\frac{\ln u}{u^{l+3}}\right), \tag{18.18}$$

where

$$a_l \equiv \frac{2^{l+1}(l-1)!(l+1)!}{(2l)!}. \tag{18.19}$$

Since $g(u)$ has a compact support, we finally arrive at the simple formula for the late-time behavior of Ψ_1 at null infinity:

$$\Psi_1^l(u>>M,\mathcal{J}^+) = a_l M \frac{I}{u^{l+2}} \quad \text{(to leading order in } 1/u\text{)}, \tag{18.20}$$

with

$$I \equiv \int_{u_0}^{0} g(u')du'. \tag{18.21}$$

We recall that the function $g(u)$ contains all the information about the shape of the initial pulse (in view of eq. (18.14), $g(u)$ can be expressed as a functional of the initial data function $\Gamma(u)$, by solving an ordinary differential equation of order l). Yet, it is only the integrated entity I which is exhibited along $\mathcal{J}^+$ at very late time.

Our calculation scheme is also applicable to the case of a nonvanishing *static* initial multipole moment of amplitude μ in the range $u < u_1$ for some u_1. This situation corresponds to $g = \text{const} = \mu$ at $u < u_1$. Applying eq. (18.18) to this case (with the integration starting at $u' = -\infty$) we find

$$\Psi_1^l(u >> M, \mathcal{J}^+) = \left(\frac{a_l}{l+1}\right) M \frac{\mu}{u^{l+1}} \quad \text{(to leading order in } 1/u\text{)}. \tag{18.22}$$

It is therefore found that along $\mathcal{J}^+$, at late time, Ψ_1 is dominated by inverse-power law tails given by eqs. (18.20) and (18.22), according to whether there is or is not an initially static wave. These power-laws are the same as those found by previous studies to characterize the overall wave Ψ along $\mathcal{J}^+$.

The essential question is now whether or not Ψ_1 dominates over the higher terms of the expansion ($\Psi_{n>1}$) at late time. By analogy to eq. (18.17), Ψ_{n+1} can be calculated from Ψ_n in a recursive way, using the Green's function method:

$$\Psi_{n+1}^l(u > 0, \mathcal{J}^+) = -\int_{u_0}^{u} du' \int_{u}^{\infty} dv' G_M^l(u, \infty, u', v')\, \delta V^l(u', v')\, \Psi_n^l(u', v'). \tag{18.23}$$

Our preliminary investigation of Ψ_n for $n > 1$ suggest that:

(i) For all $n \geq 1$, Ψ_n exhibits the the same late-time power-law tail at $\mathcal{J}^+$, $\Psi_n \propto u^{-l-2,1}$, as does Ψ_1.

(ii) The *coefficients* of this leading-order power of Ψ_n decrease with u_0 like $\left(\frac{M}{u_0} \ln \frac{u_0}{M}\right)^n$:

$$\Psi_{n+1}^l(\mathcal{J}^+) \propto \left[\frac{M}{u_0} \ln\left(\frac{u_0}{M}\right)\right]^n u^{-l-2,1}, \tag{18.24}$$

to the leading order in M/u and in M/u_0.

(So far we have shown this in full for $n = 2$, and only vaguely for $n > 2$.) We are thus led to the following conclusion: If our expansion converges, then, at $\mathcal{J}^+$, $\Psi = \Psi_1$ to the leading order in $1/u$ and in $1/u_0$. Phrased differently, the sum of the first few terms in our expansion describes correctly the 'complete' solution at $\mathcal{J}^+$, at late time, provided only that $|u_0/M|$ is large enough. [We note, however, that eq. (18.24) does not necessarily imply convergence of the expansion, even when $\left|\frac{M}{u_0} \ln\left(\frac{u_0}{M}\right)\right|$ is small. This is because we were not able to calculate the proportion factor, which apparently depends on the index n. Numerical experiments, however, strongly indicate convergence—see below]

To conclude this section, we present a few examples of our numerical results. These are in agreement with the analytic results we have been describing so far (figures (18.6), (18.7)). They also strongly suggest a convergence of the sum over Ψ_n, and, moreover, that in the case of large $|u_0|$, the convergence is fast, so that Ψ is well approximated by Ψ_1 at late time (see figures (18.7), (18.8)).

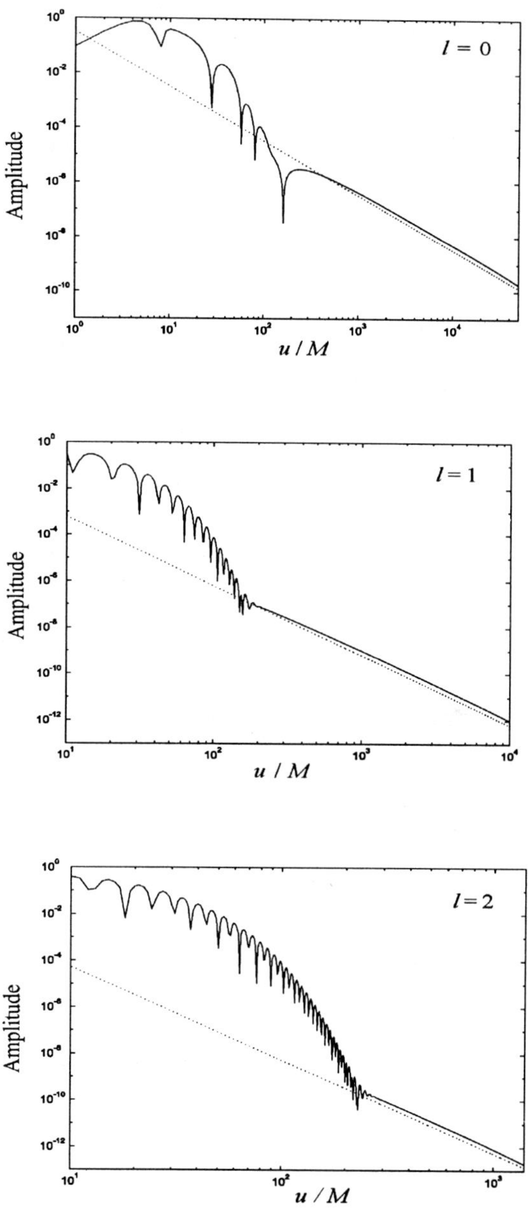

Figure 18.6: Log-Log graphs representing the complete wave functions $\Psi^{l=0,1,2}$ at large constant v (representing $\mathcal{J}^+$). Quasi-normal ringing characterizes the early evolution, while power-law tails dominate at late time. Shown for reference (dotted) are the lines $\Psi \sim u^{-l-2}$. The Initial data is a compact outgoing pulse at $u_0 = 0$.

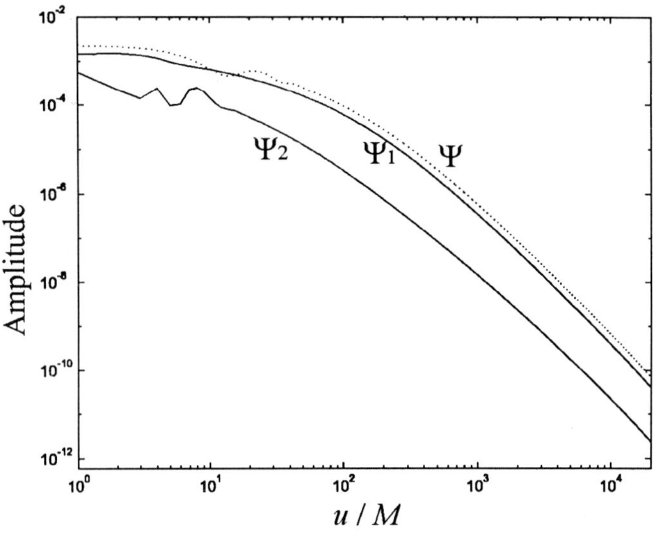

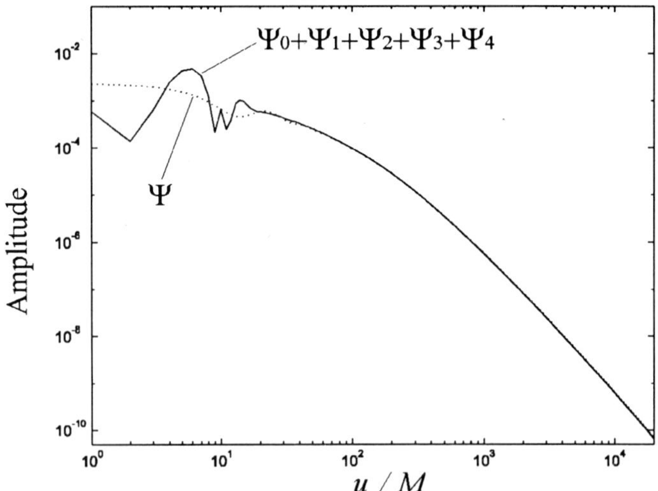

Figure 18.7: Log-Log graphs comparing the first terms in the iterative expansion with the complete wave Ψ at large v (representing $\mathcal{J}^+$), for $l = 1$ and $u_0/M = -150$. The upper figure shows Ψ_1 and Ψ_2 (solid) compared to Ψ (dotted). The bottom figure compares the sum $\Psi_0 + \Psi_1 + \Psi_2 + \Psi_3 + \Psi_4$ (solid) with Ψ (dotted), showing a good fit at late time. Present at the early stage of evolution are remainders of the quasi-normal ringing.

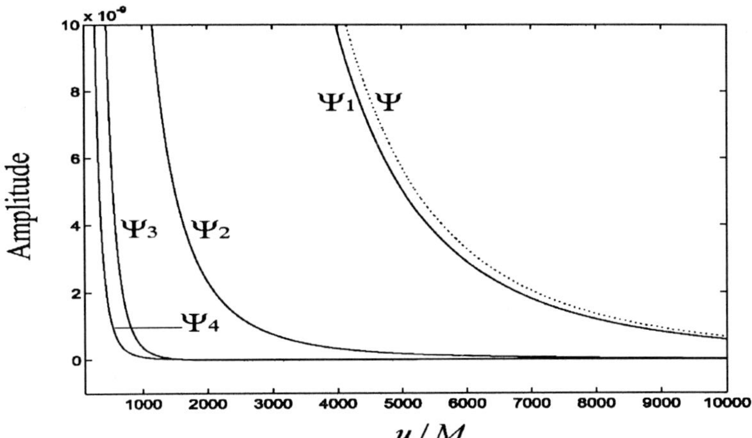

Figure 18.8: Late time tails of the four leading terms in the iterative expansion at large constant v (representing $\mathcal{J}^+$), shown in linear scale for $l = 1$ and $u_0/M = -150$ Also shown is the full wave function Ψ (dotted). The graphs illustrate the expected convergence of the expansion for large $|u_0/M|$ values.

18.2.4 Late time tails at constant r

So far, we have focused attention on the asymptotic behavior at $\mathcal{J}^+$, and found that at late time, $\Psi \propto u^{-l-2}$ (for a compact initial pulse). Based on this result, however, we can now use a simple method to determine the correct power-law index along lines of r =const> $2m$ [20]. Here we shall briefly outline this method.

We make the Anzats that at late time, the wave function admits the expansion

$$\Psi(r,t) = \sum_{k=0}^{\infty} F_k(r) t^{-k_0-k}, \tag{18.25}$$

to which we shall refer as the *late time expansion.* The primary goal is to determine the parameter k_0 and the leading-order function $F_0(r)$, which dominates the asymptotic behavior at late time ($t \gg M$).

Substituting eq. (18.25) in the field equation (18.4), one obtains a hierarchy of *ordinary* differential equations for the functions $F_k(r)$. It is straightforward to solve the ordinary equation for $F_{k=0}$. It is also easy to obtain the asymptotic form of all functions F_k at $r \to \infty$.

In order to determine k_0, we compare expression (18.25) to the previous results at $\mathcal{J}^+$. The comparison is made at the late-time section of $\mathcal{J}^+$, where both analyses are valid. Formally summing over k in eq. (18.25) and taking the limit $r \to \infty$ in an appropriate way, one finds the late-time asymptotic behavior at $\mathcal{J}^+$ to be

$$\Psi(\mathcal{J}^+) \propto \frac{1}{u^{k_0-l-1}}. \tag{18.26}$$

Comparing with eq. (18.20), we find

$$k_0 = 2l + 3 \tag{18.27}$$

Therefore, from eq. (18.25) we conclude that, when observed at fixed radii, the late time radiation is dominated by a tail of the form

$$\Psi \propto \frac{1}{t^{2l+3}} \quad \text{to leading order in } 1/t. \tag{18.28}$$

This coincides with the form originally obtained by Price [2]. It is supported by numerical analysis, as demonstrated in figure 18.9.

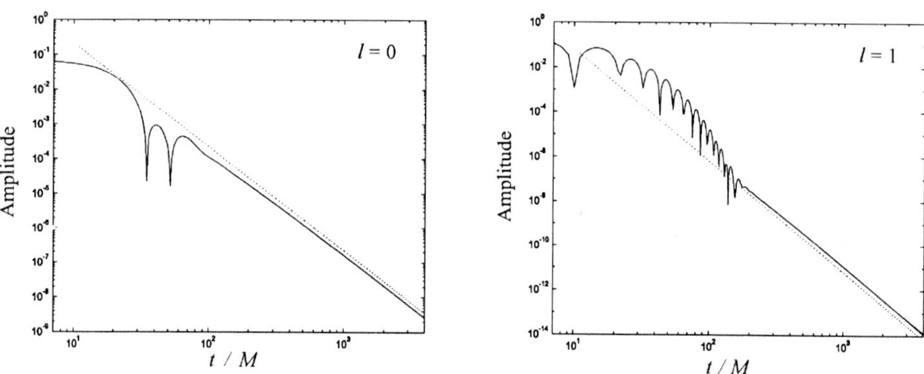

Figure 18.9: Late time tails along r=const lines. Shown (on a Log-Log scale) are numerical results at $r_* = 10M$ for the monopole and dipole modes, with compact initial pulse at $u_0 = -20M$. The lines $\sim t^{-2l-3}$ (dotted) are given for comparison. Quasi-normal ringing characterizes the early evolution.

18.3 The Analytic Approach: Kerr Black Holes

18.3.1 Scalar field in Kerr

As we have mentioned in the introduction, the main objective of this research is to analyze the late-time behavior of perturbations on the background of a spinning black hole. This requires the generalization of the above calculation scheme from

Schwarzschild to Kerr. Here we shall briefly describe the main stages in this generalization.

The Klein–Gordon equation for the scalar field $\Phi(t,r,\theta,\varphi)$ in Kerr spacetime reads (in Boyer-Lindquist coordinates)

$$\begin{aligned}\Phi_{;\mu}^{\ ;\mu} &= \left[\frac{(r^2+a^2)}{\Delta} - a^2(1-x^2)\right]\frac{\partial^2\Phi}{\partial t^2} + \frac{4Mar}{\Delta}\frac{\partial^2\Phi}{\partial t\partial\varphi} + \left(\frac{a^2}{\Delta} - \frac{1}{1-x}\right)\frac{\partial^2\Phi}{\partial\varphi^2} \\ &\quad - \frac{\partial}{\partial r}\left(\Delta\frac{\partial\Phi}{\partial r}\right) + (1-x^2)\frac{\partial^2\Phi}{\partial x^2} + 2x\frac{\partial\Phi}{\partial x} = 0, \qquad (18.29)\end{aligned}$$

where a is the black hole's angular momentum per unit mass, $x \equiv \cos\theta$, and $\Delta \equiv r^2 - 2Mr + a^2$.

In Schwarzschild, the first step of the analysis included the separation of the scalar field into its spherical harmonics components. However, a similar separation of variables is not possible in the case of Kerr, due to the lack of spherical symmetry. While the azimuthal dependence of Φ can still be described by $e^{im\varphi}$, a full separation of variables is possible only in the frequency domain. In this case, the field equation (18.29) admits separable solutions of the form

$$\Phi = \sum_{l,m}\int d\omega e^{-i\omega t}e^{im\varphi}S_\omega^{lm}(\theta)R_\omega^{lm}(r), \qquad (18.30)$$

where $S_\omega^{lm}(\theta)$ are the *spheroidal harmonics.*

This exact separation of the scalar field in Kerr, which is based on the Fourier decomposition, is not suitable for the propose of studying the asymptotic late-time behavior. In particular, we cannot apply it in our iterative scheme, in which the time dependence of the fields is kept during the calculation.

We note, however, that although the scalar field is composed of all frequencies, it is only the very small frequencies which dominate at very late time (the effective frequency of the waves decreases to zero as $1/t$). In this limit ($\omega \to 0$), the spheroidal harmonics reduce to the spherical harmonics. This implies that, to first approximation, it is natural to apply the spherical harmonics angular decomposition in order to study the late-time behavior. The deviations from such a decomposition may then be expressed as correction terms in the field equation: "interaction" terms between the various spherical modes. We shall take the effect of those correction terms into account using our iterative calculation scheme.

To that end, we proceed as follows. We first decompose Φ in spherical harmonics, as in eq. (18.2). The field equation then reads

$$\Phi_{;\mu}^{\ ;\mu} = \sum_{l,m} e^{im\varphi}P_{lm}(\cos\theta)\left\{D^{lm}(t,r)\phi^{lm}(t,r) - a^2\sin^2\theta\frac{\partial^2\phi^{lm}}{\partial t^2}\right\} = 0, \qquad (18.31)$$

where $P_{lm}(\cos\theta)$ are the Legendre polynomials, $\phi^{lm}(t,r)$ are the time-radial components of Φ, and D^{lm} is a second order differential operator. We observe that the second term in the curved brackets, proportional to $a^2\sin^2\theta$, prevents the separation of variables. The presence of this term is a manifestation of the lack of spherical symmetry in Kerr.

Next, we expand $P_{lm}(\cos\theta)\sin^2\theta$ in the Legendre polynomials, and rewrite (18.31) in terms of the new variable $\Psi(r,t)\equiv\sqrt{r^2+a^2}\,\phi(r,t)$. Reordering the terms in the sum, we can then write the field equation (for a given magnetic number m) in the form

$$\Psi^l_{,uv}+V^l_K(r)\Psi^l+i\frac{mMar}{(r^2+a^2)^2}\Psi^l_{,t}+\frac{a^2\Delta}{(r^2+a^2)^2}\left[C_1\Psi^l_{,tt}+C_2\Psi^{l-2}_{,tt}+C_3\Psi^{l+2}_{,tt}\right]=0. \tag{18.32}$$

Here $C_{1,2,3}$ are coefficients that depend on l and m, and $V^{lm}_K(r)$ is Kerr's effective potential, analogous to Schwarzschild's potential, eq. (18.5).

Equation (18.32) represents an infinite set of coupled equations. The last two terms in the square brackets in eq. (18.32) describe *interaction* between modes of different multipole numbers. From a technical point of view, this interaction presents a difficult challenge. It is here where the iterative scheme we introduced earlier becomes powerful. We hereby show how this technique decouples the equations.

Following the same procedure as in Schwarzschild, we define $\delta V_K\equiv V_K-V_0$ (V_0 is defined in eq. (18.8)) and decompose $\Psi=\Psi_0+\Psi_1+\Psi_2+\cdots$, such that

$$\Psi^{lm}_{n,uv}+V^l_0\Psi^{lm}_n=\begin{cases}0 & ,n=0\\ -(\delta V_K)\Psi^{lm}_{n-1}-i\frac{mMar}{(r^2+a^2)^2}\Psi^{lm}_{n-1,t}-\frac{a^2\Delta}{(r^2+a^2)^2}\left[\ \right]_{n-1} & ,n>0\end{cases} \tag{18.33}$$

where empty square brackets symbolize the terms in square brackets in eq. (18.32).

Eq. (18.33) is a hierarchy of equations for the functions Ψ^l_n, which, in principle, we may solve one by one. We shall assume that the initial pulse is made of a single mode l,m [21]. Now, Ψ^l_0 obeys the same equation as in Schwarzschild, and was already calculated above. Then, Ψ^l_1 may be calculated, having Ψ^l_0 as its source, using the same Green's function method described above (with the only difference that δV is now replaced by δV_K). We find that two new modes, Ψ^{l-2}_0 and Ψ^{l+2}_0, are also generated, due to the interaction (provided that $l-2\geq m$). In principle, we may proceed and calculate each of the modes of the next Ψ_n's, for any n, in a similar manner.

The above considerations imply that if a mode (l,m) is present in the initial data, then all modes (l',m) with $l'\geq m$ and even $|l'-l|$ will be present at $\mathcal{J}^+$ (see figure (18.10)). Our preliminary investigation suggests that the mode with smallest l' which is present at $\mathcal{J}^+$, will dominate the late-time behavior there, and its decay rate will be $u^{-(l'+2)}$, as in Schwarzschild.

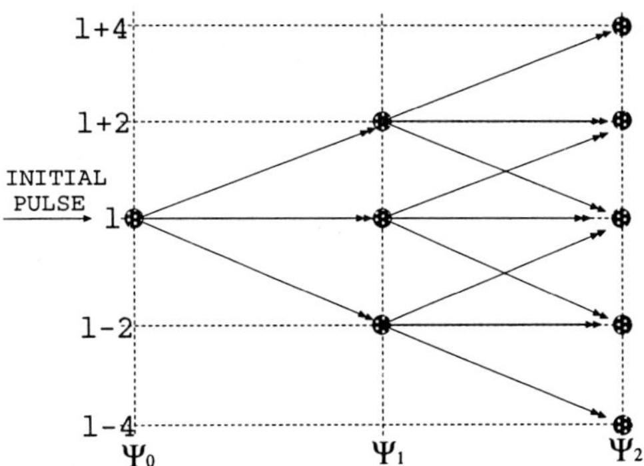

Figure 18.10: 'Modes generation' in Kerr. The diagram shows the first two steps of the mechanism, for a single-mode initial pulse.

We hereby summaries the preliminary results of the analysis in Kerr in a more quantitative way. Details of the calculation shall be given in [15].

For a compact initial pulse of a given magnetic number m and a given multipole number l, the following results are valid to leading order in $1/u$ and in $1/u_0$:

- Any l'-mode of even $|l'-l|$ and $l' \geq m$ will be observed along $\mathcal{J}^+$.

- Each of these modes will be dominated by a tail of the form

$$\psi^{l'}_{\delta/2+1} \propto \frac{a^\delta}{M^{2\delta-1}} \left(\frac{M}{u_0}\right)^{5\delta/2} \frac{1}{u^{l'+2}} \quad [\text{for } \delta \equiv (l-l') \geq 0],$$

  and

$$\psi^{l'}_{\delta/2} \propto M a^\delta \left(\frac{M}{u_0}\right)^{\delta/2-1} \frac{1}{u^{l'+2}} \quad [\text{for } \delta \equiv (l'-l) \geq 0]. \tag{18.34}$$

- Thus the dominant spherical mode is the $|m|^{th}$ (or the $|m|+1^{th}$ if $l-m$ is odd).

- This dominant mode decays at $\mathcal{J}^+$ with a tail of the form

$$\sim \frac{1}{u^{|m|+2,3}}, \tag{18.35}$$

  corresponding to $l-m$ even, odd.

- The case where the initial pulse is a mixture of modes may be inferred using a simple superposition.

18.3.2 Extension to gravitational perturbations in Kerr

So far we have considered a scalar field, as a toy-model for physical perturbations. Obviously, our ultimate goal would be to extend the analysis to the realistic physical fields, and especially to gravitational perturbations.

There are two different approaches to the study of linearized gravitational perturbations: by considering metric perturbations, or ,alternatively, by considering perturbations of the Weyl scalars. Equations governing metric perturbations of the Schwarzschild black hole were derived by Regge and Wheeler [22] (for axial perturbations) and by Zerilli [23] (for polar perturbations). Both equations are of the same form as the scalar wave equation (18.4), where this time the wave function represents certain linear combinations of entities characterizing the metric perturbation. In both Regge-Wheeler's and Zerilli's equations one also finds that the effective potential is similar in shape to that of the scalar model (eq.(5) and figure 1); in particular, they all possess the same asymptotic behavior.

We do not have similar separable equations for metric perturbations of Kerr black holes. Rather, a second approach, based on the Newman–Penrose tetrad formalism, was used by Teukolsky [13] to derive a separable wave equation for perturbations of the Weyl scalars. These five (complex) scalars, constructed from the Weyl tensor in a given tetrad basis, completely characterize the gravitational field in vacuum. They allow a more convenient approach to the study of gravitational perturbations, due to their scalar nature. The equations governing the Weyl scalars perturbations take, again, the same form as eq. (18.4), and may be treated similarly as problems of wave scattering off an effective potential (these equations are discussed, for example, in [24], pp. 174-182 (for Schwarzschild) and pp. 430-434 (for Kerr)).

In the near future we intend to explore gravitational perturbations outside the Kerr black hole, by using the Newman-Penrose formalism. To that end, we shall use the master equation governing this kind of fields [13]. In view of the remarkable similarity in the mathematical characters of scalar and gravitational fields, we expect most of the parts of the analysis to be very similar. In particular, we expect to observe a similar late-time behavior, due to the similarity of the asymptotic form of the effective potential in both cases.

Acknowledgements

This research was done in collaboration with Amos Ori. I thank Lior Burko for carefully reading the manuscript and for helpful comments.

Bibliography

[1] P. M. Wald, *General relativity* (University of Chicago Press, Chicago, 1984), pp. 322-324.

[2] R. H. Price, Phys. Rev. D**5**, 2419 (1972); **5**, 2439 (1972).

[3] C. Gundlach, R. H. Price, and J. Pullin, Phys. Rev. D**49**, 883 (1994).

[4] K. S. Thorne, in *300 Years of Gravitation*, edited by S. W. Hawking and W. Israel (Cambridge University Press, Cambridge, England, 1987).

[5] It should be noted, however, that the predicted intensity for the *late-time* tails of radiation, whose exploration is the main objective of this research, is far below the detection capability of any present and near-future experiment.

[6] A. Ori, Phys. Rev. Lett. **67**, 789 (1991); **68**, 2117 (1992).

[7] C. Gundlach, R. H. Price, and J. Pullin, Phys. Rev. D **49**, 890 (1994).

[8] L. M. Burko and A. Ori, Phys. Rev. D **56**, 7820 (1997).

[9] E. S. C. Ching, P. T. Leung, W. M. Suen, and K. Young, Phys. Rev. Lett. **74**, 2414 (1995).

[10] R. Gómez, J. Winicour, and B. G. Schmidt, Phys. Rev. D **49**, 2828 (1994).

[11] E. Leaver, J. Math. Phys (N.Y.) **27**, 1238 (1986); Phys. Rev. D **34**, 384 (1986).

[12] P. R. Brady, C. M. Chambers, W. Krivan, and P. Laguna, Phys. Rev. D **55**, 7538 (1997).

[13] S. A. Teukolsky, Phys. Rev. Lett. **29**, 1114 (1972).

[14] W. Krivan, P. Laguna, and P. Papadopoulos, Phys. Rev. D **54**, 4728 (1996).

[15] L. Barack and A. Ori, in preparation.

[16] We note, however, that other wave equations are also possible. For example, to assure conformal invariance, one should rather use the equation $\Box\Phi + \frac{1}{6}R\Phi = 0$, where R is the Ricci scalar. This equation reduces to eq. (18.1) when considering waves in vacuum.

[17] In [7], the purely spherical collapse of a self-gravitating minimally coupled scalar field is studied. It is demonstrated numerically that in this case late-time tails are formed even when the collapse fails to create a black hole. However, quasi-normal ringing are observed only when a black hole is formed.

[18] It is easy to see that in the model with $r_0 \to 0$, if Ψ_n (for any n) is finite at $r_* = 0$ then it must vanish there. This result leads to $\Psi_n \equiv 0$ in the whole region $r_* < 0$.

[19] This somewhat surprising feature of the scattering off the purely centrifugal potential may be easily understood in the corresponding 3+1 picture.

[20] In principle, the behavior along lines of r =const can be also deduced directly from the iterative scheme, for $r_0 > 0$. This method yields the correct power-law index at each order n: $\Psi_n \propto t^{-(2l+3)}$. However, even for values of r_0 for which the sum seems to converge (at finite r), this convergence is rather slow, which makes the method inefficient for the determination of the correct coefficient.

[21] This causes no loss of generality, due to the linearity of the field equation.

[22] T. Regge and J. A. Wheeler, Phys. Rev. **108**, 1063 (1957).

[23] F. J. Zerilli, Phys. Rev. D**2**, 2141 (1970).

[24] S. Chandrasekhar, *The Mathematical Theory of Black Holes* (Oxford University Press, New York, 1983.

BLACK HOLE ENTROPY IN INDUCED GRAVITY

V. P. Frolov

Theoretical Physics Institute, Department of Physics, University of Alberta, Edmonton, Canada T6G 2J1
and
P. N. Lebedev Physics Institute, Leninskii Prospect 53, Moscow 117924, Russia

Abstract

We demonstrate how Sakharov's idea of induced gravity allows one to explain the statistical-mechanical origin of the entropy of a black hole. According to this idea, gravity becomes dynamical as the result of quantum effects in the system of heavy constituents of the underlying theory. The black hole entropy is related to the properties of the vacuum in the induced gravity in the presence of the horizon. We obtain the Bekenstein-Hawking entropy by direct counting the states of the constituents.

19.1 Introduction

There are physical phenomena that allow a simple description but require tremendous efforts to explain them. Without doubts the origin of black hole entropy is such a problem. A black hole of mass M radiates as a heated body [1] with temperature $T_H = (8\pi GM)^{-1}$ and has entropy [2, 3]

$$S^{BH} = \frac{\mathcal{A}}{4G}, \tag{19.1}$$

where $\mathcal{A}$ is the surface area of the black hole ($\mathcal{A} = 16\pi G^2 M^2$), G is Newton's constant, and $c = \hbar = 1$. Statistical mechanics relates entropy to the measure of disorder and qualitatively it is the logarithm of the number of microscopically different states available for given values of the macroscopical parameters. Are there internal degrees of freedom that are responsible for the Bekenstein-Hawking entropy S^{BH}? This is the question that physicists were trying to answer for almost 25 years.

What makes the problem of black hole entropy so intriguing? Before discussing more technical aspects let us make simple estimations. Consider for example a supermassive black hole of 10^9 solar mass. According to (19.1), its entropy is about 10^{95}. This is seven orders of magnitude larger than the entropy of the other matter in the visible part of the Universe. What makes things even more complicated, a black hole is simply an empty space-time with a strong gravitational field and ... nothing more. If we put this in a more physical way, the phenomenon we are dealing with is a vacuum in the strong gravitational field. This conclusion leaves us practically no other choice but to try to relate the entropy of the black hole to properties of the physical vacuum in the strong gravitational field.

The black hole entropy is of the same order of magnitude as the logarithm of the number of different ways to distribute two signs + and − over the cells of Planckian size on the horizon surface. This estimation suggests that a reasonable microscopical explanation of the Bekenstein-Hawking entropy must be based on the quantum gravity. The string theory is the best what we have and what is often considered as the modern version of quantum gravity. Recent observation that the Bekenstein-Hawking entropy can be obtained by counting of string (D-brane) states is a very interesting and important result [see e.g [4]].

Still there are questions. The string calculations essentially use supersymmetry and are mainly restricted to extremal and near-extremal black holes. Moreover each model requires new calculations. The string calculations do not give answer to the question what regions of black hole are responsible for its entropy. And the last but not the least it remains unclear why the entropy of a black hole is universal and does not depend on the details of the theory at Planckian scales. Note that the black hole thermodynamics follows from the low-energy gravitational theory. That is why one can expect that only a few fundamental properties of quantum gravity but not its concrete details are really important for the statistical-mechanical explanation of the black hole entropy.

19.2 Where the entropy of a black hole is hidden?

We start by discussing the following question: where the black hole entropy might be hidden. In principle, there might be three different possibilities: (i) the entropy is connected with the black-hole surface physics, (ii) the physical effects responsible for the entropy takes place in the space of black hole interior, and (iii) it is the property of a singularity hidden inside the black hole.

In order to distinguish between these possibilities we consider the gedanken experiment proposed in paper [5]. Namely we assume that there exists a traversable wormhole, and its mouths are freely falling into a black hole [Figure 19.1]. If one of the mouths (say mouth 1) crosses the gravitational radius earlier than the other (mouth 2), then rays passing through the first mouth can escape from the region lying inside the gravitational radius. Such rays would go through the wormhole and enter the outside region though the second mouth. As the result the shadowed region inside the gravitational radius becomes visible by an external observer. The position of the event horizon is changed. A dashed line in Figure 19.1 shows a new position of the event horizon. During the period of time when the first of the mouth is inside and

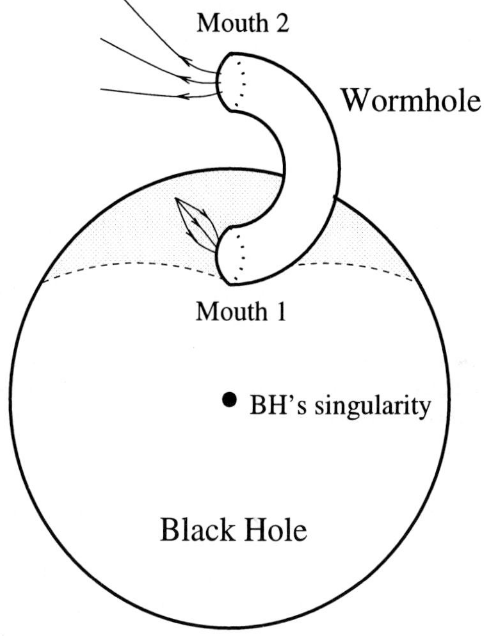

Figure 19.1: Wormhole as the device for study "black-hole interior".

the other mouth is outside the gravitational radius the surface area of the horizon decreases.

The above described device (wormhole) can be used to extract information from the spacetime regions located inside the gravitational radius (apparent horizon). The remarkable feature of this gedanken experiment is that the change of the area of the event horizon remains finite even in the limit when the cross-section of the wormhole becomes negligibly small. If we assume that the black hole entropy is related to the surface area of a black hole, then the only possibility to escape contradictions with the second law is to assume that this entropy is really connected with physics of region in the vicinity of the event horizon so that during the experiment the decrease of entropy is related to the possibility to get access to some new information concerning black hole internal states located near the horizon.

19.3 Mechanism of the entropy generation

At first sight it looks like a puzzle. We know that (at least in the classical theory of general relativity) a black hole at late time is completely specified by a finite number of parameters. For a non-rotating uncharged black hole one need to know only one parameter (mass M) to describe all its properties. It is true not only for the exterior where this property is the consequence of no-hair theorems, but also for the black hole interior [6].

The reason why it is impossible for an isolated black hole at late time to have non-trivial excitations in its interior in the vicinity of the horizon is basically the same as for the exterior regions. These states might be excited only if a collapsing body emits the pulse of fields or particles immediately after it crosses the event horizon. Due to red-shift effect the energy of the emitted pulse must be exponentially large in order to reach the late time region with any reasonable energy. Short time after the formation of a black hole (say $100M$) it is virtually impossible because it requires the energy of emission much greater than the black hole mass M.

In quantum physics the situation is completely different due to the presence of zero-point fluctuations. To analyze states of a quantum field it is convenient to use its decomposition into modes. Besides the positive-frequency modes which have positive energy, there exist also positive-frequency modes with negative energy. In a non-rotating uncharged black holes such modes can propagate only inside the horizon where the Killing vector used to define the energy is spacelike. It is possible to show that at late time for any regular initial state of the field these states are thermally excited and the corresponding temperature coincides with the black hole temperature $T_H = (8\pi M)^{-1}$. It is natural to identify the dynamical degrees of freedom of a black hole with the states of all physical fields inside the black hole [7]. The set of the fields must include the gravitational one.

Figure 19.2 gives a schematic picture of a black hole at some "instant of time". Quanta of different fields are created in the Hawking process. The quanta are created

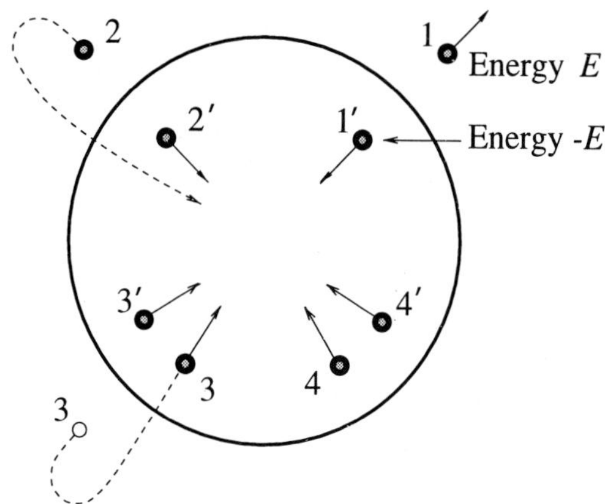

Figure 19.2: Illustration to the Hawking process.

in pairs. The energy conservation does not allow for the both of the components of the pair to be created outside the black hole. Different possibilities are shown in Figure 19.2. A particle 1 has enough energy to penetrate the potential barrier and to escape to infinity. Such particles form the flux of Hawking radiation of the black hole. Particle 2 has no enough energy to escape to infinity and will fall down into a black hole later, but at the moment of consideration it is located outside the black hole. Particle 3 was created earlier outside the horizon but after reflection by the potential barrier it had fallen down into the black hole. Particle 4 as well as its companion $4'$ are created inside the black hole.

After averaging of the given initial state over the states of the particles located outside the black hole at the given moment of time one obtains the density matrix describing the state of the black hole interior. It can be identified with the density matrix of the black hole itself. In accordance with this definition the density matrix of a black hole is

$$\hat{\rho}_H = \underbrace{Tr}_{'visible'} (\hat{\rho}_{total}) , \tag{19.2}$$

where $\hat{\rho}_{total}$ is the total density matrix for quantum fields, and the trace operation is performed over the modes located outside the horizon (that are 'visible' by a distant observer).

The corresponding statistical-mechanical entropy is

$$S^{SM} = - \underbrace{Tr}_{'invisible'} (\hat{\rho}_H \ln \hat{\rho}_H) , \tag{19.3}$$

where the trace operation is performed over the modes located inside the horizon (that are 'invisible' by a distant observer). Really, pairs 3-$3'$ and 4-$4'$ do not contribute to the entropy because both of the components are located inside, and their state is pure. It means that only pairs 1-1′ and 2-2′ are responsible for the entropy of the black hole. A particle belongs to the type 1 if its momentum l is small: $l \lesssim \sqrt{27}ME$. Since particles with energies greater that $T_H = (8\pi M)^{-1}$ practically are not created, one can conclude that particles of type 2 (trapped by the potential barrier) give the main contribution to the statistical-mechanical entropy of the black hole. (For more details see, e.g. [7].)

By using the mode decomposition and the fact, that the internal modes are thermally populated with respect to the Killing energy, one gets

$$S^{SM} = \underbrace{\Sigma}_{'visible'\ J} s_J . \tag{19.4}$$

Here

$$s_J = \frac{\beta\omega}{\exp(\beta\omega) - 1} - \ln[1 - \exp(-\beta\omega)], \tag{19.5}$$

is the entropy of a quantum oscillator of frequency ω at the temperature $T = \beta^{-1}$. It is important to stress that in our model the dynamical degrees of freedom of a black hole are connected with 'invisible' modes, nevertheless the summation in Eq.(19.4) is performed over the 'visible' states. The reason is simple: only those pairs, for which one of the components is 'visible' do contribute to the entropy. The fact that the summation in (19.4) is performed over the 'visible' states is in an agreement with the 't Hooft's [8] brick wall proposal. It should be also emphasized that the entropy S^{SM} can be considered as a special case of a so-called entanglement entropy [9, 10]. The remarkable feature of the statistical-mechanical entropy (19.4) for a black hole is that it can be written as the entropy of 'thermal excitations'. This property is connected with the fact that the boundary of the black hole coincides with the Killing horizon.

There is an elegant possibility of description the above introduced internal degrees of freedom in the framework of so called the *no boundary wave function of the black hole* introduced in Ref. [11].

It can be shown that for any chosen field the number of the modes infinitely grows as one considers the regions located closer and closer to the horizon. For this reason the contribution of a field to the statistical-mechanical entropy of a black hole calculated by counting the internal modes of a black hole is formally divergent. In order to make it finite one might restrict himself by considering only those modes which are located at the proper distance from the horizon greater than some chosen value l. For this choice of the cut-off the contribution of a field to the statistical-mechanical entropy of a black hole is

$$S^{SM} = \alpha \frac{\mathcal{A}}{l^2}, \tag{19.6}$$

where $\mathcal{A}$ is the surface area of a black hole, and a dimensionless parameter α depends on the type of the field.

The statistical-mechanical entropy S^{SM} in this as well as other 'dynamical' approaches possesses the following main properties:

1. $S^{SM} \sim \mathcal{A}$, where $\mathcal{A}$ is the surface area of a black hole;
2. S^{SM} is divergent and requires regularization $S^{SM} \sim \mathcal{A}/l^2$, where l is the cut-off parameter;
3. S^{SM} depends on the number of fields, which exist in Nature;
4. For $l \sim l_{Pl}$ one has $S^{SM} \sim S^{BH}$.

19.4 Covariant calculations and renormalization

It might be suggestive to use a special choice of the cut-off and to identify S^{BH} with S^{SM}. Unfortunately the situation is not so simple. A black hole really is a thermodynamical system, but it also possesses a very special property which singles it out. They are connected with the presence of the horizon. The origin of the divergence of S^{SM} is connected with the fact that in the close vicinity of the horizon there exists infinite number of modes that contribute to S^{SM}. In expression (19.6) this infinity is suppressed by assuming that the summation over the modes is performed only over the modes located at the proper distance larger than l from the horizon. This procedure is known as the *volume cut-off method.* A similar result can be obtained by using other schemes of regularization. For the discussion of the relation between S^{BH} and S^{SM} it is more convenient to use a *covariant regularization.*

Pauli-Villars regularization is an example of such covariant regularization which is often used for calculations (see e.g. [12]). In this method besides a given field of mass m one introduces additional fields. In the simplest case the number of these fields is five. Two of them with masses M_k with the same statistics as the field m, and the other three with masses M_r' have the "wrong" statistics (i.e., they are fermions for scalars and bosons for spinors). The masses are restricted by relations

$$m^2 + \sum_{k=1}^{2} M_k^2 - \sum_{r=1}^{3} {M'}_r^2 = 0\,, \qquad m^4 + \sum_{k=1}^{2} M_k^4 - \sum_{r=1}^{3} {M'}_r^4 = 0\,. \tag{19.7}$$

To be concrete we chose a special one-parameter solution of these equations

$$\begin{aligned} M_1 = M_2 = \sqrt{3\mu^2 + m^2}\,, \qquad {M'}_1 &= {M'}_2 = \sqrt{\mu^2 + m^2}\,, \\ {M'}_3 &= \sqrt{4\mu^2 + m^2}\,. \end{aligned} \tag{19.8}$$

Statistical-mechanical entropy calculated in the Pauli-Villars regularization is

$$S^{SM} = \frac{\mathcal{A}}{48\pi} \sum_i h(s_i) B(m_i^2, \mu^2)\,, \tag{19.9}$$

where $h(s_i)$ is the number of polarization of the field i of spin s_i, and

$$B(m^2, \mu^2) = m^2 \ln m^2 + \sum_k M_k^2 \ln M_k^2 - \sum_r M'^2_r \ln M'^2_r . \tag{19.10}$$

In the limit $\mu \to \infty$ quantity B diverges as $\sim \mu^2$. This quadratic divergence μ^2 stands instead of the quadratic divergence l^{-2} in the volume cut-off method of the regularization. One can interprete this by saying that the covariant cut-off which removes divergences at high energies $\sim \mu$ at the same time gives the cut-off at the proper distance $l \sim \mu^{-1}$.

What is the relation between S^{SM} and S^{BH}? To compare these quantitites first we need to take into account that G which enters (19.1) is an observable value of the Newtonian coupling constant. It differes from the bare coupling G_{bare} by contibutions due to quantum fields

$$\frac{1}{G} = \frac{1}{G_{bare}} + \frac{1}{G_{div}} , \tag{19.11}$$

where for the same Pauli-Villars regularization used above one has

$$\frac{1}{G_{div}} = \frac{1}{12\pi} \sum_i h(s_i) B(m_i^2, \mu^2) - \frac{1}{2\pi} \sum_s \xi_s B(m_s^2, \mu^2) . \tag{19.12}$$

Here ξ_s is the parameter of non-minimal coupling of the scalar field ϕ_s with the scalar curvature R. By comparing $\mathcal{A}/4G$ with (19.9) one can conclude that S^{BH} can be equal to S^{SM} only if $G_{bare}^{-1} = 0$ and $\xi = 0$. The first relation indicates that there is no initial (or bare) gravity, so that all the gravitational action is induced as the result of quantum (one-loop) effect of a system of quantized fields. The second relation implies that the scalar fields which might be present among the quantum fields must be necessary minimally coupled with $\xi = 0$. Under this assumption and in the absence of initial infinite (and negative) coupling constant G_{bare}^{-1} it is impossible to have a finite value of G^{-1} because the contribution of all (boson and fermion) fields to G_{div}^{-1} is positive and divergent. In other words, one cannot have $S^{BH} = S^{SM}$ in the theory where G^{-1} is one-loop finite.

Does this mean that it is impossible to explain the Bekenstein-Hawking entropy by counting quantum excitations of a black hole in the scheme described above? This conclusion is too hasty. There exists a class of theories where S^{BH} is directly related to the statistical-mechanical entropy. They are so-called theories of *induced gravity*.

19.5 Induced gravity

A theory of induced gravity was proposed by Sakharov in 1968 [13] (for the review see [14]). According to the Sakharov's idea general relativity can be considered as a low energy effective theory where the metric $g_{\mu\nu}$ becomes dynamical variable as the result of quantum effects in the system of heavy constituents propagating in the external gravitational background. Gravitons in this picture up to some extend are analogous to the phonon field describing collective excitations of a lattice in low-temperature limit of the theory.

According to Sakharov's idea the background fundamental theory is described by the action $I[\Phi_i, g_{\mu\nu}]$ of fields Φ_i propagating in the external geometry with the metric $g_{\mu\nu}$. The metric originally is not dynamical so that the original gravitational action is simply put equal to zero. By averaging over the constituent fields Φ_i one gets the dynamical effective action for the metric $g_{\mu\nu}$

$$\exp(-W[g_{\mu\nu}]) = \int \mathcal{D}\Phi_i \exp(-I[\Phi_i, g_{\mu\nu}]) \,. \tag{19.13}$$

An observable gravitational field is an extremum of the effective action $W[g_{\mu\nu}]$. The idea is to relate the Bekenstein-Hawking entropy to the statistical mechanics of ultraheavy constituents that induce the gravity in the low energy limit of the theory [15].

This idea was recently realized in Refs. [16, 17]. The model consists of Dirac fermion and scalar boson field constituents. This model is built of N_s free scalar bosons ϕ_s with masses m_s and of N_d free fermion fields ψ_d with masses m_d. The scalar fields have non-minimal coupling with constants ξ_s and their classical action $I[\phi_s, g_{\mu\nu}]$ is

$$I[\phi_s, g_{\mu\nu}] = -\frac{1}{2}\int (\phi_s^{;\mu}\phi_{s,\mu} + m_s^2\phi_s^2 + \xi_s R\phi_s^2)\sqrt{-g}\, d^4x \,. \tag{19.14}$$

The fermion fields are the Dirac spinors ψ_d with the Dirac action

$$I[\psi_d, g_{\mu\nu}] = \int \bar{\psi}_d\,(i\gamma^\mu\nabla_\mu + m_d)\,\psi_d\sqrt{-g}\, d^4x \,. \tag{19.15}$$

Thus effective gravitational action $W[g_{\mu\nu}]$ is defined by Eq. (19.13) where the classical action for the constituent fields $\Phi_i = \{\phi_s, \psi_d\}$ is the sum

$$I[\Phi_i, g_{\mu\nu}] = \sum_s I[\phi_s, g_{\mu\nu}] + \sum_d I[\psi_d, g_{\mu\nu}] \,. \tag{19.16}$$

Consider now the following two functions

$$p(z) = \sum_s m_s^{2z} - 4\sum_d m_d^{2z} \,, \qquad q(z) = \sum_s m_s^{2z}(1-6\xi_s) + 2\sum_d m_d^{2z} \tag{19.17}$$

constructed from the parameters of the constituents. Direct calculations show that the induced cosmological constant vanishes and the induced gravitational coupling constant G is finite if the following constraints are satisfied

$$p(0) = p(1) = p(2) = p'(2) = 0 \,, \qquad q(0) = q(1) = 0 \,. \tag{19.18}$$

For $N_d > 1$ and $N_s > 4$ the the first set of equations (19.18) which gives restrictions on the masses of the constituents is consistent. The second set of equations (19.18) is a linear system defining ξ_s. The condition $p(0) = 0$ requires that $N_s = 4N_d$ and is always satisfied in supersymmetric theories.

The induced Newton constant G is the function of the parameters of the constituents

$$\frac{1}{G} = \frac{1}{12\pi} q'(1) = \frac{1}{12\pi}\left(\sum_s (1-6\xi_s)\, m_s^2 \ln m_s^2 + 2\sum_d m_d^2 \ln m_d^2\right) . \tag{19.19}$$

In the induced gravity $G_{bare}^{-1} = 0$, and the expression for the induced gravitational coupling constant can be easily obtained by using relation (19.12). Under constraints (19.18) the contributions of Pauli-Villars auxiliary fields cancell each other in the limit $\mu \to \infty$. As the result one get (19.19). The non-minimality of the scalar fields plays an important role since it allows one to make G^{-1} finite. The value of the induced Newton constant is dominated by the masses of the heaviest constituents and at least some of the constituents must have Planck mass $m \sim m_{Pl}$.

The heavy constituents are unobservable at low energies and their effective action $W[g_{\mu\nu}]$ is reduced in the low energy regime to the Einstein-Hilbert action

$$W[g_{\mu\nu}] = -\frac{1}{16\pi G}\left(\int_{\mathcal{M}} dV\ R + 2\int_{\partial\mathcal{M}} dv\ K\right) + \dots . \tag{19.20}$$

The dots in r.h.s. of (19.20) indicate higher curvature terms which are suppressed by the power factors of m_i^{-2} when the curvature is small. In order to make finite the terms which are quadratic in curvature one must consider a more general set of constituent fields with additional constrains imposed on them. For our problem, since we will be interested in Ricci-flat geometries, possible local $\mathcal{R}^2$-terms give a pure topological contribution to the action which is irrelevant for our discussion. For this reason in what follows we omit such terms.

The variation of $W[g_{\mu\nu}]$ gives the Einstein equations

$$\frac{\delta W}{\delta g_{\mu\nu}} \sim R^{\mu\nu} - \frac{1}{2} g^{\mu\nu} R = 0 . \tag{19.21}$$

According to the Sakharov's equality (19.13) these equations in the induced gravity are equivalent to the relation

$$\left\langle \hat{T}^{\mu\nu}(x) \right\rangle = 0 , \tag{19.22}$$

where $\hat{T}^{\mu\nu}(x)$ is the total stress-energy tensor of the constituents.

Using equations (19.1), (19.9), (19.11)–(19.12) one can rewrite the expression for the Bekenstein-Hawking entropy in the induced gravity as

$$S^{BH} = S^{SM} - Q , \tag{19.23}$$

where

$$Q = \frac{\mathcal{A}}{2\pi}\sum_s \xi_s B(m_s^2, \mu^2)\,. \tag{19.24}$$

To summarize, using the induced theory of gravity one can directly relate the Bekenstein-Hawking entropy to the statistical-mechanical entropy of heavy constituents. The non-minimal coupling of scalar constituents which is required to make G^{-1} finite is responsible for the additional term Q in (19.23). At first glimpse this term spoils the game since it has no evident statistical-mechanical meaning. In fact it is not true. We explain now why it happens [17].

19.6 Non-minimal coupling and Noether charge

Let us start with a simple observation. Namely, in the Pauli-Villars regularization

$$\langle\hat{\phi}_s^2\rangle = \frac{1}{4\pi^2}B(m_s^2, \mu^2)\,. \tag{19.25}$$

Hence the quantity Q can be written as

$$Q = \sum_s 2\pi\xi_s \int_H \langle\hat{\phi}_s^2\rangle\sqrt{\gamma}d^2x\,, \tag{19.26}$$

where the integral is taken over the surface of the black hole.

Let us recall that the complete background spacetime of an eternal black hole contains two Rindler-like wedges bounded by the Killing horizons. The Killing horizons intersect at the two dimensional bifurcation surface Σ. We shall use a foliation of the hypersurfaces t=const orthogonal to the Killing vector ζ^μ that intersect each other at Σ.

Quantum fields propagate on the complete space-time manifold, however in our statis-
tical-mechanical calculations we restrict ourselves by considering only a part of the system located in one of the wedges. It is instructive first to discuss how this procedure manifests in the classical theory. We focus on a classical scalar field ϕ with a non-minimal coupling with the scalar curvature R described by the action

$$I[\phi] = -\frac{1}{2}\int(\phi^{,\mu}\phi_{,\mu} + m^2\phi^2 + \xi R\phi^2)\sqrt{-g}\, d^4x\,. \tag{19.27}$$

The field obeys the equation

$$\Box\phi - (m^2 + \xi R)\phi = 0\,, \tag{19.28}$$

where $\Box$ is the D'Alambert operator. The stress-energy tensor resulting from the variation of the action (19.27) with respect to the metric is

$$\begin{aligned} T_{\mu\nu} &= \phi_{,\mu}\phi_{,\nu} - \frac{1}{2}g_{\mu\nu}\left(\phi_{,\rho}\phi^{,\rho} + m^2\phi^2\right) \\ &+ \xi\left[(R_{\mu\nu} - \frac{1}{2}g_{\mu\nu}R)\phi^2 + g_{\mu\nu}(\phi^2)^{;\rho}_{;\rho} - (\phi^2)_{;\mu\nu}\right] . \end{aligned} \tag{19.29}$$

Denote by $\mathcal{B}$ a space-like hypersurface orthogonal to the Killing vector ζ^μ. The energy E of the system is defined in terms of the stress-energy tensor (19.29)

$$E = \int_{\mathcal{B}} T_{\mu\nu}\zeta^\mu d\sigma^\nu = -\int_{\mathcal{B}} T_0^0\sqrt{-g}\, d^3x\,, \tag{19.30}$$

where $d\sigma^\nu$ is the future directed vector of the volume element on $\mathcal{B}$. In the general case E differs from the *canonical* energy H that coincides with the Hamiltonian. The latter is expressed in terms of the Hamiltonian density $\mathcal{H}$

$$H = \int_{\mathcal{B}} \mathcal{H}\sqrt{-g}\, d^3x\,, \tag{19.31}$$

where

$$\mathcal{H} = \frac{1}{2}\left(-g^{00}\phi_{,0}^2 + g^{ij}\phi_{,i}\phi_{,j} + (m^2 + \xi R)\phi^2\right) . \tag{19.32}$$

To compare E and H we note that in a static space-time

$$-T_0^0 = \mathcal{H} - \xi\left(R_0^0\phi^2 + g^{ij}(\phi^2)_{;ij}\right) . \tag{19.33}$$

The last term in r.h.s. of this equation can be rewritten as

$$\begin{aligned} (\phi^2)_{;ij} &= \Box\phi^2 - g^{00}(\phi^2)_{;00} \\ &= g^{00}\left((\phi^2)_{,0,0} - (\phi^2)_{;00}\right) + \frac{1}{\sqrt{-g}}\partial_i\left(\sqrt{-g}g^{ij}\partial_j\phi^2\right) \\ &= \frac{1}{\sqrt{-g}}\partial_i\left[\sqrt{-g}g^{ij}((\phi^2)_{,j} - \phi^2 w_j)\right] + \nabla_\mu w^\mu\,\phi^2 . \end{aligned} \tag{19.34}$$

Here $w^\mu = \frac{1}{2}\nabla^\mu \ln|g_{00}|$ is the four-acceleration of the Killing observer. It can be shown that $\nabla_\mu w^\mu = -R_0^0$, so that for static space-times relation (19.33) takes the form

$$-T_0^0 = \mathcal{H} - \xi\frac{1}{\sqrt{-g}}\partial_i\left(\sqrt{-g}g^{ij}((\phi^2)_{,j} - \phi^2 w_j)\right) . \tag{19.35}$$

Then substitution of (19.35) into (19.30) gives the required relation between the energy E and the canonical energy H

$$E = H - \xi\int_{\partial\mathcal{B}} ds^k\ |g_{00}|^{1/2}((\phi^2)_{,k} - \phi^2 w_k) . \tag{19.36}$$

Here ds^k is a three dimensional vector in $\mathcal{B}$ normal to the boundary $\partial\mathcal{B}$ and directed outward with respect to $\mathcal{B}$. Thus two energies differ by a surface term given on the boundary $\partial\mathcal{B}$ of the hypersurface $\mathcal{B}$.

Obviously, when one considers a complete Cauchy surface the boundary term in (19.36) contains only a contribution from the spatial infinity, or from the external spatial boundaries if they are present. For a field falling off at infinity or obeying suitable conditions at the boundary one can get rid of the boundary term and make E and H be equal.

However, the situation is qualitatively different when we consider the theory only in one of the wedges. Then the integration region in E is restricted by the bifurcation surface Σ of the Killing horizon, where the field ϕ can take arbitrary finite values. By assuming that contribution from the spatial infinity or external boundary is absent one can write for the field regular at the horizon the following relation

$$E = H - \beta_H^{-1} Q \, , \tag{19.37}$$

were β_H^{-1} is the Hawking temperature determined by the surface gravity κ of the black hole as $\beta_H^{-1} = \kappa/(2\pi)$ and

$$Q = 2\pi\xi \int_\Sigma \phi^2 \sqrt{\gamma} d^2x \, . \tag{19.38}$$

It is easy to show that the quantity Q is directly related to the Noethor charge for a non-minimally coupled scalar field. The energy E computed for a domain restricted by the horizon differs from the canonical energy H by the quantity proportional to Q.

19.7 Statistical-mechanical explanation of the Bekenstein-Hawking entropy in induced gravity

Two energies, $\hat{E}$ and $\hat{H}$, play essentially different role. The canonical energy is the value of the Hamiltonian $\hat{H}$ which is the generator of translations of the system along ζ^μ and which enters definition (19.5) of the statistical-mechanical entropy S^{SM}. The energy E is the contribution of the constituents to the black hole mass. The number $\nu(E)\Delta E$ of microscopically different physical states of the constituents in the energy interval ΔE near $E = 0$ determines the degeneracy of the black hole mass spectrum.

Since the Killing vector ζ^μ vanishes at the bifurcation surface Σ of the Killing horizons, the Hamiltonian $\hat{H}$ is degenerate. One can add to the system an arbitrary number of *soft modes*, i.e. modes with zero Rindler frequencies, without changing the canonical energy. On the other hand, only soft modes contribute to the average $\bar{Q}$. According to (19.37), this removes the degeneracy of the energy $\hat{E}$. As the result the infinite number of thermal states of the constituents reduces to the finite number of physical states of the black hole and S^{SM} reduces to the Bekenstein-Hawking value. It is possible to show that the corresponding number density $\nu(E = 0)$ of physical states is exactly equal to $\exp S^{BH}$ [17]. There is a similarity between this mechanism and gauge theories, soft modes playing the role of the pure gauge degrees of freedom.

Let us note that the concrete model of the induced gravity may differ from the one considered here, and may contain, for example, finite or infinite number of fields of higher spins. However our consideration indicates that it is quite plausible that the same mechanism of black-hole entropy generation still works.

Our discussion can be summarized as follows. The entropy S^{BH} of a black hole in induced gravity is the logarithm of the number of different states of the constituents obeying the condition $\langle \hat{T}_{\mu\nu} \rangle = 0$. The statistical-mechanical explanation of the entropy becomes possible only when one appeals to a more deep underlying theory in which vacuum is equipped with an additional "fine" structure. The vacuum in the induced gravity is a ground state of "heavy" constituents. "Light" particles are just low energy collective excitations over this vacuum. The black hole entropy S^{BH} acquires its statistical-mechanical meaning because of this additional fine structure of the vacuum of the underlying theory. Two important elements that make statistical-mechanical picture self-consistent and universal are representation of gravity as an induced phenomenon and the ultraviolet finiteness of the induced gravitational couplings. These are main lessons the induced gravity teaches us.

Acknowledgements

This work was partially supported by the Natural Sciences and Engineering Research Council of Canada. The author is grateful to the Killam Trust for its financial support.

Bibliography

[1] S. W. Hawking, Comm. Math. Phys. **43**, 199 (1975).

[2] J. D. Bekenstein, Nuov. Cim. Lett. **4**, 737 (1972); Phys. Rev. D **7**, 2333 (1973).

[3] S. W.Hawking, Phys. Rev. D **14**, 2460 (1976).

[4] G. Horowitz, "Quantum states of black holes", preprint gr-qc/9704072.

[5] V. P. Frolov and I. D. Novikov, Phys. Rev. D **48** 1607 (1993).

[6] A. G. Doroshkevich and I. D. Novikov, Soviet Phys. JETP. **63**, 1538 (1978).

[7] V. P. Frolov and I. D. Novikov, Phys. Rev. D **48**, 4545 (1993).

[8] G. 't Hooft, Nucl. Phys. **B 256**, 727 (1985).

[9] L. Bombelli, R. K. Koul, J. Lee, and R. Sorkin, Phys. Rev. D **34**, 373 (1986).

[10] M. Srednicki, Phys. Rev. Lett. **71**, 666 (1993).

[11] A. I. Barvinsky, V. P. Frolov, and A. I. Zelnikov, Phys. Rev. D **51**, 1741 (1995).

[12] J. G. Demers, R. Lafrance, and R. C. Myers, Phys. Rev. D **52**, 2245 (1995).

[13] A. D. Sakharov, Sov. Phys. Doklady, **12**, 1040 (1968); Theor. Math. Phys. **23**, 435 (1976).

[14] S. L. Adler, Rev. Mod. Phys. **54**, 729 (1982).

[15] T. Jacobson, *Black Hole Entropy and Induced Gravity*, preprint gr-qc/9404039.

[16] V. P. Frolov, D. V. Fursaev, and A. I. Zelnikov, Nucl. Phys. **B 486**, 339 (1997).

[17] V. P. Frolov and D. V. Fursaev, Phys. Rev. D **56**, 2212 (1997).

INTERNAL STRUCTURE OF NONSINGULAR SPHERICAL BLACK HOLES

Irina Dymnikova

Institute of Mathematics and Physics, Pedagogical University of Olsztyn, Zolnierska 14, 10-561 Olsztyn, Poland

Abstract

We analyze the internal structure of a black hole whose singularity is replaced by a de Sitter-like core which could arise as a result of symmetry restoration.

De Sitter-Schwarzschild black hole has two horizons. Global structure of the de Sitter-Schwarzschild spacetime contains an infinite sequence of asymptotically flat regions, black and white holes and regular de Sitter-like cores which mark a way from black holes into baby universes asymptotically de Sitter at the beginning. The interior of de Sitter-Schwarzschild black hole contains an infinite sequence of such structures spreading into the future.

Two horizons degenerate at the critical value of mass parameter M_{cr1}. Dependence of the Hawking temperature on the mass shows that the specific heat is broken and changes sign at $M_{cr2} > M_{cr1}$, which means that a second-order phase transition occurs in the course of Hawking evaporation. The Hawking temperature drops to zero when the mass approaches the critical value M_{cr1} which puts the lower limit for a black hole mass. Beyond M_{cr1} there is no black hole, and de Sitter-Schwarzschild solution describes a recovered self-gravitating particlelike structure. At $M > M_{cr1}$ it is hidden inside a black hole.

20.1 Introduction

A black hole whose singularity is replaced by a de Sitter-like core can result from restoration of symmetry to the de Sitter group when super-high density is achieved in the course of a gravitational collapse. The idea goes back to the middle 1960's ideas that the de Sitter vacuum can be the limiting state at super-high densities [1, 2], and thus can be both a final state in gravitational collapse [2] and an initial state for an expanding universe [3].

Since such a state is achieved deeply inside a black hole, while outside spacetime is asymptotically Schwarzschild, de Sitter-Schwarzschild black hole gives an example of dynamically changed cosmological constant. The cosmological constant was introduced by Einstein in 1917 as a universal repulsion balancing gravitational attraction of matter and making the Universe stationary. The solution of the Einstein equations with cosmological constant and without matter

$$R_{\mu\nu} - \frac{1}{2}Rg_{\mu\nu} - \Lambda g_{\mu\nu} = 0 \tag{20.1}$$

has been found by de Sitter in the same year. It is given by

$$ds^2 = \left(1 - \frac{r^2}{r_0^2}\right)c^2dt^2 - \left(1 - \frac{r^2}{r_0^2}\right)^{-1}dr^2 - r^2(d\theta^2 + sin^2\theta d\varphi^2), \tag{20.2}$$

where $r_0 = \sqrt{3/\Lambda}$ is the de Sitter event horizon (photon emitted at the origin needs an infinite time to arrive at the surface $r = r_0$).

During several decades the physical essence of this solution remained obscure. In modern physics it has been mainly used as a simple testing ground for developing quantum field techniques in curved spacetime.

In 1965 the paper "Algebraic properties of stress-energy tensor and vacuum-like states of matter" by Gliner [2] appeared, in which the cosmological constant was welcomed to full membership in the family of different kinds of matter as characterized macroscopically by equation of state and algebraic structure of stress-energy tensor.

Putting $\Lambda g_{\mu\nu}$ to the right side of the Eq. (20.1), one can see that the cosmological constant corresponds to the stress-energy tensor

$$T_{\mu\nu} = \varepsilon g_{\mu\nu}, \tag{20.3}$$

where

$$\varepsilon = \frac{c^4}{8\pi G}\Lambda, \tag{20.4}$$

and to the equation of state

$$p = -\varepsilon. \tag{20.5}$$

The main algebraic property of the stress-energy tensor (20.3) is that any reference frame is comoving for it ($\Lambda g_{\mu\nu}$ is diagonal in any reference frame). Assume now that we have a test body moving through the medium characterized by the stress-energy tensor (20.3). Its comoving reference frame is also comoving for the medium. Therefore it is impossible in principle, by any experiments, to determine the velocity of a test body with respect to the medium (20.3). The relativity principle is valid in this case as for the motion in the vacuum. So, from the point of view of macroscopic description, the medium (20.3) satisfies the definition of vacuum as such a kind of matter which does not allow any preferred comoving reference frame connected with it. Therefore the stress-energy tensor (20.3) and the equation of state (20.5) describe vacuum with nonzero energy density [2].

The next conclusion Gliner has drawn from analysis of the equation of state (20.5) is as follows: The negative pressure implies that the internal volume forces in a medium are not repulsive (as for well known media made up from particles whichever they are), but rather attractive forces. This makes us think that equation (20.5) describes a superdense state of matter such that a continual medium is formed with attraction between its elements. Such a state, he concluded, could be the final state in a gravitational collapse [2].

Indeed, assume that the state (20.3) is achieved. Then from contracted the Bianchi identity $T^{\nu}_{\mu;\nu} = 0$ it follows that $\varepsilon = const$, and ε is not growing further. On the other hand, in de Sitter world gravitational acceleration is positive: $a = -4\pi G(\varepsilon+3p)/3c^4 = 8\pi G\varepsilon/3c^4 > 0$. We see, that internal volume attraction corresponding to negative pressure, instead to lead to unlimited contraction, creates, as source term for the Einstein equations, the geometry in which gravity is repulsive! This repulsion prevents further contraction.

The first candidate for a superdense vacuum was proposed by Zel'dovich in 1967 [4]: Quantum vacuum is full of virtual particles and there is no rule demanding the energy of their gravitational interaction to be zero. The effective number density of the virtual particles in vacuum is $n \sim 1/\lambda_c^3$, where $\lambda_c = \hbar/mc$ is the Compton wavelength of a particle with a mass m. The energy of the gravitational interaction of virtual particles is of order of Gm^2/λ_c for one pair. It gives energy density $\varepsilon \sim Gm^6c^4/\hbar^4$ which is related to *gravitational vacuum polarization*.

Taking into account that de Sitter geometry has the event horizon

$$r_0 = \sqrt{\frac{3}{\Lambda}} = \sqrt{\frac{3c^4}{8\pi G\varepsilon}}, \tag{20.6}$$

we can estimate a limiting density imposing the causality condition [3]: The region of localization of a quantum particle confined by its Compton wavelength λ_c, must be contained inside a causally connected region r_0. From the inequality $\lambda_c \leq r_0$ follows the limiting mass of a particle and the limiting density of a medium. They are given by

$$m_{(lim)} = \sqrt{\frac{3}{8\pi}}\sqrt{\frac{\hbar c}{G}} = \sqrt{\frac{3}{8\pi}}M_{Pl} \tag{20.7}$$

$$\rho_{(lim)} = \left(\frac{9}{64\pi^2}\right)\frac{c^5}{\hbar G^2} = \left(\frac{9}{64\pi^2}\right)\rho_{Pl}. \tag{20.8}$$

At $\lambda_c > r_0$ there are no causally connected regions of localization, which imply the impossibility of interactions. At this scale it is also difficult to speak about further contraction.

Since 1981 we know that de Sitter vacuum state can arise at the scale of symmetry breaking of Grand Unified theories [5, 6], and it makes it natural to expect the symmetry restoration to the de Sitter group when the scale of Grand Unification is achieved in the course of a gravitational collapse [7].

De Sitter-Schwarzschild spacetime has been constructed by several authors by direct matching the Schwarzschild metric to de Sitter interior through massive thin shell [8, 9, 10, 11]. The solution obtained in such a way appears to have a jump in the metric at the junction surface. Matching imposes $\rho(r) = \Theta(r_m - r)\rho_{Pl}$ and $p_r = -\rho$ for radial pressure. Tangential pressure $p_t = -\rho - r\rho'/2$, has, as a result, a δ-like behavior. Integrating the Einstein equations in the junction layer gives $g_{00} = 1 - 2M(r)/r$ with $M(r) = M_{Schw}\Theta(r - r_m) + (r^3/2r_0^2)\Theta(r_m - r)$ which has a jump at $r = r_m$ [12].

The conformal diagram for matched solution represents a black hole interior as connecting asymptotically flat region to collapsing and then reexpanding baby universe [11].

Analyzing Schwarzschild-de Sitter transition, Poisson and Israel stated that it is necessary to introduce a layer of "noninflationary material at the interface" in which the geometry remains effectively classical and governed by the Einstein field equations with a source term representing vacuum polarization effects. The characteristic scale for these effects is determined as radius $r \sim r_g^{1/3}$ where the curvature r_g/r^3 grows to order of unity [13]. (Here r_g is gravitational radius $r_g = 2GM/c^2$.)

In Ref. [15] this idea has been applied to describe smooth transition to de Sitter regime without *a priori* fixing the depth of a transitional layer . The simple semiclassical model has been based on two assumptions: First is the same as for matching: $g_{00} = -g_{11}^{-1}$, i.e. for radial pressure $p_r = -\varepsilon$. It results in algebraic structure for stress-energy tensor $T_t^t = T_r^r$, $T_\vartheta^\vartheta = T_\varphi^\varphi$ corresponding to spherically symmetric vacuum [15]. Second assumption concerns particular form for a density profile T_t^t. De Sitter-Schwarzschild spacetime is characterized by two parameters: mass M and limiting density ρ_0, which would be achieved at the scale of symmetry restoration and thus would depend on some unified coupling and would not depend on specific parameters characterizing particular fields. On the other hand, all particular fields contribute to the gravitational field tension. Therefore we can expect that the resulting effect would be vacuum polarization in the resulting gravitational field. To find the analytic solution and use it to investigate the behavior of de Sitter-Schwarzschild configuration, we applied the Schwinger formula for the expectation value of energy density T_t^t due to gravitational vacuum polarization [14].

The analytic solution obtained in such a model allows us to investigate the global structure of de Sitter-Schwarzschild spacetime, the internal structure of nonsingular spherically symmetric black holes, and the physical properties of the de Sitter-Schwarzschild self-gravitating configurations.

The conformal diagram for de Sitter-Schwarzschild spacetime contains, in place of both future and past Schwarzschild singularities, *an infinite sequence* of new structures: asymptotically flat regions, black and white holes, and regular cores. The interior of de Sitter-Schwarzschild black hole contains an infinite sequence of such a structures spreading into the future.

The de Sitter-Schwarzschild black hole has two horizons which coincide at the critical value of the mass parameter M_{cr1} wich puts the lower limit for the black hole mass. Dependence of the Hawking temperature of the mass parameter shows that at the certain value $M_{cr2} > M_{cr1}$ the specific heat is broken and changes sign. It means that second-order phase transition occurs at this scale in accordance with the underlying hypothesis concerning symmetry restoration. The Hawking temperature drops to zero when the mass approaches the lower limit $M = M_{cr1}$. Beyond M_{cr1} there is no black hole, and the de Sitter-Schwarzschild solution describes recovered particle-like structure, hidden at $M > M_{cr1}$ inside a black hole.

This paper is organized as follows. In Section 20.2 we outline the model and the de Sitter-Schwarzschild analytic solution. In Section 20.3 we describe thermodynamics of de Sitter-Schwarzschild black hole. In Section 20.4 we investigate the properties of a particle-like core which exists either inside a black hole for $M > M_{cr1}$, or separately as recovered self gravitating particle-like structure for $M < M_{cr1}$. Section 20.5 contains discussion of the results.

20.2 De Sitter-Schwarzschild Analytic Solution

In the spherically symmetric static case a line element has the form

$$ds^2 = e^{\nu} c^2 dt^2 - e^{\mu} dr^2 - r^2(d\theta^2 + sin^2\theta d\varphi^2), \tag{20.9}$$

and the Einstein equations reduce to the system

$$\frac{8\pi G}{c^4} T_t^t = -e^{-\mu}\left(\frac{1}{r^2} - \frac{\mu'}{r}\right) + \frac{1}{r^2} \tag{20.10}$$

$$\frac{8\pi G}{c^4} T_r^r = -e^{-\mu}\left(\frac{1}{r^2} + \frac{\nu'}{r}\right) + \frac{1}{r^2} \tag{20.11}$$

$$\frac{8\pi G}{c^4} T_\vartheta^\vartheta = \frac{8\pi G}{c^4} T_\varphi^\varphi = -\frac{e^{-\mu}}{2}\left(\nu\prime\prime + \frac{\nu'^2}{2} + \frac{(\nu' - \mu')}{r} - \frac{\nu'\mu'}{2}\right) \tag{20.12}$$

The boundary conditions are the Schwarzschild behavior at $r \to \infty$

$$ds^2 = \left(1 - \frac{r_g}{r}\right) c^2 dt^2 - \left(1 - \frac{r_g}{r}\right)^{-1} dr^2 - r^2(d\theta^2 + sin^2\theta d\varphi^2), \tag{20.13}$$

$$T_{\mu\nu} = 0 \tag{20.14}$$

and de Sitter behavior at $r \to 0$

$$ds^2 = \left(1 - \frac{r^2}{r_0^2}\right) c^2 dt^2 - \left(1 - \frac{r^2}{r_0^2}\right)^{-1} dr^2 - r^2(d\theta^2 + sin^2\theta d\varphi^2), \tag{20.15}$$

$$T_{\mu\nu} = \varepsilon_0 g_{\mu\nu}. \tag{20.16}$$

The limiting density $\varepsilon_0 = \rho_0 c^2$ is fixed and connected with the de Sitter horizon parameter r_0 by the relation

$$r_0{}^2 = \frac{3c^4}{8\pi G\varepsilon_0}. \tag{20.17}$$

For both asymptotics $\mu = -\nu$, and this condition is imposed for de Sitter-Schwarzschild solution to guarantee the proper asymptotic behavior (20.13) and (20.15). Then field equations (20.10) and (20.11) become identical, and

$$T_t^t = T_r^r. \tag{20.18}$$

The metric is given by

$$g_{00} = e^{-\mu} = 1 - \frac{R_g(r)}{r}; \qquad R_g(r) = \frac{8\pi G}{c^4}\int_0^r T_t^t(r) r^2 dr \tag{20.19}$$

and the components $T_\theta^\theta = T_\varphi^\varphi$ are obtained from the Eq. (20.12).

The stress-energy tensor with the algebraic structure

$$T_t^t = T_r^r; \quad T_\theta^\theta = T_\varphi^\varphi \tag{20.20}$$

describes spherically symmetric vacuum [15]. It satisfies general definition of vacuum by absence of a preferred comoving reference frame connected with it. (Such a frame can be defined uniquely if, and only if, none of eigenvalues of stress-energy tensor does coincide with T_t^t, otherwise there is an infinite set of comoving reference frames.) It has more restrictive symmetry than isotropic de Sitter vacuum (2.8) and describes a smooth transition from Minkowski vacuum at infinity to de Sitter vacuum at $r \to 0$ through anisotropic vacuum in between, representing algebraic properties of "non-inflationary material at the interface".

The natural requirement on a particular form of a density profile $T_t^t(r)$ is quick vanishing at infinity to ensure finiteness of the mass

$$M = \frac{4\pi}{c^2}\int_0^\infty T_t^t(r) r^2\, dr. \tag{20.21}$$

To calculate a density profile T_t^t describing vacuum polarization effects in a collapse, we would need a renormalizable quantum theory of all interactions including gravity. In default of such a theory one can try some qualitative approach to guess a possible form for T_t^t. It can be evaluated qualitatively applying intuitive semiclassical approach [16]: In accordance with the uncertainty relation, lifetime of virtual pair of particles with energy mc^2 is $\tau \sim \hbar/mc^2$. In a time τ particles can move apart each other at the distance $l_0 \sim \hbar/mc$. Probability w to find particles separated by a distance l is proportional to $\exp(-l/l_0)$. To create real particles with a charge g, a field F should produce work $gFl \sim mc^2$. As a result vacuum polarization and particle creation effects in a field F are described by the Schwinger formula

$w \sim \exp(-\alpha m^2c^3/\hbar gF) \sim \exp(-F_{crit}/F)$, where constants depend on particular properties of a field F. In the case $F < F_{crit}$ this qualitative formula agrees with results of detailed calculations [16, 17, 18].

Gravitational field tension in spherically symmetric case is characterized by the curvature term $F \sim r_g/r^3$, and the critical value—by the de Sitter curvature $F_{crit} \sim 1/{r_0}^2$.

The energy density of the form

$$T_t^t(r) = \varepsilon_0 \exp\left(-r^3/r_0^2 r_g\right) \tag{20.22}$$

leads, with using the Eq. (20.19), to the metric [15]

$$ds^2 = \left(1 - \frac{R_g(r)}{r}\right)c^2dt^2 - \left(1 - \frac{R_g(r)}{r}\right)^{-1} dr^2 - r^2(d\theta^2 + \sin^2\theta d\varphi^2) \tag{20.23}$$

$$R_g(r) = r_g(1 - \exp(-r^3/r_0^2 r_g)), \tag{20.24}$$

which satisfies the boundary conditions (20.13) and (20.15).

The whole stress-energy tensor has the form

$$T_t^t = T_r^r = \varepsilon_0 \exp\left(-r^3/r_0^2 r_g\right); \quad T_\theta^\theta = T_\varphi^\varphi = \varepsilon_0\left(1 - \frac{3r^3}{2r_0^2 r_g}\right)\exp\left(-r^3/r_0^2 r_g\right). \tag{20.25}$$

The case $T_t^t = T_r^r = T_\theta^\theta = T_\varphi^\varphi$ corresponding to the inflationary vacuum $p = -\varepsilon$ is achieved at $r \to 0$.

The quadratic invariant of the Riemann curvature tensor $\Re^2 = R_{\alpha\beta\gamma\delta}R^{\alpha\beta\gamma\delta}$ has the form

$$\Re^2(r) = 4\frac{R_g^2(r)}{r^6} + 4\left(\frac{3}{r_0^2}e^{-r^3/r_0^2 r_g} - \frac{R_g(r)}{r^3}\right)^2 + \left(\frac{2R_g(r)}{r^3} - \frac{9r^3}{r_0^4 r_g}e^{-r^3/r_0^2 r_g}\right)^2 \tag{20.26}$$

For $r \to 0$, $\Re^2$ remains finite and tends to the de Sitter value $\Re^2 = 24/r_0^4$, which naturally appears to be the limiting value of spacetime curvature [15, 11].

The metric (20.23) has two horizons (see Fig. 20.1). In the case $r_g \gg r_0$ they are given by

$$r_+ \approx r_g[1 - O(\exp(-r_g^2/r_0^2))]; \quad r_- \approx r_0[1 + O(r_0/4r_g)]. \tag{20.27}$$

The global structure of spacetime can be seen from the Penrose diagram shown in Fig. 20.2. There is an infinite sequence of asymptotically flat regions, black and white holes, and future and past de Sitter-like cores. Interior of a black hole contains, in place of a singularity, an infinite chain of structures. Each single segment shows that an asymptotically flat region is connected through black hole interior to a baby universe asymptotically de Sitter at the beginning.

Horizons degenerate at the critical value of the mass parameter

$$M = M_{cr1} \simeq 0.3M_{Pl}(\rho_{Pl}/\rho_0)^{1/2}, \tag{20.28}$$

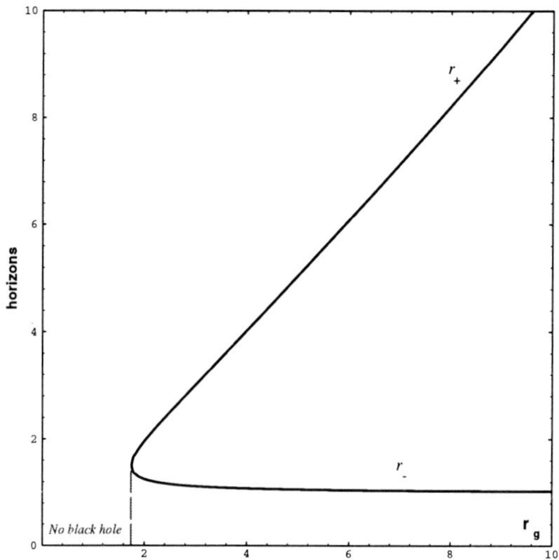

Figure 20.1: The horizon-mass diagram for the solution (20.24) in the units normalized to r_0. Horizons are plotted as the function of the Schwarzschild radius r_g. Degenerate horizon $r_+ = r_- \approx 1.4957$ exists at the critical value $r_g \approx 1.7576$; $M = M_{cr1} \approx 0.3 M_{Pl}(\rho_{Pl}/\rho_0)^{1/2}$.

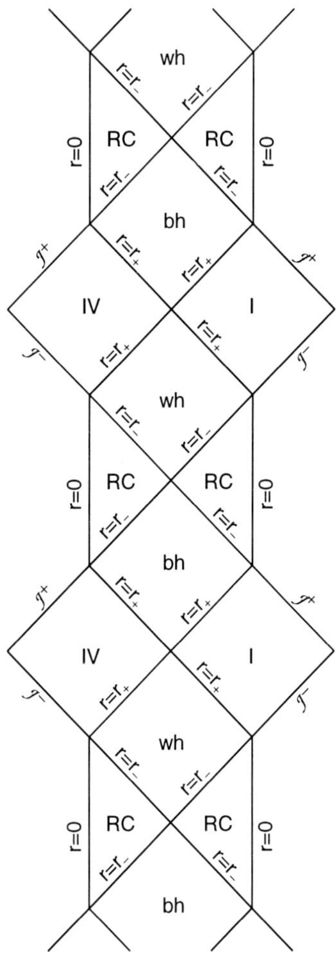

Figure 20.2: Conformal diagram for de Sitter-Schwarzschild spacetime. There is an infinite sequence of asymptotically flat regions (I, IV), black and white holes, and regular cores. The Killing vector $K = \partial/\partial t$ is timelike in regions RC, I and IV, and spacelike in bh and wh regions.

corresponding to the extreme black hole. Beyond M_{cr1} there is no black hole (see Fig. 20.3, and the solution (20.23) describes a particlelike structure which has soliton-like properties: it is globally regular and has finite energy determined by (20.21).

It follows, that de Sitter-Schwarzschild black hole of arbitrarily small mass can not exist. Nontrivial particlelike structure exists for any nonzero mass (see Fig. 20.3. At $M > M_{cr1}$ it is hidden under the external event horizon r_+.

20.3 Thermodynamics and hair

De Sitter-Schwarzschild black holes emit Hawking radiation from both horizons, with the Gibbons-Hawking temperature [19]

$$T = \frac{\hbar\kappa}{2\pi kc}, \tag{20.29}$$

where k is the Boltzmann constant. The surface gravity κ satisfies the equation

$$K_{a;b}K^b = \kappa K_a \tag{20.30}$$

on the horizons, where K_a is the Killing vector normalized to have unit magnitude at the origin. It is uniquely defined by the conditions that it should be null on both horizons and have unit magnitude at $r \to \infty$. For the de Sitter-Schwarzschild black hole the surface gravity is given by

$$\kappa = \frac{c^2}{2}\left[\frac{R_g(r_h)}{{r_h}^2} - \frac{R_g'(r_h)}{r_h}\right] \tag{20.31}$$

with r_h for both r_+ and r_-. It has a bifurcation point at $M = M_{cr1}$ (extreme black hole) determined by the condition

$$g_{00}(r_\pm) = g_{00}'(r_\pm) = 0. \tag{20.32}$$

For $R_g(r)$ defined by (20.24) the Hawking temperature is [20]

$$T_h = \frac{\hbar c}{4\pi k r_0}\left[\frac{r_0}{r_h} - \frac{3r_h}{r_0}\left(1 - \frac{r_h}{r_g}\right)\right]. \tag{20.33}$$

The temperature-mass diagram is shown in Fig. 20.4.

For $r_g \gg r_0$ the temperature tends to the Schwarzschild value

$$T_{Schw} = \frac{\hbar c^3}{8\pi kGM} \tag{20.34}$$

on the external horizon and to the de Sitter value

$$T_{deS} = -\frac{\hbar c}{2\pi k r_0} \tag{20.35}$$

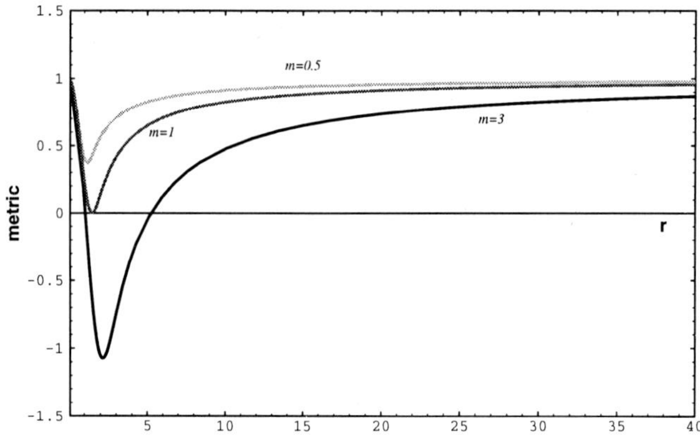

Figure 20.3: De Sitter-Schwarzschild metric (2.16). Mass parameter is normalized to $M_{cr1} \simeq 0.3 M_{Pl}(\rho_{Pl}/\rho_0)^{1/2}$. The extreme black hole ($m = 1$) is depicted by the red line, and a particlelike solution without horizon is shown by the green line.

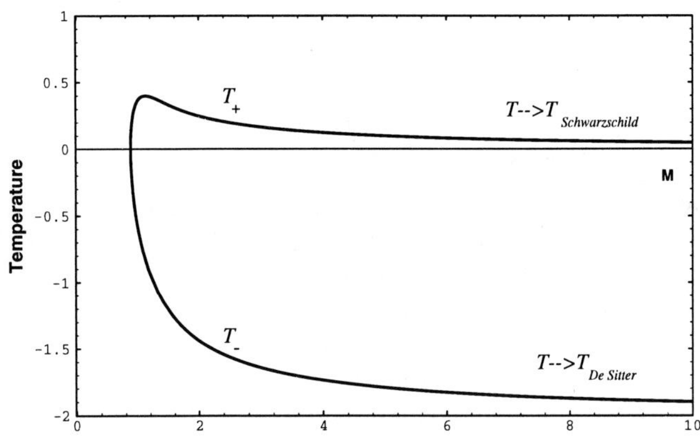

Figure 20.4: The temperature-mass diagram for de Sitter-Schwarzschild black hole in units $\hbar = c = G = 8\pi k = 1$, for the external and internal horizons. The specific heat on the external horizon is positive for $M_{cr1} \leq M \leq M_{cr2}$, where $M_{cr1} \approx 0.3 M_{Pl}(\rho_{Pl}/\rho_0)^{1/2}$ and $M_{cr2} \approx 0.38 M_{Pl}(\rho_{Pl}/\rho_0)^{1/2}$. A second-order phase transition occurs at $M = M_{cr2}$.

on the internal horizon. (The surface gravity characterizes the force that must be exerted at infinity to hold a unit test mass in place at the horizon. It has opposite signs for r_+ and r_- due to the repulsive character of gravity near r_-. Therefore the temperature on the internal horizon is negative.)

The temperature has the maximum at

$$M_{cr2} \simeq 0.38 M_{Pl}(\rho_{Pl}/\rho_0)^{1/2}, \tag{20.36}$$

and drops to zero when r_h approaches $r_+ = r_-$.

The specific heat is negative for $M > M_{cr2}$. As M decreases within the range of black hole masses $M_{cr2} \geq M \geq M_{cr1}$, the temperature T also increases. The specific heat is positive in this mass range. It changes sign as the mass decreases beyond M_{cr2}, hence a second-order phase transition occurs at this point. The temperature of a phase transition is

$$T_{tr} \simeq 0.2 T_{Pl}(\rho_{Pl}/\rho_0)^{1/2} \tag{20.37}$$

For the case of $\rho_0 \sim \rho_{GUT}$ and $M_{GUT} \sim 10^{15} GeV$,

$$T_{tr} \sim 0.2 \times 10^{11} GeV \tag{20.38}$$

$$M_{cr2} \sim 0.4 \times 10^{11} GeV \tag{20.39}$$

$$M_{cr1} \sim 0.3 \times 10^{11} GeV. \tag{20.40}$$

Due to existence of two horizons, the lower limit for a black hole mass M_{cr1} exists, independently on the particular form of the stress-energy tensor describing Schwarzschild-de Sitter transition: The temperature drops to zero when the horizons merge at a certain value of the mass M_{cr1}, and vanishes as $T \sim M^{-1}$ for $M \gg M_{cr1}$. It is nonzero and positive for $M_{cr1} < M < \infty$, since the surface gravity is nonzero and positive on the external horizon, in accordance with the Zeroth and Fourth laws of black hole thermodynamics. Therefore a curve representing the temperature-mass dependence on the external horizon must have a maximum at a certain value of mass $M_{cr2} > M_{cr1}$, which means a second-order phase transition at $M = M_{cr2}$.

The vacuum energy outside of de Sitter-Schwarzschild black hole is not zero and given by

$$E_{vac} = 4\pi \int_{r_+}^{\infty} T_t^t(r) r^2 dr = Mc^2 \exp\left(-\frac{r_+^3}{r_0^2 r_g}\right) = Mc^2 \exp\left(-\frac{4\pi r_+^3}{3M}\rho_0\right). \tag{20.41}$$

One can say that the de Sitter-Schwarzschild black hole has "Schwinger hair" vanishing only asymptotically at $(r_g/r_0) \to \infty$.

20.4 Particle-like structure

The De Sitter-Schwarzschild spacetime has two characteristic surfaces at the characteristic scale $r \sim r_g^{1/3}$.

The surface of zero gravity is the characteristic surface at which the strong energy condition of the singularities theorems [21] $(T_{\mu\nu} - g_{\mu\nu}T/2)u^\mu u^\nu \geq 0$ (where u^ν is any timelike vector) is violated, which means that the gravitational acceleration changes sign. The radius of this surface satisfies the condition $\varepsilon + \sum p_k = 0$, where $p_k = -T_k^k$, and is given by

$$r = r_b = (2r_0^2 r_g/3)^{1/3}. \tag{20.42}$$

The globally regular configuration with the de Sitter-like core, instead of a singularity, arises as a result of balance between attractive gravitational force outside, and repulsive gravitational force inside of the region bounded by the surface of zero gravity (20.42).

The surface of zero scalar curvature is determined by the condition $R = 0$ and is given by

$$r = r_c = (4r_0^2 r_g/3)^{1/3}. \tag{20.43}$$

Beyond this surface the scalar curvature changes sign.

Choosing r_c as the characteristic size of a regular core, we can estimate its mass as

$$m_{core} = 4\pi \int_0^{r_c} \rho r^2 dr \simeq 0.74M. \tag{20.44}$$

Let us note that this mass is contained in the region

$$r_c \simeq 10^{-20} cm (\rho_{Pl}/\rho_0)^{1/3} (M/3M_\odot)^{1/3}. \tag{20.45}$$

For a black hole of several solar masses it means that $\sim 5 \times 10^{33} g$ is contained in the core of a size $(10^{-20} - 10^{-15}) cm$.

Four characteristic surfaces r_+, r_-, r_c and r_b exhibit nontrivial behavior (see Fig. 20.5): while the mass parameter is decreasing, they change their places before the horizons merge – suggesting nontrivial dynamics of evaporation.

For nonextreme black hole the characteristic surfaces r_b and r_c are located between the horizons r_+, r_-, in such a sequence that

$$r_- < r_b < r_c < r_+. \tag{20.46}$$

At the value of mass parameter, $M_b \simeq 0.41 M_{Pl}(\rho_{Pl}/\rho_0)^{1/2}$, the surface of zero gravity $r = r_b$ merges with the internal horizon r_-, and then for masses $M < M_b$, the sequence becomes

$$r_b < r_- < r_c < r_+. \tag{20.47}$$

At the value of mass parameter $M_c \simeq 0.31 M_{Pl}(\rho_{Pl}/\rho_0)^{1/2}$, the surface of zero scalar curvature $r = r_s$ merges with the internal horizon, and for masses $M < M_c$

$$r_b < r_c < r_- < r_+. \tag{20.48}$$

For the extreme black hole

$$r_b < r_c < r_- = r_+, \tag{20.49}$$

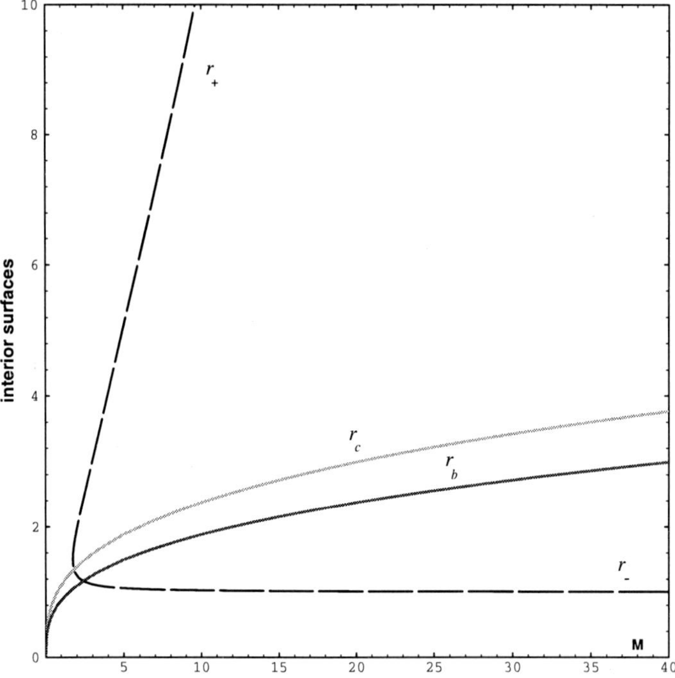

Figure 20.5: Surface of zero scalar curvature $r = r_c$ and of zero gravity $r = r_b$ (strong energy condition of singularity theorems is violated) are shown together with the horizons.

and for recovered particlelike structure without horizons

$$r_b < r_c. \tag{20.50}$$

In accordance with the results of Ref. [22], the de Sitter-like core can approximate the non-Abelian structure with mass $M = \sqrt{2}gM_0$, where M_0 is the vacuum expectation value of the relevant non-Abelian (Higgs) field with self-interaction $\lambda = g^2$.

We can estimate the coupling constant g, setting the Compton wavelength equal to the characteristic size r_c of a particle-like structure. It results in the formula for the mass $M = \sqrt{2}gM_0 = M_{Pl}(2\pi\rho_0/\rho_{Pl})^{1/4}$. Taking into account that $(\rho_0/\rho_{Pl}) = (M_0/M_{Pl})^4$, we obtain $g = (\pi/4)^{1/4}$. The corresponding fine-structure constant $\alpha = g^2/4\pi$ is $\alpha = 1/(8\sqrt{\pi}) \approx 1/14$.

20.5 Discussion

Such properties of The de Sitter-Schwarzschild spacetime as

1. algebraic structure of the stress-energy tensor as describing spherically symmetric vacuum,

2. limiting de Sitter value for the spacetime curvature,

3. global structure of spacetime,

4. existence of zero curvature and zero gravity surfaces at the characteristic scale $r \sim r_g^{1/3}$

5. of the lower limit for a black hole mass

result, in fact, from the boundary conditions and from the condition $g_{00} = -g_{11}^{-1}$ imposed for a metric. They do not depend on a particular form of $T_t^t(r)$ if it is a smooth function decreasing monotonically and quickly. This requirement does not seem to be too strong if we look at de Sitter-Schwarzschild solution as at the non-singular modification of the Schwarzschild solution which implies δ-like distribution of $T_t^t(r)$. Any particular form of $T_t^t(r)$ satisfying this requirement would result in a metric $g_{00}(r)$, whose shape would be qualitatively the same as shown in Fig. 20.3.

The form of the temperature-mass diagram is determined by the existence of two horizons whichever would be the particular form for the density profile. The temperature drops to zero when the horizons merge and tends to zero as $T \sim M^{-1}$ when the mass is increasing, being nonzero and positive for $M_{cr1} < M < \infty$. Therefore, a curve representing temperature-mass dependence must have a maximum, which means that a second-order phase transition occurs in the course of evaporation.

Following Schwarzschild, we can say that "it is always nice to have analytic solution of a simple form" [23]. It allows us to investigate the behavior qualitatively, and gives a chance to reveal features which otherwise could escape from analysis. For example, numerical analysis of non-Abelian black holes revealed existence, among them, a "neutral" type for which a non-Abelian structure can be approximated as a vacuum core of uniform density [22, 24] (suggesting existence of de Sitter-Schwarzschild black holes), however two essential features of a black hole of this type – existence of the lower limit for a black hole mass M_{cr1} and vanishing the temperature in the course of evaporation – remained concealed.

The results obtained above allow one to speculate about possible end point for evaporation of a de Sitter-Schwarzschild black hole. While it emits thermal radiation and loses mass, the horizons come together until they merge. The temperature drops to zero, and the evaporation stops. There are three further possibilities depending on the stability of the extreme configuration. If it were stable, then a black hole could settle down as an extreme black hole. Another possibility is connected with back-reaction effects. If they diminished the mass, then a black hole could disappear leaving recovered particle-like structure in its place. The third possibility is suggested by the global structure of the extreme black hole. In this case the Killing vector K is never spacelike and in principle nothing could prevent the de Sitter-like core from exploding. If it occurred, the black hole would evaporate completely. Let us note, that in any case information about a baby universe inside a de Sitter-Schwarzschild black hole is always lost in the course of evaporation.

Acknowledgments

I am very grateful to the organizers of the "Workshop on the Internal Structure of Black Holes and Spacetime Singularities" for their kind hospitality.

This work was supported by the Polish Committee for Scientific Research through the Grant 2.P03D.017.11.

Bibliography

[1] A. D. Sakharov, Sov. Phys. JETP **22**, 241 (1966).

[2] E. B. Gliner, Sov. Phys. JETP **22**, 378 (1966).

[3] E. B. Gliner, Sov. Phys. Dokl. **15**, 559 (1970).

[4] Ya. B. Zel'dovich, Sov. Phys. Lett. **6**, 883 (1967).

[5] A. Guth A., Phys. Rev. D **23**, 389 (1981).

[6] E. W. Kolb and M. S. Turner, *The Early Universe*, (Addison-Wesley, 1990).

[7] A. Strominger, Phys. Rev. D **46**, 4396 (1992) has demonstrated a natural (not *ad hoc*) arising a de Sitter-like core inside a black hole, in the model of two-dimensional dilaton gravity conformally coupled to N scalar fields.

[8] M. R. Bernstein, Bull Amer. Phys. Soc. **16**, 1016 (1984).

[9] E. Fahri and A. Guth, Phys. Lett. **B 183**, 149 (1987).

[10] W. Shen and S. Zhu, Phys. Lett. **A 126**, 229 (1988).

[11] V. P. Frolov, M. A. Markov and V. F. Mukhanov, Phys. Lett. **B 216**, 272 (1989); Phys. Rev. D **41**, 3831 (1990).

[12] D. Morgan D., Phys. Rev. D **43**, 3144 (1991), has considered a black hole in a simple model for quantum gravity in which quantum effects are represented by an upper cutoff on the curvature, and obtained de Sitter-like past and future cores (it appeared impossible to splice them together exactly).

[13] E. Poisson and W. Israel, Class. Quantum Grav. **5**, L201 (1988).

[14] J. Schwinger, Phys. Rev. **82**, 664 (1951).

[15] I. G. Dymnikova, Gen. Rel. Grav. **24**, 235 (1992).

[16] I. D. Novikov and V. P. Frolov, *Physics of Black Holes*, (Kluwer, Dordrecht, 1989).

[17] I. K. Affleck and N. S. Manton, Nucl. Phys. **B194**, 38 (1982).

[18] D. Garfinkle and A. Strominger, Phys. Lett. **B 256**, 146 (1991).

[19] G. W. Gibbons and S. W. Hawking, Phys. Rev. D **15**, 2738 (1977).

[20] I. G. Dymnikova, Int. Journ. Mod. Phys. **D 5**, 529 (1996).

[21] S. W. Hawking and G. F. R. Ellis, *The Large Scale Structure of Space-Time*, (Cambridge University Press, Cambridge, 1973).

[22] K. Maeda, T. Tashizawa, T. Torii, M. Maki M., Phys. Rev. Lett. **72**, 450 (1994).

[23] K. Schwarzschild K., Sitzber. Deut. Akad. Wiss. Berlin, Kl. Math.–Phys. Tech., 189 (1916).

[24] T. Torii and K. Maeda, Phys. Rev. D **48**, 1643 (1993).

NUMERICAL STUDY OF COSMOLOGICAL SINGULARITIES

Beverly K. Berger [a], David Garfinkle [a] and Vincent Moncrief [b]

a) Physics Department, Oakland University, Rochester, MI 48309 USA
b) Physics Department, Yale University, New Haven, CT 06520 USA

Abstract

The spatially homogeneous, isotropic Standard Cosmological Model appears to describe our Universe reasonably well. However, Einstein's equations allow a much larger class of cosmological solutions. Theorems originally due to Penrose and Hawking predict that all such models (assuming reasonable matter properties) will have an initial singularity. The nature of this singularity in generic cosmologies remains a major open question in general relativity. Spatially homogeneous but possibly anisotropic cosmologies have two types of singularities: (1) velocity dominated—(reversing the time direction) the universe evolves to the singularity with fixed anisotropic collapse rates ; (2) Mixmaster—the anisotropic collapse rates change in a deterministically chaotic way. Much less is known about spatially inhomogeneous universes. Belinskii, Khalatnikov, and Lifshitz (BKL) claimed long ago that a generic universe would evolve toward the singularity as a different Mixmaster universe at each spatial point. We shall report on the results of a program to test the BKL conjecture numerically. Results include a new algorithm to evolve homogeneous Mixmaster models, demonstration of velocity dominance and understanding of evolution toward velocity dominance in the plane symmetric Gowdy universes (spatial dependence in one direction), demonstration of velocity dominance in polarized U(1) symmetric cosmologies (spatial dependence in two directions), and exploration of departures from velocity dominance in generic U(1) universes.

21.1 Introduction

We shall describe a series of numerical studies of the nature of singularities in cosmological models. Since the interiors of black holes can be described locally as cosmological models, it is possible that our methods and results may be useful to the participants in this conference.

The generic singularity in spatially homogeneous cosmologies is reasonably well understood. The approach to it asymptotically falls into two classes. The first, called asymptotically velocity term dominated (AVTD) [1, 2], refers to a cosmology that approaches the Kasner (vacuum, Bianchi I) solution [3] as $\tau \to \infty$. (Spatially homogeneous universes can be described as a sequence of homogeneous spaces labeled by τ. Here we shall choose τ so that $\tau = \infty$ coincides with the singularity.) An example of such a solution is the vacuum Bianchi II model [4] which begins with a fixed set of Kasner-like anisotropic expansion rates, and, possibly, makes one change of the rates in a prescribed way (Mixmaster-like bounce) and then continues to $\tau = \infty$ as a fixed Kasner solution. In contrast are the homogeneous cosmologies which display Mixmaster dynamics such as vacuum Bianchi VIII and IX [5, 6, 7] and Bianchi VI_0 and Bianchi I with a magnetic field [8, 9, 10]. Jantzen [11] has discussed other examples. Mixmaster dynamics describes an approach to the singularity which is a sequence of Kasner epochs with a prescription, originally due to Belinskii, Khalatnikov, and Lifshitz (BKL) [5], for relating one Kasner epoch to the next. Some of the Mixmaster bounces (era changes) display sensitivity to initial conditions one usually associates with chaos and in fact Mixmaster dynamics is chaotic [12]. The vacuum Bianchi I (Kasner) solution is distinguished from the other Bianchi types in that the spatial scalar curvature ${}^3R$, (proporional to) the minisuperspace (MSS) potential [6, 13], vanishes identically. But ${}^3R$ arises in other Bianchi types due to spatial dependence of the metric in a coordinate basis. Thus an AVTD singularity is also characterized as a regime in which terms containing or arising from spatial derivatives no longer influence the dynamics. This means that the Mixmaster models do not have an AVTD singularity since the influence of the spatial derivatives (through the MSS potential) never disappears—there is no last bounce.

In the late 1960's, BKL claimed to show that singularities in generic solutions to Einstein's equations are locally of the Mixmaster type [5]. This means that each point of a spatially inhomogeneous universe could collapse to the singularity as the Mixmaster sequence of Kasner models. (It has been argued that this could generate a fractal spatial structure [15, 16, 17].) In contrast, each point of a cosmology with an AVTD singularity evolves asymptotically as a fixed Kasner model. Although the BKL result is controversial [14], it provides a hypothesis for testing. Our ultimate objective is to test the BKL conjecture numerically.

21.2 Numerical Methods

The work reported here was performed by using symplectic ODE and PDE solvers [18, 19]. While other numerical methods may be used to solve Einstein's equations for the models discussed here, symplectic methods have proved extremely advantageous for Mixmaster models and have also worked quite well in the Gowdy plane symmetric and polarized $U(1)$ symmetric cosmologies [20, 21, 22, 23]. Consider a system with one degree of freedom described by $q(t)$ and its canonically conjugate momentum $p(t)$

with a Hamiltonian

$$H = \frac{p^2}{2m} + V(q) = H_K + H_V. \tag{21.1}$$

Note that the subhamiltonians H_K and H_V separately yield equations of motion which are exactly solvable no matter the form of V. Variation of H_K yields $\dot{q} = p/m$, $\dot{p} = 0$ with solution

$$p(t + \Delta t) = p(t) \quad , \quad q(t + \Delta t) = q(t) + \frac{p(t)}{m}\Delta t. \tag{21.2}$$

Variation of H_V yields $\dot{q} = 0$, $\dot{p} = -dV/dq$ with solution

$$q(t + \Delta t) = q(t) \quad , \quad p(t + \Delta t) = p(t) - \left.\frac{dV}{dq}\right|_t \Delta t. \tag{21.3}$$

Note that the absence of momenta in H_V makes (21.3) exact for any $V(q)$. One can then demonstrate that to evolve from t to $t + \Delta t$ an evolution operator $\mathcal{U}_{(2)}(\Delta t)$ can be constructed from the evolution sub-operators $\mathcal{U}_K(\Delta t)$ and $\mathcal{U}_V(\Delta t)$ obtained from (21.2) and (21.3). One can show that [18]

$$\mathcal{U}_{(2)}(\Delta t) = \mathcal{U}_K(\Delta t/2)\mathcal{U}_V(\Delta t)\mathcal{U}_K(\Delta t/2) \tag{21.4}$$

reproduces the true evolution operator through order $(\Delta t)^2$. Suzuki has developed a prescription to represent the full evolution operator to arbitrary order [24]. For example

$$\mathcal{U}_{(4)}(\Delta t) = \mathcal{U}_{(2)}(s\Delta t)\mathcal{U}_{(2)}[(1 - 2s)\Delta t]\mathcal{U}_{(2)}(s\Delta t) \tag{21.5}$$

where $s = 1/(2 - 2^{1/3})$. The advantage of Suzuki's approach is that one only needs to construct $\mathcal{U}_{(2)}$ explicitly. $\mathcal{U}_{(2n)}$ is then constructed from appropriate combinations of $\mathcal{U}_{(2n-2)}$.

The generalization of this method to N degrees of freedom and to fields is straightforward. In the latter case, $V[\vec{q}(t)] \to V[\vec{q}(\vec{x}, t)]$ so that dV/dq becomes the functional derivative $\delta V/\delta q$. On the computational spatial lattice, the derivatives that are obtained in the expression for the functional derivative must be represented in differenced form. We note that, to preserve nth order accuracy in time, nth order accurate spatial differencing is required. Some discussion of this has been given elsewhere [21].

21.3 Application to Mixmaster Dynamics

(Diagonal) Bianchi Class A cosmologies [13] are described by the metric

$$ds^2 = -\,e^{3\Omega}\,dt^2 + \left(e^{2\beta}\right)_{ij}\,\sigma^i\,\sigma^j \tag{21.6}$$

where $\beta_{ij} = \mathrm{diag}(-2\beta_+, \beta_+ + \sqrt{3}\,\beta_-, \beta_+ - \sqrt{3}\,\beta_-)$, $d\sigma^i = C^i_{jk}\sigma^j \wedge \sigma^k$ defines the Bianchi type and t is the BKL time coordinate. Einstein's equations can be obtained by variation of the Hamiltonian

$$2H = -p_\Omega^2 + p_+^2 + p_-^2 + V(\beta_+, \beta_-, \Omega). \tag{21.7}$$

The logarithmic anisotropic scale factors α, ζ, and γ are given in terms of the logarithmic volume Ω and anisotropic shears $\beta_\pm$ as

$$\begin{aligned} \alpha &= \Omega - 2\beta_+, \\ \zeta &= \Omega + \beta_+ + \sqrt{3}\,\beta_-, \\ \gamma &= \Omega + \beta_+ - \sqrt{3}\,\beta_-. \end{aligned} \tag{21.8}$$

The momenta p_Ω, $p_\pm$ are canonically conjugate to Ω, $\beta_\pm$ respectively. For vacuum Bianchi IX or magnetic Bianchi VI_0, the MSS potential V has the form [20]

$$V = c^2 e^{2b\alpha} + e^{4\zeta} + e^{4\gamma} - 2\left(a\,e^{2(\alpha+\zeta)} + a\,e^{2(\alpha+\gamma)} + d\,e^{2(\zeta+\gamma)}\right) . \tag{21.9}$$

Here $a = 1$, $b = 2$, $c = 1$, and $d = 1$ for vacuum Bianchi Type IX, while $a = 0$, $b = 1$, $c = \sqrt{\xi}$, and $d = -1$ for magnetic Bianchi Type VI_0, and ξ is a constant that depends on the strength of the magnetic field. In these models, the singularity occurs at $\tau = -\Omega = \infty$. From (21.8) and (21.9), we see that $V \to 0$ as $\tau \to \infty$ unless one of α, ζ, or γ is ≈ 0. When the potential itself is small ($V \approx 0$), (21.7) describes a "free particle" in MSS and is in fact approximately the Kasner solution.

The standard algorithms for solving ODE's [25] often employ an adaptive time step. The idea is to take large time steps where nothing much happens (e.g. in the Kasner regime) while taking shorter time steps when the forces are large (e.g. at a bounce). Unfortunately, in Mixmaster dynamics, the duration of the Kasner epochs increases exponentially in τ as $\tau \to \infty$ [26] while the duration of the bounce itself is in some sense fixed [27]. This means that, although huge time steps may be taken in the Kasner segments, the time step must become very small at the bounces. Ideally, one would like the time step also to grow exponentially but this cannot be done with standard approaches. Thus, with a great deal of effort, standard methods can yield about 30 bounces for $0 < \tau < 10^8$ in about an hour on a supercomputer. The application of the symplectic method to Mixmaster dynamics is straightforward. From (21.7), we have

$$H_K = -p_\Omega^2 + p_+^2 + p_-^2 \quad ; \quad H_V = V(\Omega, \beta_+, \beta_-). \tag{21.10}$$

With an adaptive step size and a 6th order version of (21.5), one can do slightly better (approximately a factor of three fewer steps) than 4th order Runge-Kutta. However, it is well known that a bounce off an exponential wall—the Bianchi II (or Taub [4]) cosmology—is exactly solvable. If we first identify the dominant exponential wall

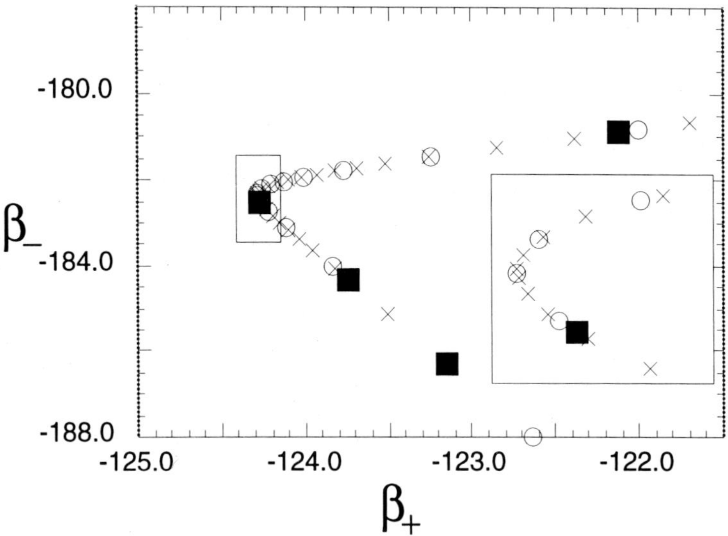

Figure 21.1: Comparison of algorithms for Mixmaster dynamics. A typical Mixmaster bounce is shown in the anisotropy plane. Crosses indicates every 10th point on a 4th order Runge-Kutta evaluation of the trajectory, while circles indicate every 10th point for a 6th order standard symplectic evaluation. The filled squares indicate *every* point using the new algorithm. The inset shows the details closest to the bounce. Note that the new algorithm does not require a point at the apex of the bounce.

(say $e^{4\alpha}$), then we find that the symplectic algorithm works for a different split of the Hamiltonian into two subhamiltonians [20]. Let

$$H = H_1 + H_2 \tag{21.11}$$

where (e.g. for Bianchi IX)

$$H_1 = -p_\Omega^2 + p_+^2 + p_-^2 + e^{4\Omega - 8\beta_+} \quad ; \quad H_2 = H_V - e^{4\Omega - 8\beta_+}. \tag{21.12}$$

Variation of H_2 is exactly solvable as before, while variation of H_1 yields equations with solution

$$\begin{aligned} p_y(t+\Delta t) &= p_y(t) \quad ; \quad p_-(t+\Delta t) = p_-(t) \quad ; \\ y(t+\Delta t) &= y(t) - 6p_y(t)\Delta t \quad ; \quad \beta_-(t+\Delta t) = \beta_-(t) + 2p_-(t)\Delta t \quad ; \end{aligned}$$

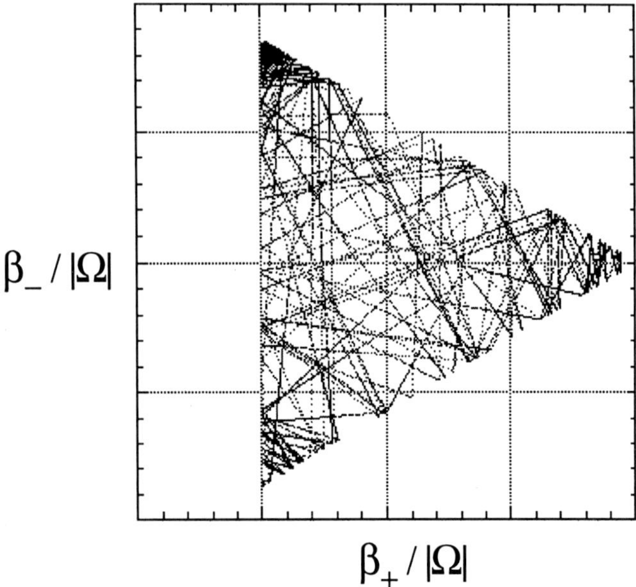

Figure 21.2: A typical trajectory consisting of 268 Mixmaster bounces is shown projected onto the anisotropy plane. In terms of the rescaled variables $\beta_+/|\Omega|$ and $\beta_-/|\Omega|$, the walls of the MSS potential and thus the bounce sites are at fixed locations.

$$\alpha(t+\Delta t) \;=\; -\frac{1}{2}\ln\left[\frac{1}{E}\cosh 4\sqrt{3}\,(t+\Delta t-t_0)\right] \quad ; \quad 6p_\alpha = \frac{d\alpha}{dt} \qquad (21.13)$$

where $y = -2\Omega + \beta_+$ and $E^2 = 3p_y^2 - p_-^2$. This new symplectic algorithm provides an enormous advantage because the bounce is built in. The time step can grow exponentially. In a few minutes on a desktop computer, one can obtain, e.g., 268 bounces for $8 < \tau < 10^{61.5}$. Fig. 21.1 shows a comparison of the new and standard algorithms while Fig. 21.2 shows a typical trajectory. Since H_1 is actually the exact Hamiltonian almost all the time, the accuracy of the method is much higher than the formal 6th order—we achieve machine precision.

The BKL parameter u characterizes the angle in the anisotropy plane of the Kasner epoch trajectory. Between the nth and $n+1$st Kasner epochs, we have

$$u_{n+1} = \begin{cases} u_n - 1 & ; \quad 2 \leq u_n < \infty \\ (u_n - 1)^{-1} & ; \quad 1 \leq u_n \leq 2 \end{cases} . \tag{21.14}$$

All numerical studies [5, 28, 29, 27] and a variety of analytic arguments [30, 20] have shown that one expects (21.14) to become ever more valid as $\tau \to \infty$. With our algorithm in a double or quadruple precision code, one can evolve until deviations from (21.14) disappear and then use the predicted values of u as a test of the accuracy of the code. For example [20], one discovers that the Hamiltonian constraint ((21.7) with $H = 0$) must be enforced although not necessarily at every time step.

For our studies of spatially inhomogeneous cosmologies, it is important to keep in mind that

(1) between bounces Mixmaster dynamics looks like the AVTD Kasner solution;

(2) in a fixed time variable such as τ, the time between bounces increases exponentially as $\tau \to \infty$;

(3) extraordinary accuracy can be achieved by symplectic methods when one of the subhamiltonians is dominant;

(4) enormous gains in accuracy and computational speed can be made by using a "custom-designed" treatment of the bounce.

21.4 The Gowdy Test Case

As the simplest example of a spatially inhomogeneous cosmology, we consider the plane symmetric vacuum Gowdy universe on $T^3 \times R$ [31, 32] described by the metric

$$\begin{aligned} ds^2 &= e^{-\lambda/2} e^{\tau/2} (-e^{-2\tau} d\tau^2 + d\theta^2) \\ &\quad + e^{-\tau} [e^P d\sigma^2 + 2 e^P Q \, d\sigma \, d\delta + (e^P Q^2 + e^{-P}) \, d\delta^2] \end{aligned} \tag{21.15}$$

where the background λ and amplitudes P and Q of the $+$ and $\times$ polarizations of gravitational waves are functions of τ and $0 \leq \theta \leq 2\pi$ and periodic in θ. There is a curvature singularity at $\tau = \infty$ [32, 2, 33]. The polarized case ($Q = 0$) has been shown to have an AVTD singularity [2]. The generic case has been conjectured to be AVTD (except perhaps at a set of measure zero) [34], which we have verified to the extent possible numerically [21, 35, 36, 37, 38, 39, 22]. A claim has been made that this model does not have an AVTD singularity [40] which we believe to be incorrect [41].

This model is especially attractive as a test case because Einstein's equations in our variables split into dynamical equations for the wave amplitudes (where $_{,a} = \partial / \partial a$):

$$P_{,\tau\tau} - e^{-2\tau} P_{,\theta\theta} - e^{2P} \left(Q_{,\tau}^2 - e^{-2\tau} Q_{,\theta}^2 \right) = 0, \tag{21.16}$$

$$Q_{,\tau\tau} - e^{-2\tau} Q_{,\theta\theta} + 2\left(P_{,\tau} Q_{,\tau} - e^{-2\tau} P_{,\theta} Q_{,\theta}\right) = 0 \qquad (21.17)$$

while the Hamiltonian and momentum constraints become respectively first order equations for λ:

$$\lambda_{,\tau} - [P_{,\tau}^2 + e^{-2\tau} P_{,\theta}^2 + e^{2P}(Q_{,\tau}^2 + e^{-2\tau} Q_{,\theta}^2)] = 0, \qquad (21.18)$$

$$\lambda_{,\theta} - 2(P_{,\theta} P_{,\tau} + e^{2P} Q_{,\theta} Q_{,\tau}) = 0. \qquad (21.19)$$

Thus two problematical aspects of numerical relativity—preservation of the constraints and solution of the initial value problem—become trivial [31]. The wave equations (21.16) and (21.17) may be obtained by variation of the Hamiltonian

$$\begin{aligned} H &= \tfrac{1}{2} \int_0^{2\pi} d\theta \left[\pi_P^2 + e^{-2P} \pi_Q^2\right] \\ &+ \tfrac{1}{2} \int_0^{2\pi} d\theta \left[e^{-2\tau} \left(P_{,\theta}^2 + e^{2P} Q_{,\theta}^2\right)\right] = H_K + H_V \end{aligned} \qquad (21.20)$$

where π_P and π_Q are canonically conjugate to P and Q respectively. Variation of H_K yields the AVTD solution

$$\begin{aligned} P &= \ln|\mu| + v(\tau - \tau_0) + \ln[1 + e^{-2v(\tau-\tau_0)}] \to v\tau \quad \text{as } \tau \to \infty, \\ Q &= Q_0 + \frac{1}{\mu} \frac{e^{-2v(\tau-\tau_0)}}{(1 + e^{-2v(\tau-\tau_0)})} \to Q_0 \quad \text{as } \tau \to \infty, \\ \pi_P &= v \frac{(1 - e^{-2v(\tau-\tau_0)})}{(1 + e^{-2v(\tau-\tau_0)})} \to v \quad \text{as } \tau \to \infty, \\ \pi_Q &= -2\mu v \end{aligned} \qquad (21.21)$$

and is thus exactly solvable in terms of four functions of θ: μ, $v > 0$, Q_0, and τ_0. H_V is also (trivially) exactly solvable so that symplectic methods can be used [21].

If the singularity is AVTD, one would expect the spatial derivative terms to go to zero exponentially as $\tau \to \infty$. However, if $P \to v\tau$, the term

$$V_2 = e^{-2\tau + 2P} Q_{,\theta}^2 \qquad (21.22)$$

in (21.16) would grow rather than decay if $v > 1$. Thus Grubišić and Moncrief (GM) [34] conjectured that, in a generic Gowdy model as $\tau \to \infty$, $0 \le v < 1$ except perhaps at isolated spatial points. If we consider generic initial data—e.g. $P = 0$,

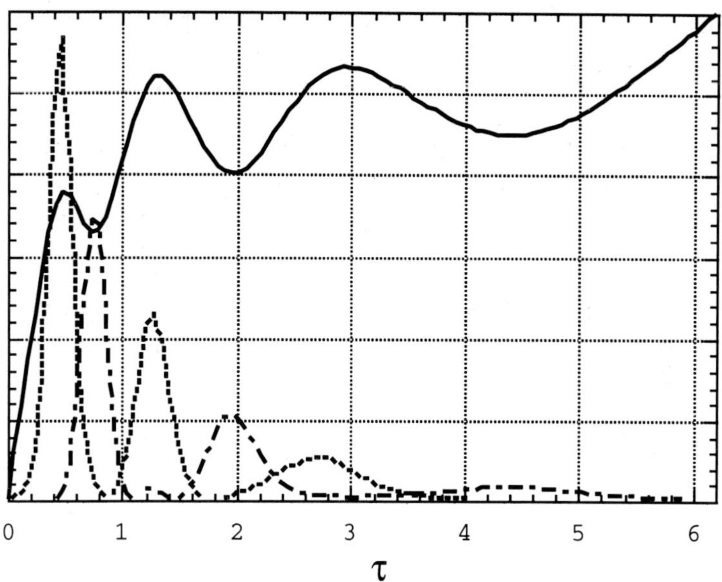

Figure 21.3: P (solid curve), V_1 (dash-dotted curve), and V_2 (dashed curve) vs τ at a fixed value of θ. Note that the slope of P, $P_{,\tau}$, decreases after each interaction with V_2 while $P_{,\tau}$ goes from negative to positive after each interaction with V_1. A continuation of this graph in τ would show that V_1 and V_2 have permanently died off and that P continues to increase with fixed positive slope $P_{,\tau} < 1$.

$\pi_P = v_0 \cos\theta$, $Q = \cos\theta$, $\pi_Q = 0$—then we must ask how a generic Gowdy solution evolves toward the AVTD solution at each spatial point and how an initial $P_{,\tau} > 1$ or $P_{,\tau} < 0$ is brought into the conjectured range. Typically, either V_2 or

$$V_1 = \pi_Q^2 \, e^{-2P} \tag{21.23}$$

(where $\pi_Q = e^{2P} \, Q_{,\tau}$) will dominate (21.16) to yield either

$$P_{,\tau}^2 + \pi_Q^2 \, e^{-2P} \approx \kappa_1^2 \tag{21.24}$$

or

$$Z_{,\tau}^2 + Q_{,\theta}^2 \, e^{2Z} \approx \kappa_2^2 \tag{21.25}$$

where $Z = P - \tau$. If $P_{,\tau} < 0$, an interaction with V_1 will occur to drive $P_{,\tau}$ to $-P_{,\tau}$ to yield $P_{,\tau} > 0$. If $P_{,\tau} > 1$, an interaction with V_2 will occur to drive $(P_{,\tau} - 1)$ to $-(P_{,\tau} - 1)$. If this yields $P_{,\tau} < 0$, a second interaction with V_1 will occur, etc.

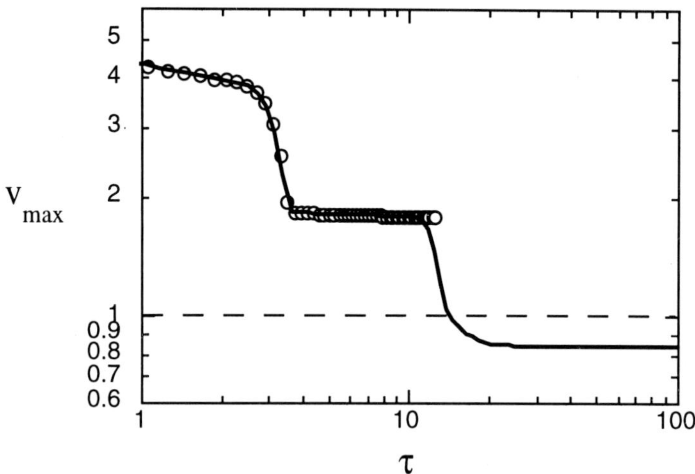

Figure 21.4: Plot of v_{max} vs τ. The maximum value of v is found for two simulations with 3200 (solid line) and 20000 spatial grid points (circles) respectively. The horizontal line indicates $v = 1$. Continuation in τ of the finer resolution simulation would show $v_{max} > 1$ for a much longer τ with $v_{max} < 1$ eventually.

When $|P,_\tau| < 1$, V_2 disappears so that, after a possible final interaction with V_1, $0 \le P,_\tau < 1$ forever. A typical sequence of bounces is shown in Fig. 21.3. Fig. 21.4 shows the maximum value of v on the spatial grid vs τ. First we see that high values of τ can be reached at which $0 \le v < 1$ everywhere.

Non-generic behavior can occur at isolated spatial points where either π_Q or $Q,_\theta$ vanishes. In the former case, the absence of V_1 where $\pi_Q = 0$ and its flatness where $\pi_Q \approx 0$ allow $P,_\tau$ and thus P to remain negative for a long time. Since $Q,_\tau = \pi_Q\, e^{-2P}$, Q will grow exponentially in opposite directions on either side of the points where $\pi_Q = 0$ producing a characteristic apparent discontinuity. On the other hand, if $Q,_\theta \approx 0$, $P,_\tau$ can remain large for a long time causing a spiky feature in P. Both types of features sharpen and narrow with time. The features and their association with non-generic points are shown in Fig. 21.5. The presence of this non-generic behavior at isolated spatial points leads to a dependence of simulation results on the spatial resolution. The finer the spatial resolution, the closer will a grid point be to the non-generic point. Near these non-generic points, the generic process of approach to $0 \le P,_\tau < 1$ will occur but slowly since either π_Q or $Q,_\theta \approx 0$. The closer one is to a non-generic site, the longer this process will take. Thus a finer resolution code will have narrower spiky features at which it takes longer for $P,_\tau$ to move into the range $[0, 1)$. Some evidence for this is seen in Fig. 21.4 where the finer spatial resolution

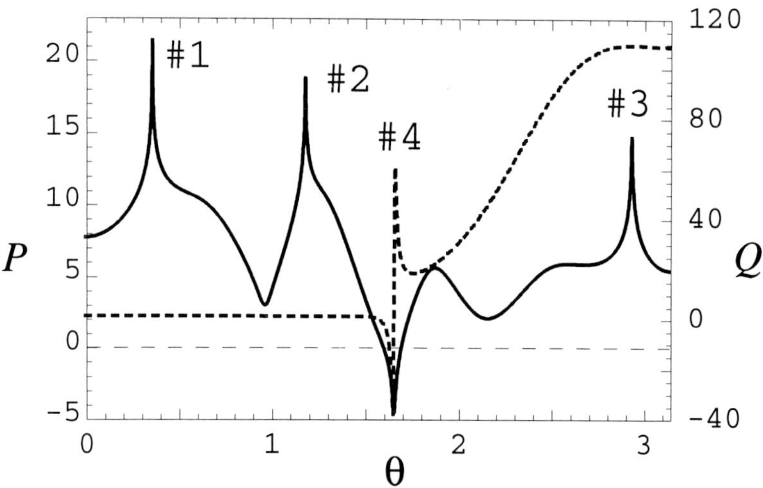

Figure 21.5: P (solid line) and Q (dashed line) vs θ at $\tau = 12.4$ for the initial data set given here with $v_0 = 5$ for $0 \le \theta \le \pi$ for a simulation containing 20000 spatial grid points in the interval $[0, 2\pi]$. The numbers on the graph refer to the most interesting features. Peaks #1, #2, and #3 in P are essentially the same in that they occur where $Q_{,\theta} \approx 0$. Peak #4 shows an apparent discontinuity in Q where $\pi_Q \approx 0$.

simulation diverges from the coarser one and will be considered in detail elsewhere [22].

Finally, we note that V_2 is analogous to the MSS potential. One may consider then application of the new Mixmaster algorithm to the Gowdy Hamiltonian (21.20) with $H = H_1 + H_2$ for

$$H_1 = \tfrac{1}{2} \oint d\theta \left(\pi_P^2 + e^{-2\tau + 2P} Q_{,\theta}^{\,2} \right) \tag{21.26}$$

and

$$H_2 = \tfrac{1}{2} \oint d\theta \left(\pi_Q^2 \, e^{-2P} + e^{-2\tau} P_{,\theta}^{\,2} \right). \tag{21.27}$$

These yield the exact solutions

$$\begin{aligned} Q(\tau + \Delta\tau) &= Q(\tau), \\ \pi_Q(\tau + \Delta\tau) &= \xi + \left[\frac{\kappa}{Q(\tau)_{,\theta}} \tanh \kappa(\tau - \tau_0) \right]_{,\theta}, \end{aligned}$$

$$
\begin{aligned}
P(\tau+\Delta\tau) &= \tau - \ln\left[\left|\frac{Q(\tau)_{,\theta}}{\kappa}\right| \cosh\kappa(\tau-\tau_0)\right], \\
\pi_P(\tau+\Delta\tau) &= 1-|\kappa|\tanh\kappa(\tau-\tau_0)
\end{aligned}
\tag{21.28}
$$

from the variation of H_1 (where ξ, κ, and τ_0 are functions of θ) and

$$
\begin{aligned}
P(\tau+\Delta\tau) &= P(\tau), \\
\pi_Q(\tau+\Delta\tau) &= \pi_Q(\tau), \\
\pi_P(\tau+\Delta\tau) &= \pi_P(\tau)+\pi_Q^2(\tau)\,e^{-2P(\tau)}\Delta\tau+\tfrac{1}{2}e^{-2\tau}\left(1-e^{-2\Delta\tau}\right)P(\tau)_{,\theta\theta}\ , \\
Q(\tau+\Delta\tau) &= Q(\tau)+\pi_Q(\tau)\,e^{-2P(\tau)}\Delta\tau
\end{aligned}
\tag{21.29}
$$

from the variation of H_2. Application of this algorithm is in progress.

From the Gowdy test case, we learn that:

(1) Since the singularity is AVTD, H_K dominates H_V asymptotically so our (current) algorithm is very accurate.

(2) Non-linear terms in the wave equations act as potentials. In the Gowdy case, they drive the system to the AVTD regime as $\tau \to \infty$ with $0 \leq v < 1$, where the potentials permanently die out.

(3) Non-generic points where $\pi_Q = 0$ or $Q_{,\theta} = 0$ lead to the growth of spiky features in P and Q.

(4) Spiky features appear narrower with finer spatial resolution.

21.5 $U(1)$ Symmetric Cosmologies

Given our understanding of the Gowdy model, we can move to spatially inhomogeneous cosmologies with one Killing field rather than two, retaining a $U(1)$ symmetry on $T^3 \times R$ [42]. These models can be described by five degrees of freedom $\{\varphi, \omega, \Lambda, z, x\}$ and their respective conjugate momenta $\{p, r, p_\Lambda, p_z, p_x\}$ which are functions of spatial variables u and v and time τ. Einstein's equations may be obtained by variation of [21, 23]

$$
\begin{aligned}
H &= \oint\oint du\,dv\,\mathcal{H} \\
&= \oint\oint du\,dv\,\left(\tfrac{1}{8}p_z^2 + \tfrac{1}{2}e^{4z}p_x^2 + \tfrac{1}{8}p^2 + \tfrac{1}{2}e^{4\varphi}r^2 - \tfrac{1}{2}p_\Lambda^2 + 2p_\Lambda\right)
\end{aligned}
$$

$$
\begin{aligned}
&+e^{-2\tau} \oint \oint du\,dv \left\{ \left(e^{\Lambda} e^{ab}\right)_{,ab} - \left(e^{\Lambda} e^{ab}\right)_{,a} \Lambda_{,b} \right. \\
&+ e^{\Lambda} \left[\left(e^{-2z}\right)_{,u} x_{,v} - \left(e^{-2z}\right)_{,v} x_{,u} \right] \\
&\left. + 2 e^{\Lambda} e^{ab} \varphi_{,a} \varphi_{,b} + \tfrac{1}{2} e^{\Lambda} e^{-4\varphi} e^{ab} \omega_{,a} \omega_{,b} \right\} \\
= \; & H_K + H_V = \oint \oint du\,dv\,\mathcal{H}_K + \oint \oint du\,dv\,V. \qquad (21.30)
\end{aligned}
$$

Here φ and ω are analogous to P and Q while Λ, x, z describe the metric $\tilde{e}_{ab}$ in the u-v plane perpendicular to the symmetry direction with

$$
\tilde{e}_{ab} = e^{\Lambda} e_{ab} = \tfrac{1}{2} e^{\Lambda} \begin{pmatrix} e^{2z} + e^{-2z}(1+x)^2 & e^{2z} + e^{-2z}(x^2-1) \\ e^{2z} + e^{-2z}(x^2-1) & e^{2z} + e^{-2z}(1-x)^2 \end{pmatrix}. \qquad (21.31)
$$

This model is sufficiently generic that local Mixmaster dynamics is allowed. The $U(1)$ Hamiltonian (21.30) has the standard symplectic form. Note that H_K consists of two Gowdy-like H_K's and a free particle term so that H_K is exactly solvable. H_V is also (again trivially) exactly solvable although spatial differencing must be performed with care. In two spatial dimensions, there are a variety of ways to represent derivatives to a given order of accuracy. We currently use a scheme provided by Norton [43] to minimize the growth of short wavelength modes.

Unlike the Gowdy model, the constraints and initial value problems must be considered. We find

$$
\mathcal{H}^0 = \mathcal{H} - 2p_\Lambda = 0 \qquad (21.32)
$$

and

$$
\begin{aligned}
\mathcal{H}^u = \; & p_z z_{,u} + p_x x_{,u} + p_\Lambda \Lambda_{,u} - p_{\Lambda,u} + p\,\varphi_{,u} + r\,\omega_{,u} \\
& + \tfrac{1}{2} \left\{ \left[e^{4z} - (1+x)^2 \right] p_x - (1+x) p_z \right\}_{,v} \\
& - \tfrac{1}{2} \left\{ \left[e^{4z} + (1-x^2) \right] p_x - x\,p_z \right\}_{,u} = 0, \qquad (21.33)
\end{aligned}
$$

$$
\begin{aligned}
\mathcal{H}^v = \; & p_z z_{,v} + p_x x_{,v} + p_\Lambda \Lambda_{,v} - p_{\Lambda,v} + p\,\varphi_{,v} + r\,\omega_{,v} \\
& - \tfrac{1}{2} \left\{ \left[e^{4z} - (1-x)^2 \right] p_x + (1-x) p_z \right\}_{,u} \\
& + \tfrac{1}{2} \left\{ \left[e^{4z} + (1-x^2) \right] p_x - x\,p_z \right\}_{,v} = 0. \qquad (21.34)
\end{aligned}
$$

While a general solution to the initial value problem is not known, we use the particular solution obtained as follows: To solve the momentum constraints (21.33) and (21.34) set $p_x = p_z = \varphi_{,a} = \omega_{,a} = 0$ to leave $p_\Lambda \Lambda_{,a} - p_{\Lambda}{}_{,a} = 0$ which may be satisfied by requiring $p_\Lambda = c\, e^\Lambda$. For sufficiently large c, the Hamiltonian constraint may be solved algebraically for either p or r. In general, this leaves as free data the four functions x, z, Λ, and either r or p. Since there are four free functions at each spatial point, we expect generic behavior.

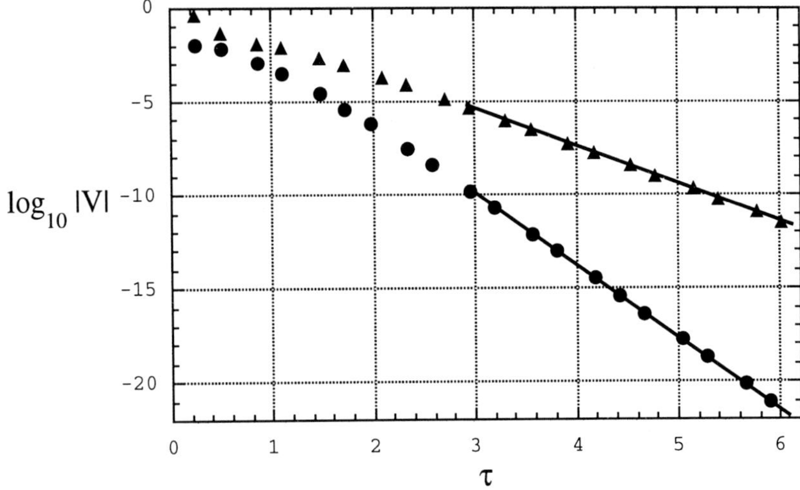

Figure 21.6: Evolution of $\log_{10}|V|$ vs τ for a polarized $U(1)$ cosmology at two representative spatial points. The solid lines are exponential fits.

As a first case, we consider polarized $U(1)$ models obtained by setting

$$r = \omega = 0. \tag{21.35}$$

This condition is preserved numerically as well as analytically. It has been conjectured [44] that polarized $U(1)$ models are AVTD. This is reasonable because the Mixmaster potential-like term

$$V_{\nabla\omega} = e^{-2\tau} e^{\Lambda} e^{ab} e^{-4\varphi} \omega_{,a}\, \omega_{,b} \tag{21.36}$$

is absent. Other spatial derivative terms decay as $e^{\Lambda - 2\tau - 2z}$ for the expected AVTD limits of the variables as $\tau \to \infty$ [23]:

$$z \;\; \to \;\; -v_z \tau \quad , \quad x \to x_0 \quad , \quad p_z \to -4\, v_z \quad ,$$

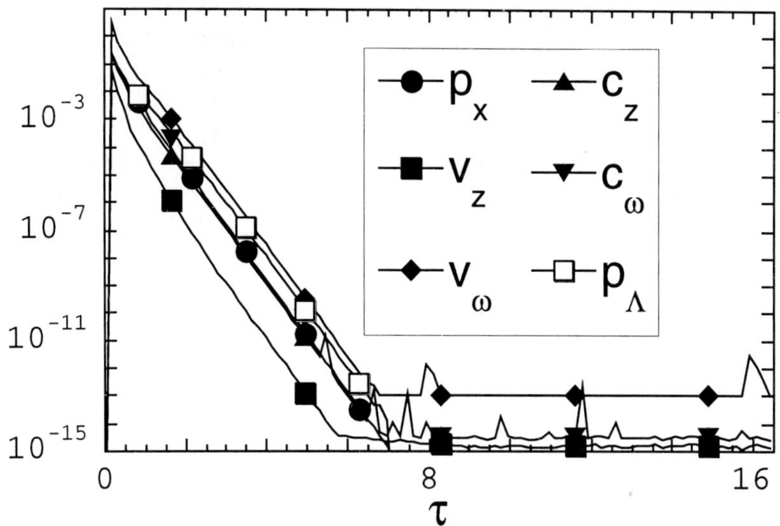

Figure 21.7: The **maximum** value of the **change** with time over the spatial grid of AVTD regime constants for a polarized $U(1)$ symmetric model. Here $v_z = \sqrt{p_z^2/8 + e^{4z}p_x^2/2}$, $v_\omega = \sqrt{p^2/8 + e^{4\varphi}r^2/2}$, $c_z = p_z/2 + p_x\,x$, and $c_\omega = p/2 + r\,\omega$. Exponential decay is observed until the maximum values of the changes reach the level of machine precision.

$$
\begin{aligned}
p_x &\rightarrow p_x^0 \quad , \quad \varphi \rightarrow -v_\varphi \tau \quad , \quad p \rightarrow -4\,v_\varphi \quad , \\
\Lambda &\rightarrow \Lambda_0 + (2 - p_\Lambda^0)\tau \quad , \quad p_\Lambda \rightarrow p_\Lambda^0,
\end{aligned}
\tag{21.37}
$$

so that an AVTD singularity is consistent.

Fig. 21.6 shows $\log_{10}|V|$ vs τ for typical spatial points. Thus we see the expected exponetial decay. In Fig. 21.7, we see that the maximum values of the changes with time of the quantitites expected to be constant in an AVTD regime also decay exponentially as expected. These results will be discussed in detail elsewhere [23].

We emphasize here that the polarized $U(1)$ models present *no* numerical difficulties. The absence of $V_{\nabla\omega}$ means that spiky features do not develop. The situation unfortunately changes for generic (unpolarized) $U(1)$ models. It appears that consistency arguments which suggest an AVTD singularity in polarized $U(1)$ models and restrict v to $[0,1]$ in the Gowdy models fail in generic $U(1)$ models. This suggests that $V_{\nabla\omega}$ will always grow exponentially if the system tries to be AVTD producing a Mixmaster-like bounce. A bounce in the opposite direction will come, as in Gowdy

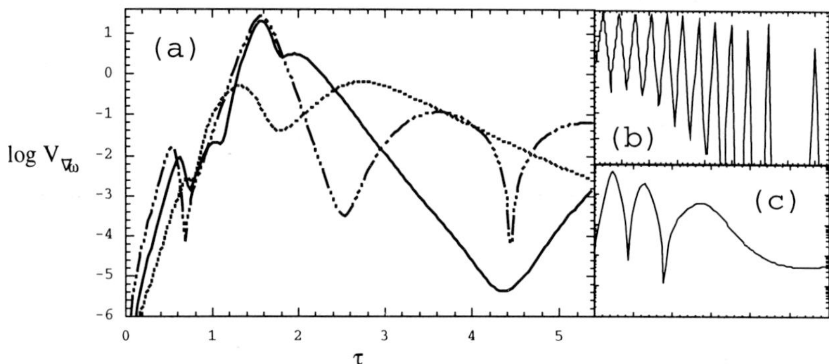

Figure 21.8: (a) $\log_{10} V$ vs τ for three representative spatial points of a generic $U(1)$ symmetric cosmology. (b) The analogous quantity for a typical Mixmaster evolution. (c) The analogous quantity for a Gowdy simulation.

models, from terms in H_K. This bouncing could continue indefinitely.

Unfortunately, the bounces off $V_{\nabla\omega}$ probably cause numerical instabilities that limit the duration of the simulations. Spatial averaging has been used to improve stability but it is known to produce numerical artifacts. Typical evolutions of $V_{\nabla\omega}$ at single spatial grid points are shown in Fig. 21.8 and compared to similar quantities in Mixmaster and Gowdy models and, of course, to Fig. 21.6. While generic $U(1)$ models are clearly different from polarized ones, it is not clear yet whether the observed bounces are Mixmaster-like or will eventually die out as in Gowdy models. Recall that Mixmaster universes are very close to the AVTD Kasner solution between bounces. One is also concerned about numerical artifacts although it is not clear that standard tests are helpful. Fig. 21.9 shows two frames from a movie of $V(u, v)$ vs τ for two different spatial resolutions. Features are narrower on the finer grid. While this usually indicates artifacts, we recall that this is precisely what happens in Gowdy models where the resolution dependence is well understood.

21.6 Future Directions

In the search for the nature of the generic cosmological singularity, we have obtained convincing evidence that both the Gowdy universes and the polarized $U(1)$ symmetric cosmologies have AVTD singularities. These results are found in simulations which present no numerical difficulties. In contrast, we cannot draw definite conclusions for generic $U(1)$ models except to say that we have not found evidence that the singularities are AVTD everywhere.

Presumably, numerical difficulties in generic $U(1)$ models are due to Gowdy-like spiky features (which are less easy to represent accurately in two spatial dimensions). It is possible that the new Mixmaster algorithm [20] which can be adapted to the Gowdy model will help in the generic $U(1)$ case where it can also be implemented. The hope is that better treatment of the bounces would give better local treatment of spiky features that arise from them.

While we solve the constraints initially in $U(1)$ models, we do nothing to preserve them thereafter. In the polarized case, they remain acceptably small (and in fact converge to zero with increasing spatial resolution). We have learned from the Mixmaster case that one must preserve the constraints [20]. In fact, it is the kinetic part of the Hamiltonian constraint which restricts the exponential factor in $V_{\nabla\omega}$. An error in the constraints could give the argument of the exponential the wrong sign leading to the observation of qualitatively wrong behavior. By starting closer to the singularity, we can supress some of the numerical instabilities and study this controlling exponential. Studies of this type have provided some evidence that it is essential to solve the constraints. Work on implementing a constraint solver is in progress.

Acknowledgements

B.K.B. and V.M. would like to thank the Albert Einstein Institute at Potsdam for hospitality. B.K.B. would also like to thank the Institute of Geophysics and Planetary Physics of Lawrence Livermore National Laboratory for hospitality. This work was supported in part by National Science Foundation Grants PHY9507313 and PHY9722039 to Oakland University and PHY9503133 to Yale University. Computations were performed at the National Center for Supercomputing Applications (University of Illinois).

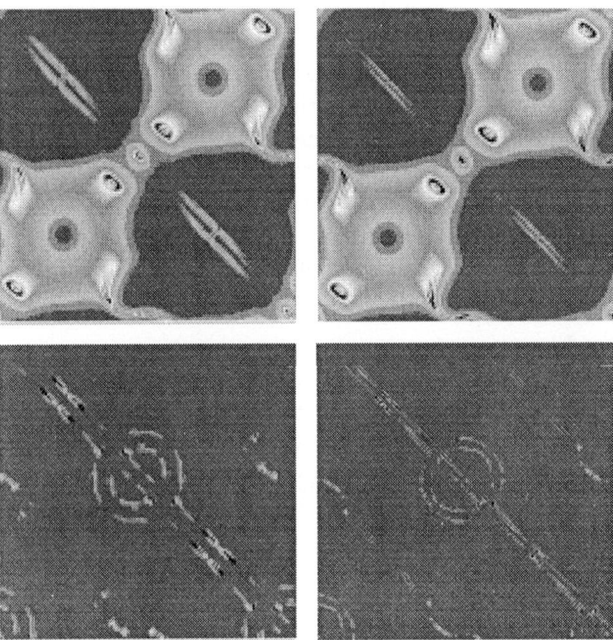

Figure 21.9: Movie frames of $V(u,v)$ for a generic $U(1)$ symmetric cosmology for two fixed values of τ. The upper frames precede the lower ones. The spatial coordinates u and v run from 0 to 2π in both directions. The frames on the left are from a simulation with 128^2 spatial grid points while the ones on the right have 256^2 spatial grid points. Both simulations have the same initial data. Note that there are features independent of spatial resolution as well as those that depend on the resolution. The background shade of gray indicates regions where $V \approx 0$.

Bibliography

[1] D. Eardley, E. Liang, R. Sachs, J. Math. Phys. **13**, 99 (1972).

[2] J. Isenberg, V. Moncrief, Ann. Phys. (N.Y.) **199**, 84 (1990).

[3] E. Kasner, Am. J. Math **43**, 130 (1921).

[4] A. Taub, Ann. Math. **53**, 472 (1951).

[5] V. A. Belinskii, E. M. Lifshitz, I. M. Khalatnikov, Sov. Phys. Usp. **13**, 745 (1971); Adv. Phys. **19**, 525 (1970).

[6] C. W. Misner, Phys. Rev. Lett. **22**, 1071 (1969).

[7] P. Halpern, Gen. Rel. Grav. **19**, 73 (1987).

[8] V. G. LeBlanc, D. Kerr, J. Wainwright, Class. Quantum Grav. **12**, 513 (1995).

[9] B. K. Berger, Class. Quantum Grav. **13**, 1273 (1996).

[10] V. G. LeBlanc, Class. Quantum Grav. **14**, 2281 (1997).

[11] R. T. Jantzen, Phys. Rev. D **33**, 2121 (1986).

[12] N. J. Cornish, J. J. Levin, Phys. Rev. Lett. **78**, 998 (1997)

[13] M. P. Ryan, Jr. and L. C. Shepley, *Homogeneous Relativistic Cosmologies* (Princeton University, Princeton,1975).

[14] J. D. Barrow, F. Tipler, Phys. Rep. **56** 372 (1979).

[15] G. Montani, Class. Quantum Grav. **12**, 2505 (1995).

[16] V. A. Belinskii, JETP Lett. **56**, 421 (1992).

[17] A. A. Kirillov, A. A. Kochnev, JETP Lett. **46**, 435 (1987); A. A. Kirillov, Sov. Phys. JETP **76**, 355 (1993).

[18] J. A. Fleck, J. R. Morris, M. D. Feit, Appl. Phys. **10**, 129 (1976).

[19] V. Moncrief, Phys. Rev. D **28**, 2485 (1983).

[20] B. K. Berger, D. Garfinkle, E. Strasser, Class. Quantum Grav. **14**, L29 (1997).

[21] B. K. Berger, V. Moncrief, Phys. Rev. D **48**, 4676 (1993).

[22] B. K. Berger, D. Garfinkle, B. Grubišić, V. Moncrief, "Phenomenology of the Gowdy Cosmology on $T^3 \times R$", unpublished.

[23] B. K. Berger, V. Moncrief, "Numerical Evidence for a Velocity Dominated Singularity in $U(1)$ Symmetric Cosmologies," unpublished.

[24] M. Suzuki, Phys. Lett. **A 146**, 319 (1990).

[25] W. H. Press, B. P. Flannery, S. A. Teukolsky, W. T. Vetterling, *Numerical Recipes: the Art of Scientific Computing (2nd edition)* (Cambridge University, Cambridge, 1992).

[26] I. M. Khalatnikov, E. M. Lifshitz, K. M. Khanin, L. N. Shchur, and Ya. G. Sinai, J. Stat. Phys. **38**, 97 (1985).

[27] B. K. Berger, Gen. Rel. Grav. **23**, 1385 (1991).

[28] A. R. Moser, R. A. Matzner, M. P. Ryan, Jr., Ann. Phys. (N.Y.) **79**, 558 (1973).

[29] S. E. Rugh, Cand. Scient. Thesis, Niels Bohr Inst. (1990); S. E. Rugh, B. J. T. Jones , Phys. Lett. **A 147**, 353 (1990).

[30] A. D. Rendall, Class. Quantum Grav. **14**, 2341 (1997).

[31] R. H. Gowdy, Phys. Rev. Lett. **27**, 826 (1971).

[32] B. K. Berger, Ann. Phys. (N.Y.) **83**, 458 (1974).

[33] V. Moncrief, Ann. Phys. (N.Y.) **132**, 87 (1981).

[34] B. Grubišić, V. Moncrief, Phys. Rev. D **47**, 2371 (1993).

[35] B. K. Berger, D. Garfinkle, B. Grubisic, V. Moncrief, V. Swamy, in *Proceedings of the Cornelius Lanczos Symposium*, edited by J.D. Brown, M.T. Chu, D.C. Ellison, R.J. Plemmons (SIAM, Philadelphia, 1994).

[36] B. K. Berger, D. Garfinkle, V. Moncrief, C. M. Swift, in *Proceedings of the AMS-CMS Special Session on Geometric Methods in Mathematical Physics*, ed. by J. Beem and K. Duggal (Americal Mathematical Society, 1994).

[37] B. K. Berger, in *Seventh Marcel Grossmann Meeting*, edited by R. Ruffini, M. Keiser (World Scientific, Singapore, 1995).

[38] B. K. Berger, in *Relativity and Scientific Computing,* edited by F. W. Hehl, R. A. Puntigam, H. Ruder (Springer-Verlag, Berlin, 1996).

[39] B. K. Berger, in *Proceedings of the 14th Internatial Conference on General Relativity and Gravitation,* edited by M. Francaviglia, G. Longhi, L. Lusanna, E. Sorace (World Scientific, Singapore, 1997).

[40] S. J. Hern, J. M. Stewart, "The Gowdy T^3 Cosmologies Revisited," gr-qc/9708038.

[41] B. K. Berger, D. Garfinkle, V. Moncrief, "Comment on 'The Gowdy T^3 Cosmology Revisited'," gr-qc/9708050.

[42] V. Moncrief, Ann. Phys. (N.Y.) **167**, 118 (1986).

[43] A. Norton, private communication

[44] B. Grubišić, V. Moncrief, Phys. Rev. D **49**, 2792 (1994).

QUANTUM MECHANICAL INSTABILITIES OF CAUCHY HORIZONS IN TWO DIMENSIONS — A MODIFIED FORM OF THE BLUESHIFT INSTABILITY MECHANISM

Éanna É. Flanagan

Cornell University, Newman Laboratory, Ithaca, NY 14853-5001

Abstract

There are several examples known of two dimensional spacetimes which are linearly stable when perturbed by test scalar classical fields, but which are unstable when perturbed by test scalar quantum fields. We elucidate the mechanism behind such instabilities by considering minimally coupled, massless, scalar, test quantum fields on general two dimensional spacetimes with Cauchy horizons which are classically stable. We identify a geometric feature of such spacetimes which is a necessary condition for obtaining a quantum mechanical divergence of the renormalized expected stress tensor on the Cauchy horizon for regular initial states. This feature is the divergence of the affine parameter length of a certain one parameter family of null geodesics which lie parallel to the Cauchy horizon, where the affine parameter normalization is determined by parallel transport along a fixed, transverse null geodesic which intersects the Cauchy horizon. (By contrast, the geometric feature of such spacetimes which underlies classical blueshift instabilities is the divergence of a holonomy operator). We show that the instability can be understood as a "delayed blueshift" instability, which arises from the infinite blueshifting of an energy flux which is created locally and quantum mechanically. The instability mechanism applies both to chronology horizons in spacetimes with closed timelike curves, and to the inner horizon in black hole spacetimes like two dimensional Reissner-Nördstrom-de Sitter.

22.1 Introduction and summary

22.1.1 Cauchy horizon instabilities: background and motivation

The physical question addressed by the Haifa workshop can be summarized as "What is the generic physical nature of black hole interiors?". Current attempts to answer this question involve using all of the laws of physics which are well understood today, including classical gravity and semiclassical gravity. As remarked by Valeri Frolov at the workshop, we theorists who study this issue are fortunate in two separate ways: First, we are protected from possibly embarrassing confrontations with experimental data by the fact that the gravitational singularities inside black holes are, very likely, always hidden behind event horizons. Second, some of the most interesting and deep issues involving the singularities inside black holes such as their possible traversability [1] and such as the information loss paradox [2] are ultimately obscured by the "Planck fog" of Planck scale physics where both classical and semiclassical gravity break down. Rather than attempt to grapple with these deep issues, we can justifiably throw up our hands once Planck scale curvatures are reached and declare the subsequent evolution to be beyond our purvue, given the absence of a well understood theory of quantum gravity.

Nevertheless, it seems to me that it is still of interest to investigate the structure of black hole interiors within the domains of validity of classical and semiclassical gravity. Black hole interiors can in principle be probed experimentally, by observers who venture inside the event horizon, and possibly even by external observers if weak cosmic censorship turns out not to be valid. Moreover, the structure of singularities even in the sub-Planckian regime is a deep and complicated question, and is interesting both in its own right and because understanding this structure is presumably a necessary prerequisite to the eventual understanding of black holes in full quantum gravity.

The study of black hole interiors is impeded by the fact that under certain circumstances general relativity breaks down; indeed, it predicts its own demise. As is well known, given initial data specified on some spacelike surface Σ, this breakdown can take two forms. First, in the maximal Cauchy evolution $D^+(\Sigma)$ of the initial data, the predicted gravitational field strength (curvature scalars or components of curvature tensors on parallel propagated bases along curves) can grow without limit. This prediction is presumably valid only up to the regime of Planckian curvatures, at which point our understanding breaks down. The prediction of infinities or singularities is not particularly worrisome; similar situations, where a continuum theory predicts its own breakdown, occur in other contexts in physics. A good example, as Amos Ori pointed out during the workshop, is the formation of shocks in fluids. There, the hydrodynamic equations predict that the fluid variables diverge; in actuality, the structure of shocks is determined by microscopic physics for which the fluid approximation is invalid. Presumably, something similar occurs at gravitational singularities: their structure is ruled by the as-yet-unknown laws of quantum gravity.

The second, well-known type of breakdown of general relativity is where the maximal Cauchy evolution $D^+(\Sigma)$ of the initial data is geodesically incomplete without any curvature singularities. For example, if the initial data on Σ were to consist of a spherically symmetric, collapsing charged star, the maximal Cauchy evolution would be a spacetime consisting of an interior solution describing the star, together with an exterior solution consisting of a portion of the Reissner-Nördstrom spacetime (the "initial globally hyperbolic region"). In such spacetimes, certain observers, after a finite amount of their proper time, come to the "edge" of the maximal Cauchy evolution; the theory fails to predict what such observers subsequently measure. Clearly this is a serious breakdown of the theory [3].

Mathematically, the breakdown is characterized by the fact that one can extend the maximal Cauchy evolution, which we will denote by (M, g_{ab}), to a larger spacetime (M', g'_{ab}). In any such larger spacetime, the future Cauchy horizon $H^+(\Sigma)$ is the boundary of M in M'; the breakdown is thus signified by the existence of a Cauchy horizon. Note, however, that the extension spacetime is not uniquely determined by the initial data and thus is not physical (even in situations where a natural extension is determined by analytic continuation).

The disturbing aspect of this second type of breakdown is that it can occur within the (apparent) domain of validity of general relativity, entirely at low curvature scales, as in the Reissner-Nördstrom example [4]. Of course, here we are assuming that classical general relativity [5] is a good approximation at any point $\mathcal{P}$ in spacetime whenever the curvature is sub-Planckian everywhere in the past lightcone of $\mathcal{P}$, which seems like a reasonable assumption. In order to highlight how disturbing such breakdowns are, consider the hypothetical analogous situation in fluid hydrodynamics. Suppose one were given a solution of the hydrodynamic equations in which all of the lengthscales and timescales determined by the solution are much larger than the relevant microscopic lengthscales and timescales (so that the continuum approximation should be good) but where the solution is nevertheless is incomplete, cannot be uniquely extended, and fails to predict the complete future evolution of the fluid. Such a situation would be paradoxical, and of course does not occur.

In the context of general relativity, if such breakdowns were ubiquitous one would apparently be forced to abandon general relativity as a viable description of Nature even in the regime of low curvatures. The traditional refuge of theorists has been to assert that such breakdowns should be *non-generic*; that is, that only "isolated" initial data give rise to geodesically incomplete spacetimes without curvature singularities. In other words, Cauchy horizons in spacetimes without singularities should always be unstable. This is the essential content of the strong cosmic censorship hypothesis [6].

The subject of this contribution to the proceedings is the instability properties of Cauchy horizons in classical and semi-classical gravity. To summarize the above discussion, one of the main motivations for studying the stability of Cauchy horizons is to show that the breakdowns of general relativity are not so serious as to render it not viable as a description of Nature. However, there are additional motivations. In attempting to determine the interior structure of black holes, one finds that the well-known analytic solutions posses Cauchy horizons in the interior, and in order to determine the generic interior structure one must investigate solutions in a neighborhood of the solution with the Cauchy horizon. (Such investigations were one of the focuses of the Haifa workshop.) Moreover, the stability of Cauchy horizons is also relevant to the question "Does Nature permit the occurrence of closed timelike curves?" [7].

22.1.2 Meaning of stability/instability

One would like to show that spacetimes with Cauchy horizons and without curvature singularities are not generic, in the sense that "generic" perturbations to the initial data $\mathcal{I}$ for the gravitational and matter fields on some initial Cauchy surface Σ will give rise to spacetimes without Cauchy horizons (i.e., inextendible spacetimes). To make the notion of genericity precise would involve defining a topology and/or measure on the set of such initial data, generic then meaning either "all initial data sets $\mathcal{I}'$ in some open set containing $\mathcal{I}$" or "all initial data sets $\mathcal{I}'$ in some set whose complement is of measure zero" [8]. It has become customary to say that a Cauchy horizon is *unstable* when this non-genericity property is satisfied.

A somewhat different notion of instability is *linear instability*: a Cauchy horizon is linearly unstable if generic perturbations to the initial data (both matter and gravitational), when evolved forward using the linearized Einstein-matter equations, yield a singularity of the perturbed metric on the Cauchy horizon. By continuity, one would expect stability to imply linear stability, and thus linear instability should be a sufficient condition for a true nonlinear Cauchy horizon instability. It need not be a necessary condition as instabilities need not show up in linearized analyses, but this has not occurred in most investigations to date in the context of black holes.

In this contribution, we will use a still weaker notion of instability. We will call a Cauchy horizon *test field unstable* if, when one evolves linearized test matter fields on the spacetime, either classical or quantum mechanical, the stress-energy tensor of the test field diverges on the Cauchy horizon (in the sense that observers who cross the Cauchy horizon measure diverging stress tensor components). One would expect that test field instability should imply linear instability, since the behavior of test matter fields should presumably be similar to the behavior of linearized metric perturbations, and also the test field's stress tensor should act as a source for the leading order metric perturbation in a coupled perturbation analysis. Again, in cases that have been examined, test field instability has been a good indicator of linear instability [9]. (For a detailed analysis of the relation between various types of test field instability and nonlinear instability, see Ref. [10] and references therein).

In the remainder of this contribution, we shall for the most part adopt a conventional abuse of terminology and abbreviate "test field unstable" as simply "unstable". Thus, by stability or instability we shall *not* mean the notion of full, nonlinear stability or instability discussed above. Our focus shall not be on understanding the effects of Cauchy horizon instabilities, for which purpose one typically needs to perform nonlinear analyses; rather we shall focus on trying to understand when and why Cauchy horizon instabilities occur. Also, we shall restrict attention to the simplified context of two dimensional spacetimes.

22.1.3 Purpose and overview of this contribution

Many investigations have found Cauchy horizons to be classically linearly unstable, for example, Cauchy horizons in black holes in asymptotically flat spacetimes [11, 12]. However, there is some evidence for the existence of stable Cauchy horizons. The first evidence for stable Cauchy horizons of which I am aware is the work of Morris, Thorne and Yurtsever, who showed that the Cauchy horizons in wormhole spacetimes with closed timelike curves were classically test field stable [13]; their analysis was later generalized by Hawking [14]. However, the spacetimes in these analyses were not solutions of Einstein's equation for a given matter model; rather they were simply posited background spacetimes. Also, the Cauchy horizon in the two dimensional version of the Reissner-Nördstrom-deSitter spacetime is classically linearly test field stable [15].

However, it has been shown in Refs. [18, 19] that the two dimensional Reissner-Nördstrom-deSitter spacetime is always semiclassically unstable (see Sec. 22.5.1 below for more details). Similarly, several researchers have shown that Cauchy horizons in spacetimes with closed timelike curves are semiclassically test field unstable [20, 21, 22, 23], even in those cases which are classically test field stable [24].

In the case of classically unstable Cauchy horizons, the physical mechanism causing the instability is well understood — it is just the blueshifting of radiation to higher and higher energies [11, 12, 15]. Moreover, one can identify a simple geometric feature of the background spacetime — whether or not the blueshift factor diverges — which allows one to predict whether or not the Cauchy horizon is classically stable.

By contrast, our understanding of the nature of semiclassical instabilities has been far less detailed. We have had no simple and general physical explanation of how or why such instabilities operate in classically stable spacetimes. The main purpose of this contribution is to show that, in the simplified context of two dimensional spacetimes, semiclassical instabilities can be understood in a simple and intuitive way, and that one can identify a geometric property of the background spacetime which allows one to predict instabilities.

We examine conformally coupled, massless, scalar test quantum fields on general two dimensional spacetimes with Cauchy horizons and without singularities. In such contexts, it is well known that the expected stress tensor can be split in a unique way into the sum of two terms [Eq. (22.5) below]: an "initial data" piece which depends just on the initial values of the stress tensor on some initial surface, and a "locally generated" piece which describes local particle creation and/or vacuum polarization effects, and which depends only on the spacetime geometry and not on the initial data. Thus, there are two different types of semiclassical instabilities: (i) Instabilities for which the "initial data" piece of the stress tensor diverges on the Cauchy horizon. This type of instability arises classically as well as quantum mechanically; it is just the blueshift instability. (ii) Instabilities for which the "locally generated" piece diverges on the Cauchy horizon. We will call such instabilities *locally created energy flux* instabilities.

We now define a particular geometric property of spacetimes which is relevant to locally created energy flux instabilities. In a neighborhood of any point $\mathcal{P}$ on the Cauchy horizon, we construct a family of null geodesics which lie parallel to the Cauchy horizon. We normalize the affine parameter λ on these geodesics by demanding that the vector field $d/d\lambda$ be parallel transported along a fixed, transverse null geodesic Λ which intersects the Cauchy horizon at $\mathcal{P}$. Let $\Delta\lambda$ denote the total affine parameter length along any of these null geodesics parallel to the Cauchy horizon, from Λ back to the initial data surface (see Fig. 22.2 below). Then, if $\Delta\lambda$ diverges as one moves closer and closer to the point $\mathcal{P}$ on the Cauchy horizon, we will say that the spacetime has the property of "divergence of affine parameter length".

The main result of this contribution is that, under suitable mild assumptions, the divergence of affine parameter length is a necessary condition for a locally created energy flux instability. Roughly speaking, if the total affine parameter length is bounded above and there are no curvature singularities, then the locally small semiclassical corrections do not have enough time to accumulate and cause a divergence of the stress tensor.

The divergence of affine parameter length is not a sufficient condition for semiclassical *test field* instabilities, as there are locally flat spacetimes such as Misner space, which satisfy the divergence of affine parameter length condition, but which do not suffer from the locally created energy flux instability. However, locally flat spacetimes are not generic; spacetimes close to Misner space but with small amounts of curvature on the chronology horizon will suffer from the locally created energy flux instability. Hence, in suitable dynamical two-dimensional versions of semiclassical gravity [25], one would expect the locally created energy flux instability to be apparent in second order semiclassical perturbation theory about locally flat backgrounds which satisfy the divergence of affine parameter length property. We therefore conjecture that, quite generally, the divergence of affine parameter length is a sufficient condition for the full nonlinear instability of Cauchy horizons in suitable dynamical two-dimensional versions of semiclassical gravity.

The properties of the blueshift and locally created energy flux instability mechanisms are summarized and contrasted in Table 22.1 below.

The concept of divergence of affine parameter length meshes nicely with our understanding of Cauchy horizon instabilities in the special case of spacetimes with closed timelike curves [20, 7, 14, 23, 26, 27, 28, 29, 30, 31]. In such spacetimes, it is easy to see intuitively that the property of divergence of affine parameter length will always be satisfied: geodesics starting from points near the Cauchy horizon (which will be a closed null geodesic in two dimensions) will circle around very close to this closed null geodesic many times before eventually making their way back to the initial data surface. We make this argument more precise in Sec. 22.5.2 below [32].

Why should the property of divergence of affine parameter length be relevant to

	Blueshift instability	Locally created energy-flux instability
Applies classically?	Yes	No
Applies semiclassically?	Yes	Yes
Necessary condition for instability	Divergence of holonomy	Divergence of affine-parameter length
Sufficient condition for instability	Divergence of holonomy (with some conditions on initial data)	Divergence of affine parameter length is probably a sufficient condition for nonlinear instability, but is not for linear instability
Spacetimes in which applies	Reissner-Nördstrom Misner space	some Reissner-Nördstrom-deSitter spacetimes
Spacetimes in which does not apply	some Reissner-Nördstrom-deSitter spacetimes	Misner space: linear instability does not apply

Table 22.1: A table contrasting the two different instability mechanisms discussed in this contribution, in the context of two dimensional spacetimes: the *blueshift instability* characterized by the divergence on the Cauchy horizon of the "initial data" piece of the expected stress tensor, and what we call the *locally created energy flux instability* mechanism, which is characterized by the divergence on the Cauchy horizon of the "locally generated" piece of the expected stress tensor.

divergences of the expected stress tensor? There are two different types of intuitive explanation which, although apparently quite different, are not necessarily incompatible. First, in spacetimes in which the curvature is low everywhere, it is well known that classical general relativity coupled to classical fields is locally a good approximation. However, in a global context, small semiclassical corrections can "accumulate" and eventually become important. A good example of this is the Hawking evaporation of macroscopic black holes, which takes place over long timescales. Now when the property of divergence of affine parameter length is satisfied, in some sense the quantum field perceives points very near the Cauchy horizon to lie at "asymptotically late times". Thus, there is enough time for semiclassical corrections which are locally small to accumulate and to become large at the Cauchy horizon.

The second intuitive explanation is that the locally created energy flux instability can be understood as a modified type of blueshift instability, a "delayed blueshift instability". In the normal blueshift instability mechanism, an energy flux that is present in the initial data propagates from the initial surface to a point near the Cauchy horizon, and in doing so suffers some total net blueshift; this net blueshift becomes larger and larger near the Cauchy horizon. However, it is possible for the energy flux to be first redshifted (frequency multiplied by some factor $F_{\text{red}} < 1$) and then blueshifted (frequency multiplied by some factor $F_{\text{blue}} > 1$) in such a way that the total blueshift factor $F_{\text{red}} F_{\text{blue}}$ is bounded above. In such situations the usual blueshift instability does not apply. Nevertheless, if an energy flux is *created* near the "turning point" where quanta stop being redshift and start becoming blueshifted, then this locally created energy flux suffers a net blueshift factor $\sim F_{\text{blue}}$ which is divergent, giving rise to a divergent stress tensor on the Cauchy horizon. In Sec. 22.4.3 below we show that the above situation always occurs when the property of divergence of affine parameter length is satisfied in two dimensional spacetimes. That is, the divergence of the stress tensor is always caused by the infinite blueshifting of a portion of the locally generated piece of the stress tensor.

In Sec. 22.4.4 below we argue that locally generated energy flux instabilities should also occur in four dimensional spacetimes. We also conjecture that, in the four dimensional context, locally generated energy flux instabilities need not always correspond to delayed blueshift instabilities.

22.1.4 Organization of this contribution

The organization of this contribution to the proceedings is as follows. Section 22.2 is devoted to general analyses that underly both the blueshift and locally created energy flux instabilities. In Sec. 22.2.1 we define the class of spacetimes which we analyze. In Sec. 22.2.3 we define a basis of null vectors which is adapted to the local geometry near a given point on the Cauchy horizon. This null basis serves as a convenient tool throughout our calculations, and it allows us to avoid having to introduce a coordinate system to describe the spacetime. Using this basis, we derive in Sec. 22.2.2 a (well-known) general formula for the expected stress tensor in terms of its initial values on some initial surface Σ [Eq. (22.10)]. That formula exhibits the split of the stress tensor, referred to above, into an "initial data" piece plus a "locally generated" piece. We show in Sec. 22.2.4 that all observers of bounded acceleration measure finite energy densities while crossing the Cauchy horizon if and only if the components of the expected stress tensor on the null basis are finite. Thus, in investigating the stability of the Cauchy horizon, we can focus attention on the components of the stress tensor on the null basis.

In Sec. 22.3 we review and discuss the well-known blueshift instability mechanism, in order to contrast it with the locally created energy flux instability mechanism. In Sec. 22.3.1 we give a very general definition of the blueshift factor, and recall that it can be interpreted in terms of a holonomy operator around a certain closed loop in spacetime, as illustrated in Fig. 22.2. A discussion of necessary and sufficient conditions for the instability is given in Sec. 22.3.2. Finally, in Sec. 22.3.3, we review the well-known fact that the instability acts only on radiation propagating parallel to the Cauchy horizon and not to radiation which crosses the Cauchy horizon; and we show that the radiation which crosses the Cauchy horizon is always finite, semiclassically as well as classically.

Section 22.4 is devoted to the locally created energy flux instability mechanism. In Sec. 22.4.1 we show that the property of divergence of affine parameter length is a necessary condition for a locally created energy flux instability, when one assumes that the blueshift factor is globally bounded and one makes some other mild assumptions about the spacetime and the initial slice Σ. In Sec. 22.4.2 we explain that the divergence of affine parameter length property is not a sufficient condition for a linear instability (giving the counterexample of Misner space), and conjecture that it should be a sufficient condition for instabilities in a full, nonlinear analyses. In Sec. 22.4.4, we argue that some of the key ideas which underly, in two dimensions, the classification of instabilities into blueshift and locally created energy flux instabilities, should also generalize to four dimensions. We speculate that the divergence of affine parameter length *might* be relevant to Cauchy horizon instabilities in four dimensions.

In Sec. 22.4.3 we show that the divergence of the stress tensor is always caused by an infinite blueshifting of a portion of the locally created piece of the stress tensor, thus showing the instability mechanism can be understood as a "delayed blueshift" instability.

In Section 22.5 we apply the general analyses of Secs. 22.2, 22.3 and 22.4 to several different spacetimes and classes of spacetimes, in order to clarify and illustrate the results. In Sec. 22.5.1, we show that the property of divergence of affine parameter length is satisfied in two-dimensional Reissner-Nördstrom-deSitter spacetimes, and reproduce the result of Marković and Poisson [19, 33] that such spacetime are semiclassically unstable. In Sec. 22.5.2 we show, by adapting an argument due to Hawking, that two dimensional spacetimes with closed timelike curves always satisfy the property of divergence of affine parameter length. A particular example of such a spacetime, Misner space, is analyzed in Sec. 22.5.3. We review the well-known fact that Misner space is blueshift unstable. We also show that Misner space does *not* suffer from the locally created energy flux instability mechanism; we argue, however, that generic spacetimes "close to" Misner space should suffer from the instability.

Finally, section 22.6 summarizes our main conclusions. Appendix A shows that the property of divergence of affine parameter length will be satisfied at a point on the Cauchy horizon whenever the generator of the Cauchy horizon through that point has infinite affine parameter length in the past direction.

We use units in which $G = c = \hbar = 1$, and we use the $(+,+,+)$ sign convention in the notation of Ref. [34].

22.2 Cauchy horizon stability: fundamental analyses

22.2.1 Class of spacetimes and matter models

We start by specifying the class of spacetimes which we shall discuss. We shall be interested in globally hyperbolic, two-dimensional spacetimes (M, g_{ab}), i.e., spacetimes in which there exists a partial Cauchy surface Σ whose domain of dependence $D(\Sigma)$ is the entire spacetime M [6]. The reason we restrict attention to such spacetimes is the following. Physically realistic spacetimes should be obtainable by specifying initial data for the gravitational and matter fields on some initial slice Σ and by solving the classical (or semiclassical) Einstein equation together with the matter equations of motion in order to recover the entire spacetime. In this way one recovers the maximal Cauchy evolution of the initial data, which will be just $D(\Sigma)$. Thus, spacetimes obtainable by this procedure will all be globally hyperbolic.

We will also assume that the spacetime (M, g_{ab}) is extendible to a larger spacetime (M', g'_{ab}), and thus is geodesically incomplete. The spacetime (M', g'_{ab}) is useful from a mathematical point of view — using it one can define the future Cauchy Horizon $H^+(\Sigma)$, which is just the boundary in M' of the future domain of dependence $D^+(\Sigma)$. Note, however, that the extension is not unique and thus not physically meaningful. Our arguments and conclusions will be independent of the choice of extension (M', g'_{ab}).

The initial surface Σ may either be spacelike, or it may consist of two null segments as indicated in Fig. 22.1 below. Also, we do not restrict the topology of Σ: it may be compact (topology of a circle) or non-compact (topology of the real line). In Sec. 22.5 below we discuss several specific examples of spacetimes to which our analyses apply; most of these spacetimes will be spatially non-compact, but one of them, Misner space, will be spatially compact.

We will also assume that the spacetime (M, g_{ab}) does not have any curvature singularities, either scalar or parallel propagated. Thus, we assume that (i) all local curvature invariants are globally bounded on (M, g_{ab}), and (ii) there are no causal geodesic curves along which the components of local curvature tensors on parallel propagated bases diverge. These assumptions are valid, for example, for the initial globally hyperbolic regions (M, g_{ab}) of the Reissner-Nördstrom and Reissner-Nördstrom-de Sitter spacetimes.

Our matter model will consist of a massless, minimally coupled scalar quantum field $\hat{\Phi}$ on (M, g_{ab}). We shall be concerned with the expected stress-energy tensor $\langle T_{ab} \rangle$ of the field $\hat{\Phi}$ in some quantum state in the vicinity of some point $\mathcal{P}$ on the Cauchy horizon. For ease of notation, we will denote this expected stress tensor simply as T_{ab}.

In this model, the test field stability of Cauchy horizons which we are investigating should provide a good indication to the full, nonlinear stability properties of two dimensional spacetimes, in the context of a suitable two dimensional version of semiclassical gravity [25]. Alternatively, one can regard two dimensional, semiclassical, test field instability results as implying that test field instabilities are also likely in similar four dimensional, spherically symmetric spacetimes [19]; such four dimensional test field instabilities would be directly relevant to nonlinear instabilities in standard four dimensional semiclassical gravity.

22.2.2 General formula for stress tensor

The stress tensor T_{ab} obeys the conservation equation

$$\nabla^a T_{ab} = 0 \tag{22.1}$$

together with the trace anomaly equation [35]

$$T^a_a = \frac{1}{24\pi} R, \tag{22.2}$$

where R is the two dimensional Ricci scalar. As is well known [35], these conservation and trace anomaly equations are sufficient to determine the evolution of T_{ab} from its initial value on Σ; one does not need to specify the details of the quantum state, unlike the situation in four dimensions. We can therefore calculate the behavior of T_{ab} near the Cauchy horizon in terms of its initial data on Σ.

We split up the stress tensor into a traceless part and a trace part:

$$T_{ab} = \hat{T}_{ab} + \frac{1}{48\pi} R g_{ab}, \tag{22.3}$$

where $g^{ab}\hat{T}_{ab} = 0$. It follows from Eq. (22.1) that the traceless part $\hat{T}_{ab}$ obeys the equation

$$\nabla^a \hat{T}_{ab} = -\frac{1}{48\pi} \nabla_b R. \tag{22.4}$$

This equation has a well posed initial value formulation; initial values of $\hat{T}_{ab}$ determine uniquely its evolution. The general solution of Eq. (22.4) can be written as

$$\hat{T}_{ab} = \hat{T}_{ab}^{(\text{initial data})} + \hat{T}_{ab}^{(\text{locally generated})}, \tag{22.5}$$

where $\hat{T}_{ab}^{(\text{initial data})}$ is the solution of the homogeneous version of Eq. (22.4) with the same initial data on Σ, and $\hat{T}_{ab}^{(\text{locally generated})}$ is the solution of Eq. (22.4) with vanishing initial conditions.

We now derive explicit formulae for these two pieces of the general solution. These formulae are now new, but are usually derived and expressed in a "double-null" coordinate systems (u, v) in which the metric is conformally flat:

$$ds^2 = -2e^{\sigma(u,v)}\, du dv; \tag{22.6}$$

see, for example, Ref. [35]. Here, however, we derive and express the formulae using a basis of null vectors $\{\vec{k}, \vec{l}\}$, which has the minor advantage that one does not need to assume the global existence of a coordinate system of the form (22.6).

Let Γ be any null geodesic in the spacetime, going from some point $\mathcal{Q}$ to some point $\mathcal{R}$. Let k^a denote the future directed, null tangent to the geodesic, with associated affine parameter λ, so that $\vec{k} = (d/d\lambda)$. Let l^a be the vector field on Γ which is null, future directed and which satisfies $l^a k_a = -1$, from which it follows that $\vec{l}$ is parallel transported along Γ. Using the relation

$$g_{ab} = -2l_{(a}k_{b)} \tag{22.7}$$

in Eq. (22.4) yields

$$-2k^{(a}l^{c)}\nabla_a \hat{T}_{cb} = -\frac{1}{48\pi}\nabla_b R. \tag{22.8}$$

Contracting this equation with l^b yields

$$-k^a l^c b^b \nabla_a \hat{T}_{bc} - k^{(c}l^{b)} l^a \nabla_a \hat{T}_{cb} = -\frac{1}{48\pi} l^b \nabla_b R. \tag{22.9}$$

The second term on the left hand side in Eq. (22.9) vanishes by Eq. (22.7). Also the first term can be written as $-d/d\lambda(\hat{T}_{bc}l^b l^c)$, since the vector $\vec{l}$ is parallel transported along the geodesic. Integrating with respect to λ and using $\hat{T}_{ab}l^a l^b = T_{ab}l^a l^b$ therefore yields

$$T_{ab}l^a l^b(\mathcal{R}) = T_{ab}l^a l^b(\mathcal{Q}) + \frac{1}{48\pi}\int_{\mathcal{Q}}^{\mathcal{R}} d\lambda\; l^a \nabla_a R. \tag{22.10}$$

Note that the first term in Eq. (22.10) clearly corresponds to the first term in Eq. (22.5), and similarly for the second term.

The formula (22.10) allows one to determine the entire stress tensor from its initial value on Σ. Namely, given any point $\mathcal{R}$ in M, one can choose a pair of future directed null vectors $\vec{k}$ and $\vec{l}$ at $\mathcal{R}$ with $k^a l_a = -1$. Then, the stress tensor at $\mathcal{R}$ can be written as

$$T_{ab}(\mathcal{R}) = \rho l_a l_b + \sigma k_a k_b + \frac{1}{24\pi} R g_{ab}, \tag{22.11}$$

where $\rho = T_{ab}k^a k^b$ and $\sigma = T_{ab}l^a l^b$. One can determine $\sigma(\mathcal{R})$ by shooting off a past-directed null geodesic Γ from $\mathcal{R}$ with initial tangent $-\vec{k}$, extending it until it intersects the initial surface Σ, parallel transporting $\vec{l}$ along Γ, and then applying the formula (22.10). Similarly one can determine $\rho(\mathcal{R})$ by shooting off a null geodesic from $\mathcal{R}$ with initial tangent $-\vec{l}$, and applying the formula (22.10) with $\vec{k}$ and $\vec{l}$ interchanged.

Note that, in the formula (22.10), one has the freedom to perform the rescaling

$$\begin{aligned} \vec{k} &\rightarrow e^{\mu}\vec{k} \\ \vec{l} &\rightarrow e^{-\mu}\vec{l} \end{aligned} \qquad (22.12)$$

where μ is a constant. Both sides of Eq. (22.10) then get multiplied by $e^{-2\mu}$. If one has a family of geodesics (which corresponds to a specification of the basis $\vec{k}$, $\vec{l}$ in an open region), then μ can be any function on the manifold M satisfying $k^a\nabla_a\mu = 0$. In terms of double null coordinates (u, v) [cf. Eq. (22.6) above], μ can be an arbitrary function of the null coordinate v, when we adopt the convention $\vec{k} \propto \partial/\partial u$ and $\vec{l} \propto \partial/\partial v$.

22.2.3 A null basis adapted to the local geometry near the Cauchy horizon

Fix a point $\mathcal{P}$ in the Cauchy horizon $H^+(\Sigma)$. We can construct a natural basis $\{\vec{k}, \vec{l}\}$ in the intersection of the past light cone of $\mathcal{P}$ with a neighborhood U of $\mathcal{P}$, in which we resolve the "gauge freedom" (22.12) where $\mu = \mu(v)$, as follows (see Fig. 22.1). Let $\vec{k}(\mathcal{P})$ be the future-directed tangent to the null generator of the Cauchy horizon at $\mathcal{P}$, with some arbitrary choice of normalization. Then there is a unique, null, future directed vector $\vec{l}(\mathcal{P})$ with $\vec{l} \cdot \vec{k} = -1$. Let Λ be the null geodesic which starts from $\mathcal{P}$ with initial tangent $-\vec{l}$ and extends into the past, and extend $\vec{l}$ and $\vec{k}$ along Λ by parallel transport. Finally, at an arbitrary point $\mathcal{R}$ on Λ in U, shoot out a geodesic Γ into the past with initial tangent $-\vec{k}$ and continue it until it reaches that initial surface Σ, and extend $\vec{k}$ and $\vec{l}$ along Γ by parallel transport [36]. The resulting dreibein (set of basis vectors) $\{\vec{k}, \vec{l}\}$ is unique up to transformations of the form (22.12) where now μ is a constant. Also, it follows from the construction that Eq. (22.7) holds in the domain of definition of the null basis.

22.2.4 Measurements made by observers crossing the Cauchy horizon

Our aim is to characterize when the stress tensor T_{ab} diverges at the Cauchy horizon. Now, as is well known, to detect divergences it is insufficient in general to examine coordinate invariant scalar quantities such as $T_{ab}T^{ab}$ near the Cauchy horizon. Instead, one must examine the behavior of components of T_{ab} with respect to parallel propagated bases along curves of bounded proper acceleration which cross the Cauchy horizon. In other words, one must examine what physical observers who cross the Cauchy horizon would measure. Now, we have constructed above a basis of null vectors which is naturally adapted to the spacetime geometry in the vicinity of a point $\mathcal{P}$

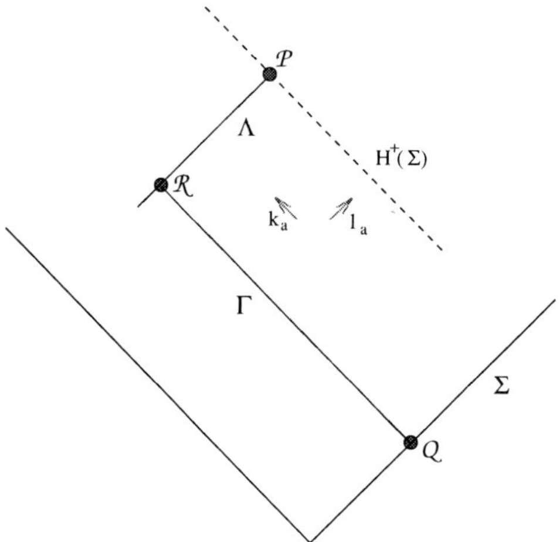

Figure 22.1: An illustration of the construction outlined in the text. The surface Σ is the initial Cauchy surface, the shaded area schematically represents the future domain of dependence $D^+(\Sigma)$, and $H^+(\Sigma)$ is the Cauchy horizon. In the vicinity of any point $\mathcal{P}$ on the Cauchy horizon, one chooses two future directed, null vectors $\vec{k}$ and $\vec{l}$ with $\vec{k} \cdot \vec{l} = -1$ and with $\vec{k}$ normal to the Cauchy horizon, and one resolves the ambiguity in the normalization of these vectors by starting at $\mathcal{P}$, parallel transporting both vectors first along the geodesic Λ and then along the each geodesic Γ starting from points $\mathcal{R}$ on Λ. The result is a dreibein $\{\vec{k}, \vec{l}\}$ which is adapted to the local geometry near $\mathcal{P}$ on the Cauchy horizon.

on the Cauchy horizon. Therefore, one would expect that the stress tensor is regular in the above parallel-propagated sense at $\mathcal{P}$ if and only if the components of T_{ab} on this basis, i.e., the quantities σ and ρ defined by Eq. (22.11), are regular at $\mathcal{P}$. In this section we give a brief proof that this is indeed the case. More precisely, we shall show that the measured energy densities will be finite for all observers of finite proper acceleration who cross the Cauchy horizon at $\mathcal{P}$ if and only if both σ and ρ are bounded above in a neighborhood of $\mathcal{P}$.

Let $\mathcal{C}$ be the timelike worldline of an observer who passes through $\mathcal{P}$. Her velocity can be expressed as

$$\vec{u} = \frac{1}{\sqrt{2}} \left[e^{\chi} \vec{l} + e^{-\chi} \vec{k} \right] \tag{22.13}$$

for some $\chi = \chi(\tau)$, where τ is her proper time with $\tau = 0$ at $\mathcal{P}$. Now it follows from our construction of the basis $\{\vec{k}, \vec{l}\}$ that

$$\begin{aligned} \nabla_{\vec{k}} \vec{k} &= \nabla_{\vec{k}} \vec{l} = 0 \\ \nabla_{\vec{l}} \vec{l} &= \kappa \vec{l} \\ \nabla_{\vec{l}} \vec{k} &= -\kappa \vec{k} \end{aligned} \tag{22.14}$$

where κ is a spin coefficient analogous to Newman-Penrose quantity γ in four dimensions. Combining Eqs. (22.13) and (22.14) we obtain for the acceleration

$$a^a = \left(\frac{1}{2} e^{2\chi} l^a - k^a\right) \left(\kappa + \sqrt{2} e^{-\chi} u^b \nabla_b \chi\right). \tag{22.15}$$

Squaring this equation yields

$$\frac{d\chi}{d\tau} = \pm \frac{1}{\sqrt{2}} a(\tau) - \frac{1}{\sqrt{2}} e^{\chi} \kappa, \tag{22.16}$$

where $a(\tau) \equiv \sqrt{a_a a^a}$. It follows from Eq. (22.16) that the boost parameter χ will be regular in a neighborhood of $\tau = 0$, since by assumption the geometry is regular in a neighborhood of $\mathcal{P}$, so κ will be regular, and also the proper acceleration $a(\tau)$ is bounded by assumption.

Next, the energy density $\rho_{\mathcal{C}} = T_{ab}\, u^a u^b$ that the observer measures is given by, from Eqs. (22.11) and (22.13),

$$\rho_{\mathcal{C}} = \frac{1}{2} \rho e^{-2\chi} + \frac{1}{2} \sigma e^{2\chi} - \frac{1}{24\pi} R. \tag{22.17}$$

The third term here is always bounded since the background geometry is regular in a neighborhood of $\mathcal{P}$. Now if both σ and ρ are finite at $\mathcal{P}$, then it follows that $\rho_{\mathcal{C}}$ will be finite for all observers of bounded acceleration, since for such observers χ is bounded. Conversely, if either ρ or σ diverged at $\mathcal{P}$, then there will be some choice of worldline $\mathcal{C}$ for which $\rho_{\mathcal{C}}$ will be divergent [37].

To summarize, the Cauchy horizon will be stable if σ and ρ are finite on the Cauchy horizon, and unstable otherwise.

22.2.5 Comparison between classical and semiclassical theories

The classical version of the theory of a massless, conformally coupled scalar field differs from the above semiclassical version in only two respects:

- In the equation of motion (22.4), the source term on the right hand side is not present in the classical theory; it describes local particle creation and/or vacuum polarization effects. Correspondingly, in the split (22.5) of the stress tensor into a "locally generated" piece and an "initial data" piece, the locally generated term is absent in the classical theory.

- The set of allowed initial data for $\hat{T}_{ab}$ on Σ is larger in the semiclassical theory than in the classical theory [38].

Therefore, the piece $\hat{T}_{ab}^{(\text{initial data})}$ of the general solution is *purely classical* — it has exactly the same behavior in semiclassical solutions as in classical solutions (except for the greater freedom in initial data in semiclassical solutions). In particular, if the Cauchy horizon is stable classically but unstable quantum mechanically, the instability must be due to the locally generated term $\hat{T}_{ab}^{(\text{locally generated})}$. In the following two sections we turn to a discussion of the two different instability mechanisms discussed in the Introduction: the well-known blueshift instability which manifests itself as a divergence of the term $\hat{T}_{ab}^{(\text{initial data})}$ and which can cause both classical and semiclassical instabilities, and the locally created energy flux instability mechanism which manifests itself as a divergence of the term $\hat{T}_{ab}^{(\text{locally generated})}$ in Eq. (22.5).

22.3 The blueshift instability

The blueshift instability mechanism is responsible for the classical instability of Cauchy horizons in charged and/or rotating black hole spacetimes [11, 12, 9, 16], which is nicely reviewed in the contributions by Poisson and Chambers to this volume [33, 39]. In this section, we review this mechanism in the language we have developed above, in order to contrast the blueshift mechanism with the locally created energy flux instability mechanism we discuss in Sec. 22.4 below. Our discussion will apply both to classical and quantum mechanical analyses.

22.3.1 Preliminary definitions and constructions

As discussed in the Introduction, when we say that the Cauchy horizon is unstable we mean that the stress tensor diverges on the Cauchy horizon for generic, regular initial data on the initial Cauchy surface Σ. We now discuss what is a suitable meaning of "regular" in our context of very general two dimensional spacetimes. To this end, we define a basis of null vectors $\{\vec{k}_\Sigma, \vec{l}_\Sigma\}$ on the initial surface Σ, according to the following prescription: pick, at some point $\mathcal{Q}_0$ on Σ, two future directed, null vectors $\vec{k}_\Sigma(\mathcal{Q}_0)$ and $\vec{l}_\Sigma(\mathcal{Q}_0)$ with $\vec{k}_\Sigma(\mathcal{Q}_0) \cdot \vec{l}_\Sigma(\mathcal{Q}_0) = -1$, and extend this basis to all of Σ by parallel transport. If Σ has the topology of a circle, as is the case for Misner space discussed in Sec. 22.5.3 below, this definition of the basis $\{\vec{k}_\Sigma, \vec{l}_\Sigma\}$ is ambiguous. We resolve the ambiguity by demanding that the the basis be continuous at all points on Σ except possibly at $\mathcal{Q}_0$; a non-unit holonomy around Σ will yield a discontinuity at $\mathcal{Q}_0$. The resulting basis $\{\vec{k}_\Sigma, \vec{l}_\Sigma\}$ is unique up to an overall constant boost of the form (22.12).

We will say that initial data for T_{ab} on Σ is *regular* if the components $T_{ab} l_\Sigma^a l_\Sigma^b$ and $T_{ab} k_\Sigma^a k_\Sigma^b$ of the stress tensor are continuous (at all points except possibly $\mathcal{Q}_0$) and bounded. It is clear that all physically reasonable initial data should be regular in this sense. In some spacetimes, the set of "physically reasonable" initial data will be a proper subset of the set of regular initial data (due, eg, to asymptotic fall-off conditions); see the discussion of various examples of spacetimes in Sec. 22.5.

We now extend the basis $\{\vec{k}_\Sigma, \vec{l}_\Sigma\}$ to the entire spacetime (M, g_{ab}) by parallel transporting along null geodesics parallel to $\vec{k}_\Sigma$. This basis will be related to the basis $\{\vec{k}, \vec{l}\}$ discussed in Sec. 22.2.3 above by a spacetime dependent boost:

$$\begin{aligned} \vec{k} &= e^{-\Psi}\vec{k}_\Sigma \\ \vec{l} &= e^{\Psi}\vec{l}_\Sigma. \end{aligned} \tag{22.18}$$

Since the vector fields $\vec{k}$ and $\vec{k}_\Sigma$ are geodesic, we have $k^a\nabla_a\Psi = 0$, *i.e.*, Ψ depends only on one of the two null coordinates (u, v). For convenience, we will always choose the boost-normalization of the basis $\{\vec{k}, \vec{l}\}$ so that $\Psi(\mathcal{Q}_0) = 0$.

Note that the quantity Ψ can be understood as characterizing a holonomy operator around a closed loop in spacetime, in the special case where the topology of Σ is that of the real line (which excludes examples like Misner space). More specifically, consider the construction of the basis $\{\vec{k}, \vec{l}\}$ described in Sec. 22.2.3 above. That construction describes a family of null geodesics Γ joining points $\mathcal{Q}$ on the initial slice Σ to points $\mathcal{R}$ on the geodesic Λ which emanates from the Cauchy horizon at the point $\mathcal{P}$ (see Fig. 22.2). Pick a particular null geodesic Γ_0 joining some points $\mathcal{R}_0$ and $\mathcal{Q}_0$. Then it follows that at any point $\mathcal{R}$ on Λ, the holonomy around the loop $\mathcal{R}\,\mathcal{R}_0\,\mathcal{Q}_0\,\mathcal{Q}\,\mathcal{R}$ indicated in Fig. 22.2 is simply the following boost with rapidity parameter $\Psi(\mathcal{R})$:

$$\alpha\,\vec{k}(\mathcal{R}) + \beta\,\vec{l}(\mathcal{R}) \rightarrow \alpha\, e^{\Psi(\mathcal{R})}\,\vec{k}(\mathcal{R}) + \beta\, e^{-\Psi(\mathcal{R})}\,\vec{l}(\mathcal{R}). \tag{22.19}$$

Here α and β are arbitrary real numbers. To derive Eq. (22.19), consider starting at $\mathcal{R}$ with the vector $\vec{k}(\mathcal{R})$. When one parallel transports this vector to $\mathcal{R}_0$ along Λ and then to $\mathcal{Q}_0$ along Γ_0, the result is just $\vec{k}(\mathcal{Q}_0)$ by the definition of the basis $\{\vec{k}, \vec{l}\}$, which equals $\vec{k}_\Sigma(\mathcal{Q}_0)$ from Eq. (22.18) and the fact that $\Psi(\mathcal{Q}_0) = 0$. When one then parallel transports this vector along Σ to $\mathcal{Q}$ and then along Γ back to $\mathcal{R}$, the result is $\vec{k}_\Sigma(\mathcal{R})$ by the definition of the basis $\{\vec{k}_\Sigma, \vec{l}_\Sigma\}$, which by Eq. (22.18) is the same as $e^{\Psi(\mathcal{R})}\,\vec{k}(\mathcal{R})$. A similar argument applies when one starts at $\mathcal{R}$ with $\vec{l}(\mathcal{R})$, and one obtains in this way Eq. (22.19).

Note also that the blueshift factor e^{Ψ} can be understood as the ratio of differential proper times of freely falling observers near $\mathcal{P}$ and on Σ, as explained in detail in Refs. [15, 39].

22.3.2 Instability mechanism

We now describe the blueshift instability mechanism. Spacetimes in which this instability operates have the property that, for some point $\mathcal{P}$ on the Cauchy horizon, the "blueshift factor" $e^{\Psi(\mathcal{R})}$ diverges as the point $\mathcal{R}$ approaches $\mathcal{P}$. From Eqs. (22.11),

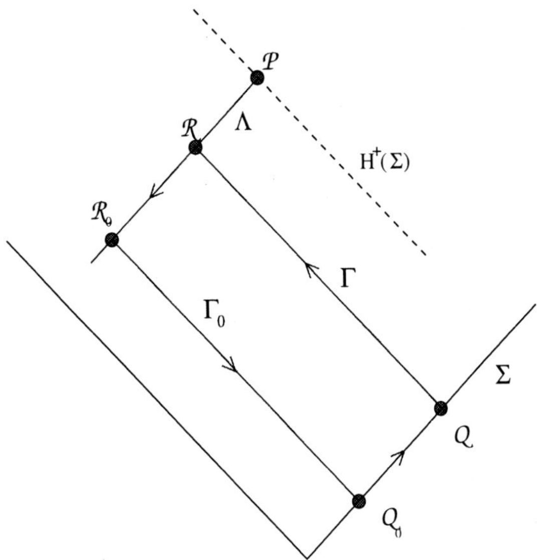

Figure 22.2: A specialization of the construction shown in Fig. 22.1 to illustrate the difference between the blueshift and locally created energy flux instability mechanisms. Here Σ is the initial Cauchy surface, $\mathcal{P}$ is a point on the future Cauchy horizon $H^+(\Sigma)$, and $\mathcal{R}$ and $\mathcal{R}_0$ are points on the null geodesic Λ emanating from the Cauchy horizon at $\mathcal{P}$, which are connected to points $\mathcal{Q}$ and $\mathcal{Q}_0$ on Σ via null geodesics. The *blueshift instability* mechanism is characterized by a divergence of the total holonomy around the loop $\mathcal{R}\mathcal{R}_0\mathcal{Q}_0\mathcal{Q}$ as $\mathcal{R} \to \mathcal{P}$. By contrast, what we call the *locally created energy flux instability* mechanism is characterized by the divergence of the affine parameter length of the geodesic $\mathcal{R}\mathcal{Q}$ as $\mathcal{R} \to \mathcal{P}$, where the normalization of the affine parameter λ is such that $d/d\lambda$ is parallel transported along the geodesic Λ. This property of divergence of affine parameter length is a necessary but not sufficient condition for a locally created energy flux instability.

(22.10) and (22.18) we find that the quantity σ (describing radiation propagating parallel to the Cauchy horizon) evaluated at the point $\mathcal{R}$ near the Cauchy horizon is given by

$$\begin{aligned}\sigma(\mathcal{R}) &= e^{2\Psi(\mathcal{R})}\sigma_\Sigma\left[\mathcal{Q}(\mathcal{R})\right] \\ &\quad +(\text{locally generated term}).\end{aligned} \tag{22.20}$$

Here, $\mathcal{Q}(\mathcal{R})$ denotes the unique point $\mathcal{Q}$ on Σ determined by $\mathcal{R}$ according to the construction described in Sec. 22.2.3, and $\sigma_\Sigma \equiv T_{ab}l^a_\Sigma l^b_\Sigma$ is the initial data on Σ which we have assumed is bounded. In this section we will ignore the locally generated term in Eq. (22.20); it will be absent in a classical treatment, and even in a semiclassical treatment, a divergence of the first term in Eq. (22.20) should be sufficient to produce an instability [40].

From Eq. (22.20) the quantity $\sigma(\mathcal{R})$ will diverge at $\mathcal{P}$ unless the initial data $\sigma_\Sigma[\mathcal{Q}(\mathcal{R})]$ goes to zero as $\mathcal{R} \to \mathcal{P}$ faster than the divergence of the blueshift factor $e^{\Psi(\mathcal{R})}$. Thus, whether or not the spacetime is unstable depends on the precise specification of the class of "physically reasonable" initial data on Σ (cf. Sec. 22.3.1 above).

There are several possibilities for the behavior of $\mathcal{Q}(\mathcal{R})$ on Σ as $\mathcal{R} \to \mathcal{P}$:

- When Σ is non-compact, it can happen that $\mathcal{Q}(\mathcal{R}) \to \infty$ as $\mathcal{R} \to \mathcal{P}$. Then the behavior of $\sigma_\Sigma[\mathcal{Q}(\mathcal{R})]$ depends on the asymptotic fall-off behavior of the initial data. In the case of classically unstable black hole spacetimes, the fall off rate of $\sigma_\Sigma[\mathcal{Q}(\mathcal{R})]$ for physically reasonable initial data is sufficiently slow that, although $\sigma_\Sigma[\mathcal{Q}(\mathcal{R})] \to 0$, the quantity (22.20) still diverges [15].

- It can happen that $\mathcal{Q}(\mathcal{R}) \to \mathcal{Q}_1$ as $\mathcal{R} \to \mathcal{P}$, where $\mathcal{Q}_1$ is some fixed point on Σ [41]. In this case, generic initial data will have $\sigma_\Sigma(\mathcal{Q}_1) \neq 0$, so the spacetime will be unstable.

- In the case where Σ is compact, it can happen that $\mathcal{Q}(\mathcal{R})$ "circles around and around" Σ as $\mathcal{R} \to \mathcal{P}$, as occurs for example in Misner space. In this case also, it follows from Eq. (22.20) that $\sigma(\mathcal{R})$ will be unbounded near $\mathcal{P}$ for all non-zero initial data, and thus the spacetime will be unstable.

To summarize, when the background spacetime (M, g_{ab}) has the property that the holonomy-like quantity $e^{\Psi(\mathcal{R})}$ diverges, and when in the non-compact case the fall-off rate of the class of "physically reasonable" initial data is sufficiently slow, then the spacetime will be (classically and quantum mechanically) unstable. The condition on the fall-off rate of the initial data depends on the spacetime and on the choice of initial surface Σ; it is satisfied in black hole spacetimes [42].

22.3.3 Radiation crossing the Cauchy horizon

So far in this section we have considered only the term $\sigma k_a k_b$ in Eq. (22.11), which describes radiation propagating parallel to the Cauchy horizon. What of the other term $\rho l_a l_b$ in that equation? This term describes radiation which crosses the Cauchy horizon. We now show that such radiation can never give rise to an instability by showing that the quantity ρ is always bounded near the point $\mathcal{P}$.

Extend the geodesic Λ into the past until it intersects the initial surface Σ at some point $\mathcal{S}$. Then, since the vectors $\vec{k}$ and $\vec{l}$ are both parallel transported along Λ, it follows from Eq. (22.10) with $\vec{k}$ and $\vec{l}$ interchanged that

$$\rho(\mathcal{R}) = (T_{ab}k^a k^b)(\mathcal{S}) + \frac{1}{48\pi}\int_{\mathcal{S}}^{\mathcal{R}} d\xi \; k^a \nabla_a R, \tag{22.21}$$

where ξ is an affine parameter along Λ such that $\vec{l} = d/d\xi$. The first term in Eq. (22.21) is just

$$e^{-2\Psi(\mathcal{S})} \rho_\Sigma(\mathcal{S}), \tag{22.22}$$

where $\rho_\Sigma \equiv T_{ab}k^a_\Sigma k^b_\Sigma$, from Eq. (22.18). Now the point $\mathcal{S}$ does not vary as $\mathcal{R} \to \mathcal{P}$, and thus the quantity (22.22) is fixed and finite as $\mathcal{R} \to \mathcal{P}$. The second term in Eq. (22.21) will converge to the finite quantity

$$\frac{1}{48\pi}\int_{\mathcal{S}}^{\mathcal{P}} d\xi \; k^a \nabla_a R \tag{22.23}$$

as $\mathcal{R} \to \mathcal{P}$, and so $\rho(\mathcal{R})$ is bounded as $\mathcal{R} \to \mathcal{P}$.

Thus, when the blueshift factor is finite, it follows from Eq. (22.20) and the analysis of this subsection that the spacetime is stable except possibly for a divergence of the semiclassical, locally generated piece of $\sigma(\mathcal{R})$. We now turn to a discussion of this possibility.

22.4 The locally created energy flux instability

In this section we discuss a second instability mechanism which can cause a divergence of the second term $\hat{T}_{ab}^{(\text{locally generated})}$ in Eq. (22.5). This mechanism is a purely quantum mechanical effect as the term $\hat{T}_{ab}^{(\text{locally generated})}$ is absent in classical analyses.

Let us focus attention on spacetimes in which the blueshift factor e^Ψ is globally bounded, and in which therefore the blueshift instability mechanism does not operate. For example, the two dimensional Reissner-Nördstrom-de Sitter spacetime has this property in a certain region of parameter space [15], when the initial surface Σ is chosen to lie outside the event horizon (see Sec. 22.5.1 below). Such spacetimes will be classically stable but may be semiclassically unstable.

From Eq. (22.10), the locally generated piece of σ is given by

$$\sigma^{(\text{locally generated})}(\mathcal{R}) = \frac{1}{48\pi}\int_{\mathcal{Q}(\mathcal{R})}^{\mathcal{R}} d\lambda \; l^a \nabla_a R. \tag{22.24}$$

Now by assumption the background geometry is nonsingular, so the Ricci scalar R is bounded; however the quantity $l^a \nabla_a R$ may still be unbounded and may give rise to a divergence of the integral (22.24) as $\mathcal{R} \to \mathcal{P}$. Alternatively, one might imagine that the integrand $l^a \nabla_a R$ could be bounded, but that the total affine parameter length in λ between $\mathcal{Q}(\mathcal{R})$ and $\mathcal{R}$ could diverge as $\mathcal{R} \to \mathcal{P}$ giving rise to a divergence of the integral. In Sec. 22.5 below, we examine several spacetimes for which the integral (22.24) diverges, and we find that in these example spacetimes, *both* the integrand and the total affine parameter length diverge.

We will say that a spacetime has the property of "divergence of affine parameter length" if, for some point $\mathcal{P}$ on the Cauchy horizon,

$$\Delta\lambda(\mathcal{R}) \to \infty \quad \text{as} \quad \mathcal{R} \to \mathcal{P}. \tag{22.25}$$

Here $\Delta\lambda(\mathcal{R})$ is the affine parameter length of the null geodesic from $\mathcal{R}$ to $\mathcal{Q}(\mathcal{R})$, where the affine parameter λ is normalized according to $\vec{k} = d/d\lambda$ (see Fig. 22.2 above). In Appendix A we show that the property (22.25) will hold if the generator of the Cauchy horizon through the point $\mathcal{P}$ has infinite affine parameter length towards the past. [Note that, although any such generator must be inextendible in the past direction [6], it may have finite affine parameter length in the past direction if the spacetime (M', g'_{ab}) is geodesically incomplete; see, for example, Fig. 8.2 of Ref. [6].]

22.4.1 A necessary condition for instability

In this subsection section we prove that when one makes certain mild assumptions about the spacetime (M, g_{ab}) and the initial slice Σ, the property of divergence of affine parameter length is a *necessary* condition for an instability of the Cauchy horizon mediated by a divergence of the quantity (22.24).

A precise statement of our result is the following. Assume that (i) The spacetime (M, g_{ab}) is non-singular in the sense discussed in Sec. 22.2.1 above; (ii) The holonomy-like quantity $\Psi(\mathcal{R})$ is globally bounded; (iii) The total affine parameter length $\Delta\lambda(\mathcal{R})$ is globally bounded by some maximum $\Delta\lambda_{\text{max}}$; (iv) The initial slice Σ is regular in the sense that the quantity $l^a_\Sigma \nabla_a R$ is bounded on Σ. Then the quantity (22.24) is bounded and thus the Cauchy horizon is stable. In other words, assuming the conditions (i), (ii) and (iv), the divergence of affine parameter length is a necessary condition for an instability.

Assumption (iv) is only relevant when Σ is non-compact; it is satisfied automatically in compact cases like Misner space. In the non-compact case, the condition (iv) is a reasonable assumption as the vector $\vec{l}_\Sigma$ is parallel transported along Σ, and moreover if Σ is asymptotically spacelike and the spacetime is asymptotically flat, then R will go to zero at large distances. The condition should thus be satisfied by slices Σ which are asymptotically null or asymptotically spacelike in spacetimes that are asymptotically flat.

We now turn to a proof of the above result. Consider the quantity $k^a\nabla_a l^b\nabla_b R$, which we can write as

$$\begin{aligned} k^a\nabla_a l^b\nabla_b R &= k^a l^b\nabla_a\nabla_b R + (k^a\nabla_a l^b)\nabla_b R \\ &= -\frac{1}{2}\Box R. \end{aligned} \tag{22.26}$$

Here the second term on the first line vanishes by Eq. (22.14), and the second equality follows from Eq. (22.7). Now integrating with respect to λ along the geodesic Γ in Fig. 22.1, using $k^a\nabla_a = d/d\lambda$ and Eq. (22.18), we obtain

$$\begin{aligned} (l^a\nabla_a R)(\mathcal{R}') &= (l^a\nabla_a R)(\mathcal{Q}) - \frac{1}{2}\int_{\mathcal{Q}}^{\mathcal{R}'} d\lambda\, \Box R \\ &= e^{\Psi(\mathcal{Q})}\,(l^a_\Sigma\nabla_a R)(\mathcal{Q}) - \frac{1}{2}\int_{\mathcal{Q}}^{\mathcal{R}'} d\lambda\, \Box R, \end{aligned} \tag{22.27}$$

where $\mathcal{R}'$ is some point on Γ between $\mathcal{Q}$ and $\mathcal{R}$. Integrating once more with respect to λ and using Eq. (22.24) yields

$$\begin{aligned} |\sigma^{(\text{locally generated})}(\mathcal{R})| \le\; & \frac{1}{48\pi}\Delta\lambda_{\text{max}} e^{\Psi_{\text{max}}} ||l^a_\Sigma\nabla_a R||_\infty \\ & + \frac{1}{96\pi}(\Delta\lambda_{\text{max}})^2 ||\Box R||_\infty, \end{aligned} \tag{22.28}$$

where Ψ_{max} is the maximum value of Ψ, $||l^a_\Sigma\nabla_a R||_\infty < \infty$ is the maximum value of $l^a_\Sigma\nabla_a R$ on Σ, and $||\Box R||_\infty < \infty$ is the maximum value of $\Box R$ on M. It follows from Eq. (22.28) that $\sigma^{(\text{locally generated})}(\mathcal{R})$ is bounded.

22.4.2 Discussion of instability

As discussed in the Introduction, one can intuitively understand the reason for the instability in the following way. The locally generated term in Eq. (22.10) describes local particle creation and/or vacuum polarization effects due to the background gravitational field. [In general dynamic spacetimes it is not meaningful to distinguish between particle creation and vacuum polarization effects.] Radiation propagating parallel to the Cauchy horizon is generated all along the geodesic Γ, and since the length of this geodesic is becoming infinite, an infinite amount of radiation can be accumulated at the Cauchy horizon.

Consider now the issue of under what conditions the divergence of affine parameter length is a sufficient condition for a locally created energy flux instability of the Cauchy horizon. First, as we discuss in Sec. 22.5.3 below, there are spacetimes such as Misner space which are locally flat so that the locally generated term (22.24) vanishes identically, but for which nevertheless the condition of divergence of affine parameter length is satisfied. Therefore, the divergence of affine parameter length is not a sufficient condition for *test field* locally created energy flux instabilities. However, the example of Misner is very special — it is locally flat. In general spacetimes, one might imagine that if the curvature were very small or vanishing near the Cauchy horizon in the background spacetime, then the locally created energy flux instability might be present in second order semiclassical perturbation if not in first order: the first order perturbation would give rise to some curvature near the Cauchy horizon, and this curvature would then act as a source and give rise to an divergence of the expected stress tensor at second order. Therefore, we conjecture that the divergence of affine parameter length is a *sufficient* condition for a full, nonlinear instability of Cauchy horizons in dynamical, two dimensional semiclassical theories with backreaction [25].

22.4.3 Explanation of instability as a "delayed blueshift" instability

Our classification of Cauchy horizon instabilities is based on the split (22.5) of the stress tensor into "initial data" and "locally generated" pieces. However, this split depends on the location of the initial data surface Σ. An instability which is a locally generated energy flux instability from the point of the initial data surface Σ can instead appear to be an initial-data-related blueshift instability from the point of view of some later surface Σ'. An example of this behavior is given in Sec. 22.5.1 below. We now show that this situation always occurs: there is always some initial data surface Σ' with respect to which the instability is a blueshift instability.

The blueshift factor $\Psi(\mathcal{R})$ defined in Sec. 22.3.1 above depends not only on the point $\mathcal{R}$ but also implicitly on the point $\mathcal{P}$ on the Cauchy horizon. In this section we will write the blueshift factor as

$$\Psi_{\mathcal{P}}(\mathcal{R})$$

to make the dependence on $\mathcal{P}$ explicit [43]. Consider now the construction illustrated in Fig. 22.2. Let $\mathcal{R}'$ be an arbitrary point on the geodesic Γ between $\mathcal{R}$ and $\mathcal{Q}$, and let $\mathcal{R}'_0$ and $\mathcal{P}'$ be corresponding points on the geodesic Γ_0 and on the Cauchy horizon respectively, so that $\mathcal{R}'_0$, $\mathcal{R}'$ and $\mathcal{P}'$ all lie on a null geodesic parallel to $\vec{l}$. Let λ_0 be the affine parameter along the geodesic Γ_0, so that $\vec{k} = d/d\lambda_0$ along Γ_0. Then it is possible to show that the affine parameter length $\Delta\lambda(\mathcal{R})$ of the geodesic Γ is given by the following integral along the geodesic Γ_0:

$$\Delta\lambda(\mathcal{R}) = \int_{\Gamma_0} d\lambda_0 \, \frac{e^{\Psi_{\mathcal{P}}(\mathcal{R})}}{e^{\Psi_{\mathcal{P}'}(\mathcal{R}')}}. \tag{22.29}$$

Now, if the total blueshift factor $\Psi_{\mathcal{P}}(\mathcal{R})$ is globally bounded, and if $\Delta\lambda(\mathcal{R}) \to \infty$ as $\mathcal{R} \to \mathcal{P}$, it follows from Eq. (22.29) that the quantity $e^{\Psi_{\mathcal{P}'}(\mathcal{R}')}$ cannot be globally bounded below. Therefore, for some point $\mathcal{P}'$ on the Cauchy horizon (passing to some conformal completion of the spacetime if necessary), there is a diverging redshift, i.e. $e^{\Psi_{\mathcal{P}'}(\mathcal{R}')} \to 0$ as $\mathcal{R}' \to \mathcal{P}'$.

Consider now an "initial data surface" Σ' for which the rightmost portion is a null geodesic parallel to $\vec{l}$ which intersects the Cauchy horizon at $\mathcal{P}'$. Denote by $\Psi'_{\mathcal{P}}(\mathcal{R})$ the blueshift factor given by the construction of Sec. 22.3.1, with respect to the surface Σ'. Then one has

$$\Psi'_{\mathcal{P}}(\mathcal{R}) = \Psi_{\mathcal{P}}(\mathcal{R}) - \Psi_{\mathcal{P}'}(\mathcal{R}'), \tag{22.30}$$

which diverges as $\mathcal{R} \to \mathcal{P}$. Therefore, the Cauchy horizon is blueshift unstable with respect to the surface Σ'. As explained in the Introduction, any energy flux present on the initial surface Σ will not give rise to a divergence of the stress tensor on the Cauchy horizon, since the net blueshift is finite. By contrast, an energy flux which is created locally near Σ' will suffer an infinite blueshift and give rise an instability.

22.4.4 Relevance to four dimensional spacetimes

We now discuss the issue of to what extent this two dimensional locally created energy flux instability mechanism is relevant to four dimensional spacetimes. Suppose one has a four dimensional spacetime on which propagates a quantized scalar field $\hat{\Phi}$. In this context, the expected stress tensor is not determined by its value on an initial surface Σ, so the split (22.5) of the expected stress tensor in any state into a "locally generated" piece plus an "initial data" piece would not seem to have an analog in four dimensions. However, we now show that this splitting, and in addition some of the features of the two dimensional semiclassical theory which underly the instability mechanism, do generalize to at least a certain class of four dimensional spacetimes.

Consider a four dimensional spacetime (M, g_{ab}) which, to the past of some Cauchy surface Σ, is isometric to a portion of Minkowski spacetime. For any quantum state, let $G(x,y) = \langle\hat{\Phi}(x)\,\hat{\Phi}(y)\rangle$ be that state's the two point function, and let $G_0(x,y)$ be the two point function of the state which to the past of Σ is just the usual Minkowski vacuum. Let

$$F(x,y) = G(x,y) - G_0(x,y), \tag{22.31}$$

which is a smooth bisolution of the wave equation. Then, as explained in detail in Ref. [44], the total expected stress tensor *can* be written in a form analogous to Eq. (22.5):

$$T_{ab} = T_{ab}^{(\text{initial data})} + T_{ab}^{(\text{locally generated})}, \tag{22.32}$$

where now

- The "initial data" piece is given by $T_{ab}^{(\text{initial data})} = \text{Lim}_{y\to x}\mathcal{D}_{ab}F(x,y)$, where $\mathcal{D}_{ab}$ is a second order differential operator. This is the same stress tensor as would be obtained in a purely *classical* theory of a statistical ensemble of scalar fields, for which the stress tensor is determined by the expected value $F(x,y) = \langle\Phi(x)\Phi(y)\rangle$ of $\Phi(x)\Phi(y)$ with respect to the classical ensemble. Moreover, the quantity $F(x,y)$ is uniquely determined (both classically and semiclassically) by its initial data and derivatives on $\Sigma \times \Sigma$ from the classical equations of motion

[44]. Just as in two dimensions, the main difference between the classical and semiclassical theories is that the class of allowed initial data (here, initial data for F and its derivatives on $\Sigma \times \Sigma$) is larger in the semiclassical case than in the classical case.

- The remaining "locally generated" piece $T_{ab}^{(\text{locally generated})}$ of the stress tensor is the same for all quantum states, and is just a unique functional of the spacetime geometry, just as in two dimensions. An explicit formula for this piece of the stress tensor is known for the special case of metrics that are linearized perturbations off Minkowski spacetime [45].

These similarities between the two dimensional and four dimensional semiclassical theories indicate that

- As in two dimensions, so also in four dimensions, one can probably classify instabilities at Cauchy horizons into two types (i) "blueshift type" instabilities that cause a divergence of the term $T_{ab}^{(\text{initial data})}$ and that operate both classically and semiclassically; and (ii) intrinsically quantum mechanical, "locally created energy flux type" instabilities that cause a divergence of the term $T_{ab}^{(\text{locally generated})}$.
- The property of divergence of affine parameter length *might* be relevant to divergences of the term $T_{ab}^{(\text{locally generated})}$ in four dimensions. On the other hand, other issues such as the focusing/defocusing along any geodesic, which enter in four dimensions [7] but not in two, probably complicate the situation.

In summary, it is an interesting open question whether or not any of the ideas and results discussed here can be extended to four dimensions; it seems conceivable that some of them could be.

22.5 Example spacetimes

In this section, in order to illustrate and clarify the above discussions and results, we examine several specific spacetimes — the two dimensional Reissner-Nördstrom-deSitter spacetime in Sec. 22.5.1, general two dimensional spacetimes with closed timelike curves in Sec. 22.5.2, and Misner space in Sec. 22.5.3.

22.5.1 The two dimensional Reissner-Nördstrom-de Sitter spacetime

The metric for a spherically symmetric black hole of mass m and charge e in de-Sitter space can be written as [46]

$$ds^2 = -f(r)dt^2 + f(r)^{-1}dr^2 + r^2 d\Omega^2, \qquad (22.33)$$

where

$$f(r) = 1 - \frac{2M}{r} + \frac{e^2}{r^2} - \frac{1}{3}\Lambda r^2, \tag{22.34}$$

and Λ is the cosmological constant. There are three positive roots of the equation $f(r) = 0$, which we will denote as r_i, r_e and r_c, following Ref. [19]. Here $r_i < r_e < r_c$, and these three roots correspond to the inner or Cauchy horizon, the event horizon, and the cosmological horizon, respectively (see Fig. 22.3). We also define $\kappa_j = |f'(r_j)|/2$, which is the surface gravity at the jth horizon, for $r_j = r_i$, r_e or r_c. We will discuss for the most part only the two dimensional version of the spacetime in which the last term in Eq. (22.33) is omitted.

In Ref. [15], Brady and Poisson showed that the Cauchy horizon is classically test field stable in the region $\kappa_i < \kappa_c$ of parameter space, when the matter model is taken to consist of an infalling null fluid (or equivalently a minimally coupled scalar field in the two dimensional context), and they argued that the stability result should also hold for more realistic matter models and for gravitational perturbations. In Ref. [19], Marković and Poisson showed that the two dimensional spacetime is semiclassically unstable, by showing that the stress tensor diverges on the Cauchy horizon in the Marković-Unruh state [47], and in Ref. [33] Poisson extended this result to show that the stress tensor must diverge on the Cauchy horizon in *any* state which is regular on the cosmological and event horizons. (See also earlier calculations by Davies and Moss [18] which indicated that the Cauchy horizon is unstable, and see Ref. [39] for a detailed overview of this literature).

In this section, we will focus attention on some of the above results, and for the most part simply verify that they are reproduced by the general formalism of Secs. 22.2 – 22.4. We will also show that in the integral (22.24), both the integrand $l^a \nabla_a R$ and the total affine parameter length diverge, as claimed in Sec. 22.4.

We take the initial surface to consist of two null half-lines, as illustrated in Fig. 22.3, where the rightmost half-line is outside the event horizon. The null basis $\{\vec{k}, \vec{l}\}$ must have the form

$$\begin{aligned} \vec{k} &= \alpha \left[\frac{1}{f}\frac{\partial}{\partial t} - \frac{\partial}{\partial r} \right] \\ \vec{l} &= \frac{f}{2\alpha} \left[\frac{1}{f}\frac{\partial}{\partial t} + \frac{\partial}{\partial r} \right], \end{aligned} \tag{22.35}$$

where α is some positive function on spacetime, since the vectors $\vec{k}$, $\vec{l}$ are future directed and null. Now the two dimensional version of the metric (22.33) can be written in the usual way as $-f(dt^2 - dr_*^2)$, and since in any spacetime of the form (22.6) the vector fields $e^{-\sigma}\partial_u$ and $e^{-\sigma}\partial_v$ are geodesic, it follows that the vector fields $\pm f^{-1}\partial_t \pm \partial_r$ are locally geodesic. Therefore since $\vec{l}$ is geodesic along the geodesic Λ in Fig. 22.3, it follows from Eq. (22.35) that f/α must be constant along Λ, and without loss of generality we can take this constant to be -1:

$$\alpha(\mathcal{R}) = -f(\mathcal{R}), \quad \mathcal{R} \text{ on } \Lambda. \tag{22.36}$$

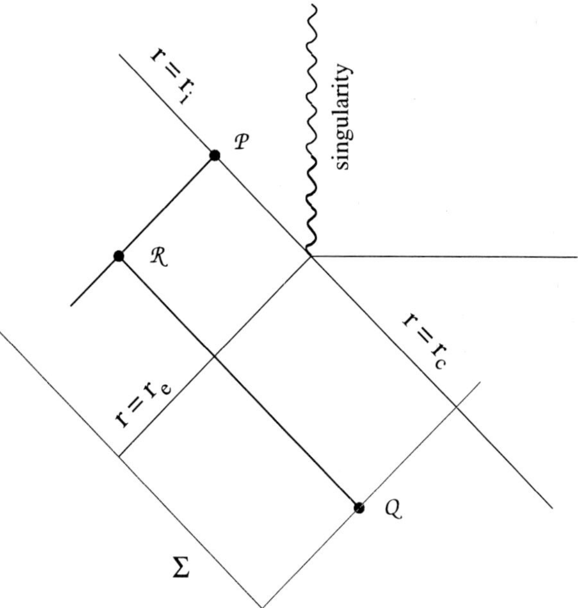

Figure 22.3: The Penrose diagram for a portion of the two dimensional Reissner-Nördstrom-deSitter spacetime which evolves from data specified on the initial Cauchy surface Σ. The lines marked $r = r_c$, $r = r_e$ and $r = r_i$ are the cosmological, event and inner or Cauchy horizons respectively. Also shown are the geodesics Γ and Λ and the points $\mathcal{P}$, $\mathcal{R}$ and $\mathcal{Q}$ of the general construction of Sec. 22.2.3.

Similarly, $\vec{k}$ is geodesic along Γ, so α is constant along Γ,

$$\alpha(\mathcal{R}') = \alpha(\mathcal{R}) \tag{22.37}$$

where $\mathcal{R}'$ is any point along Γ. Combining Eqs. (22.35) – (22.37) we find that the integrand in Eq. (22.24) is given by

$$(l^a \nabla_a R)(\mathcal{R}') = -\frac{f(\mathcal{R}')}{2f(\mathcal{R})} R'(r), \tag{22.38}$$

where $R = R(r)$ is the Ricci scalar which depends only on r.

Next, from Eq. (22.35) we find that the relationship between the affine parameter λ along Γ and the coordinate r is given by $d\lambda = -dr/\alpha$, which from Eq. (22.37) gives

$$\lambda = \frac{1}{f(\mathcal{R})} \left[r + \text{const}\right]. \tag{22.39}$$

Combining Eqs. (22.24), (22.38) and (22.39) now gives

$$\sigma^{(\text{locally generated})}(\mathcal{R}) = \frac{1}{96\pi f(\mathcal{R})^2} \int_{r(\mathcal{R})}^{r[\mathcal{Q}(\mathcal{R})]} dr f(r) R'(r). \tag{22.40}$$

Here as before $\mathcal{Q} = \mathcal{Q}(\mathcal{R})$ denotes the unique point $\mathcal{Q}$ on the initial surface Σ determined by $\mathcal{R}$ according to the construction of Sec. 22.2.3. Using the formula $R(r) = -f''(r)$ and integrating by parts gives

$$\int_{r(\mathcal{R})}^{r[\mathcal{Q}(\mathcal{R})]} dr\, f(r) R'(r) = \left[\frac{1}{2}(f')^2 - f f''\right]_{r(\mathcal{R})}^{r[\mathcal{Q}(\mathcal{R})]}. \tag{22.41}$$

Finally, using that fact that as $\mathcal{R} \to \mathcal{P}$, the limits of integration behave as $r(\mathcal{R}) \to r_i$, $r[\mathcal{Q}(\mathcal{R})] \to r_c$ and that $f(r_i) = f(r_c) = 0$ gives

$$\sigma^{(\text{locally generated})}(\mathcal{R}) \approx (\text{const}) \frac{1}{f(\mathcal{R})^2} \left[\kappa_c^2 - \kappa_i^2\right]. \tag{22.42}$$

Equation (22.42) reproduces the result of Refs. [19, 33] that σ is divergent on the Cauchy horizon in the classically stable region $\kappa_c > \kappa_i$ of parameter space, since $f(\mathcal{R}) \to 0$ as $\mathcal{R} \to \mathcal{P}$. The agreement with Ref. [19] can be made more explicit by noting that in this limit $f(\mathcal{R}) \approx (\text{const}) \exp[-\kappa_i v(\mathcal{R})]$, where v is the advanced time coordinate $t + r_*$ which goes to infinity as $\mathcal{R} \to \mathcal{P}$ [15].

Several points should be noted about the above derivation. First, from Eq. (22.39) it can be seen that the affine parameter length is divergent:

$$\Delta\lambda(\mathcal{R}) \approx \frac{1}{f(\mathcal{R})}(r_c - r_i) \approx e^{\kappa_i v}\,(r_c - r_i); \tag{22.43}$$

and that the integrand (22.38) itself also has an overall factor of $1/f(\mathcal{R})$ which is divergent as $\mathcal{R} \to \mathcal{P}$ [48]. Second, the instability result applies to *all* quantum states for which the initial data on Σ is regular. The quantity $\sigma(\mathcal{R})$ will be a sum of the divergent, locally generated piece (22.42), together with the piece (22.20) which depends on the initial data and which will be finite as the blueshift factor e^{Ψ} is finite. [It is easy to verify that up to an overall constant factor the blueshift factor is

$$\begin{aligned} e^{\Psi(\mathcal{R})} &= \frac{f[\mathcal{Q}(\mathcal{R})]}{f(\mathcal{R})} \\ &\approx \frac{e^{-\kappa_c v(\mathcal{R})}}{e^{-\kappa_i v(\mathcal{R})}} \quad \text{as } \mathcal{R} \to \mathcal{P}. \end{aligned} \tag{22.44}$$

This is finite for $\kappa_c > \kappa_i$, as first noted in Ref. [15]]. Third, the derivation allows us to understand why the divergent piece of the quantity $\sigma(\mathcal{R})$ in semiclassical analyses does not contain a cosmological redshift factor $e^{-2\kappa_c v}$ as it does in classical analyses [19]. The reason is that the radiation giving rise to the divergence (22.42) is produced in the vicinity of the event horizon and propagates from there inward to the Cauchy horizon, and thus is not affected by the redshift factor describing propagation from the cosmological horizon to the event horizon.

Note that it is straightforward to show that in the classically stable region $\kappa_c > \kappa_i$ of parameter space, the blueshift factor $e^{\Psi(\mathcal{R})}$ is globally bounded, not just inside the black hole but also along the event horizon. Thus, the class of spacetimes to which the general result of Sec. 22.4.1 applies is not empty.

We have explained the above instability as a locally created energy flux instability and not as a blueshift instability. However, it can also be thought of as a "delayed blueshift" instability, as discussed in Sec. 22.4.3 above. If one chooses the right-most portion of the initial data surface Σ to lie on the event horizon, then the corresponding blueshift factor given by the construction of Sec. 22.3.1 is proportional to $e^{2\kappa_i v}$ and so is unbounded. In classical analyses, the initial data on the event horizon must fall off like $e^{-2\kappa_c v}$ (due to having propagated inward from some initial surface outside the event horizon), thus giving rise to a finite value of σ on the Cauchy horizon. However, in a semiclassical analysis, things are different. Consider an ingoing mode of the quantum field whose mode function is concentrated near some geodesic that is very close to the Cauchy horizon. Any ingoing quanta in this mode from outside the black hole experience first a large redshift (from the outside to the event horizon), then a large blueshift (from the event horizon to the interior) which is smaller than the redshift. The crucial feature in the semiclassical analysis is that ingoing quanta are created in the vicinity of the event horizon; these propagate inwards and suffer only the large blueshift. Thus, the "initial data" for the stress tensor on the event horizon need not fall off like $e^{-2\kappa_c v}$ in the semiclassical theory. Correspondingly, near the Cauchy horizon σ diverges like $e^{2\kappa_i v}$ by the usual blueshift effect.

Thus, the distinction between blueshift and locally created energy flux instabilities is dependent on the choice of location of the initial data surface, as discussed in

Sec. 22.4.3 above

22.5.2 General spacetimes with closed timelike curves

A special case of the general class of spacetimes described in Sec. 22.2.1 above is when the spacetime (M', g'_{ab}) contains closed timelike curves, so that the Cauchy horizon $H^+(\Sigma)$ is also a chronology horizon. A simple example of such a spacetime is given by Yurtsever [28], in which the lightcones on a cylinder "tip over" to produce a closed null geodesic around the cylinder [49]. This closed null geodesic coincides with the chronology horizon. [In the four dimensional context, closed null geodesics or "fountains" will form a small subset of the full chronology horizon [7].]

In Ref. [14], Hawking gives a general argument, in the context of chronology horizons in four dimensional spacetimes which are compactly generated, that the total affine parameter length of all fountains should be infinite in the past direction (although generically finite in the future direction): If $\vec{k}$ is a future directed, null vector which is tangent to the geodesic at some point, then the total holonomy around the fountain in the future direction will map $\vec{k}$ to $e^h\vec{k}$ for some constant h. Hawking shows that if $h < 0$, then there must exist a closed timelike curve to the past of the chronology horizon, which is a contradiction. Therefore $h \geq 0$ and so the the total affine parameter length in the past direction of the closed null geodesic is proportional to

$$\sum_{n=0}^{\infty} e^{nh} = \infty, \tag{22.45}$$

(although the total affine parameter length in the future direction is proportional to $\sum e^{-nh} < \infty$ as long as $h \neq 0$). It can be checked that this argument applies equally well to two dimensional spacetimes. In the two dimensional context, the generator of the Cauchy horizon through any point $\mathcal{P}$ will therefore have infinite affine parameter length in the past direction, and it follows from the argument of appendix A that all such spacetimes satisfy the property (22.25) of divergence of affine parameter length.

However, it does not follow that all such spacetimes are subject to the locally created energy flux instability. We now turn to a specific spacetime containing closed timelike curves which illustrates this point.

22.5.3 Misner space

Misner space is a well-known locally flat spacetime with topology $S^1 \times R$ [50, 51]. As is well known, Misner space is both classically and semiclassically unstable [52]. These instability properties of the spacetime are reviewed in Ref. [53]. In this section we show how the construction of Secs. 22.3 and 22.4 applies to Misner space. The construction reproduces the well-known fact that Misner space suffers from the blueshift instability, which explains both the classical and semiclassical instabilities. We also show that Misner space does *not* suffer from the locally created energy flux instability, despite the fact that the spacetime does satisfy the property of divergence of affine parameter length.

Misner space can be constructed as follows [53]. In two dimensional Minkowski spacetime with coordinates (t, x), identify the worldline $x = 0$ with the worldline $x = L - \beta t$, for some $L > 0$, $\beta > 0$ (see Fig. 22.4). Let τ be the proper time measured by a clock on one of these world lines; $\tau = t$ for the leftmost worldline, whereas $\tau = t/\gamma$ for the rightmost worldline, where γ is the usual special relativistic time dilation factor. An observer crossing the rightmost worldline at coordinate time $t = t_0$ therefore emerges from the leftmost line at coordinate time $t = t_0/\gamma$. Hence, at large t, there will be closed timelike curves in the spacetime, after the closed null geodesic marked in Fig. 22.4. This closed null geodesic is the Cauchy horizon or chronology horizon of Misner space.

Consider now how the constructions of Secs. 22.2.3 and 22.3.1 apply to Misner space. First, note that if any vector is parallel transported through the leftmost worldline to emerge from the rightmost worldline, it undergoes the boost given by $\partial_u \to e^{\xi}\partial_u$, $\partial_v \to e^{-\xi}\partial_v$, where $\cosh \xi = \gamma$ and $v = t + x$, $u = t - x$. From the diagram of Misner space shown in Fig. 22.4, we can fairly easily see that (i) If we take the initial surface Σ to be the surface $t = 0$, then Σ has the topology of a circle. The basis $\{\vec{k}_\Sigma, \vec{l}_\Sigma\}$ obtained by parallel transport along Σ will be discontinuous at some point Q_0, because of the overall boost when passing through the identified worldlines. (ii) The null geodesic Γ through a point $\mathcal{R}$ near the Cauchy horizon passes through the identified worldlines some number n of times before it reaches the initial surface, where n grows without limit as $\mathcal{R} \to \mathcal{P}$. Thus, the initial point $\mathcal{Q}(\mathcal{R})$ circles around and around Σ, as claimed in Sec. 22.3.2 above. (iii) When one parallel transports the vector $\vec{k}$ along Γ to Σ, it is boosted n times, and thus the redshift factor is

$$e^{\Psi(\mathcal{R})} = e^{n\xi}, \tag{22.46}$$

which diverges as $\mathcal{R} \to \mathcal{P}$. This redshift factor changes discontinuously by a factor of e^{ξ} every time $\mathcal{Q}(\mathcal{R})$ crosses $\mathcal{Q}_0$ on Σ, as the basis $\{\vec{k}_\Sigma, \vec{l}_\Sigma\}$ is discontinuous there. (iv) The total affine parameter length $\Delta\lambda(\mathcal{R})$ of Γ also diverges as $\mathcal{R} \to \mathcal{P}$. [This also follows from the general argument of Sec. 22.5.2 above].

Since the blueshift factor diverges, the spacetime is classically and quantum mechanically unstable, from the arguments of Sec. 22.3 above. In the classical case, this means that the stress tensor diverges on the Cauchy horizon for generic initial data of the scalar field [54]. Of course, for the exceptional case of vanishing initial data on Σ, the stress tensor vanishes identically. Nevertheless, one still regards the Cauchy horizon as being unstable as vanishing initial data is not generic or stable in any physical sense.

The situation is very similar in the semiclassical theory. Originally it was conjectured in Ref. [14] that the stress tensor must diverge on the Cauchy horizon for *all* quantum states. The argument of Sec. 22.3 shows that the stress tensor must diverge unless the initial data σ_Σ on Σ is vanishing; thus, the conjecture was effectively that

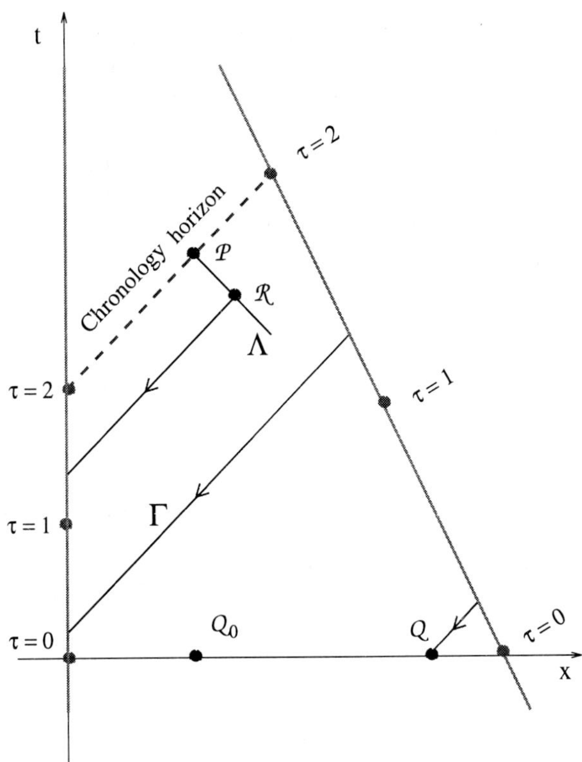

Figure 22.4: A spacetime diagram of Misner space (adapted from Ref. [51]). The two thick black lines are identified. Shown are a representative point $\mathcal{P}$ on the chronology horizon, the null geodesic Λ emanating from $\mathcal{P}$, a point $\mathcal{R}$ on Λ near $\mathcal{P}$, and the null geodesic Γ joining $\mathcal{R}$ to the initial surface $t = 0$ at the point $\mathcal{Q}$. It can be seen that the spacetime satisfies the property of divergence of affine parameter length (as must all spacetimes with closed timelike curves) since the number of times the geodesic Γ winds around the spacetime diverges as $\mathcal{R} \to \mathcal{P}$. Moreover the blueshift factor $e^{\Psi(\mathcal{R})}$, which is obtained from the Lorentz transformation got by parallel transporting along Γ between $\mathcal{Q}$ and $\mathcal{R}$, also diverges as $\mathcal{R} \to \mathcal{P}$, because all vectors are boosted when they pass through the identified worldlines.

there is no state on Misner space for which the expected stress tensor vanishes identically on the initial surface. However, it is now known that there are quantum states on both two dimensional and four dimensional Misner space for which the stress tensor does vanish identically [29, 30, 31]. Nevertheless, states for which the initial data for the stress tensor is vanishing are clearly in some sense non-generic, and therefore one is still justified in regarding Misner space as unstable in the semiclassical theory, just as in the classical theory.

Finally, we note that the locally created energy flux instability mechanism does not operate in Misner space, since it is locally flat and so the integrand in Eq. (22.24) is identically vanishing. This does not contradict the fact that the condition (22.25) is satisfied by Misner space, since we have only shown that it is a necessary condition for the instability, and not a sufficient condition. Note, however, that all spacetimes which are "close" to Misner space but which differ from it by having an arbitrarily small amount of spacetime curvature on the Cauchy horizon should suffer from the locally created energy flux instability. See also the related discussion in Sec. 22.4.2 above.

22.6 Conclusions

We have shown that in the context of semiclassical gravity in two dimensions, there are two different types of instabilities of Cauchy horizons. The first is a divergence of the piece of the stress tensor which is determined by the initial data; it operates classically as well as semiclassically, and is just the well-understood blueshift instability. The second, which we call the locally created energy flux instability, is a divergence of the locally generated piece of the expected stress tensor. This instability is characterized by the divergence of the affine parameter lengths of null geodesics parallel to and close to the Cauchy horizon.

It is natural to conjecture that for all Cauchy horizons in two dimensional spacetimes without singularities, one or other of these two instability mechanism always applies, if not in linear perturbation theory, then at least in nonlinear analyses.

acknowledgements

It is a pleasure to thank Lior Burko, Amos Ori and Liz Youdim for organizing the very enjoyable workshop on "The Internal Structure of Black Holes and Spacetime Singularities" at Haifa, and for the generous hospitality offered during our stay there. I also wish to thank Patrick Brady, Chris Chambers, Ian Moss, Amos Ori, Eric Poisson, Kip Thorne, and Robert Wald for helpful discussions on Cauchy horizon stability issues. This research was supported in part by NSF grant PHY 9722189 and by a Sloan Foundation fellowship.

A Diverging affine parameter length condition follows from Cauchy horizon generator being of infinite affine parameter length in the past

In this appendix we show that if the generator of the Cauchy horizon through the point $\mathcal{P}$ has infinite affine parameter length in the past, then the condition (22.25) must be satisfied by the spacetime.

Let TM' be the tangent bundle over the spacetime (M', g'_{ab}), and let

$$\exp : D \subset TM' \to TM'$$

be the exponential map, where the domain of definition D of the exponential map is an open subset of TM'. [If (M', g'_{ab}) were geodesically complete we would have $D = TM'$]. In other words, for any point $\mathcal{B}$ in M' and any vector $\vec{v}$ at $\mathcal{B}$, $\exp[\mathcal{B}, \vec{v}]$ will be the pair $(\mathcal{B}', \vec{v}')$ in TM' such that the geodesic starting at $\mathcal{B}$ with initial tangent $\vec{v}$ reaches $\mathcal{B}'$ after one unit of parameter length, and such that $\vec{v}'$ is the tangent to the geodesic at $\mathcal{B}'$.

Let $U_+ = M' - \overline{D^-(\Sigma)}$, the complement of the closure of the past domain of dependence of Σ, and let $W_+ \subset TM'$ be the tangent bundle over U_+. It is clear that the set W_+ is open in TM'. Consider now the construction outlined in Sec. 22.2.3. It can be seen that, for any $l > 0$ and for any $\mathcal{R}$ on Λ,

$$\exp[\mathcal{R}, -l\,\vec{k}(\mathcal{R})] \in W_+ \quad \text{implies} \quad \Delta\lambda(\mathcal{R}) \geq l, \tag{22A.1}$$

by the definition of $\Delta\lambda(\mathcal{R})$ given in Sec. 22.4 above and using the fact that Σ is a Cauchy surface for (M, g_{ab}).

Suppose now that the affine parameter length towards the past of the generator of the Cauchy horizon through $\mathcal{P}$, normalized with respect to $\vec{k}(\mathcal{P})$, is greater than some number β. It follows that $[\mathcal{P}, -\beta\,\vec{k}(\mathcal{P})]$ lies in the domain of definition D of the exponential map. Also it follows that $\exp[\mathcal{P}, -\beta\,\vec{k}(\mathcal{P})]$ lies in W_+, and thus by continuity of the exponential map there is some open neighborhood U of $[\mathcal{P}, -\beta\,\vec{k}(\mathcal{P})]$ in D whose image under the exponential map lies in W_+. Hence there is some neighborhood V of $\mathcal{P}$ in M' such that for all points $\mathcal{R}$ on Λ and in V, $\exp[\mathcal{R}, -\beta\vec{k}(\mathcal{R})]$ is defined and lies in W_+ and so $\Delta\lambda(\mathcal{R}) \geq \beta$. Since this is true for all β the result follows.

Note that the converse of this result if not true: if the diverging affine parameter condition holds for a given point $\mathcal{P}$ on the Cauchy horizon, it does not follow that the generator of the Cauchy horizon through $\mathcal{P}$ has infinite affine parameter length in the past. This is because one always has the freedom to redefine the spacetime (M', g'_{ab}) by excising points on the Cauchy horizon.

We have chosen to express the condition in terms of the behavior of the spacetime (M, g_{ab}) [divergence of affine parameter length of null geodesics parallel to and close to the Cauchy horizon] rather than the behavior of the larger spacetime (M', g'_{ab}) [generator of Cauchy horizon having infinite affine parameter length in the past] in order that the condition be manifestly independent of which extension we pick.

Bibliography

[1] A. Ori, *The singularity inside realistic black holes: Major open questions*, talk given at the workshop on The Internal Structure of Black Holes and Spacetime Singularities, June 29 – July 3, 1997 Haifa, Israel (unpublished).

[2] For a review of the information loss paradox, see, for example, J. Preskill, Report No. hep-th/9209058 (unpublished).

[3] We assume that the initial surface is a complete Riemannian manifold to exclude examples, such as the region $|u| < 1$, $|v| < 1$ of two dimensional Minkowski spacetime, where the predicted spacetime is incomplete simply because the initial data is incomplete.

[4] Of course, it could also happen that one can have maximal Cauchy evolutions with both curvature singularities and Cauchy horizons. However, this possibility is not so disturbing. Suppose we define the "classical region" of spacetime to consist of all points $\mathcal{P}$ for which the curvature is sub-Planckian everywhere to the past of $\mathcal{P}$; then if the Cauchy horizon is outside the "classical region" there is less cause for concern, as one would not believe the predictions of the classical theory outside the classical region anyway.

[5] By classical general relativity we mean that both the metric and matter are classical, as opposed to semiclassical general relativity.

[6] R. M. Wald, *General relativity* (University of Chicago, Chicago, 1984).

[7] K. S. Thorne, in *Proceedings of the 13th International Conference on General Relativity and Gravitation*, Ed. R. J. Gleiser, C. N. Kozameh and O. M. Moreschi (Institute of Physics Publishing, Bristol, 1993) p. 295.

[8] R. M. Wald, *Gravitational Collapse and Cosmic Censorship*, Report No. gr-qc/9710068 (unpublished).

[9] F. Mellor and I. Moss, Class. Quantum Grav. **9**, L43 (1992).

[10] D. A. Konkowski and T. M. Helliwell, Phys. Rev. D **54**, 7898 (1996).

[11] M. Simpson and R. Penrose, Int. J. Theor. Phys. **7**, 183 (1973).

[12] S. Chandrasekhar and J.B. Hartle, Proc. R. Soc. London **A284**, 301 (1982).

[13] M. S. Morris, K. S. Thorne, U. Yurtsever, Phys. Rev. Lett. **61**, 1446 (1988).

[14] S. W. Hawking, Phys. Rev. D **46**, 603 (1992).

[15] P. R. Brady and E. Poisson, Class. Quantum Grav. **9**, 121 (1992).

[16] C. M. Chambers and I. G. Moss, Class. Quantum Grav. **11**, 1034 (1994).

[17] G. T. Horowitz and H. J. Sheinblatt, Phys. Rev. D **55**, 650 (1997) (also gr-qc/9607027).

[18] P. C. W. Davies and I. G. Moss, Class. Quantum Grav. **6**, L173 (1989).

[19] D. Marković and E. Poisson, Phys. Rev. Lett. **74**, 1280 (1995).

[20] S. W. Kim and K. S. Thorne, Phys. Rev. D **43** 3929 (1991).

[21] V. P. Frolov, Phys. Rev. D **43**, 3878 (1991).

[22] N. N. Gnedin, unpublished; D. A. Kompaneets, unpublished.

[23] A general theorem that the expected stress tensor must either diverge or be ill-defined at some points on compactly generated chronology horizons in four dimensional spacetimes was proved in B. S. Kay, M. J. Radzikowski and R. M. Wald, Commun. Math. Phys. **183**, 533 (1997).

[24] Ref. [20] showed that the expected stress tensor diverges on the Cauchy horizon for quantum states for which the two point function is dominated by a series of singular contributions which can be calculated using geometric optics. Recently, examples have been recently found of states of automorphic fields on four dimensional spacetimes with closed timelike curves for which the stress tensor does not diverge on the chronology horizon [29], and similar examples have been found for massless scalar fields in two dimensions [30]. Thus, the above property of two point functions used in Ref. [20] must not be valid for all states; however, it is presumably valid for generic states and thus the stress tensor should diverge on the chronology horizon for generic states.

[25] As is well known, it is necessary to couple in additional scalar fields as part of the gravitational sector of the theory to get a non-trivial, dynamical theory of gravity in two dimensions. One can then add quantum scalar fields to get a coupled semiclassical theory. See, for example, T.M. Fiola, J. P. Preskill, A. D. Strominger and S. P. Trivedi, Phys. Rev. D **50**, 3987 (1994) (also hep-th/9403137).

[26] G. Klinkhammer, Phys. Rev. D **46**, 3388 (1992).

[27] M. Visser, *Lorentzian wormholes, from Einstein to Hawking* (American Institute of Physics, Woodbury, 1995).

[28] U. Yurtsever, Class. Quantum Grav. **3**, 1127 (1991).

[29] S. V. Sushkov, Class. Quantum Grav. **14**, 523 (1997).

[30] S. V. Krasnikov, Phys. Rev. D **54**, 7322 (1996).

[31] C. R. Cramer and B. S. Kay, *The thermal and two-particle stress-energy must be ill-defined on the 2-d Misner space chronology horizon*, Report No. gr-qc/9708028 (unpublished).

[32] It follows from our analyses that the instabilities of chronology horizons in two dimensions (and probably also in four dimensions) can be split into two categories: (i) blueshift instabilities, which apply both classically and semiclassically, as in Misner space, and (ii) locally created energy flux instabilities, which can be understood as the "piling up of vacuum fluctuations", as explained in Ref. [7]. For locally created energy flux instabilities, a crucial feature of this pile-up is that each time a wavepacket of a particular mode of a quantum field propagates around an almost-closed null curve near the chronology horizon, the total expected value of propagating energy is augmented by a small amount. [See Eq. (22.24) above and associated discussion.] The instability cannot be explained just from the fact that the wavepackets from each successive trip around almost-closed null curves tend to overlap more and more as the horizon is approached; that overlap is also present classically.

[33] E. Poisson, *Black-hole interiors and strong cosmic censorship*, contribution to this volume (also gr-qc/9709022).

[34] C. W. Misner, K. S. Thorne, and J. A. Wheeler, *Gravitation* (Freeman, San Francisco, 1973).

[35] N. D. Birrell and P. C. W. Davies, *Quantum fields in curved space* (Cambridge University Press, Cambridge, 1982).

[36] The resulting basis can be expressed as

$$\vec{l}(u,v) = e^{-\sigma(u_{\mathcal{P}},v)} \frac{\partial}{\partial v}$$

and

$$\vec{k}(u,v) = e^{\sigma(u_{\mathcal{P}},v)-\sigma(u,v)} \frac{\partial}{\partial u},$$

for any double null coordinates (u,v) for which the metric takes the form (22.6) and which are regular at $\mathcal{P}$, where $u_{\mathcal{P}}$ is the value of the u coordinate at $\mathcal{P}$.

[37] One might worry that if the behavior of, say, ρ near $\mathcal{P}$ were very direction dependent, then it might happen that $\rho[x^a(\lambda)]$ would diverge for some curves $x^a(\lambda)$ while nevertheless remaining always bounded for all timelike curves of bounded proper acceleration. This possibility is excluded by the equation of motion (22.4) for the stress tensor: since the geometry is regular near $\mathcal{P}$, in a sufficiently small neighborhood of $\mathcal{P}$ the effect of the source term will be unimportant and to a good approximation ρ will depend on only one of the two null coordinates u and v.

[38] Specifically, consider the special case where Σ consists of two intersecting null lines as illustrated in Fig. 22.1. Then, the free data on one of these two half-lines consists of the function $\sigma(\lambda) \equiv T_{ab}(\partial/\partial\lambda)^a(\partial/\partial\lambda)^b$, where λ is an affine parameter along the line. In the classical theory, $\sigma(\lambda)$ can be any non-negative smooth function. In the semiclassical theory in the special case of two dimensional Minkowski spacetime, it follows from the analyses of Ref. [55] that the class of allowed functions $\sigma(\lambda)$ consists of those smooth functions which satisfy

$$\int d\lambda\, f(\lambda)\,\sigma(\lambda) \geq -\frac{1}{48\pi}\int d\lambda\,\frac{f'(\lambda)^2}{f(\lambda)},$$

for all smooth non-negative functions $f(\lambda)$ [where $f'(\lambda)^2/f(\lambda)$ is interpreted to mean zero when $f(\lambda) = 0$]. This characterization also applies to spacetimes which are globally conformal to two dimensional Minkowski spacetime, as can be seen from Eq. (6.133) of Ref. [35].

[39] C. M. Chambers, *The Cauchy Horizon in Black Hole-de Sitter Spacetimes*, contribution to this volume (also gr-qc/9709025).

[40] This is not quite true as one could envisage a situation where the "locally generated" and "initial data" terms in Eq. (22.5) both diverge but where they cancel each other out so that that the sum is finite. We ignore this possibility here. See however Ref. [18] for a scenario where such a cancelation could occur.

[41] An example of a spacetime where this occurs is the following: Let M_0 be the region $|u| < 1$, $|v| < 1$ of Minkowski spacetime, where u and v are the usual null coordinates, and let M be $M_0 - \overline{I_+(\mathcal{P})}$, where $\mathcal{P}$ is the point $(u, v) = (1/2, 1/2)$.

[42] This statement assumes that the initial slice is outside the event horizon. When considering classically stable Reissner-Nördstrom-de Sitter spacetimes ($\kappa_c > \kappa_i$) in Sec. 22.5.1, we shall take the rightmost half of our initial data slice Σ to be a null line outside the event horizon as depicted in Fig. 22.3; in this case the blueshift factor is finite. If one took the rightmost half of the initial slice to be the event horizon itself, then the blueshift factor would be infinite, but the spacetime would still be (classically) stable because of the falloff rate of the initial data on the event horizon.

[43] Suppose that one uses null coordinates (u, v) such that $\vec{l} \propto \partial/\partial v$ and $\vec{k} \propto \partial/\partial u$. Then one might think that the quantity $\Psi_{\mathcal{P}}(\mathcal{R})$ would depend on the four coordinate values $u(\mathcal{P})$, $v(\mathcal{P})$, $u(\mathcal{R})$ and $v(\mathcal{R})$. However, $v(\mathcal{P})$ is fixed by the requirement that $\mathcal{P}$ lie on the Cauchy horizon, and $u(\mathcal{R}) = u(\mathcal{P})$ by construction. Thus, $\Psi_{\mathcal{P}}(\mathcal{R})$ depends only on two independent real parameters.

[44] É. É. Flanagan and R. M. Wald, Phys. Rev. D, **54**, 6233 (1996) (gr-qc/9602052).

[45] G. T. Horowitz, Phys. Rev. D **21** 1445 (1980).

[46] B. Carter, in *Black Holes*, edited by C. DeWitt and B.S. DeWitt (Gordon and Breach, New York, 1973).

[47] D. Marković and W.G. Unruh, Phys. Rev. D **43**, 332 (1991).

[48] This does not contradict the fact that the integrand is finite on the Cauchy horizon itself near $\mathcal{P}$. As a function of the affine parameter λ and the retarded time coordinate v, the integrand in Eq. (22.24) at large v is $l^a \nabla_a R \propto f[r(\lambda)]\, f'''[r(\lambda)]\, \exp[\kappa_i v]$, with $r(\lambda) \approx r_i + \exp[-\kappa_i v]\lambda$. Therefore in the limit $v \to \infty$, $l^a \nabla_a R \to f'(r_i) f'''(r_i)\, \lambda$, which is finite at each fixed λ but which grows linearly without limit as one moves down the Cauchy horizon. A similar result is found on the cosmological horizon.

[49] Note that in four dimensional contexts, one needs to distinguish between smoothly closed null geodesics, and self-intersecting null geodesics in which the direction of the tangent to the geodesic is discontinuous at some point. In the two dimensional context, however, no such distinction is necessary, if we assume that the spacetime (M', g'_{ab}) is orientable and time orientable. This is because given a future directed, null, tangent vector k^a at a point on a self-intersecting null geodesic, when one parallel transports this vector around the geodesic, in two dimensions the result must be proportional to one of the four null directions at that point. The three possibilities other than k^a are excluded by orientability and time orientability.

[50] C. W. Misner, "Taub-NUT space as a counterexample to almost anything", in *Relativity Theory and Astrophysics I. Relativity and Cosmology*, Ed. J. Ehlers (American Mathematical Society, Providence, 1967) pp. 160-169.

[51] S. W. Hawking and G. F. R. Ellis, *The large scale structure of space-time* (Cambridge University Press, Cambridge, 1973).

[52] W. A. Hiscock and D. A. Konkowski, Phys. Rev. D **26**, 1225 (1982).

[53] "Misner Space as a Prototype for Almost Any Pathology", K. S. Thorne in *Directions in General Relativity* Eds. B. L. Hu *et al* (Cambridge University Press, Cambridge, 1993).

[54] The classical instability of Misner space follows from the general arguments of Sec. 22.3. However, there is in addition a well-known direct proof that Misner space is classically unstable [28]: The general solution of the wave equation for a massless scalar field in two dimensional Minkowski space can be written as $\Phi(u,v) = F(u)+G(v)$ for some functions F and G, and thus the general solution on Misner space is of this form where F and G satisfy

$$F(u) + G(v) = F(e^{\xi}u) + G(e^{-\xi}v),$$

where ξ is a fixed parameter. By differentiating this equation with respect to u it can be shown that the general solution is of the form

$$\Phi(u,v) = A + B\ln(uv) + f[\ln(-v)] + g[\ln(-u)],$$

where A and B are constants, and f and g are periodic functions with period ξ. All of these solutions give rise to a diverging stress tensor at $u = 0$ or $v = 0$ except for the constant solution.

[55] É. É. Flanagan, Phys. Rev. D **56**, 4922 (1997); gr-qc/9706006.

TRAVERSABLE WORMHOLES IN GEOMETRIES OF CHARGED SHELLS IN A DE SITTER COSMOS

F. Schein

Institut für Theoretische Physik, Universität Wien
Boltzmanngasse 5, A–1090 Wien, Austria

Abstract

We construct axisymmetric wormholes which connect distant parts of an asymptotic de Sitter universe. Spacetime consists of two parts: the exterior region corresponds to the gravitational field of two charged shells obtained by using the multiblack hole solutions on a de Sitter background found by Kastor and Traschen [1]. We cut out the interior of these shells and match them to a (mass M equals charge Q) Reissner Nordström de Sitter black hole.

The transition between the exterior and interior spacetime can by made continuous but enforces the introduction of two surface layers of charged matter. It turns out that the parameters of the models can be chosen in a way such that the surface energy density is positive and all energy conditions are satisfied.

We restrict the discussion of the causal structure to wormhole models containing layers built from positive energy densities. We prove that for small values of the (positive) cosmological constant Λ there exist causal curves originating at past infinity $\mathcal{J}^-$, traversing the wormhole, and running back to future infinity $\mathcal{J}^+$. Moreover, we prove that for small values of Λ these wormhole models contain closed timelike curves (CTCs). Nevertheless, there are no CTCs in the neighborhood of $\mathcal{J}^-$ and $\mathcal{J}^+$.

23.1 Introduction

In this note we study hypothetical shortcuts for interstellar travel between distant parts of a de Sitter universe. As wormholes between the different regions we use Reissner-Nordström de Sitter (RNdS) black holes. We take advantage of the fact that its complete manifolds consist of infinite chains of universes connected successively in time by wormhole tunnels.

This issue was discussed by an number of authors before: e.g. Carter [2], Novikov [3], and Hawking and Ellis [4] in their book, when discussing the Reissner-Nordström spacetime, mention the possibility to travel to other universes by passing through wormholes made by charges. An identification of successive asymptotic regions was suggested in order to allow for observers who enter the black hole to return to their universe. Such an identification of asymptotically flat regions was also considered for example by Morris and Thorne [5], and Frolov and Novikov [6] when studying static wormholes. However, no explicit way of achieving this was given.

We have already addressed this problem in [7] where we give a scheme which allows for the construction of wormholes providing shortcuts to arbitrary distant regions in an asymptotically flat universe. Here we show that this scheme can also be applied to the case of a non-vanishing, positive, cosmological constant.

In Section 23.2 we shortly describe the ingredients for this construction: the axisymmetric exterior region is based on the multi-black hole solutions on a de Sitter background found by Kastor and Traschen [1]. We apply the image charge method well known from electrostatics to obtain a special solution of Einstein-Maxwell equations which allows for matching a spherically symmetric interior region. This interior region - the wormhole tunnel - is chosen to be a $M = Q$ RNdS black hole. We cut these spacetimes both at two timelike hypersurfaces, shells S_1 and S_2, and match these surfaces in order to obtain a topologically non-trivial spacetime.

In Section 23.3 we prove that the matching can be made continuous. In addition, we calculate the distributional stress energy tensor on these hypersurfaces by using the standard thin shell formalism [8]. It turns out that the shells consist of charged dust and it is possible to choose the various parameters of the model in a way such that the surface energy density is positive. Consequently, all energy conditions can be satisfied.

In Section 23.4 we discuss the causal structure of the considered wormhole models requiring that the energy density on the shells is positive. Concerning the traversability of the wormholes we address the following question: Is it possible to enter the black hole region, traverse the wormhole, and return to the asymptotic de Sitter region? In order to make this question more precise we will define the notion of the asymptotic de Sitter region. We prove that for sufficient small values of the cosmological constant this question can be answered affirmativly. Thus, adapting the terminology of Friedman, Schleich, and Witt [9] the topology is actively probable.

In addition, we prove that for sufficient small values of the cosmological constant Λ there exist closed timelike curves (CTCs). In contrast to the asymptotically flat wormhole models ($\Lambda = 0$) discussed in [7] the region containing CTCs is bounded and there do not exist CTCs near past and future infinity.

In the Questionary we comment on the classical (in-)stability of the inner horizon of charged black holes in the presence of a positive cosmological constant.

23.2 Wormhole geometry

In cosmological or isotropic coordinates $(\tau_+, \vec{x} = (x, y, z))$ the multi-black hole solutions on a de Sitter background have the form

$$ds^2 = -\frac{d\tau_+^2}{U_+^2} + U_+^2(dx^2 + dy^2 + dz^2), \tag{23.1}$$

$$U_+(\tau_+, \vec{x}) = H\tau_+ + f(\vec{x}), \qquad H = \sqrt{\frac{\Lambda}{3}} \tag{23.2}$$

where Λ is the positive cosmological constant. The function $U_+(\tau_+, \vec{x})$ consists of a term due to the cosmological expansion and a purely spacial part $f(\vec{x})$ which we call the potential function. Together with the electromagnetic potential $A_{\tau_+} = \pm\frac{1}{U_+}$ Einstein-Maxwell field equations reduce to Laplace's equation,

$$\left(\frac{\partial^2}{\partial x^2} + \frac{\partial^2}{\partial y^2} + \frac{\partial^2}{\partial z^2}\right) f(\vec{x}) = 0. \tag{23.3}$$

The multi-black hole solutions found by Kastor and Traschen [1] and discussed for example by Brill at al. [10] are obtained by inserting the (negative of the) Newtonian potential of an arbitrary number of point masses for the function $f(\vec{x})$.

23.2.1 Geometry of charged shells

For our wormhole construction we use a special solution of Laplace's equation. We choose the potential function to be constant on two spheres S_1^+ and S_2^+ (see Fig 23.1). For simplicity we restrict ourselves to the symmetric case, i.e we center the spheres at $z = \pm d_1$ and assume that both have the same Euclidean coordinate radius, $S_i^+ : |\vec{x} \pm \vec{d_1}| = R^+$ (i=1,2). The boundary values for the potential are fixed to be

$$f(\vec{x})|_{S_i^+} = \frac{m_1}{R^+}, \tag{23.4}$$

the same on both shells S_i^+. In addition, we assume that the function $f(\vec{x})$ tends to zero for $|\vec{x}| \to \infty$. This choice ensures that for large distances of the two spheres the field is that of two particles with mass equals charge m_1.

In order to solve this boundary value problem we apply the image charge method. It turns out [7] that the image masses (charges) m_n have to be located on the z-axis at $z = \pm d_n$, where

$$d_n = d_1 - \frac{(R^+)^2}{d_1 + d_{n-1}} \quad (n > 1) \tag{23.5}$$

$$m_n = -\frac{m_{n-1}R^+}{d_1 + d_{n-1}} \quad (n > 1) \tag{23.6}$$

The resulting expression for the function $U(\tau_+, \vec{x})$ reads

$$U_+(\tau_+, \vec{x}) = H\tau_+ + \sum_{n=1}^{\infty} \left(\frac{m_n}{|\vec{x} + \vec{d}_n|} + \frac{m_n}{|\vec{x} - \vec{d}_n|} \right) . \tag{23.7}$$

In what follows we cut out the interior of the shells S_i^+. To calculate the induced metric on these timelike hypersurfaces we introduce spherical polar coordinate systems $(r_+, \vartheta, \varphi)$ centered either at $z = -d_1$ or at $z = d_1$. We choose a left handed spherical polar cooordinate system with polar axis pointing to the positive z-axis near shell S_1^+ but a right handed spherical polar cooordinate system with polar axis pointing to the negative z-axis near shell S_2^+.

In these coordinates the shells S_i^+ are given by $r_+ = R^+$ and the induced metric reads

$$ds^2|_{S_i^+} = -\frac{d\tau_+^2}{(U_0^+)^2} + (U_0^+ R^+)^2 (d\vartheta^2 + sin(\vartheta)^2 d\varphi^2) \tag{23.8}$$

$$\text{with} \quad U_0^+(\tau_+) := H\tau_+ + \frac{m_1}{R^+}$$

Notice that the spacial part of the metric (23.8) is proportional to the standard metric on the two sphere. This is the crucial fact which allows for matching a spherically symmetric interior region in a continuous way.

23.2.2 $M = Q$ Reissner-Nordström de Sitter black holes

Inserting the Newtonian potential of a single point mass into Eqs.(23.1,23.3), we obtain a $M = Q$ RNdS black hole. In spherical polar, cosmological coordinates $(\tau_-, r_-, \vartheta, \varphi)$ the metric reads,

$$ds^2 = -\frac{d\tau_-^2}{U_-^2} + U_-^2 \left(dr_-^2 + r_-^2 (d\vartheta^2 + sin(\vartheta)^2 d\varphi^2) \right) , \tag{23.9}$$

$$\text{with} \quad U_-(\tau_-, r_-) = H\tau_- + \frac{M}{r_-}$$

We add the suffix "-" to quantities which should not be confused with their analogues in the exterior region. On the other hand we do not distinguish angle coordinates because we want to identify points on the shells with equal values of (ϑ, φ).

Figure 23.2 shows the Penrose diagram of a mass equals charge, undermassive ($M < \frac{1}{4H}$) RNdS black holes. The complete manifold contains three different types of Killing horizons, de Sitter horizons R_{dS}, black (white) hole horizons R_{BH}, and inner (Cauchy) horizons R_{IN}. The dependence of these horizons on the parameters (M, H) $(H > 0)$ can be most conveniently expressed using the Schwarzschild-type radial coordinate R defined by

$$R := U_-(\tau_-, r_-) r_- = H\tau_- r_- + M \tag{23.10}$$

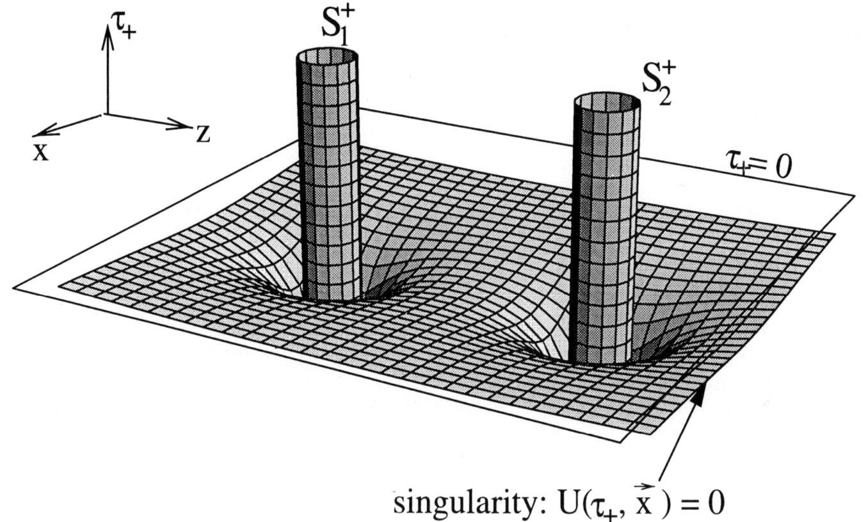

Figure 23.1: Coordinate diagram of the exterior multi-black hole region. In comological coordinates $(\tau_+, \vec{x})$ the curvature singularity lying at $U_+(\tau_+, \vec{x}) = 0$ appears as a surface that bounds spacetime to negative values of cosmological time. The spheres S_1^+ and S_2^+ emerge from the singularity and expand to future infinity $\mathcal{J}^+$.

In Section 23.4 it will serve useful to have at hand formulas for these horizons,

$$R_{IN}(M,H) = \frac{1}{2H}(\sqrt{1+4MH}-1) \tag{23.11}$$

$$R_{BH}(M,H) = \frac{1}{2H}(1-\sqrt{1-4MH}) \tag{23.12}$$

$$R_{dS}(M,H) = \frac{1}{2H}(1+\sqrt{1-4MH}). \tag{23.13}$$

The hypersurfaces S_i are chosen to be comoving with respect to the interior black hole region as well, i.e. in Figure 23.2 the hypersurface S_1^- is given by $r_- = R^-$. We cut off the region with larger coordinate radius and match it to the surface S_1^+ of the outer multi black hole spacetime, i.e. $S_1 := S_1^+ \equiv S_1^-$. We repeat this procedure in another asymptotic de Sitter region (successing in time as shown in Fig. 23.2) to obtain shell $S_2 := S_2^+ \equiv S_2^-$.

23.3 Continuity of matching and surface energy tensor

In order to make sure that we get a well defined spacetime we have to check that the induced metric on the hypersurfaces S_i can be made continuous. Calculating the induced metric on the surfaces S_i^-,

$$ds^2|_{S_i^-} = -\frac{d\tau_-^2}{(U_0^-)^2} + (U_0^- R^-)^2(d\vartheta^2 + sin(\vartheta)^2 d\varphi^2) \tag{23.14}$$

$$\text{where} \quad U_0^-(\tau_-) := H\tau_- + \frac{M}{R^-} \tag{23.15}$$

and comparing it with Eq. (23.8) leads to the equations

$$U_0^+(\tau_+)R^+ = U_0^-(\tau_-)R^- \tag{23.16}$$

$$\frac{d\tau_+}{d\tau_-} = \frac{R^-}{R^+} \tag{23.17}$$

These two relations can be satisfied since we have chosen equal values of the cosmological constant with respect to the interior and exterior region. Prescribing the mass parameters (m_1, M) and the coordinate radii $R^\pm$ Eqs. (23.16,23.17) determine the identification of events on the shells $S_i^\pm$ uniquely.

Although the induced metric can be made continuous on the shells, the extrinsic curvature K_{ab} will be discontinuous in general. (Latin letters a,b... are used for intrinsic coordinates (s, ϑ, φ) on the shells, where s denotes proper time along streamlines of constant angles (ϑ, φ).) Denoting by $e^\alpha_{(a)}$ the three basis vector fields associated to the intrinsic coordinates on the hypersurface S_i and by n^α the unit normal vector

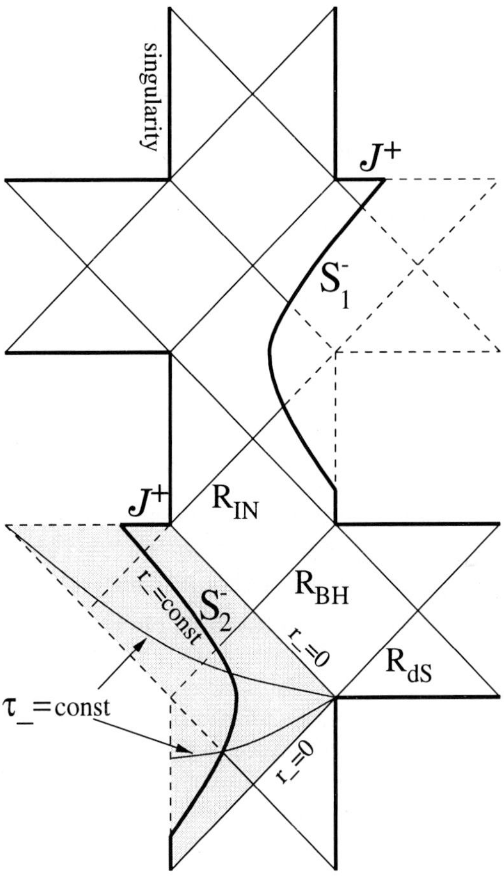

Figure 23.2: Penrose diagram of a $M = Q$, undermassive RNdS black hole. The shaded wedge covering shell S_2^- corresponds to one patch of cosmological coordinates $(\tau_-, r_-, \vartheta, \varphi)$. The shells are given by constant values of the cosmological radial coordinate, $r_- = R^-$, thus they emerge from singularities and expand to $\mathcal{J}^+$. We cut off the regions to the right of shell S_1^- and to the left of shell S_2^- and match them to the exterior region. Also the de Sitter horizons R_{dS}, black hole horizons R_{BH}, and inner (Cauchy) horizons R_{IN} are indicated.

pointing from the black hole region to the exterior region, the extrinsic curvature can be calculated by $K_{ab} := -n_\alpha e^\beta_{(b)} \nabla_\beta e^\alpha_{(a)}$. The standard thin shell formalism links the jump in the extrinsic curvature, $[K_{ab}]^\pm_{S_i} := K_{ab}|_{S_i^+} - K_{ab}|_{S_i^-}$, to the distributional stress energy tensors S_{ab} produced by the matching. For the considered models S_{ab} turns out to be diagonal: the energy density $\sigma(\vartheta)$ is angle-depend but surface pressures vanish identically,

$$\sigma = -\frac{1}{4\pi}[K^\vartheta{}_\vartheta]_{S_i} \tag{23.18}$$

$$= \frac{1}{4\pi(U_0^+)^2}\left(-\frac{\partial U_+}{\partial r_+}|_{r_+=R^+} - \frac{M}{(R^+)^2}\right) \tag{23.19}$$

$$p = \frac{1}{8\pi}\left([K^s{}_s] + [K^\vartheta{}_\vartheta]\right)_{S_i} \equiv 0. \tag{23.20}$$

This result allows the interpretation that the surfaces layers comprise of dust. It is convenient to define the "effective" mass m_1^{eff} of the shells S_i by

$$m_1^{eff}(\vartheta) := -(R^+)^2 \frac{\partial U}{\partial r_+}|_{S_i}, \tag{23.21}$$

being proportional to the normal derivative of the potential function on the shells. The condition of positivity of the energy density reads $m_1^{eff} > M$. For large coordinate separations of the shells, $\frac{R_+}{2d_1} < \frac{1}{3}$, it can be shown explicitly [7] that the effective mass m_1^{eff} is strictly positive. A stronger result can be obtained by applying the classical maximum principles developed for linear elliptic partial differential operators [11]. For our special symmetric boundary value problem the weak maximum principle tells us that the maximum of the potential function $f(\vec{x})$ is given by its value on the shells. The strong maximum principle enables us to prove that the outward pointing normal derivative on the shells (pointing to smaller radial coordinate values) is strictly positive. As a consequence the effective mass is strictly positive for any values of the separation $2d_1$, radius R^+, and angle ϑ on the shells.

By (independently) diminishing the value of the mass parameter M of the RNdS black hole we can ensure that the surface energy density $\sigma(\vartheta)$ is positive. Because pressures vanish identically all energy conditions will be satisfied.

23.4 Causal structure

In this note we solely discuss the causal structure of wormhole models obtained by cutting the extended RNdS back hole in a diagonal way (see Fig. 23.2). Furthermore, we restrict the discussion to the part of parameter space $(d_1, R^\pm, M, m_1)$ such that the energy density σ on the surface layers is positive. The cosmological constant has to be chosen such that the black hole is undermassive, i.e. $M < M_{ext} := \frac{1}{4H}$, in order to obtain the presumed topology.

As mentioned in the Introduction the considerations are guided by two issues: Are there causal curves originating in the asymptotic de Sitter region, running through the back hole, returning to the asymptotic de Sitter region and escaping to future infinity? We will define below which part of spacetime we want to denote by "asymptotic de Sitter region." Guided by the terminology of Friedman, Schleich and Witt [9] we call the topology to be "actively probable" whenever this property is satisfied.

Secondly, we ask whether and in what regions of spacetime there exist closed timelike curves.

In order to answer these questions we first make a few remarks: Far away from the shells S_i^+ the metric (23.1) tends to that of a spherically symmetric RNdS black hole with mass M^{tot}. M^{tot} is given by the sum of all image masses, $M^{tot} = 2\sum_{n=1}^{\infty} m_n$, and is finite for any value $\frac{R^+}{d_1} < 1$. If $M^{tot} < M_{ext}$ one expects that the region of spacetime described by Eq.(23.1) is incomplete, even away from the curvature singularity.

Brill at al. [10] have established this incompleteness and the existence of a de Sitter horizon for multi-black hole solutions. They have also shown that the horizon has only finite differentiability and spacetime cannot be analytically continued. Nevertheless, they prove that one can match essentially any solution of the form (23.1) with the same value of the cosmological constant and the same total mass.

In the case that $M^{tot} < M_{ext}$ we conclude that the considered wormhole spacetimes contain a past de Sitter horizon which corresponds to $(\tau_+, |\vec{x}|) \to (0, \infty)$. Spacetime can be extended in such a way that there exists a past infinity $\mathcal{J}^-$. Therefore causal curves which can reach the past de Sitter horizon can be extended to past infinity $\mathcal{J}^-$.

The boundary of events that can be causally connected to the past de Sitter horizon (disregarding topological identifications), may be denoted as the (past) white hole horizon. The region outside the white hole horizon extended across the shells S_1 and S_2 up to the black (and white) hole horizon of the inner RNdS region (see Fig. 23.2) will be called the "asymptotic de Sitter region".

23.4.1 Actively probable topology

We prove that for sufficiently small values of Λ the topology of spacetime is actively probable in the defined sense. We first establish the existence of past directed null geodesics starting at the outer pole of shell S_2, $z^I = d_1 + R^+$, which run outwards along the axis of symmetry (z-axis) but do not hit the singularity.

Because of symmetry the worldline of a null geodesic running along the z-axis can be obtained by integrating the null condition

$$\frac{d\tau_+}{dz} = -(H\tau_+ + f(\vec{x}))^2. \tag{23.22}$$

We estimate solutions of Eq.(23.22) by giving a lower bound for $\frac{d\tau_+}{dz}$,

$$\frac{d\tau_+}{dz} \geq -(H\tau_+ + \frac{M^{pos}}{(z-d_1)})^2. \tag{23.23}$$

The second term on the right hand side is obtained from (23.7) by moving all positive image masses to the center of shell S_2, $z = d_1$, and dropping all negative image masses. (Notice that the sum M^{pos} of all positive masses is still finite for $\frac{R^+}{d_1} < 1$.)

For equality in (23.23) the equation can be integrated explicitly. The solution is the same as for radial null geodesics in the spherically symmetric RNdS black hole. It is obvious that for $H < \frac{1}{4M^{pos}}$ there exist null geodesics running along the z-axis which do not hit the singularity. Therefore we have to choose the initial value τ_+^I on shell S_2^+ sufficiently late such that $R_{S_2}^{pos}(\tau_+) := (H\tau_+ R^+ + M^{pos}) > R_{BH}(M^{pos}, H)$. Taking into account that the physical radius of shell S_2 satisfies $R_{S_2}(\tau_+) := U_0^+(\tau_+)R^+ < R_{S_2}^{pos}(\tau_+)$ and having in mind the following line of inequalities,

$$R_{BH}(M^{pos}, H) < R_{dS}(M^{pos}, H) < R_{dS}(M, H), \tag{23.24}$$

we see that at the same time it is possible to choose the initial point on shell S_2 inside the future de Sitter horizon $R_{dS}(M, H)$ of the RNdS black hole. As a consequence, such a curve can be extended to the future by a causal curve traversing the black hole region and finally reaching future infinity $\mathcal{J}^+$.

23.4.2 Wormholes containing closed timelike curves

Constructing the wormhole by pasting the interior region as shown in Fig. 23.2 reduces the problem of the existence of CTCs to a question concerning only the exterior region: Is there a causal curve starting from the inner pole of shell S_1 at a physical radius $R_{S_1}(\tau_+) = U_0^+(\tau_+)R^+$ larger than $R_{in}(M, H)$, traversing the exterior region along the axis of symmetry (z-axis) and reaching shell S_2 at a radius $R_{S_2}(\tau_+) = U_0^+(\tau_+)R^+$ smaller than the de Sitter horizon $R_{dS}(M, H)$? This condition is sufficient for the existence of CTCs. (Note that it is important here how we have chosen the angle coordinates and done the identification of the hypersurfaces $S_i^{\pm}$.)

We can answer the above question affirmativly by studying the null geodesic starting at shell S_1 at cosmological time $\tau_+ = 0$. With this choice of the initial time we ensure that the radius $R_{S_1}(\tau_+ = 0) = m_1$ is larger than the inner horizon $R_{in} \leq M$. (Notice that assuming the energy density σ to be positive enforces $m_1 > M$.)

Intuitively, it is clear that the larger the coordinate distance of the shells is chosen the slower the expansion of the universe has to be to allow for causal curves reaching shell S_2 from a given starting point on S_1. Hence, in the first step we prove that for sufficient small values of the constant H the two shells are in causal contact at early times.

We estimate the required null geodesic by calculating an upper limit curve $\tau^{up}(z)$. Therefore we insert into to null condition the absolute maximum value of the potential function which is attained on the shells, $f_{max} = \frac{m_1}{R^+}$,

$$\frac{d\tau_+^{up}}{dz} = (H\tau_+ + f_{max})^2 \tag{23.25}$$

Notice that the estimate $U_+(\tau_+, \vec{x})^2 \leq (H\tau_+ + f_{max})^2$ is valid at least for positive cosmological times. The limit curve reads

$$\tau_+^{up}(z) = \frac{1}{H^2(z_0 - z)} - \frac{1}{H} f_{max} \quad \text{where} \quad z_0 = -d_1 + R^+ + \frac{1}{H f_{max}}. \tag{23.26}$$

Here the pole z_0 has been determined by the initial values $(\tau_+^I = 0, z_I = -d_1 + R^+)$. Now we see that by diminishing the value of H we can move the pole to larger positive values on the z-axis, especially beyond the location of the second shell. Therefore it is possible to choose H sufficiently small such that the limit curve $\tau_+^{up}(z)$ and consequently the studied null geodesic remain at finite values of cosmological time on the coordinate interval $z \in [-d_1 + R^+, d_1 - R^+]$.

In the second step we have to answer the question at what radius of shell S_2 the studied null geodesic arrives. Having at hand the upper limit curve we are able to give upper bounds of the time of arrival τ_+^{ar} of the null geodesic and consequently of the radius $R_{S_2}(\tau_+^{ar})$ of shell S_2^+,

$$R_{S_2}(\tau_+^{ar}) < \frac{m_1}{(1 - 2Hm_1(\frac{d_1}{R^+} - 1))}. \tag{23.27}$$

From this we see that the physical radius of shell S_2 at the time of arrival gets smaller by diminishing H and tends to m_1 when approaching the limit $H \to 0$. On the other hand by diminishing H the radius of the de Sitter horizon $R_{dS}(M, H)$ increases and even tends to infinity (23.11). Hence, also the requirement $R_{S_2}(\tau_{ar}) < R_{dS}$ can be contented. This concludes our proof of the possible existence of CTCs.

Past and future chronology horizons

We have proven above that for small H there exist CTCs in the neighborhood of the shells S_1 and S_2 at early times.

At late cosmological times, whenever the radius of shell S_2 is larger than the $R_{dS}(M, H)$ and causal curves crossing the shell cannot traverse the wormhole anymore, CTCs cease to exist through points of the asymptotic de Sitter region. On the other hand no (locally) future directed causal curve starting at shell S_1 can reach past infinity $\mathcal{J}^-$ of the extended spacetime. As a consequence, there cannot exist CTCs in the neighborhood of $\mathcal{J}^-$.

Hence, there appear past and future chronology horizons which separate the regions containing CTCs from the regions without CTCs. Especially, there are no CTCs in the neighborhood of $\mathcal{J}^+$ and $\mathcal{J}^-$. This is an important difference to the wormhole models on an asymptotically flat background discussed in [7].

23.5 Questionary

Are the considered wormhole spacetimes stable against the accretion of infalling matter?

In 1989, Brady and Poisson [13] studied the effect of a radial flux onto charged black holes immersed in de Sitter space. From their results we conclude that in the $M = Q$ case the energy density of ingoing radiation as measured by free-falling observers approaching the inner horizon inevitably will grow without bound. But there is a way out of the problem of mass inflation: by fine tuning the mass and charge parameters and the cosmological constant one can achieve that the surface gravity κ_{IN} of the inner horizon is less than two times the surface gravity κ_{dS} of the de Sitter horizon, $\kappa_{IN} < 2\kappa_{dS}$. Consequently, no mass inflation occurs. This is the case for $M = Q$ RNdS black holes and for sufficiently small values of the cosmological constant.

On the other hand Brady, Núñez and Sinha [14] have shown that especially in the $M = Q$ case there appears a curvature singularity even without mass inflation. However, we do not know if this ultimately excludes the possibility to use charged black holes as wormholes.

All these results come from the analysis of spherically symmetric black holes. Whether they also apply to the axisymmetric case, such as ours, is unknown.

Acknowledgements

It is a great pleasure to thank the organizers for the invitation and kind hospitality during the workshop. The author is also grateful to P.C. Aichelburg, P.T. Chruściel, and C. Chambers for fruitful discussions and helpful comments. This work was supported by Fundacion Federico.

Bibliography

[1] D.Kastor and J. Traschen, Phys. Rev. D47, 12, 5370 (1993)

[2] B. Carter, Phys. Lett. 21, 4, 423 (1966)

[3] I. Novikov, JETP Lett.3, 142 (1966)

[4] S.W. Hawking and G.F.R. Ellis, The large scale structure of space-time, Cambridge University Press 1973

[5] M. Morris and K. Thorne, Am. J. Phys. 56, 395 (1988)

[6] V.P. Frolov and I.D. Novikov, Phys. Rev. D 42, 4, 1057 (1990)

[7] F. Schein and P.C. Aichelburg, Phys. Rev. Lett. 77, 20, 4130 (1996)

[8] C. Barrabès and W. Israel, Phys. Rev. D43, 4, 1129 (1991)

[9] J.L. Friedman, K. Schleich, D.M. Witt, Phys. Rev. Lett. 71, 1486 (1993)

[10] D. Brill, Horowitz, Kastor and J. Traschen, Phys Rev D49, 2, 840 (1994)

[11] D. Gilbarg and N.N. Trudinger, Elliptic Partial Differential Equations of Second Order, Springer Verlag (1977)

[12] E. Poisson and W. Israel, Phys. Rev. Lett. 63, 16, 1663 (1989); Phys. Rev. D41, 6, 1796 (1990)

[13] P. Brady and E. Poisson, Class. Quantum Grav. 9, 121 (1992)

[14] P. Brady, D. Núñez and S. Sinha, Phys. Rev. D47, 4239 (1993)

ON THE TRAVERSABILITY OF THE CAUCHY HORIZON: HERMAN AND HISCOCK'S ARGUMENT REVISITED

Amos Ori

Department of Physics, Technion—Israel Institute of Technology, 32000 Haifa, Israel

Abstract

It has been argued by Herman and Hiscock that even if geometry extends classically beyond the Cauchy horizon (CH), any physical object traversing the CH will be destroyed due to unbounded increase of internal kinetic energy and entropy. In this article we re-examine this issue. We first consider the situation up to the CH and find that no such phenomenon occurs on the approach to the CH. Then we discuss the applicability of the above argument to the region beyond the CH. For a C^0 extension of the metric tensor, there exists a unique C^1 extension of timelike geodesics beyond the CH. If one assumes that the physical geodesic motion proceeds along this C^1 extension, then no catastrophic growth of internal kinetic energy and entropy occurs beyond the CH either.

24.1 Introduction

In the last decade the issue of the inner structure of realistic black holes (BHs) underwent interesting developments which can almost be described as a "phase transition": Until about a decade ago, the main challenge was to predict the evolution of geometry *up to* the singularity (based on the classical Einstein theory). At present, although many details and subtle points are still to be fixed, it appears that to a large extent this goal has already been achieved, and the (classical) inner structure of realistic spinning black holes has been uncovered up to the curvature singularity at the Cauchy horizon (CH) [1]. The research has now arrived at a new phase, in which the main challenge is to try understand how physics extends *beyond* the singularity. This is obviously a much more difficult goal.

Without going into any details, we can generally divide the various options into three categories:

1. Physics just "ends" at the CH singularity;

2. In principle physics carries on, but the classical notion of geometry does not: Instead, it is replaced by the notion of "quantum geometry" (whatever it means);
3. Classical geometry extends beyond the CH singularity. In this last option, the weak classical curvature singularity at the CH (together with the quantum effects which possibly take place there) functions as a thin transition layer between two patches of classical spacetimes – pretty much like a shock wave in fluid dynamics.

At present it is extremely hard to make a progress in this problem, due to the combination of two facts: On the one hand, the classical Einstein theory loses its predictive power at the CH (mathematically there are infinite C^0 extensions of the metric at the CH). On the other hand, the quantum theory of gravity, which is believed to provide the answer to the above question of extension, has not been formulated yet.

The analysis of the evolution of classical geometry *up to* the CH revealed that the curvature singularity is weak and non-destructive [2]; Namely, the tidal distortion experienced by extended physical objects is finite (and, typically, extremely small) as the CH is approached. It has been suggested that at the semiclassical level quantum effects will be extremely strong at the singularity and will cause the ultimate destruction of any macroscopic object. Detailed calculations on various simplified models [see e.g. [3, 4]] indeed indicated that formally such an unbounded growth of tidal distortion is to be anticipated. However, these calculations also show that this large deformation only takes place if the semiclassical calculation is extrapolated far beyond its domain of validity; Namely, whenever the calculation is restricted to the region of sub-Planckian curvature, no significant increase in the tidal distortion is obtained. [This situation is to be contrasted with another important phenomenon involving semiclassical gravity – the evaporation of black holes. The major part of the evaporation process (for macroscopic BHs) occurs already at the stage where the curvature at the horizon is much smaller than Planckian.] Thus, the semiclassical considerations do not provide a strong evidence in favor of any of the possibilities mentioned above. (Though, the semiclassical effects do provide a clue that interesting quantum effects may take place at the CH. This is not surprising, of course, because the pure classical theory anyway loses its predictive power there.)

In this article I would like to discuss another argument, due to Herman and Hiscock (HH) [5], concerning the issue of traversability of the CH. HH employed simple entropy considerations to argue that macroscopic physical objects will not be able to survive the travel through the CH. According to HH, the divergence of the rate of deformation means that the entropy of the macroscopic object diverges, which in turn implies that the infalling object will necessarily be "melted" (or "evaporated") while traversing the CH. This argument was raised several times in the discussions during the workshop (see also [6]), and I therefore found it useful to give a careful consideration to it.

Let me first briefly explain Herman and Hiscock's argument: Consider an infalling extended object which moves toward the singularity at the CH. For simplicity, let us model this extended object by two particles connected by a string. Let us define the *"internal velocity"* as the rate of change of the proper distance with proper time. This "internal velocity" is known to diverge at the CH singularity [7]. According to HH, this implies a divergence of the object's *temperature*, *kinetic energy*, and *entropy*. In turn, this will cause a total destruction of the body, and also the lose of all information concerning its initial state.

From the paper by Herman and Hiscock it is not completely clear whether the anticipated loss of information should occur on the approach to the CH (in concert with the growth of the "internal velocity"), or immediately after the CH was crossed. In the next section I will discuss the validity of this argument to the first stage, i.e. the approach to the CH. I will argue that no such growth of kinetic energy, entropy, etc. occurs at this stage. Then, in Section 24.3 I will discuss the applicability of this argument to the next stage, i.e. after crossing the CH. The analysis of this stage is more speculative, as we do not yet have the fundamental theory for describing the physics in this region. I will therefore base the discussion on two assumptions:

1. Classical geometry extends continuously (i.e. C^0 metric) beyond the CH, and the geometry extension is prescribed (this working assumption was also made by HH; without this assumption the whole discussion is meaningless);

2. The physical geodesics extend through the CH as smooth as possible. Mathematically, we find that a timelike geodesic approaching the singular CH has a unique C^1 extension, and we shall thus make the assumption that the physical geodesic orbits will indeed proceed along these C^1-extended geodesics. Making this assumption, we find that the catastrophic effects predicted by HH do not occur beyond the CH either.

Applying the Raychaudhuri equation to a congruence of such C^1-extended geodesics, one finds that the singular CH is endowed with infinite negative energy density. In Section 24.4 I will briefly discuss this issue, and argue that this is not as pathological or unexpected as one might naively think. Finally, in Section 24.5 I summarize the results.

24.2 Before the CH

The validity of the argument by HH can be considered from two different points of view: (i) phenomenological considerations, and (ii) more fundamental considerations, based on concrete (even though simplified) thought experiments. Although the second approach is much more efficient and decisive, I would like to start the discussion from the first point of view, and to discuss the applicability of the standard flat-space terminology of thermodynamics to the situation near the CH singularity: One can easily arrive at erroneous conclusions by inappropriate usage of standard terminology out of its domain of validity. Specifically, I would like to make the following points:

1. The notion of "internal velocity", and its usage as a measure of kinetic energy, entropy, temperature, etc, must be handled very carefully if spacetime is curved. In flat space, the rate of change of internal distances is an appropriate measure of internal velocities. In curved spacetime, however, this naive definition can only be applied to spatial regions whose diameter is small compared to the radius of curvature. Otherwise, one inevitably faces the issue of how to compare two four-velocity vectors is two remote points. (The procedure of parallel transport generally yields ambiguous results, because it depends on the choice of orbits for the transport.) The radius of curvature goes to zero at the singular CH. The rate of change of proper distances inside a (steadily decreasing) sphere whose diameter is bounded by the radius of curvature can be shown to *vanish* on the approach to the singularity. It is therefore clear that in our case the naive association of the notions of internal velocities, entropy, temperature, etc. to the rate of change of proper distance would be false [8].

2. The association of entropy (and hence "loss of information") to this "internal velocity" (as defined above) is incorrect from yet another reason: Entropy is a measure of the *disordered* motion inside the matter. Here we are dealing with a *coherent* "motion", caused by the coherent tidal force. No loss of information is associated with such a coherent motion!

3. The notion of "internal kinetic energy" may be defined in an unambiguous way through the energy-momentum tensor associated with the extended object. The latter is a well-defined covariant notion. Clearly, the "internal kinetic energy" is not larger than the total mass-energy density, as measured by a comoving observer. This entity is finite at the CH. To discuss this issue, let us model the extended object as made of a collection of point-like particles, which move along almost-parallel timelike geodesics, with some initial thermal dispersion velocities, on the background geometry of the CH singularity. In this model, the calculation of the energy-density (for a comoving observer) is reduced to evaluating the scalar product of two four-velocity vectors at an intersection point of two geodesics (e.g. the "center of mass" geodesic and another one with some thermal dispersion velocity). A straightforward calculation (e.g. for the mass-inflation model) shows that this scalar product is finite, and is rather insensitive to the growth of curvature (and rate of deformation) at the CH. This implies that the internal kinetic energy of the extended object does not diverge (or significantly grow) in any invariant sense.

However, the most effective way to understand the internal features of a physical object is by carrying out basic thought experiments, and not by semantic manipulations. Consider, first, an infalling piece of solid matter, which falls into the BH at late time ($v \gg M$) and moves toward the early portion of the CH ($|u| \gg M$); Assume, for example, that this object is a monatomic crystal, initially at reasonably-low temperature (i.e. $T \ll T_{melting}$). Question: would the crystal melt on the approach to the

CH singularity? The answer is definitely NO: The deformation in the inter-atomic distance will be of the typical order of magnitude $(v/M)^{-n}$ [9] (with e.g. $n = 6$). It is therefore obvious that by the time the CH is approached each atom still sits in its original site (with respect to the other atoms), so no melting has occurred. For the same reason, if any information is etched on the surface of the solid crystal, it will still be there when the CH is approached. So, no loss of information has occurred either.

Herman and Hiscock also raised the issue of the *dissipation* of the internal kinetic energy, and the contribution of this phenomenon to the growth of entropy. They argued that this dissipation will cause the divergence of entropy. A priori it is not obvious what will be the strength of the dissipation effects, especially because of the extremely short period of time on which the "internal velocity" (i.e. the rate of contraction) is large. Thus, this issue cannot be decided without explicit calculations. No such calculations are mentioned in Ref. [5]. It would be dangerous to use the standard phenomenological formalism of, e.g., viscosity to evaluate the dissipation, because of the extreme conditions at the CH singularity: First, the smallness of radius of curvature, and second, the rapid changes in "internal velocity". The more safe way to evaluate the effect of dissipation is, again, by considering simple thought experiments.

Consider, for example, the dissipation in a gas of atoms (e.g. an ideal gas). What will be the amount of dissipation caused by the large "internal velocity" field? In order to avoid the need in detailed complicated calculations (like trying to struggling with the Boltzmann equation on a curved mass-inflation background), let us design a simple thought experiment: Consider a congruence of timelike geodesics γ_n which represents the bulk motion of the extended object. Consider now another timelike geodesic λ which is located inside the world-tube and is almost (but not exactly) parallel to the congruence. The deviation in the direction of λ from that of γ_n represents the thermal dispersion velocity, and it may be used as a measure for the true intrinsic internal velocity and temperature. By evolving the system (using the geodesic equation) we can estimate how the temperature evolves on the approach to the CH. In addition, by evaluating the energy transfer in collisions between the particle λ and the particles of the congruence, one can evaluate the effect of dissipation on the growth of internal energy (and entropy). (Recall that in ordinary hydrodynamics, too, the dissipative effects like viscosity may be expressed in terms of momentum transfer in collisions.)

More specifically, the calculation proceeds as follows: Assume that at a moment τ_1 the particle λ intersects a timelike geodesic γ_1, and at a later moment τ_2 it intersects another timelike geodesic γ_2. (For concreteness we use here the proper-time τ along the geodesic λ as a measure of the system's time. Note that since λ and γ_n are almost parallel, their proper times are approximately the same.) The energy of the particle λ, as measured by a comoving observer (following a congruence geodesic γ) is $\mu(u_\lambda \cdot u_\gamma)$, where μ is the particles' rest mass, and the dot denotes the scalar product of the two four-velocity vectors. [The signature is $(+ - --)$.] Hence, the particle's kinetic energy is $\mu(u_\lambda \cdot u_\gamma - 1)$. The temperature of the ideal gas is proportional to the kinetic energy per particle. Let T_1 and T_2 denote the temperatures at the two moments, τ_1 and τ_2, respectively. The ratio of the two temperatures is

$$\frac{T_2}{T_1} = \frac{u_\lambda \cdot u_2 - 1}{u_\lambda \cdot u_1 - 1} , \tag{24.1}$$

where u_1 and u_2 denote the four-velocities of the geodesics γ_1 and γ_2, respectively. (Of course, the term in the numerator is evaluated at time τ_2, and that at the denominator is evaluated at τ_1, so the scalar product always involves two vectors at the same spacetime event.) Now, we take the initial moment τ_1 to be just before the curvature starts to rise near the CH (but nevertheless in a short proper distance from the CH). Then, we shall take τ_2 to be at the CH singularity itself. (Alternatively, we could consider the *limit* in which τ_2 approaches the CH. This is not needed, however, because when regular coordinates are used, all the entities in the above equation turn out to be well-defined at the CH singularity.) The dimensionless entity

$$H \equiv (T_2/T_1) - 1 \tag{24.2}$$

represents (in a true local sense) the amount of heating due to the increase of internal kinetic energy.

It is straightforward to calculate H from the geodesic equation, once the background geometry is prescribed. In a calculation which I carried out some time ago [10], I used Hiscock's model [11] of the charged Vaidya (CV) solution as a background geometry. (As HH point out, this model should properly capture the essence of the problem, and should give the same qualitative results as the more realistic models of the CH singularity.) To simplify the analysis as much as possible, I took γ_n to be a congruence of radial timelike geodesics, and the geodesic λ was taken to be slightly non-radial. As explained above, I took τ_2 to be just at the CH, and τ_1 such that $\tau_2 - \tau_1 \ll M$. This also implied $r_2 - r_1 \ll r_2 \propto M$, where r represents the area coordinate. A straightforward calculation then yields

$$\frac{T_2}{T_1} \cong (r_1/r_2)^2 , \tag{24.3}$$

and hence

$$H \propto (r_1 - r_2)/r_2 \ll 1 . \tag{24.4}$$

Thus, the heating H is extremely small, of the same order as the tidal deformation [the latter can be estimated to be of order $(v/M)^{-n}$, presumably with $n = 6$]. It should be pointed out that H represents the total heating, includes both the adiabatic and the dissipative heating.

We conclude that even when the dissipative effect of viscosity is taken into account, no significant heat production occurs up to the CH. Intuitively, this should be understood as a result of two factors, which mark the fundamental difference between our problem and standard flat-space hydrodynamics:

1. Although the rate of change of distances diverges, the relative velocity calculated by parallelly-transporting the four-velocity vector from one geodesic of the congruence to another one (along λ) is negligibly small (this drastic difference between the two notions of "relative velocity" is only possible because of the extreme curvature).

2. The stage of large deformation rate is extremely short.

24.3 Beyond the CH

So far we focused attention on the internal processes which may take place on the approach to the CH. One can raise another question, however: What will happen to the infalling extended object immediately after it crosses the CH? In the rest of this article we shall discuss this issue.

Obviously, when considering the evolution of any physical system beyond the CH, we inevitably run into a serious conceptual difficulty: We do not yet have the fundamental theory for describing the physics beyond the CH. We must realize, therefore, that any statement concerning the physics beyond the CH is merely a speculation. Nevertheless, we may try assess the level of *likelihood* of various possibilities.

If geometry itself does not extend classically beyond the CH, then the entire notion of extended object becomes meaningless. Therefore, for the present discussion we must make the working assumption (like HH) that classical geometry does extend beyond the CH. The spacetime is thus made of two patches, P_1 (before the CH) and P_2 (beyond the CH), matched together at the CH (see Fig. 24.3). We denote by P the union of P_1 and P_2 (P is not necessarily the entire spacetime – in fact, we shall view P as a piece of the overall spacetime which includes the neighborhood of a portion of the CH. The consideration of this neighborhood is sufficient for the present discussion). Each of the two patches P_1 and P_2 is smooth, except at the boundary (the CH) where the curvature diverges (at least in P_1). It is well known that the CH singularity is characterized by a well-defined continuous (C^0) metric, even at the CH singularity itself [12, 2]. Motivated by this, we shall demand that throughout P the geometry is C^0 in terms of the metric functions; Namely, coordinates x^μ exist such that the metric tensor is

1. smooth inside both P_1 and P_2;

2. continuous at the CH;

3. nowhere degenerate.

Whereas the geometry in P_1 is uniquely determined by the field equations, the geometry of P_2 is not. In fact, there are infinite number of possible C^0 extensions P_2 (this is the case even when the field equations for P_2 are prescribed). For the sake of the present discussion, we shall assume that the geometry in P_2 is prescribed.

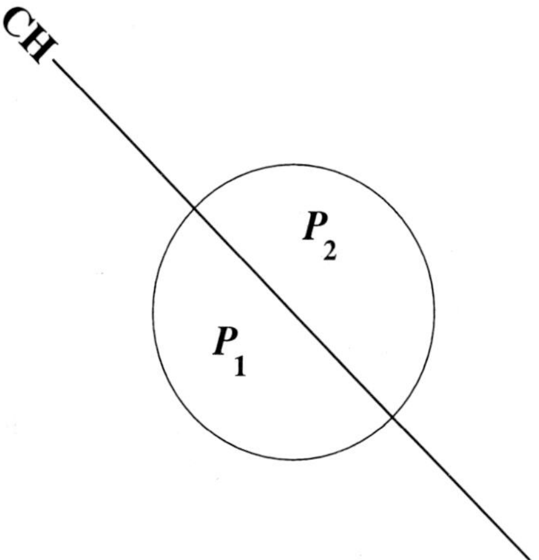

Figure 24.1: The spacetime in the vicinity of the CH: The patch P_1 before the CH, and the patch P_2 beyond the CH. The union of P_1 and P_2 is P (including the thin transition layer, namely, the CH).

As a simple example, let us assume that P_1 is the same CV solution considered above, with a curvature singularity at the CH (this is also the example considered by HH). Then, we can take P_2 to be a piece of the Reissner-Nordström (RN) geometry. The continuity of the metric implies the continuity of the area coordinate r at the CH. We shall further demand that the affine parameter Λ along the CH is the same at both sides (it is not completely obvious to me whether this requirement is dictated by physical considerations, but anyway we have enough freedom to make this assumption). Also, whenever dealing with spherical electrovac geometries, we shall demand $Q_2 = Q_1$, which amounts to continuity of the electric field at the CH. The parameters of the RN geometry at P_2 are then uniquely determined by the matching conditions: Namely, $Q_2 = Q_1$, and the continuity of r at the CH implies $M_2 = M_1$, where M_1 is the final mass of the CV geometry in P_1 (i.e. the value of the mass function at the CH).

[This example can be immediately generalized in several ways: First, still for a CV geometry in P_1, we can take P_2 to be any other CV geometry, provided that $Q_2 = Q_1$ and $M_2 = M_1$, where M_2 is now the value of the mass function of P_2 at the CH. Second, we can take P_1 and P_2 to be mass-inflation [13] geometries. Then the matching will be uniquely determined by (i) the charge Q_1 and the initial mass parameter of P_1, which we denote m_1 (i.e. the parameter m_1 in Fig. 3 of Ref. [13]); (ii) the function $L_{out}(v)_1$, describing the outflux rate in P_1; (iii) the two functions $L_{in}(v)_{1,2}$, describing the influx rate in P_1 and P_2. These entities, combined with the above continuity conditions (i.e. the continuity of r, Λ, and the electric field), will completely determine the geometry in both patches. Third, we can take P_1 to be the interior of a generic vacuum spinning BH. Then, we shall naturally take P_2 to be a vacuum geometry. The discussion below holds for all these cases.]

After we have specified the geometry throughout P, we are in a position to consider the evolution of an extended object which moves in P and crosses the CH. As before, we model the extended object as a collection of point-like particles moving along timelike geodesics. (As before, it is assumed that in view of the extremely strong tidal force near the CH, the internal forces are unimportant.) The physical problem is thus reduced to understanding the evolution of a congruence of timelike geodesics in P. In principle (and in more usual spacetimes), such a setup uniquely determines the evolution of the object's shape (and internal state) from the geometry solely (i.e. by solving the geodesic equation). In our case, however, in order to predict the evolution we must first answer the following question: *How do the geodesics extend beyond the CH?* This question is not trivial, because the connections $\Gamma^{\mu}_{\nu\eta}$, which appear in the geodesic equation, diverge at the CH singularity (the metric tensor fails to be C^1 there). Here we face the same conceptual difficulty mentioned above: Since we do not have the fundamental theory for the physics in region P_2, we cannot be sure about the "right" extension. We can, however, try to guess what might be the reasonable (or "natural") possibilities.

The essence of the geodesic equation is that, the free-fall orbits should be as "straight" as allowed by the curved background spacetime. We may also interpret this as a demand for maximal possible smoothness of the free-fall orbits. This motivates the suggestion that the most natural extension of geodesics through the CH is the one with the maximal level of smoothness. One is therefore led to the following question: What is the maximal level of smoothness allowed for geodesics in P at the CH singularity? Clearly, this level of smoothness will be determined (or at least restricted) by the behavior of geodesics in P_1 as they approach the CH singularity. It is possible to show that the timelike geodesics in P_1 are C^1 at their intersection with the CH (that is, the functions $x^\alpha(\tau)$ are C^1, which also implies that the four-velocity $u^\alpha(\tau) \equiv dx^\alpha/d\tau$ is C^0 there) [14]. (Generically, however, geodesics fail to be C^2 at the CH singularity.)

This observation raises the following questions:

1. Can any timelike geodesic intersecting the CH be extended into P_2 in a C^1 manner?

2. Is this C^1 extension into P_2 unique?

I carried out an analysis [15] of the geodesic equation in the case where both P_1 and P_2 are the non-symmetric local vacuum geometries constructed by Flanagan and myself [16]. The conclusions are the following:

1. Any timelike geodesic in P_1 that intersects the CH singularity is C^1 there;

2. Any such geodesic has a *unique* C^1 extension into P_2.

(Although the full rigorous analysis was carried out only for the spacetimes of Ref. [16], the same conclusions seem to hold also in the other cases mentioned in [14].)

These observations have an obvious interpretation: There is a C^0 metric tensor [with $det(g) \neq 0$] throughout P. Therefore, there is a well-defined tangent space at each point – even at the CH. Thus, if a timelike geodesic arrives at a point p on the CH, it has a well-defined tangent vector u^μ there. And naturally, a well-defined tangent vector u^μ at a point p uniquely defines a future-directed geodesic. From the calculational point of view, this result should be understood as follows. The metric tensor fails to be C^1 at the CH; that is, for any coordinates x^μ which yield a continuous non-degenerate metric tensor throughout the CH, first derivatives of some of the metric functions will diverge at the CH. As a consequence, the connections $\Gamma^\mu_{\nu\eta}$ diverge at the CH. By virtue of the geodesic equation, this implies that $\ddot{x}^\mu$ will blow up there in a similar manner, where the overdot denotes a proper-time derivative. Now, the divergence of the connections (which is proportional to the first-order derivatives of the metric functions with respect to the distance from the CH) disappears upon integration across the CH. In the same way, a proper-time integration removes the divergence of $\ddot{x}^\mu$, and yields finite and well-defined four-velocity components $\dot{x}^\mu$.

The above observation immediately suggests that the natural extension of geodesics throughout the CH is simply the unique C^1 extension! In what follows we shall *assume* that this in indeed the physical extension of geodesics.

It is not difficult to show that with this extension, the rate of deformation (which was infinite at the CH), attains finite values immediately after the CH was crossed. For simplicity, consider a radial congruence of timelike geodesics in one of the above spherically-symmetric constructions of P (e.g. a CV geometry in P_1 and a RN geometry in P_2). Then, the metric in P can be expressed as $g_{uv}(u,v)dudv - r^2(u,v)d\Omega^2$, where r and g_{uv} are both continuous and nonvanishing throughout P. At the CH (which we take to be at $v = 0$), $r_{,v}$ diverges; however, throughout P_2 (like P_1) both r and g_{uv} are smooth. Consider now two neighboring radial geodesics of the congruence which hit $v = 0$ at the same u, and at $\theta = \pi/2$ (i.e. the equatorial plane of the spherical coordinate system), but with a slight (infinitesimal) difference $\Delta\varphi$ in the azimuthal coordinate φ. The C^1 character of the geodesics means that at $v = 0$ the geodesics have well-defined values of $du/d\tau$ and $dv/d\tau$. Therefore, the rate of change of the proper distance $L(\tau) = r(\tau)\Delta\varphi$ is

$$dL/d\tau = \Delta\varphi(r_{,u}du/d\tau + r_{,v}dv/d\tau)\,. \tag{24.5}$$

This entity is clearly well-defined and finite at any point of P_2, meaning that the expansion is finite after the CH is crossed; and, more importantly, L is finite throughout P (moreover, the relative change in L may be arbitrarily small throughout P) [18]. It is straightforward to extend this result to a nonspherical construction of P, and the results are the same. Thus, the extended object crossing the CH will be in a good physical state, in terms of both its shape (i.e. deformation) and its internal state [19].

24.4 A remark on the energy-momentum content of the CH singularity

One may be concerned about the consistency of the above conclusion with the Raychaudhuri equation. According to this equation, in a vacuum spacetime (or one with positive energy densities), the expansion of any (twist-free) congruence on timelike geodesics must decrease monotonically. Naively, this would mean that as the expansion diverges to $-\infty$ at the CH, there is no way to turn it around to regular values. This is obviously disturbing. A closer look reveals, however, that there is no inconsistency here: The Raychaudhuri equation simply tells us that in the above geometric construction of P the CH singularity functions as a thin layer with infinite negative energy density. Indeed, the presence of this surface layer there should in principle be deduced directly from the Einstein equations (when applied to the CH in a distributional sense); The Raychaudhuri equation is just an alternative probe for the energy-momentum content of this thin layer.

Although the subject of this article is Herman and Hiscock's argument concerning the traversability of the CH, and *not* the issue of physical extension of the geometry itself, I would like to make a few remarks concerning this thin layer and its implications to the issue of geometric extension. The presence of this peculiar thin layer may be viewed as a disturbing feature of the geometric extension, because of several reasons: First, consider for example the CH singularity of a generic spinning BH. Then, the field equations are the vacuum Einstein equations, so we shall naturally demand that P_2 will be a vacuum solution (like P_1). However, the *matching layer itself* fails to be vacuum; And, as a consequence, the entire region P will have certain global non-vacuum features (the overall non-monotonic character of the expansion is one example) [20]. Second, the energy-density content of this layer is negative rather than positive. Third, not only does the energy-density diverge, but also its integral through the CH diverges. (In this sense, the CH differs from the standard concept of thin layers in GR.)

In my opinion, although the above features are certainly unusual, they are not so unreasonable, after all: We are dealing here with the conditions inside the microscopic thin layer, where (classically) curvature diverges. It is almost clear that any attempt to describe the physics *inside* this thin layer in terms of classical GR would be meaningless; and, the association of energy-momentum is obviously such an attempt: Energy-momentum is nothing but the outcome of applying the classical Einstein operator to the geometry (the effective geometry, in this case). One of course has the right to perform this mathematical operation (in terms of limiting integrated values only; the pure differential operator itself is meaningless at the CH, where the effective metric is not even C^1), but one should not be surprised if the results of this formal operation do not conform with our daily experience with "energy-momentum". In fact, one may expect that any features of matter in low-curvature physics (except those resulting from first principles) may be violated inside the singular matching layer at the CH. This situation is analogous to the phenomenon of shock formation in fluid mechanics, in a compressible perfect fluid: Again, we have a thin transition layer (the shock) which connects the two smooth regions (let us call them again P_1 and P_2, and their union is P). Although the original theory conserves entropy, and indeed entropy is conserved in each of the individual patches P_1 and P_2, it is *not* conserved throughout P. This violation of entropy conservation has an obvious reason: Inside the transition region the dynamics is dominated by viscosity (and other dissipative effects), so the theory of perfect fluid is not valid there. (In this analogy, perfect-fluid dynamics is analogous to classical GR, and the more general viscous-fluid theory is analogous to quantum gravity.)

In this regard, it is interesting to note that all the above features of the energy-momentum content of the CH were in fact obtained in simplified models in which the semiclassical theory is applied to BHs with a CH. One such example is Hiscock's model of a 2-dimensional charged BH [21]: The semiclassical effects break the (electro-)vacuum character of the geometry, the energy-density is negative (at least in some cases), and the integrated energy throughout the region of the CH is infinite (e.g. when the energy density is integrated over the last Planckian proper time along the orbit of a typical free-falling orbit that intersects the CH.) More recent analyses of semiclassical effects (e.g. [3, 4]) yield results which are are not inconsistent with Hiscock's. Obviously we cannot take these results too seriously, because in these models the significant semiclassical effects appear at the stage where curvature is Planckian and the semiclassical approximation is thus invalid . Nevertheless, these analyses do teach us one important lesson: If indeed the physics at the CH and its immediate neighborhood is dictated by Quantum Gravity, one should not be surprised if the (effective) energy-momentum content of this thin layer will have the above exotic features!

24.5 Summary

We considered the fate of a macroscopic extended object which falls into a spinning or charged BH and moves towards the CH, focusing attention on the object's internal state. Our results can be divided into two parts:

1. *Up to* the CH: We have shown that no "melting", or significant growth or entropy, or loss of information, occurs on the approach to the CH.

2. *Beyond* the CH: Here the situation is more subtle, because in principle the current theories do not provide a unique prediction concerning the evolution beyond the CH. One can therefore raise various speculations, and there is no absolute way to test their validity. Nevertheless, we can try to assess how *reasonable* the various possibilities are.

   For the sake of this discussion we made the working assumption (also made by HH) that geometry extends classically beyond the CH, and this extension is prescribed – otherwise the whole issue of the internal state of a classical physical object beyond the CH would be meaningless. We have thus assumed that the metric tensor extends continuously (i.e. C^0) beyond the CH. Then, a timelike geodesic intersecting the CH has a *unique* C^1 extension beyond the CH. Assuming that the physical geodesic motion should proceed along this C^1 extension, one arrives at the following conclusion: No catastrophic growth of entropy, internal kinetic energy, etc. occurs beyond the CH either. The rate of deformation (interpreted by HH as "internal velocity"), which diverges at the CH, retains finite values at the other side of the CH.

Our conclusion is, therefore, the following: If the geometry extends classically beyond the CH, then physical objects may traverse it peacefully. [22] We therefore return to the major question: *Does geometry extend classically beyond the CH singularity?* Unfortunately, at present, with the lack of a valid theory, we don't have an answer to this question, and any attempt to exclude one of the possibilities would be a mere speculation. We shall have to await further developments in Quantum Gravity (or perhaps in other approaches to black-hole geometrodynamics beyond the standard Einstein theory?) before this issue may be resolved.

Bibliography

[1] We should emphasize, though, that the structure of the singular hypersurface at the "deeper" part of the BH (i.e. the part of the singular hypersurface located to the left of the CH on the standard Penrose diagram) is still unclear. It may include a BKL type of spacetime singularity, but this has not been verified yet.

[2] A. Ori, Phys. Rev. Lett. **68**, 2117 (1992).

[3] W. G. Anderson, P. R. Brady, W. Israel, S. M. Morsink, Phys. Rev. Lett. **70**, 1041 (1993).

[4] R. Balbinot and E. Poisson, Phys. Rev. Lett. **70**, 13 (1993).

[5] R. Herman and W. A. Hiscock, Phys. Rev. D **46**, 1863 (1992).

[6] C. M. Chambers, These proceedings.

[7] This is known to be the case in asymptotically-flat models like Hiscock's model [11] based on the Charged Vaidya solution, the mass-inflation model, and also in the CH singularity of a generic spinning BH. This is not necessarily the case, however, for solutions with a cosmological constant – see [6].

[8] One may still attempt to take the rate of change of distances inside a sphere of fixed (small) diameter as the basis for defining "internal velocities", overcoming the curvature- related problems by performing a parallel transport. As we mentioned above, however, the results will depend on the choice of the orbits for the transport. Particularly, in our case there will be a drastic difference between (i) transporting along spacelike orbits along an angular direction, or (ii) transporting along almost-radial timelike orbits.

[9] A. Ori, Gen. Rel. Grav, **29**, 881 (1997).

[10] A. Ori, unpublished.

[11] W. A. Hiscock, Phys. Lett. **A 83**, 110 (1981).

[12] A. Ori, Phys. Rev. Lett. **67**, 789 (1991).

[13] E. Poisson and W. Israel, Phys. Rev. D **41**, 1796 (1990).

[14] I showed this rigorously [15] for the case in which P_1 is the non-symmetric local vacuum geometry constructed in Ref. [16], and also, in a less rigorous form, in other types of CH singularity, namely: (i) a CV or mass-inflation geometry, (ii) the CH of a generic spinning vacuum BH, (iii) the plane-symmetric geometry constructed in Ref. [17]. Note, however, that if there is a cosmological constant, there may be regions in the space of parameters for which the geodesics are even C^2 at the CH.

[15] A. Ori, unpublished.

[16] A. Ori and É. É. Flanagan, Phys. Rev. D **53**, R1754 (1996).

[17] A. Ori, submitted to Phys. Rev. D.

[18] In the above example of a CV geometry in P_1 and a RN geometry in P_2, the deformation rate (e.g. the expansion) jumps to small values immediately after the CH was crossed. If P_2 is also a CV geometry, however, then the deformation rate may diverge at the CH on the approach from P_2 also.

[19] If one insists on interpreting the large deformation rate near the CH as "internal velocity", then one would have to interpret the situation as follows. The gravitational force "accelerates" this "internal velocity" to infinite values at the CH, but immediately afterwards it "de-accelerates" it back to regular values. From the statistical-mechanical point of view, this is only possible because of the *coherent* nature of the above "internal velocity" field, which means that no entropy (or information loss) is associated with it (see the discussion in the previous section).

[20] The failure to construct a pure vacuum matching is related to the nonlinearity of the Einstein equation.

[21] W. A. Hiscock, Phys. Rev. D **15**, 3054 (1977).

[22] Note, however, that the journey through the CH may be fatal for an unshielded human being, due to the fatal dose of radioactive γ-rays [23]. (This is essentially the issue raised by Penrose a long time ago [24]).

[23] L. M. Burko, Phys. Rev. D **55**, 2105 (1997).

[24] R. Penrose, in *Battelle Rencontres, 1967 lectures in mathematics and physics*, edited by C. M. DeWitt and J. A. Wheeler (Benjamin, New York, 1968), p. 222.

LIST OF PARTICIPANTS

- Shai Ayal
  Racah Institute of Physics, Hebrew Univarsity, 91904 Jerusalem, Israel
  shai@shemesh.fiz.huji.ac.il

- Leor Barack
  Department of Physics, Technion, 32000 Haifa, Israel
  leor@tx.technion.ac.il

- Claude Barrabès
  Physics Department, Faculte des Sciences, 37200 Tours, France
  barrabes@chopin.phys.univ-tours.fr

- Eran Ben-Shahar
  Department of Physics, Technion, 32000 Haifa, Israel
  eran@physics.technion.ac.il

- Beverly Berger
  Department of physics, Oakland Univeristy, Rochester, MI 48309, USA
  berger@Oakland.edu

- Alfio Bonanno
  Instituto di Astronomica, Universita di Catania, Viale Andrea Doria 6, 95125 Catania, Italy
  abonanno@alpha4.ct.astro.it

- Lior Burko
  Department of Physics, Technion, 32000 Haifa, Israel
  burko@pegasus.technion.ac.il

- Aleksander Burinskii
  Nuclear Safety Institute, Russian Academy of Sciences, B. Tulskaya 52, Moscow 113191, Russia
  grg@ibrae.ac.ru

- Liora Dori
  Department of Physics, Technion, 32000 Haifa, Israel
  ismael@aluf.technion.ac.il

- Chris Chambers
  Department of Physics, Montana State University, Bozeman, MT 59717-3840, USA
  chambers@physics.montana.edu

- Irina Dymnikova
  Institute of Mathematics and Physics, Pedagogical University of Olsztyn, Zolnierska 14, Olsztyn 10-561, Poland
  irina@tufi.wsp.olsztyn.pl

- Éanna Flanagan
  Newman Laboratory, Cornell University, Ithaca, NY 14853-5001, USA
  flanagan@spacenet.tn.cornell.edu

- Valeri Frolov
  Department of Physics, University of Alberta, Edmonton, Canada T6G 2J1
  frolov@phys.ualberta.ca

- Dmitri Gal'tsov
  28-21 Leninsky prospect, 117971 Moscow, Russia
  galtsov@grg.phys.msu.su

- Eduardo Guendelman
  Department of Physics, Ben-Gurion University of the Negev, 84105 Beer Sheva, Israel
  guendel@bgumail.bgu.ac.il

- Tom Helliwell
  Department of Physics, Harvey Mudd College, Claremont, CA 91711, USA
  helliwell@thuban.ac.hmc.edu

- Shahar Hod
  Racah Institute of Physics, Hebrew University, 91904 Jerusalem, Israel
  hod@shemesh.fiz.huji.ac.il

- Ted Jacobson
  Department of Physics, University of Maryland, College Park, MD 20742-4111, USA
  jacobson@nscpmail.physics.umd.edu

- Sanjay Jhingan
  Tata Institute of Fundamental Research, Homi Bhabha Road, Mumbai, India
  sanju@tifrvax.res.in

- Burkhard Kleihaus
  Fachbereich Physik, University of Oldenburg, Postfach 2503, D-26111 Oldenburg, Germany
  kleihaus@darkstar.physik.uni-oldenburg.de

- Deborah Konkowski
  Department of Mathematics, U.S. Naval Academy, Annapolis, MD 21402, USA
  dak@nadn.navy.mil

- George Lavrelashvili
  Institute of Theoretical Physics, University of Bern, Sidlerstrasse 5, CH–3012 Bern, Switzerland
  lavrela@butp.unibe.ch

- Robert Mann
  University of Waterloo, Waterloo, Ontario, Canada N2L 3G1
  mann@avatar.uwaterloo.ca

- Ian Moss
  Physics Department, University of Newcastle, Newcastle upon Tyne NE1 7RU, England
  ian.moss@newcastle.ac.uk

- Amos Ori
  Department of Physics, Technion, 32000 Haifa, Israel
  amos@physics.technion.ac.il

- Renaud Parentani
  Laboratoire de Mathematique et Physique Theoretique, Faculate des Sciences, 37200 Tours, France
  parenta@parenta.phys.univ-tours.fr

- Eric Poisson
  Deaprtment of physics, University of Guelph, Guelph, Ontario N1G 2W1, Canada
  poisson@terra.physics.uoguelph.ca

- Friedrich Schein
  Institut fur Theoretische Physik, Universitat Wien, Boltzmanngasse 5, A–1090 Wien, Austria
  schein@galileo.thp.univie.ac.at

- Abha Sood
  Fachbereich Physik, University of Oldenburg, Postfach 2503, D–26111 Oldenburg, Germany
  sood@cygnus.physik.uni-oldenburg.de

- Alexei Starobinsky
  Landau Institute for Theoretical Physics, Russian Academy of Sciences, Kosigina2, V–334, Moscow, GSP–1, 117334, Russia
  alstar@landau.ac.ru

- Roberto Sussman
  Instituto de Ciencias Nucleares, UNAM, Apartade Pastal 70–543, Mexico D.F. 04510, Mexico
  sussman@nucleco.unam.mx

- Matt Visser
  Department of Physics, Washington University, One Brookings Drive, St. Louis, MO 63130-4899, USA
  visser@kiwi.wustl.edu

- Michael Volkov
  Institute of Theoretical Physics, University of Zurich–Irchel, Winterthurerstrasse 190, CH-8057 Zurich, Switzerland
  volkov@physik.unizh.ch

- Gilbert Weinstein
  Department of Mathematics, University of Alabama at Birmingham, 452 Campbell Hall, Birmingham, AL 35294-1170, USA
  weinstei@vorteb.math.wab.edu